SHRUBS OF ONTARIO

James H. Soper

Curator Emeritus
National Museum of Natural Sciences
Ottawa

and

Margaret L. Heimburger

Professor Emeritus
University of Toronto

With drawings by Leslie A. Garay
and Ronald A. With

A Life Sciences Miscellaneous Publication
of the Royal Ontario Museum, Toronto

ROYAL ONTARIO MUSEUM
PUBLICATIONS IN LIFE SCIENCES

The Royal Ontario Museum publishes three series in the Life Sciences:

LIFE SCIENCES CONTRIBUTIONS, a numbered series of original scientific publications including monographic works.

LIFE SCIENCES OCCASIONAL PAPERS, a numbered series of original scientific publications, primarily short and usually of taxonomic significance.

LIFE SCIENCES MISCELLANEOUS PUBLICATIONS, an unnumbered series of publications of varied subject matter and format.

All manuscripts considered for publication are subject to the scrutiny and editorial policies of the Life Sciences Editorial Board, and to review by persons outside the Museum staff who are authorities in the particular field involved.

The Royal Ontario Museum gratefully acknowledges the generous financial assistance of the National Museum of Natural Sciences, Ottawa, in the production of this volume.

Canadian Cataloguing in Publication Data

Soper, James H., 1916–
Shrubs of Ontario

(Life sciences miscellaneous publication, ISSN 0082-5093)
Bibliography: p.
Includes index.
ISBN 0-88854-283-6

1. Shrubs—Ontario—Identification. I. Heimburger, Margaret L. II. Royal Ontario Museum. III. Title. IV. Series: Life sciences miscellaneous publications.

QK485.2.O6S66 582.1′7′09713 C82-094182-4

First printing, 1982
Second printing (with revisions), 1985
Third printing, 1990
Fourth printing, 1994

Cover: *Physocarpus opulifolius* (p. 193). Drawing by Ronald With.

100 Queen's Park, Toronto, Canada M5S 2C6
ISBN 0-88854-283-6
ISSN 0082-5093
Printed and bound in Canada by Best Book Manufacturers

CONTENTS

SHRUBS OF ONTARIO

FOREWORD

It is now twenty years since the publication, by the provincial Department of Commerce and Development, of Soper and Heimburger's *100 Shrubs of Ontario*, a useful book which has long been out of print. *Shrubs of Ontario* should prove to be an even more useful successor to the earlier work.

It is appropriate that *Shrubs of Ontario* should be published by the Royal Ontario Museum, since much of the basic research was done in Toronto on herbarium collections soon to be transferred from the University of Toronto to the ROM. When the authors began their collaborative studies on shrubby plants of Ontario, they were both professors of Botany at the University of Toronto. Prof. Soper left Toronto in 1967 to become the Chief Botanist of the Museum of Natural Sciences of the National Museums of Canada; Prof. Heimburger retired in 1972 and went to live in Victoria, B.C. The collaboration begun more than twenty years ago in Toronto has been continued from Ottawa and Victoria and has culminated in the completion of *Shrubs of Ontario*, a significant contribution towards an understanding of the flora of this extremely large and diverse province.

Dr. James H. Soper's first interest in shrubs as a group stems from his association with the late Dr. W. Sherwood Fox, former President of the University of Western Ontario. Soon after his arrival in Toronto in September 1946 as Special Lecturer and Curator of the Vascular Plant Herbarium, Dr. Soper became involved with the Federation of Ontario Naturalists. He was named Chairman of a Checklist Committee of the F.O.N., charged with the preparation of a checklist of vascular plants of southern Ontario. The other members of the committee were W. Sherwood Fox, Monroe Landon, Fred H. Montgomery, and Edmund M. Walker, and their work concluded with the distribution of the checklist in March 1949. In November 1948 Dr. Fox and Dr. Soper began a collaborative effort to map the distribution of certain Carolinian trees and shrubs in southern Ontario. This association culminated in a three-part joint publication in the Transactions of the Royal Canada Institute between 1952 and 1955.

At about this same time Dr. Margaret Landes Heimburger, in collaboration with Dr. Soper, began the recording and mapping of a group of about 650 species selected from the 1949 Checklist. In 1953 Dr. Soper wrote an article on flowering dogwood for the F.O.N. Bulletin, and then the two of them published a short series on shrubs in the same bulletin (button-bush and witch-hazel in 1953; leatherwood and hobble-bush in 1954). It was perhaps Dr. Heimburger who suggested a specialization on shrubs, since the trees of Ontario had been covered by J.H. White in 1925 and 1946.

These articles on Ontario shrubs came to the attention of Mr. A.H. Richardson of the Conservation Branch of the Ontario Department of Planning and Development. Dr. Soper and Dr. Heimburger prepared descriptions and maps for a group of 25 of southern Ontario's shrubs, with illustrations by Dr. Leslie A. Garay, and this material appeared in Chapter 8 of the Forestry Section of the Credit Valley Conservation Report of May 1956. A similar article entitled "Another 30 Shrubs of Ontario", also by Soper, Heimburger, and Garay, appeared in Chapter 10 of the Forestry Section of the Napanee Valley Conservation Report of September 1957. These two groups of descriptions comprising 55 species were subsequently brought together in *Shrubs of Ontario, Part I*, published in 1957.

During this same period Dr. Soper published articles on the distribution in Ontario of bayberry, poison ivy, and pawpaw and, with Dr. Heimburger, an article on winterberry. The two of them then collaborated to produce descriptions of a third group comprising 45 species. These were combined with *Shrubs of Ontario, Part I* and the whole collection was issued as *100 Shrubs of Ontario*, in 1961.

Mr. Richardson encouraged a continuation of the collaborative study by Dr. Soper and Dr. Heimburger. When Dr. Garay left Toronto to take a position at the Oakes Ames Orchid Herbarium at Harvard University, it was Mr. Richardson who arranged for Mr. Ronald A. With to prepare the drawings for the next lot of shrubs. Mr. With has continued with the preparation of the drawings, the latest ones having been done since his retirement from the Conservation Branch, and with the financial support of the National Museum of Natural Sciences, Ottawa. He now lives in Victoria, B.C.

Although Mr. Richardson died before *Shrubs of Ontario* was completed, his interest was continued by Mr. A.S.L. Barnes, who became Director of the Conservation Authorities Branch, Ontario Ministry of Energy and Resources. After his retirement, Mr. Barnes became Executive Secretary of the Quetico Foundation and President of the Royal Canadian Institute. He died in December 1976.

Shrubs, including woody vines and subshrubs, are an important component of any vascular flora. They are of value in the conservation of soil, water, and a wide range of wildlife species. Their importance, however, is not only ecological but aesthetic, and their study both practical and intellectually challenging. Many different plant families are represented by shrubby species in the indigenous flora of Ontario.

No one is better qualified than the authors to produce a definitive work on the shrubs of Ontario. Dr. Soper and Dr. Heimburger are mature scientists who have developed an entirely original work of great significance. For many decades to come this book will be an essential companion of botanists, foresters, naturalists, conservationists, and others interested in our natural environment.

Toronto
September 1981

J.E. Cruise
Director
Royal Ontario Museum

ACKNOWLEDGEMENTS

The authors wish to express their thanks to all who have helped in the gathering of data for this book by acting as assistants in the field and the herbarium, making facilities available during visits to herbaria, sending specimens on loan for study, or providing information from personal field notes, observations, and private collections. Financial support over the years for field work, visits to herbaria, and the hiring of assistants has come from various sources: the National Research Council of Canada, the Ontario Research Foundation, the University of Toronto, and, since 1967, the National Museum of Natural Sciences in Ottawa.

Dr. G.W. Argus has identified a large proportion of the collections of *Salix* used in the plotting of maps for that genus and has provided much constructive comment during the preparation of descriptions, illustrations, and the key to species of *Salix*. Mrs. Sheila Kuja (*née* McKay) kindly allowed us to examine the large series of specimens of *Amelanchier* on loan at the herbarium of the University of Toronto (TRT) during her study of the Ontario material for a Master's degree thesis in the Department of Botany. Dr. P.F. Maycock made available the records (species lists and unmounted collections) from the many sites in Ontario where he has examined the vegetation by stand analysis and quadrat studies. His "sight records" have been especially useful for northern Ontario, where the available collections do not provide a sufficiently detailed sample for the vast area around James Bay and Hudson Bay.

Our appreciation is also extended to Mr. John Riley for specimen records from the Hudson Bay Lowlands; to Dr. A.A. Reznicek for maps of shrub distribution in Simcoe County; to Mr. R.E. Whiting for data from the Muskoka District and northern Ontario; to Mr. W.G. Stewart for new records from Elgin County; to Dr. Larry B. Morse for distribution records of *Hudsonia tomentosa*; and to Dr. S.P. Vander Kloet for unpublished data on *Vaccinium*.

We are particularly grateful to Mrs. Joyce Watts for her careful editing of the original manuscript, for checking consistency in style and language, and for offering numerous suggestions which have resulted in improvements in the final text.

Mr. John Campsie provided assistance in tracing the origin and meaning of the scientific names for some of the shrubs. We are greatly indebted to Mr. Campsie for handling the technical details of layout of text, illustrations, and maps, and for the supervision of all stages of production of this book.

James H. Soper

Margaret L. Heimburger

INTRODUCTION

The purpose of this book is to provide, in the form of a field guide or manual, a means of identifying the shrubs that can be found in Ontario growing outside cultivation. The term *shrub* has been variously defined. As interpreted in this book it includes perennial plants with usually more than one low-branching woody stem, also woody vines and a number of low or trailing perennials which are sometimes called subshrubs. Since certain trees occasionally develop shrubby growth with several stems from the ground, some of these have been included.

Many shrubs are of horticultural interest for their attractive flowers and fruits or for the fine autumn coloration of their foliage. A few have brightly coloured stems and persistent fruits which are decorative in winter. Others produce edible nuts and berries and lend themselves to cultivation as sources of food in the home garden. Some are excellent for hedges or foundation plantings. Various species can be used in conservation projects either for erosion control on slopes and bare ground or for the provision of food and shelter for wildlife along fencerows and borders of fields.

The information on the shrubs covered by this manual has been obtained through the examination of living plants in the field, through the study of specimens in the herbarium, and by reference to books and articles. Measurements are based chiefly on specimens examined in the herbarium at the University of Toronto (TRT) and in the National Herbarium of Canada (CAN) at Ottawa. Data plotted on distribution maps have come from specimens seen in the major herbaria that have holdings of Ontario plants, supplemented by field observations (sight records), reports from published literature, and other sources considered as acceptable evidence for the occurrence of the species concerned. Much of the distribution data is on file in the Catalogue of Vascular Plants at the Department of Botany, University of Toronto, and in a similar catalogue in the National Herbarium. Recently some specimen records have been stored in the Botany databank under the National Inventory of Collections Programme operated by the National Museums of Canada.

Seventy of the drawings were prepared by Leslie A. Garay who participated as a co-author and illustrator at the University of Toronto during the first stages of the project. The rest of the drawings have been made by Ronald A. With of Victoria, B.C., formerly on the staff of the Conservation Branch, Ontario Department of Commerce and Development, Toronto. The illustrations have been prepared from pressed herbarium specimens, as it was felt that line drawings would emphasize the diagnostic features, such as leaf shape, venation, and marginal toothing, more clearly than would photographs. The leafy branch that forms the main portion of most drawings was sketched to life size and has been reduced to approximately 60 per cent of natural size for publication. The accessory drawings of details of leaf surface, margin, flower, and fruit are shown at various magnifications or reductions as indicated by the scales placed nearby.

Scientific and Common Names

Scientific names have been used throughout the text. In addition, vernacular names have been included when they appear to be in common use. For anyone not familiar with the make-up of scientific names, it may be helpful to give a few explanatory notes.

In an area such as the province of Ontario, the vascular plant flora is well known, and identification can be made by keys available in regional floras and manuals covering the area.

The keys in this book have been constructed to cover all the shrubs known to occur in Ontario outside of cultivation. Names for these shrubs are already available because most of our species were first collected elsewhere and subsequently described, named, and classified. Scientific names date from the time of Carolus Linnaeus (1707 – 1778), and their use is governed by the International Rules of Botanical Nomenclature.

Basically, each botanical name requires a minimum of two words that indicate the genus and species, and these are usually followed by an authority which is a reference for tracing the origin and place of publication of that name. The basic two-part name (the genus and species) is known as a *binomial*, and this system of naming plants is roughly comparable to the way in which we give names to people belonging to different families. In plants, however, the first part of the name tells the genus (group) to which the plant belongs, for example, *Rosa*, which means a rose. This is the same as the surname or family name for a person, such as *Smith*, *Bentham*, or *Darwin*, and is always capitalized. The second part of the botanical name indicates the species, that is, the particular member of the group, for example, *palustris*, which means "of swamps". This, in turn, may be compared to the given name of an individual, such as *James*, *Joseph*, or *Charles*, to accompany the examples of surnames given above. In botany, however, this second part of the name is not capitalized. Scientific names for shrubs are based on Latin words unless otherwise stated and both parts of the name (genus and species) are italicized.

The botanical name *Rosa palustris* is specific for a particular kind of rose even though thousands of individuals (separate plants) of that kind may occur in nature. Two-part scientific names (binomials) are also used in zoology for the naming of animals, including man (*Homo sapiens*), and these too are governed by a set of international rules. To complete the example of a binomial for a plant, we must add the authority, which is usually abbreviated. In our present example it would give us the name *Rosa palustris* Marsh. It was Humphry Marshall who in 1785, in a catalogue of American trees and shrubs called *Arbustum Americanum*, first described the swamp rose and proposed the appropriate specific name *palustris*.

It will be noted that in many cases the authority is not a single name but consists of a name in parentheses followed by another name, for example, *Spiraea latifolia* (Ait.) Borkh. The use of two names of authors in such a sequence indicates that a transfer of names has taken place. Actually this shrub was first described as a variety of the species *Spiraea salicifolia* L., so the name was then *Spiraea salicifolia* L. var. *latifolia* Ait. Later, study by Borkhausen showed that the plants so named could be considered as a separate species, distinct from *Spiraea salicifolia*. The original name *latifolia* used by Aiton for the plant as a variety was retained in the new status as a species (following traditional practice), and the new name is therefore *Spiraea latifolia* (Ait.) Borkh. Thus the credit for the new binomial is given to Borkhausen, but the fact that he made use of Aiton's name and description is shown by retaining "Aiton" (abbreviated) in parentheses. All published names proposed for plants are recorded in certain indexes, for example, *Index Kewensis*, *Index Genericorum*, *Gray Herbarium Index*. It is possible therefore to trace the history of the naming of a plant, as in this example.

No attempt has been made to distinguish between the subspecies level and the variety level. Names below the level of species have been accepted as they have been published in the botanical literature and used at the rank assigned by the author, thus avoiding the necessity of creating new combinations.

The term "family" is also used for purposes of classification within the Plant Kingdom. It is a higher category or level at which a number of related groups (genera) are brought together. For example, the genus *Rosa* (roses) is put with many other genera, including *Prunus* (cherries and plums), *Pyrus* (pears), *Sorbus* (mountain ash), *Malus* (apples), and *Rubus* (raspberries and blackberries), to form a family called *Rosaceae* (rose family). This simply means that these plants share more attributes with one another than they do with the plants in certain other genera, for example, peas, beans, vetches, clovers, and lupines, which in turn make up another family, *Leguminosae* (pea family).

The botanical names used for genera and species are not always descriptive. Some are commemorative, for example, *Forsythia*, a genus including attractive spring-flowering shrubs used in gardens and foundation plantings, which was named after William Forsyth (1737–1800), a prominent English horticulturist, and *Pinus banksiana*, jack pine, named in honour of Sir Joseph Banks (1743–1820), a famous English naturalist, traveller, and philanthropist.

Botanists use scientific (technical) names for plants not to confuse or impress the amateur but because such names are more precise than common names and are understood by botanists all over the world. In contrast, common names are often confusing because one name may be used for two or more different kinds (species) of plants; for example, *Acer pensylvanicum*, *Dirca palustris*, and *Viburnum alnifolium* have all been called moosewood, but botanically they belong to three different families. Also, more than one name may be applied to a single kind of plant; for example, *Viburnum edule* is known variously as squashberry, mooseberry, and pembina. Some plants have no common names. Common names in English are meaningless to someone whose everyday language is French, German, Russian, or Chinese.

Authors of Botanical Names of Shrubs

Most of the native shrubs of Ontario were named by European botanists who received for study the specimens collected in North America during the early exploration of this continent. The official nomenclature for all vascular plants, including all our shrubs, started with the publication of *Species Plantarum* (ed. 1) by Carolus Linnaeus in 1753. Over eighty of the shrub names in the present book were given by Linnaeus in that work, although certain specific names have since been transferred to different genera from the ones under which they were originally described. For example, the ninebark was described as *Spiraea opulifolia* by Linnaeus in 1753 but was transferred to the genus *Physocarpus* by Maximovich in 1879. The correct name *Physocarpus opulifolius* (L.) Maxim. gives credit to Linnaeus for the specific epithet. Frequently it is necessary to alter the ending of the specific epithet to agree with the gender of the genus to which it is being transferred. Hence, *opulifolia* (feminine ending after *Spiraea*) becomes *opulifolius* (masculine form) to agree with *Physocarpus*. Most of the generic names of trees and shrubs are either feminine (e.g., *Spiraea*) or so treated (e.g., *Prunus americana*), but there are numerous exceptions such as *Physocarpus* and *Rubus* (masculine) and *Acer*, *Ribes*, and *Sassafras* (neuter).

The persons who have described the shrubs of Ontario, as well as those who have transferred specific names to a different genus or changed the rank (species, subspecies, variety, or form), are listed separately at the end of the book (p. 476). The list includes, for each author, the abbreviation (if one is in use), the birth and death dates, and the country where major work was done. Most of those listed were responsible for only one, or a few, of the shrub names in Ontario, but several whose contribution has been outstanding are given special mention below:

William Aiton (1731–1793) was a prominent horticulturist in England who published many new species in 1789 in a work called *Hortus Kewensis*.

Nils Johan Andersson (1821–1880), a professor at the University of Stockholm, Sweden, named four species and one variety of willow (*Salix*) which are found in Ontario.

Humphry Marshall (1722–1801), a Pennsylvanian who is credited with the first book devoted entirely to botany to be printed in America, *Arbustum Americanum: The American Grove* (1785), described nine of our native shrubs.

André Michaux (1746–1802) came to North America in 1785 at the request of the French government to study the trees and shrubs and assess their suitability for use in France. He was one of the pioneer explorers and described about a dozen of our shrubs.

Frederick Traugolt Pursh (1774-1820) came to Philadelphia from Saxony at an early age and later managed a botanical garden near Baltimore. Several of our shrubs were described in his publication entitled *Flora Americae Septentrionalis* (1814).

The naming of most of our shrubs had been completed by the middle of the 19th century so that even Asa Gray (1810–1888) and M.L. Fernald (1873–1950), both of whom worked on the entire flora of northeastern North America, found no undescribed shrubs for the Ontario part of that region. Fernald's *Viburnum recognitum* is essentially the northern representative of the southern species of arrow-wood *Viburnum dentatum* L., and may be treated as *V. dentatum* var. *lucidum* Ait., using the earlier name proposed at the varietal rank by Aiton. Fernald's 1950 revision of *Gray's New Manual of Botany Illustrated* resulted in one of the most useful books available on the flora of eastern North America, and it is still widely used for identification of vascular plants of that region.

Arrangement of Materials

For each species treated fully there is a description, a brief field check, an illustration, and a distribution map. A few species that are rare or should be looked for because they occur in adjacent regions have been included in the keys and are mentioned in a note under a closely related species. The note is also used for other information about related species, name changes, edibility, poisonous properties, and so on. When a genus is represented by more than one species, a key is provided.

Families are arranged in the taxonomic sequence found in *Gray's Manual of Botany* (Fernald, 1950), but within each family the genera are presented in alphabetical order as are the species within each genus. The detailed treatment of species is preceded by a key to all the genera included in the book. This introductory key eliminates the necessity of having keys to the many families and, within each family, to the genera represented. Normally it is the genus that one learns to recognize through familiarity with one or more of its species, and the family relationships are of secondary interest.

Distribution Patterns

One of the most interesting aspects of the study of our native shrubs has been the mapping of the known occurrences of each species. Analysis of their distribution patterns has shown them to be highly individualistic. Some species occur almost everywhere in Ontario from Lake Erie to Hudson Bay and from the Quebec border to the Manitoba border. Others are limited to the southern or the northern part of the province, and still others are found mostly in particular places, such as along the shores of rivers and lakes or only in bogs and fens. A few are known so far only from a single site or from two or three localities within the province.

The present-day ranges of our shrubs are the results of many factors. Because Ontario was completely covered by ice during the Wisconsin period, the most recent advance of the Pleistocene Ice Age, all the plants and animals now living in the area arrived in the wake of the retreating ice. The first type of plant cover is thought to have been an open tundra, composed of hardy pioneer species which invaded the newly exposed terrain. The development of forests in Ontario took between five and ten thousand years and the northernmost area along Hudson Bay still supports a tundra or treeless arctic barren (Figure 1).

Depending on the efficiency of seed dispersal, availability of suitable habitats, and other factors, some shrubs have migrated more rapidly into the area than others. Also, they have come into Ontario from various directions, although most came from the south. Their ability to grow and establish themselves on different types of soil and bedrock (Figure 2) is sometimes reflected in their current distribution. Other factors that have influenced their spread or survival are soil moisture, topography (especially exposure), climate (rainfall, length of the growing season, extremes of temperature in summer and winter), natural fires, and competition with other plants. Finally, the impact of man has modified existing plant ranges at a steadily increasing rate. The extensive clearing of forests for agriculture, industry, and housing developments has all but eliminated the original stands of trees in the Deciduous Forest Region of southern Ontario. Pollution of the land, air, and water is affecting many habitats so that more and more species are becoming rare or endangered. Many species of shrubs have been introduced into Ontario from Eurasia and some from other parts of North America. Most of these do not stray from gardens, parks, or highway borders where they have been planted, but others have spread freely and a few have become common weeds, for example, the barberry and buckthorn.

The current boundaries of several counties are different from those shown on the base map used for plotting distributions in southern Ontario, but the changes have been indicated on the new introductory map (Figure 3). The following summary is an arbitrary grouping of species of shrubs that indicates some of the main patterns of distribution in Ontario. An attempt has been made to relate these patterns to the vegetation zones in Ontario, namely, the tundra and three main forest regions as outlined by Rowe (1972), or to preferences for certain habitats. It must be remembered, however, that vegetation boundaries that appear clearly defined on a map are not so well defined in nature. The forest regions often pass freely into one another, sometimes in an inter-digitating fashion, sometimes by a gradual transition. Within any region, the variations in topography, soil, and local climate provide a corresponding variety of sites for plant growth.

Phytogeographic Groups

Group 1. Species of widespread range.

1A. Throughout Ontario: in the Deciduous Forest Region and northward into the Tundra. (6)
Arctostaphylos uva-ursi, Ribes triste, Salix candida, S. pedicellaris, Shepherdia canadensis, Vaccinium oxycoccos.

1B. In the Deciduous Forest Region and northward into the forest and barren portion of the Boreal Forest Region (frequently also in the Tundra). (23)
Alnus incana ssp. *rugosa, Amelanchier humilis, Andromeda glaucophylla, Chamaedaphne calyculata, Cornus canadensis, C. stolonifera, Gaultheria hispidula, Juniperus communis* var. *depressa, J. horizontalis, Linnaea borealis* ssp. *longiflora, Potentilla fruticosa, Prunus pensylvanica, Rhamnus alnifolia, Ribes hirtellum, Rosa acicularis, Rubus pubescens, R. idaeus* var. *strigosus, Salix bebbiana, S. exigua, S. lucida, S. serissima, Vaccinium angustifolium, V. myrtilloides.*

1C. In the Deciduous Forest Region and northward into the forest and barren portion of the Boreal Forest Region, where mainly along river valleys. (24)
Acer spicatum, Amelanchier laevis, A. sanguinea, Chimaphila umbellata, Corylus cornuta, Diervilla lonicera, Epigaea repens, Gaultheria procumbens, Lonicera canadensis, L. dioica, L. hirsuta, L. oblongifolia, Prunus pumila, P. virginiana, Rhus radicans var. *rydbergii, Rosa blanda, Salix discolor, S. humilis, S. petiolaris, Sambucus pubens, Spiraea alba, Symphoricarpos albus, Taxus canadensis, Viburnum trilobum.*

1D. Widespread in the Canadian Shield Region and northward, occasionally farther south in bogs. (12)
Alnus viridis ssp. *crispa, Amelanchier bartramiana, Betula pumila* var. *glandulifera, Kalmia poliifolia, Ledum groenlandicum, Myrica gale, Potentilla tridentata, Ribes glandulosum, R. lacustre, Salix pellita, S. pyrifolia, Sorbus decora.*

Group 2. Species of northern distribution (primarily Arctic, sub-Arctic or Boreal).

2A. Confined to the Arctic Tundra. (5)
Ledum decumbens, Phyllodoce coerulea, Salix arctica var. *kophophylla, S. lanata* ssp. *calcicola, S. reticulata.*

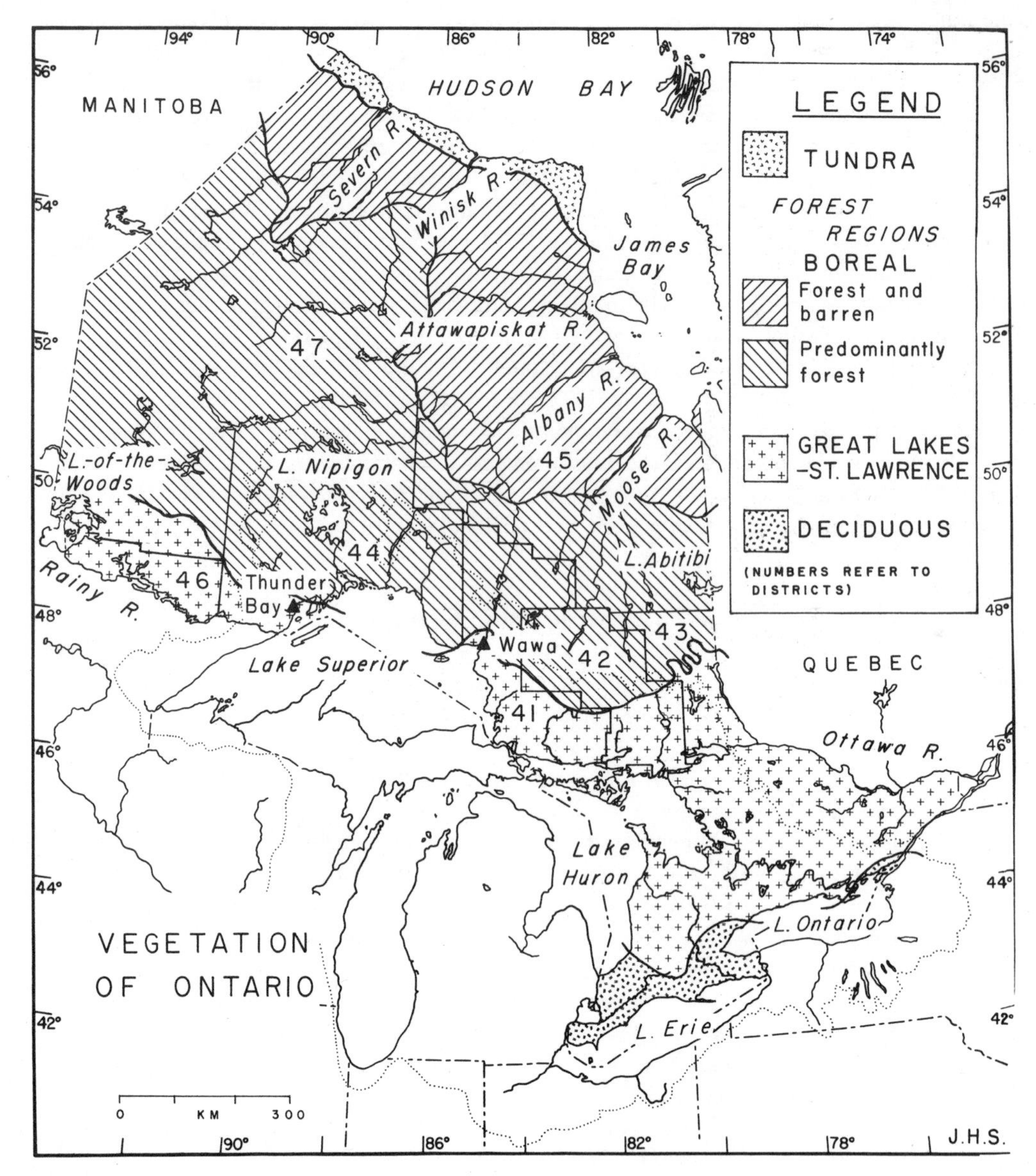

Figure 1. The vegetation of Ontario.

Names of forest regions and boundaries after Rowe (1972), with minor changes in the southern limit of the Tundra and the northern limit of the Deciduous Forest Region.

Numbers identify the districts in northern Ontario: 41 — Algoma; 42 — Sudbury; 43 — Timiskaming; 44 — Thunder Bay; 45 — Cochrane; 46 — Rainy River; 47 — Kenora. (For counties, districts, regional municipalities, and the one district municipality in southern Ontario, see Figure 3.)

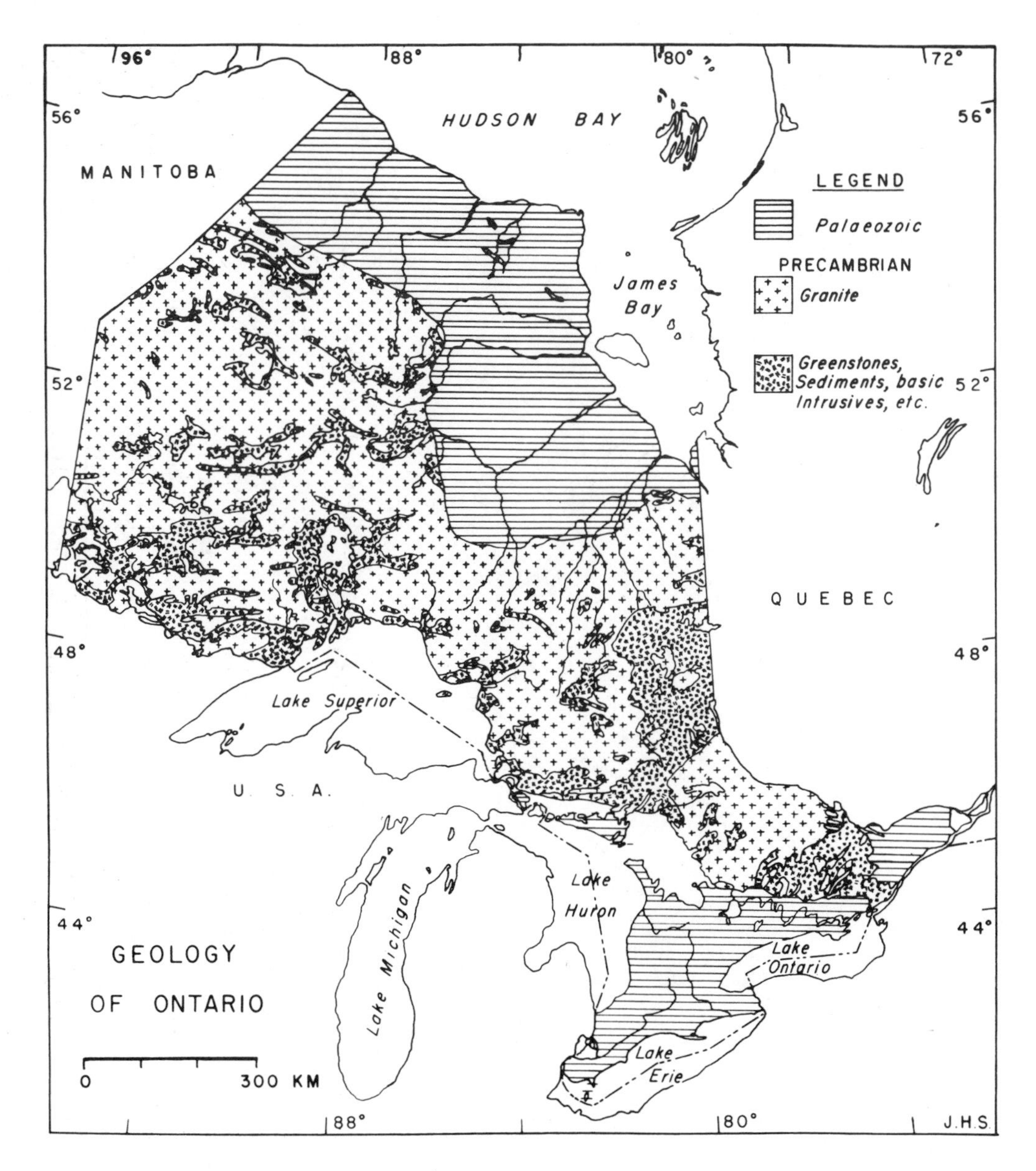

Figure 2. The geology of Ontario.
Redrawn from Map 1250A, Geological Map of Canada, published by the Geological Survey of Canada, Department of Energy, Mines & Resources, 1969.

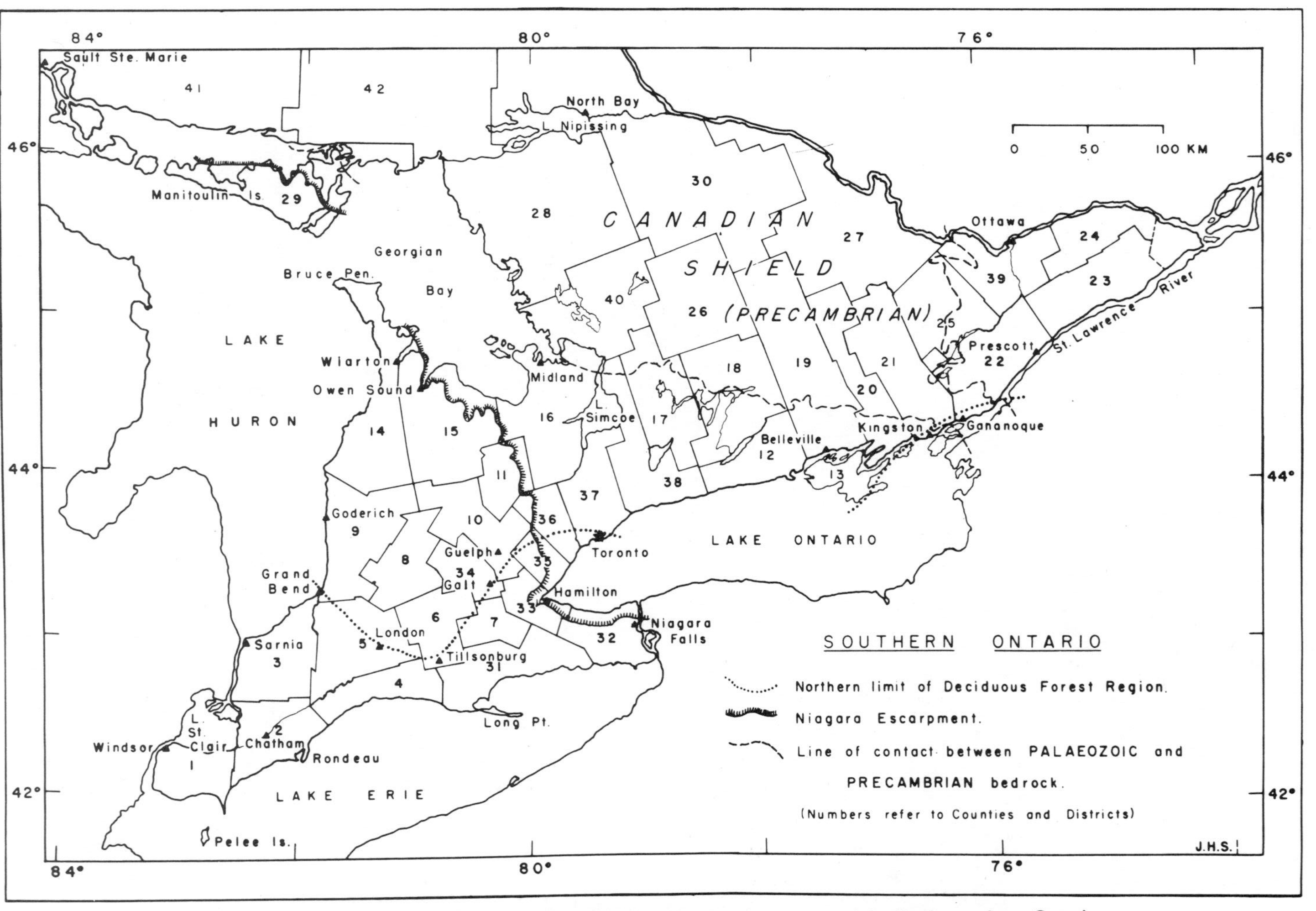

Figure 3. Counties, districts, regional municipalities, and the one district municipality in southern Ontario.

Counties

1 Essex
2 Kent
3 Lambton
4 Elgin
5 Middlesex
6 Oxford
7 Brant
8 Perth
9 Huron
10 Wellington
11 Dufferin
12 Northumberland
13 Prince Edward
14 Bruce
15 Grey
16 Simcoe
17 Victoria
18 Peterborough
19 Hastings
20 Lennox & Addington
21 Frontenac
22 Leeds & Grenville
23 Stormont, Dundas & Glengarry
24 Prescott & Russell
25 Lanark
26 Haliburton
27 Renfrew
28 Parry Sound
29 Manitoulin
30 Nipissing

Regional Municipalities

31 Haldimand-Norfolk
32 Niagara (Formerly Welland & Lincoln Counties)
33 Hamilton-Wentworth
34 Waterloo
35 Halton
36 Peel
37 York
38 Durham
39 Ottawa-Carleton

District Municipality

40 Muskoka

2B. In the Tundra and adjacent parts of the forest and barren portion of the Boreal Forest Region. (9)
Andromeda polifolia, Arctostaphylos alpina, Betula glandulosa, Rhododendron lapponicum, Salix arbusculoides, S. arctophylla, S. brachycarpa, S. glauca ssp. *callicarpaea, S. vestita.*

2C. In the Tundra and the Boreal Forest Region, rarely also in the adjacent portion of the Great Lakes - St. Lawrence Forest Region. (10)
Lonicera involucrata, Rubus acaulis, R. chamaemorus, Salix maccalliana, S. myrtillifolia, S. planifolia, S. pseudomonticola, Vaccinium uliginosum, V. vitis-idaea, Viburnum edule.

2D. In the Tundra, Boreal Forest Region and west of Sault Ste. Marie in the Lake Superior portion of the Great Lakes - St. Lawrence Forest Region. (5)
Amelanchier alnifolia, Betula occidentalis, Elaeagnus commutata, Ribes oxyacanthoides, Vaccinium caespitosum.

Group 3. Species of the Great Lakes - St. Lawrence Forest Region.

3A. Widespread from the St. Lawrence River to the Manitoba border and in adjacent parts of the Boreal Forest Region. (26)
Aralia hispida, Aronia melanocarpa, Ceanothus herbaceus, Celastrus scandens, Clematis occidentalis, C. virginiana, Comptonia peregrina, Cornus alternifolia, C. racemosa, C. rugosa, Corylus americana, Crataegus chrysocarpa, C. succulenta, Ilex verticillata, Parthenocissus vitacea, Prunus nigra, Rhus glabra, Ribes americanum, Rubus flagellaris, Salix amygdaloides, Sorbus americana, Spiraea latifolia, Vaccinium macrocarpon, Viburnum lentago, V. rafinesquianum, Vitis riparia.

3B. Generally confined to the portion of the Great Lakes - St. Lawrence Forest Region eastward from the eastern shore of Lake Superior, rarely in the southern part of the Boreal Forest Region. (15)
Acer pensylvanicum, Amelanchier arborea, Dirca palustris, Gaylussacia baccata, Hypericum kalmianum, Mitchella repens, Nemopanthus mucronata, Rhus typhina, Rosa palustris, Rubus allegheniensis, R. canadensis, R. hispidus, Spiraea tomentosa, Viburnum alnifolium, V. cassinoides.

3C. In the Great Lakes - St. Lawrence Forest Region and the Boreal Forest Region northward to James Bay, but rare or absent between Lake Superior and the Manitoba border. (4)
Amelanchier stolonifera, Kalmia angustifolia, Physocarpus opulifolius, Salix eriocephala.

Group 4. Southern species, confined to southern Ontario.

4A. Restricted to the Deciduous Forest Region or nearly so. (25)
Asimina triloba, Campsis radicans, Celtis tenuifolia, Chimaphila maculata, Cornus drummondii, C. florida, Euonymus atropurpureus, E. obovatus, Hamamelis virginiana, Hypericum prolificum, Malus coronaria, Myrica pensylvanica, Parthenocissus quinquefolia, Prunus americana, Ptelea trifoliata, Quercus prinoides, Rhus copallina, Rosa carolina, R. setigera, Sassafras albidum, Smilax rotundifolia, Vaccinium pallidum, V. stamineum, Vitis aestivalis, V. labrusca.

4B. Generally not found north of Lake Nipissing. (12)
Ceanothus americanus, Cephalanthus occidentalis, Cornus obliqua, Rhododendron canadense, Rhus aromatica, Ribes cynosbati, Rubus odoratus, R. setosus, Sambucus canadensis, Smilax hispida, Vaccinium corymbosum, Viburnum acerifolium.

4C. Rare or lacking on the Canadian Shield, mostly calciphiles but some on sand dunes or in swamps. (13)

Crataegus crus-galli, C. flabellata, C. punctata, C. mollis, Lindera benzoin, Menispermum canadense, Rhus radicans var. *radicans, R. vernix, Rubus occidentalis, Salix nigra, Staphylea trifolia, Viburnum recognitum, Zanthoxylum americanum.*

Group 5. Disjunct species.

5A. Species of western affinity found in the Upper Great Lakes Region. (5)
Crataegus douglasii, Dryas drummondii, Oplopanax horridus, Rubus parviflorus, Vaccinium ovalifolium.

5B. Species of arctic affinity found in the Lake Superior Region but common around Hudson Bay and James Bay. (2)
Dryas integrifolia, Empetrum nigrum.

5C. Species with two or more disjunct areas in Ontario. (5)
Hudsonia tomentosa, Lonicera villosa, Ribes hudsonianum, Salix cordata, S. myricoides.

Rare and Endangered Species

Thirty-three of the species included in this study have been listed by Argus and White (1977) as rare in the province of Ontario.

Andromeda polifolia
Arctostaphylos alpina
Asimina triloba
Campsis radicans
Celtis tenuifolia
Chimaphila maculata
Cornus florida
Crataegus douglasii
Dryas drummondii
Dryas integrifolia
Euonymus atropurpureus
Hypericum prolificum
Ledum decumbens
Myrica pensylvanica
Oplopanax horridus
Prunus americana
Ptelea trifoliata
Quercus prinoides
Rhus copallina
Rosa setigera
Salix arctica var. *kophophylla*
Salix arctophila
Salix brachycarpa
Salix lanata ssp. *calcicola*
Salix maccalliana
Salix reticulata
Salix vestita
Smilax rotundifolia
Vaccinium ovalifolium
Vaccinium pallidum
Vaccinium stamineum
Viburnum recognitum
Vitis aestivalis

The following eight species are proposed additions based on their restricted range (see maps) or scarcity of known locations. Three are known from a single site (marked *).

Betula glandulosa
Betula occidentalis
Parthenocissus quinquefolia
**Phyllodoce coerulea*
**Rhododendron canadense*
Rubus setosus
**Salix arbusculoides*
Vitis labrusca

The indigenous status of two of these may be questioned, namely, *Parthenocissus quinquefolia* and *Vitis labrusca*, but we consider that at least some of the early collections were from native stands in the province.

Keys for the Identification of Shrubs

The keys are based mainly on vegetative characters as this is the condition most frequently encountered in the field. The use of vegetative attributes alone may not always yield unequivocal results, and floral and fruiting characters have therefore been used to supplement vegetative characters when necessary.

The type of leaf (evergreen vs. deciduous, simple vs. compound) and the arrangement of leaves on the stem (alternate vs. opposite or whorled) were selected as main characters to divide all Ontario shrubs into four artificial groups. If the shrub being examined is completely unknown to the reader, the first step is to use the first of the general keys (below) which will indicate to which of the four groups the sample belongs. Then, using the corresponding key for that group, it should be possible to identify the specimen to a genus or, in some cases, directly to a species. In the former case there will be a key to the species in the genus and it will be located in that part of the text where the species are described and illustrated. Once having reached a species name in the last key used, it is wise to examine the illustration and read the description of that species if it is not obvious that the correct identification has been made. Since hybrids between species occur in nature in many plant groups, some specimens may be intermediate between two of the species in the key. Difficulties can therefore be expected with some specimens. The notes which follow descriptions may be useful in indicating variations to be encountered in some genera.

General Keys

A. Leaves evergreen (including conifers with persistent needlelike leaves) **Key I**, p. xxii
A. Leaves deciduous
  B. Leaves opposite, subopposite, or whorled **Key II**, p. xxiv
  B. Leaves alternate
    C. Leaves compound **Key III**, p. xxvi
    C. Leaves simple **Key IV**, p. xxvi

Key I. Leaves evergreen (including conifers with persistent needlelike leaves)

A. Leaves needlelike, scalelike, or somewhat fleshy, less than 5 mm wide
  B. Stems ascending and forming clumps or trailing, the branches elongate and spreading along the ground; leaves green, fleshy, scalelike or needlelike; seeds borne singly and surrounded by a red pulpy aril *or* several in blue berrylike cones (conifers) or in red or blue berries (angiosperms)
    C. Leaves scalelike or needlelike, sharp-pointed or, if blunt, terminating in a short sharp tip; seeds borne singly in red berrylike cups or several in dark blue berrylike cones
      D. Stems ascending, curving upwards, forming clumps
        E. Leaves linear, alternate, short-stalked, with a minute short sharp tip; seed solitary, in a red, pulpy, berrylike cup ***Taxus canadensis***, p. 3
        E. Leaves awl-shaped, in whorls of three around the stem; seeds 1–3 in a blue to blackish, berrylike cone ***Juniperus communis* var. *depressa***, p. 7
      D. Stems creeping or trailing, often forming mats
        F. Leaves decussate, scalelike, closely appressed and overlapping, clothing the stem; true flowers lacking and seeds in bluish berrylike cones ***Juniperus horizontalis***, p. 9
        F. Leaves alternate, narrowly ovate to triangular and flat, borne singly and widely spaced along the slender wiry stem; flowers pink, 4-parted; fruit an edible red berry ***Vaccinium***, p. 394
    C. Leaves fleshy, blunt-tipped, crowded around the numerous small branches; fruit a blackish berry ***Empetrum nigrum***, p. 267
  B. Stems short, erect, much-branched; fruit a capsule
    G. Leaves 2–4 mm long, grayish, clothing the stem; flowers yellow ***Hudsonia tomentosa***, p. 327
    G. Leaves 4–10 mm long, green, divergent; flowers purple ***Phyllodoce coerulea***, p. 387

A. Leaves broad, usually more than 5 mm wide
  H. Leaves opposite
    I. Stems erect; leaves elliptic-oval, 1–5 cm long; flowers in clusters *Kalmia*, p. 381
    I. Stems trailing; leaves round-oval to obovate, 1–2.5 cm long; flowers in pairs
      J. Leaves entire, with light-coloured midrib; flowers in terminal or axillary pairs; fruit a red berry *Mitchella repens*, p. 427
      J. Leaves crenate; flowers in pairs on erect slender peduncles; fruit a capsule *Linnaea borealis* ssp. *longiflora*, p. 431
  H. Leaves alternate
    K. Leaves compound, with 3 leaflets in a fanlike cluster *Potentilla tridentata*, p. 199
    K. Leaves simple
      L. Lower surface of mature blades with dense covering of white to brownish hairs or scales or with scattered bristlelike hairs or glands
        M. Lower surface of leaves with a dense coating of white to grayish or brownish hairs
          N. Stems decumbent, trailing or mat-forming; leaves less than 2 cm long
            O. Leaves with distinct slender petioles up to 10 mm long; blades with whitish to grayish pubescence beneath; flowers solitary on erect bractless peduncles; calyx with long dark glands *Dryas*, p. 189
            O. Leaves sessile or nearly so; blades with rusty-brown woolly pubescence beneath; flowers in clusters on the leafy branches; calyx without dark glands *Ledum decumbens*, p. 385
          N. Stems erect; leaves 2–6 cm long
            P. Leaves covered beneath with a dense layer of rusty-woolly hairs; flowers creamy-white, not urn-shaped, in upright clusters *Ledum groenlandicum*, p. 385
            P. Leaves covered beneath with a close layer of short white hairs; flowers white to pinkish, urn-shaped, in drooping clusters *Andromeda glaucophylla*, p. 365
        M. Lower surface of leaves with a coating of small scales or with scattered hairs or glands
          Q. Lower surface of leaves with a coating of small scales; flowers 5-parted
            R. Low shrubs, less than 3 dm high; leaves 1–2 cm long, crowded towards the ends of the branches; flowers deep purple, 1–2 cm broad, very fragrant *Rhododendron lapponicum*, p. 393
            R. Taller shrubs, up to 1 m high; leaves 1–4.5 cm long, all along the stems and becoming gradually smaller towards the ends; flowers whitish, urn-shaped and drooping, not fragrant *Chamaedaphne calyculata*, p. 371
          Q. Lower surface of leaves with scattered hairs or glands; flowers 4-parted
            S. Stems clothed with appressed or incurved brownish hairs; lower surface of leaf with scattered brown bristlelike hairs; fruit a translucent white berry *Gaultheria hispidula*, p. 375
            S. Stems glabrous or with short crisp curly white hairs; lower surface of leaf black-dotted with bristlelike glands; fruit a firm red berry *Vaccinium vitis-idaea*, p. 417
      L. Lower surface of mature blades pale green or glaucous and glabrous or nearly so
        T. Leaves with rounded or cordate base, ciliate margins and petiole to one half the length of the blade or longer *Epigaea repens*, p. 373
        T. Leaves tapered to the base, margins not as above and petioles short or none

U. Leaves up to 7 cm long, crowded towards the ends of the branches; margins with obscure to prominent teeth

V. Leaves obscurely toothed; flowers bell-shaped, hanging singly below the leaves from the leaf axils; fruit red, fleshy, with wintergreen flavour **Gaultheria procumbens**, p. 377

V. Leaves with prominent teeth; flowers saucer-shaped, in clusters on a long peduncle above the leaves; fruit a capsule **Chimaphila**, p. 361

U. Leaves less than 3 cm long, all along the stems; margins entire

W. Stems and branches stiff and woody; leaves 1 – 3 cm long

X. Leaves narrowly linear-lanceolate to oblong, strongly revolute; lower surface with white to bluish waxy coating; tundra habitats along Hudson Bay shores **Andromeda polifolia**, p. 363

X. Leaves oblong or oval, only slightly revolute, green above and slightly paler beneath; widely distributed **Arctostaphylos uva-ursi**, p. 369

W. Stems and branches slender and flexible; leaves up to 1.5 cm long **Vaccinium**, p. 394

Key II. Leaves deciduous; opposite, subopposite, or whorled

A. Leaves compound

B. Upright shrubs

C. Leaves pinnately compound with 5 – 11 leaflets; flowers in erect clusters; fruit a berrylike drupe **Sambucus**, p. 449

C. Leaves trifoliolate; flowers in drooping clusters; fruit an inflated 3-angled pod **Staphylea trifolia**, p. 295

B. Vines with trailing, twining or climbing stems

D. Slender vine climbing by its leaf petioles which act as tendrils; leaves trifoliolate; flowers white or bluish to pinkish-purple **Clematis**, p. 113

D. Coarse, vigorous vine climbing by aerial roots; leaves pinnately compound; flowers trumpet-shaped, orange-red **Campsis radicans**, p. 421

A. Leaves simple

E. Leaves mostly opposite but some subopposite

F. Leaves oblanceolate, spatulate, or narrowly elliptic to linear-lanceolate with short straight veins; flowers in catkins, appearing before leaves **Salix purpurea**, p. 75

F. Leaves broadly elliptic to ovate with long curved veins; flowers on slender stalks in clusters from axils of lower leaves, appearing after the leaves **Rhamnus cathartica**, p. 309

E. Leaves all opposite or in whorls or 3 or 4

G. Leaves with prominently lobed margins

H. Leaves with 3 or 5 pointed lobes, palmately 3–5-veined from the base

I. Leaves usually with stipules; petioles with glands near the junction with blade or lower surface of leaf pubescent and with brown or blackish resinous dots; flowers creamy-white in clusters broader than long; fruit a berrylike drupe **Viburnum**, p. 457

I. Leaves without stipules; petioles lacking glands; flowers greenish-yellow in clusters longer than broad; fruit a pair of winged nutlets (keys) **Acer**, p. 297

H. Leaves with few to several rounded lobes, especially near the base; pinnately veined **Symphoricarpos**, p. 453

G. Leaves not lobed

J. Margins of leaves with coarse or fine teeth

K. Teeth relatively few (less than 25 to a side), coarse and prominent **Viburnum**, p. 457

K. Teeth numerous (more than 25 to a side) and fine

L. Stems round in section; flowers in terminal clusters or both terminal and axillary

M. Leaves ovate or elliptic, longer than broad; flowers white or yellow, all alike

N. Large shrub (up to 2–5 m tall); petioles winged; flowers white, in open cymes; fruit a blue drupe **Viburnum lentago**, p. 465

N. Low shrub (less than 1.5 m high); petioles not winged; flowers yellow, in few-flowered clusters; fruit a slender capsule **Diervilla lonicera**, p. 429

M. Leaves broadly oval or almost round; flowers white, of two kinds, the outer ones enlarged and showy **Viburnum alnifolium**, p. 459

L. Stems 4-angled or winged; flowers on slender stalks from the leaf axils, not terminal **Euonymus**, p. 289

J. Margins of leaves entire, wavy, or with low rounded teeth, sometimes ciliate

O. Branchlets and leaves with numerous small, circular, brown scales; fruit a red berry **Shepherdia canadensis**, p. 337

O. Branchlets and leaves without brown scales; fruit a capsule, nutlet, drupe, or berry

P. Leaves with translucent glandular dots on the upper surface; stamens numerous **Hypericum**, p. 323

P. Leaves without translucent dots; stamens 4 or 5 per flower

Q. Leaves opposite or in whorls of 3 (rarely 4); flowers very numerous (up to 100 or more) in dense, spherical, long-stalked heads; fruits brown nutlets; in wet habitats **Cephalanthus occidentalis**, p. 425

Q. Leaves always opposite (paired); flowers not as above; fruits white or coloured berrylike drupes; in dry or wet habitats

R. Lateral veins of the leaf curved prominently towards the tip; flowers 4-parted, in terminal stalked clusters or, if sessile, surrounded by four white bracts **Cornus**, p. 342

R. Lateral veins of the leaf not curved prominently towards the tip; flowers mostly 5-parted, in terminal or axillary clusters or few from axils of upper leaves

S. Bark of older stems shredding; petioles not grooved; fruit a several-seeded berry or a berrylike drupe with 2 nutlets

T. Flowers yellow to orange-red (or rarely pink or white in introduced species up to 3 m tall), more than 12 mm long; fruit a blue, purplish, or orange-red several-seeded berry; shrubs or vines **Lonicera**, p. 433

T. Flowers white to pinkish, less than 10 mm long; fruit a white or greenish, spongy, berrylike drupe with 2 nutlets; low shrubs, usually less than 1 m high **Symphoricarpos**, p. 453

S. Bark of older stems not shredding; petioles grooved; fruit a blue ellipsoidal or nearly spherical drupe with a single flat stone **Viburnum cassinoides**, p. 461

Key III. Leaves deciduous; alternate and compound

A. Leaflets 3 or 5, palmately or pinnately arranged
  B. Margins of leaflets sharply toothed
    C. Stipules prominent, tendrils lacking
      D. Stipules fused with the petiole for about half their length; flowers rose-coloured, rarely white; fruit a red hip **Rosa**, p. 212
      D. Stipules not fused with the petiole; flowers white to pink or rose-coloured; fruit an aggregation of tiny, red to purple-black, fleshy drupelets (dewberries, raspberries, blackberries) **Rubus**, p. 226
    C. Stipules lacking, tendrils present **Parthenocissus**, p. 313
  B. Margins of leaflets entire, wavy or with remote fine teeth, coarse irregular teeth or lobes
    E. Low shrub or vine climbing by aerial roots; fruit a capsule or drupe
      F. Leaflets usually 5 (sometimes 3 or 7), narrow; fruit a capsule; shrub **Potentilla**, p. 195
      F. Leaflets 3, broad; fruit a drupe; shrub or vine **Rhus**, p. 269
    E. Tall shrub or small tree; fruit a roundish waferlike samara **Ptelea trifoliata**, p. 263
A. Leaflets 6–31, pinnately arranged
  G. Leaves twice compound; stems bristly at the base; flowers in umbels; fruit a purple-black berry **Aralia hispida**, p. 339
  G. Leaves once compound
    H. Stems usually prickly
      I. Tall aromatic shrubs up to 3 m or more; leaves up to 25 cm long, the leaflets entire or with shallowly crenulate margins; flowers small, in axillary clusters **Zanthoxylum americanum**, p. 265
      I. Medium-sized shrubs usually less than 2 m high; leaves less than 10 cm long, the leaflets with sharply toothed margins; flowers large, pink or rarely white **Rosa**, p. 212
    H. Stems not prickly
      J. Low shrub usually less than 1 m high; fruit a capsule
        K. Leaves less than 1 dm long with 3–7 leaflets; flowers yellow, 1–2.5 cm across **Potentilla fruticosa**, p. 197
        K. Leaves 1–4 dm long, with 13–21 leaflets; flowers less than 1 cm across, white, numerous in dense, terminal, pyramidal clusters **Sorbaria sorbifolia**, p. 257
      J. Coarse shrubs or trees up to 10 m high; fruit a pome or berrylike drupe
        L. Stipules present; flowers white, showy, in clusters broader than long; fruit a small red pome **Sorbus**, p. 253
        L. Stipules absent; flowers small, greenish-yellow in elongate clusters; fruit a berrylike drupe **Rhus**, p. 269

Key IV—Leaves deciduous; alternate and simple

A. Leaves deeply or shallowly cut or lobed, the segments or lobes blunt, rounded, or sharp-pointed (For second A, see p. xxviii)
  B. Trailing, scrambling, or climbing vines
    C. Tendrils present; leaf margins toothed **Vitis**, p. 315
    C. Tendrils lacking; leaf margins entire
      D. Leaves about as broad as long; petioles long, attached to lower surface of blade; flowers creamy-white; fruit a blue berrylike drupe **Menispermum canadense**, p. 121

D. Leaves longer than broad, often with one or more earlike lobes at the base; petiole short, attached to the base of the blade; flowers purple; fruit a translucent red berry **Solanum dulcamara**, p. 419

B. Shrubs or trees

E. Leaves palmately lobed

F. Stems unarmed

G. Stems slender, creeping, with low upright branches 1 - 3 dm high; leaves few, reniform, with rounded lobes and shallow sinuses; flowers solitary; north of 48° N. **Rubus chamaemorus**, p. 235

G. Stems stout, trailing, ascending or erect, up to 3 m high; leaves numerous, not reniform, the lobes blunt or sharp-pointed; flowers several to many, in clusters

H. Leaves large (10 - 20 cm across); stems and petioles usually clammy with glandular hairs; calyx lobes caudate-tipped; flowers 3 - 5 cm across

I. Flowers rose-purple; fruits pink to red, dry and insipid; southern and eastern **Rubus odoratus**, p. 245

I. Flowers white; fruits orange to red, edible; northern and western **Rubus parviflorus**, p. 247

H. Leaves smaller (3 - 10 cm long); stems glandless or with a few scattered glands; calyx lobes not caudate-tipped; flowers less than 1 cm across

J. Low to medium-sized shrubs usually less than 1 m high (except *Ribes odoratum* with fragrant yellow tubular flowers); stems sparingly branched, the bark close or, if peeling, the inner bark exposed; flowers variously coloured, solitary or few, short-stalked in axillary or lateral clusters; fruit a many-seeded berry **Ribes**, p. 130

J. Tall shrubs up to 3 m high; stems much branched, with conspicuously peeling and persistent papery bark; flowers white, numerous, long-stalked in showy terminal clusters; fruit a papery brown pod up to 1 cm long **Physocarpus opulifolius**, p. 193

F. Stems armed with thorns or prickles

K. Mature leaves less than 2 dm across; shrubs and trees

L. Low shrubs with trailing, ascending, or erect stems usually less than 1 m high; stems with long, stout, nodal prickles or bristles or both; flowers solitary or few in axillary or lateral clusters; corolla bell-shaped or saucer-shaped with small erect petals; fruit a many-seeded berry **Ribes**, p. 130

L. Coarse shrubs or trees with stout erect stems and spreading or ascending branches; stems with stout straight or curved thorns or spine-tipped branches; flowers in clusters on short shoots or at the ends of leafy branches, showy, with conspicuous spreading petals; fruit a pome or a berrylike drupe

M. Thorns usually leafy or with lateral buds; flowers few (2 - 6) on long slender stalks in umbel-like clusters on short shoots, sweet-scented; seeds naked within the papery-walled cavities of the ovary **Malus coronaria**, p. 191

M. Thorns lacking lateral buds; flowers numerous on stout pedicels in a branched inflorescence at the ends of leafy shoots, usually ill-smelling; seeds enclosed in a bony covering and surrounded by the pulp of the fruit **Crataegus**, p. 170

K. Mature leaves 2 - 4 dm across; petioles and lower surface of leaves with strong stout prickles; large shrub up to 2 m tall; on islands off the north shore of Lake Superior **Oplopanax horridus**, p. 341

E. Leaves pinnately lobed or of three different shapes and margins entire

N. Leaves trilobed, mitten-shaped, or unlobed; margins entire; large shrub or tree with aromatic bark and foliage; flowers greenish-yellow; fruit a red-stalked, blue, berrylike drupe **_Sassafras albidum_**, p. 129

N. Leaves all of similar shape, the margins with 10 or more blunt to rounded short lobes, gland-dotted, and aromatic; fruit a bur **_Comptonia peregrina_**, p. 87

A. Leaves not lobed

O. Margins of leaves entire

P. Vines with bristly or prickly stems; leaves ovate or rounded, with 1–3 main veins on each side of the midrib, running from base to apex **_Smilax_**, p. 11

P. Shrubs and trees without bristles or prickles

Q. Veins arising from the midrib and extending towards the leaf tip **_Cornus alternifolia_**, p. 345

Q. Veins arising from the midrib and extending towards the margin

R. Leaves with golden-yellow resinous dots or brown scales on the lower surface

S. Leaves green, with resinous dots on both sides (more noticeable below); flowers yellowish-orange to reddish **_Gaylussacia baccata_**, p. 379

S. Leaves silvery gray-green with small brown scales on the lower surface (at least along the veins); flowers yellow, with a heavy scent **_Elaeagnus commutata_**, p. 335

R. Leaves lacking resinous dots and brown scales

T. Leaves large (15–30 cm long), drooping from the branches; flowers reddish-brown, 3–4 cm across, 3-parted, the fleshy perianth conspicuously veiny **_Asimina triloba_**, p. 123

T. Leaves smaller (less than 15 cm long), not drooping; flowers smaller, white, greenish-white, yellow, or pinkish

U. Plants spicy-aromatic, the branchlets, leaves, and bright red drupes with citronellalike odour when bruised **_Lindera benzoin_**, p. 127

U. Plants not aromatic

V. Stems with soft wood, tough flexible bark, and conspicuously swollen nodes; buds hidden under the dome-shaped base of the petiole **_Dirca palustris_**, p. 331

V. Stems and buds not as above

W. Buds covered with a single scale; flowers in catkins, the male and female on separate plants **_Salix_**, p. 14

W. Buds naked or with two or more scales; flowers solitary or in few- to many-flowered, terminal or lateral clusters

X. Fruit a capsule; flowers rose-purple, showy, 2–3 cm long; known only from Alfred Bog, Prescott County **_Rhododendron canadense_**, p. 391

X. Fruit a berry or a berrylike drupe; flowers less than 1.5 cm long

Y. Fruit a blue to blue-black berry with many small seeds; flowers cylindric to bell-shaped **_Vaccinium_**, p. 394

Y. Fruit a berrylike drupe with 1–5 stones; flowers not as above

Z. Flowers opening before the leaves, fragrant, rosy-purple; drupes bright red **_Daphne mezereum_**, p. 329

Z. Flowers opening with the leaves, inconspicuous, not fragrant

a. Drupes orange to salmon-coloured; leaves asymmetrical, with 3 main veins from the base **_Celtis tenuifolia_**, p. 111

a. Drupes purple-red or purple-black; leaves symmetrical, with pinnate venation

b. Leaves elliptic to oval, mucronate at the tip; drupe solitary, purple-red on a slender stalk much longer than the petiole of the subtending leaf **_Nemopanthus mucronata_**, p. 285

b. Leaves elliptic to obovate, abruptly pointed but not mucronate; drupes solitary or few, purplish-black, nearly sessile on stalks barely exceeding the petiole **_Rhamnus frangula_**, p. 311

O. Margins of leaves toothed or wavy

c. Vines with tough twining stems, scrambling or high-climbing; fruits orange and red **_Celastrus scandens_**, p. 287

c. Shrubs with trailing or erect stems, or trees

d. Low shrubs with creeping, trailing, or ascending stems

e. Buds covered with a single scale; leaves not persistent; flowers in catkins **_Salix_**, p. 14

e. Buds with 2 or more scales; leaves persistent after withering; flowers perfect, in clusters from scaly terminal buds **_Arctostaphylos alpina_**, p. 367

d. Erect shrubs or trees

f. Stems armed with thorns

g. Thorns less than 2 cm long, in clusters of 3 at the nodes; wood bright yellow; leaf margins prominently spiny-toothed; fruit a drooping cluster of oblong scarlet berries **_Berberis vulgaris_**, p. 119

g. Thorns borne singly along the stem, simple or rarely branched; wood not yellow; leaf margins not spiny

h. Margins of leaves finely and regularly toothed; fruit a juicy drupe 2-3 cm long (Plums) **_Prunus_**, p. 200

h. Margins of leaves coarsely and irregularly toothed or lobed; fruit a pome

i. Thorns usually leafy or with lateral buds; flowers few (2-6) on long slender stalks in clusters on short shoots, sweet-scented; seeds naked within the papery-walled cavities of the ovary **_Malus coronaria_**, p. 191

i. Thorns lacking lateral buds; flowers numerous, on stout pedicels in a branched inflorescence at the ends of leafy shoots, usually with disagreeable odour; seeds enclosed in a bony covering and surrounded by the pulp of the fruit **_Crataegus_**, p. 170

f. Stems unarmed

j. Leaves mostly oblique at the base, i.e., the leaf tissue joined to the petiole at different points on the two sides

k. Leaves broadly oval or roundish, the margins wavy with irregular broad teeth; flowers in autumn; fruit a persistent woody capsule **_Hamamelis virginiana_**, p. 151

k. Leaves ovate, tapered to a pointed tip, the margins dentate with low sharp teeth, at least above the middle; flowers in spring; fruit a berrylike drupe **_Celtis tenuifolia_**, p. 111

j. Leaves mostly not oblique at the base, i.e., the leaf tissue joined to the petiole at about the same level on both sides

l. Buds covered by a single scale; flowers in catkins; fruit a capsule **_Salix_**, p. 14

l. Buds with 2 or more scales; flowers solitary or in various types of inflorescences (or if in catkins, the fruit a nut or nutlet)

m. Foliage aromatic when crushed; leaves tapered towards the base and toothed near the apex; fruit a cluster of tiny nutlets **_Myrica_**, p. 89

m. Foliage not aromatic; leaves various

n. Leaves with a midvein and two additional veins from the base
 o. Low shrubs (less than 1 m high); leaves serrate with gland-tipped teeth; flowers white in long-stalked axillary or terminal clusters; fruit a 3-lobed brown capsule ***Ceanothus***, p. 301
 o. Taller shrubs (up to 4 m high); teeth not gland-tipped; flowers small, in groups of 2 or 3 on short pedicels along the branches; fruit an orange berrylike drupe ***Celtis tenuifolia***, p. 111
n. Leaves with one main vein (midrib) and lateral branches
 p. Branchlets and leaves gland-dotted ***Betula***, p. 97
 p. Branchlets and leaves not gland-dotted
  q. Leaves elliptic with tapered base and 3–7 widely spaced rounded teeth on each side; fruit an acorn ***Quercus prinoides***, p. 109
  q. Leaves of various shapes, the margins with fine teeth or doubly serrate; fruit not an acorn
   r. Margins coarsely doubly serrate or crispy-wavy
    s. Stems with conspicuous pale lenticels; pith 3-sided in section; nutlets small, in persistent woody cones less than 2.5 cm long ***Alnus***, p. 93
    s. Stems without conspicuous lenticels; pith oval in section; nuts about 1 cm long in leafy-bracted clusters or at the base of long-beaked bristly involucres ***Corylus***, p. 105
   r. Margins neither doubly-toothed nor crispy-wavy
    t. Glands on upper midrib of leaf or on petioles or leaf margins near junction with the petiole
     u. Glands on petiole and margin of blade near junction with the petiole; lenticels prominent; medium to tall shrubs up to 12 m high ***Prunus***, p. 200
     u. Small dark glands on midrib of upper surface of leaf near the petiole; lenticels not prominent; medium-sized shrubs up to 2.5 m high ***Aronia***, p. 169
    t. Glands lacking on leaf blades and petioles
     v. Leaves sessile or on very short petioles (less than 0.5 cm long)
      w. Flowers bell-shaped, solitary or in few-flowered clusters; fruit a many-seeded berry ***Vaccinium***, p. 394
      w. Flowers minute, roselike, in many-flowered, erect, terminal clusters; fruit a persistent capsule ***Spiraea***, p. 257
     v. Leaves with petioles 0.5–3 cm long
      x. Base of leaf rounded to cordate
       y. Margins crenate-serrate; flowers yellowish-green, about 3 mm in diameter, in small clusters from the axils of the lower leaves; fruit a 1–3-seeded berrylike drupe ***Rhamnus alnifolia***, p. 307

y. Margins finely serrate to coarsely dentate; flowers white, showy, 1 - 3 cm in diameter, in terminal racemes or from the axils of upper leaves; fruit a many-seeded berrylike pome *Amelanchier*, p. 152

x. Base of leaf blade tapered

z. Petiole grooved (narrowly winged); leaves 2 - 5 cm wide; flowers greenish, nearly sessile or short-stalked and borne along the branches; fruit a red or purple-black berry-like drupe with 1 - 3 stones

AA. Leaf margins crenate-serrate; stipules linear, conspicuous, but falling before the fruit ripens; fruit a stalked purple-black drupe
Rhamnus alnifolia, p. 307

AA. Leaf margins sharply serrate with incurved teeth; stipules small, narrow, nearly black, persistent; fruit a nearly sessile bright red drupe
Ilex verticillata, p. 283

z. Petiole neither grooved nor winged; leaves 1 - 3 cm wide; flowers white to rose-coloured, in terminal pyramidal clusters, or on short lateral branchlets, or one to several on long pedicels from the leaf axils; fruit a capsule, a pome, or a drupe with a large stone

BB. Flowers small (less than 10 mm across), in dense terminal clusters; fruit a capsule, often persisting over winter ***Spiraea***, p. 257

BB. Flowers larger (10 mm or more in diameter), long-stalked, and solitary or few in axillary or lateral clusters; fruit a pome or a drupe

CC. Leaf margins finely and remotely toothed at least above the middle; fruit a purple to blackish drupe with a single pit
Prunus pumila, p. 209

CC. Leaf margins finely and closely sharp-toothed; fruit a several-seeded purple-black pome
Amelanchier bartramiana, p. 159

DESCRIPTIONS OF SPECIES

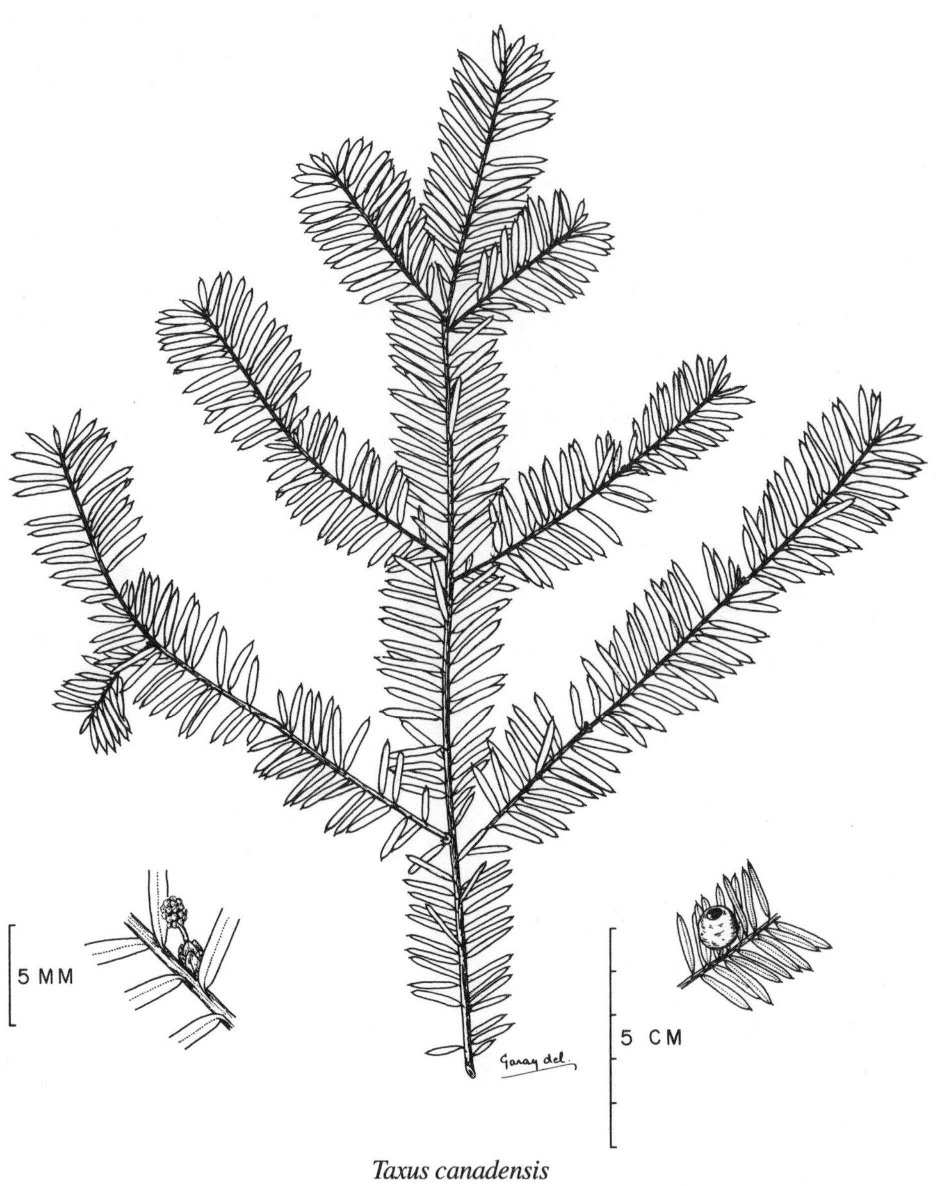

Taxus canadensis

TAXACEAE — YEW FAMILY

Five genera and about 20 species in North America, Eurasia, and New Caledonia. Evergreen shrubs and trees with spirally arranged, needlelike to linear-lanceolate leaves. Stems and leaves lacking the resin ducts that are common to conifers generally. Plants dioecious; male strobili solitary or in small spikes or scaly cones in the leaf axils; female strobili on small axillary shoots that bear several pairs of scales; ovule 1, terminal. The single bony seed produced by the female strobilus matures in one season and is surrounded by a fleshy disc (aril). Wind-pollinated; seed dispersal by birds and small mammals.

The name of the family is based on the genus *Taxus*.

Taxus L. — Yew

About 10 species in the northern hemisphere. Shrubs or trees with the spirally attached leaves arranged in two rows along the upper surface of the branches in a nearly flat plane; dioecious. Strobili unisexual, solitary in the axils of the leaves of the previous season; male strobilus with a few scales and 8 - 10 stamens; female strobilus with a single ovule at the end of a short shoot bearing three pairs of scales; mature seed surrounded by a fleshy aril.

Species commonly used for hedge and foundation plantings are English yew (*T. baccata* L.) and Japanese yew (*T. cuspidata* Sieb. & Zucc.). Wood of the former was used for the making of bows in the British Isles and continental Europe. (*Taxus*—Greek *taxos*, a yew tree; *baccata*—bearing berries; *cuspidata*—sharp-pointed)

Taxus canadensis Marsh. — American yew, Ground hemlock

A low spreading shrub of damp and shaded places, seldom reaching a height of more than 2 m, but with branches up to 2 m long spreading from the base of the plant for about one-third their length before curving upwards. Branchlets slender and green at first but becoming brownish and scaly.

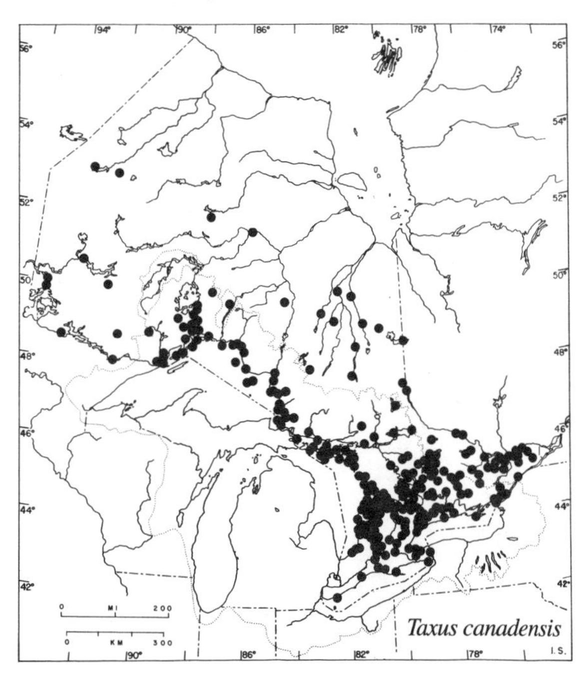

Taxus canadensis

Leaves evergreen, numerous; the blade linear, flat, 1 - 2.5 cm long, 1 - 3 mm wide, dark green above and paler beneath, abruptly narrowed to a short sharp tip (mucronate) and arising in a close spiral around the stem; the petioles twisted so that the leaves form flat sprays along the branches.

Plants dioecious; the male strobilus a small stalked cluster of pollen sacs projecting from a basal cuplike group of tiny scales and forming a minute cone, most readily seen in the spring as pollen is being shed, the female a single minute ovule on a short stalk which bears 3 pairs of scales, its presence revealed in spring by the shining pollination droplet, but otherwise inconspicuous until maturity in midsummer when the fertilized ovule has become a brown bony seed surrounded by a bright red pulpy aril 5 - 10 mm across, and open at the free end so that it resembles a red berry with a hole at the top. Strobilar buds arise on the lower side of the branches in the axils of leaves of the previous year's growth and open in early spring before new leaves have appeared. (*canadensis* — Canadian)

In swampy thickets, coniferous or mixed woods, on ravine slopes and rocky banks.

Widespread in southern Ontario but becoming rare and local north of 50° N; absent from the Boreal Forest and Barren Region. (Nfld. to s.e. Man., south to Ky. and W. Va.)

Note The fleshy red aril is edible, sweet-tasting, but slimy. Seeds, branchlets, and leaves are reported to be poisonous, but this reputation may be based on the known toxicity of related species of Eurasia.

Hemlock and balsam fir, trees with flat needles, may also occur as low spreading plants in response to injury or other damage, but in both the needles are blunt, whitened beneath, and lacking the mucronate tip characteristic of the yew. Yew is the only non-resinous conifer in Ontario.

Field check Low spreading evergreen shrub with flat sprays of abruptly sharp-pointed flat needles, green on both sides; red "berries" with a single seed visible at the open end.

CUPRESSACEAE – CYPRESS FAMILY

Evergreen shrubs and trees, the former sometimes prostrate in growth habit; leaves needlelike or scaly, opposite or whorled; strobili unisexual, the male and female on the same plant (monoecious) or on separate plants (dioecious); seeds borne in scaly or berrylike cones. About 19 genera and 150–200 species, which may be divided into two groups: one with cone scales small, woody, and distinct—*Cupressus* (cypress), *Chamaecyparis* (yellow cedar), *Thuja* (arborvitae); the other with fleshy cone scales uniting during maturation to form a berrylike mass (cone)—*Juniperus* (juniper).

Juniperus L. – Juniper

About 60 species in the northern hemisphere; dioecious, occasionally monoecious. The wood of red cedar, *Juniperus virginiana* L., is used for lining cupboards and linen chests and for making pencils. The berrylike cones are eaten by birds and small mammals. (*Juniperus*—the Latin name for the European juniper)

Key to *Juniperus*

a. Stems stiffly erect or arching upwards; leaves awl-shaped, 10–20 mm long, in whorls of three, usually well spaced; each leaf jointed at its rounded base and tapered to a spinelike tip, a white stripe running down the middle ***J. communis*** **var.** ***depressa***

a. Stems prostrate and trailing, with numerous ascending short side branches; leaves scalelike, less than 5 mm long, opposite, mostly crowded and overlapping; leaf decurrent at the base (even if sharp-pointed at the tip in juvenile or infected foliage), without a median white stripe ***J. horizontalis***

Juniperus communis var. *depressa*

Juniperus communis L. var. *depressa* Pursh

Common juniper
Ground juniper

A low shrub, erect or spreading, rarely over 1.5 m in height, with erect or ascending stiff branches but usually lacking a main central stem or "leader" and often forming broad patches which gradually die off from the centre. Branchlets smooth and greenish at first, soon becoming ridged and pale brown and eventually dark brown with scaly bark.

Leaves (needles) evergreen, awl-shaped, and usually shallowly keeled, 12–20 mm long and 2 mm or less in width at the rounded base, gradually tapered to a slender spinelike tip, with a broad bluish-white stripe down the middle of the upper surface, at first in stiff erect whorls of three along the branchlets, becoming more open on the older branches.

Male and female strobili in dissimilar cones on separate plants; the male a small catkinlike cluster of stamens in twos or threes around a central stalk with several small scales at the base; the female cone consisting of 3–8 small, pointed, tightly clustered scales, some or all of which bear 1–3 ovules; both kinds of cones arise in the axils of the leaves of the previous year's growth; May and June. After fertilization the upper scales of the female cone become fleshy and grow together around the ovules, forming a round mass which matures in the third season into an aromatic, blue-black, berrylike cone 6–10 mm across, covered with a waxy bluish-white bloom, and enclosing 1–3 seeds. (*communis*—common; *depressa*—depressed or low)

Along sandy or rocky shores and banks, in open woods, clearings, and old fields and pastures.

From the shores of Lake Erie, Lake Ontario, and the St. Lawrence River north to Lake Superior and James Bay and west to the Ontario-Manitoba boundary. (Nfld. to Alaska, south to Calif. and Va.)

Note Juniper "berries" are used to give gin its characteristic flavour.

Dispersal is by means of birds and small mammals which eat the "berries" and excrete the seeds, with the germination capability improved by passage through the digestive tract.

Field check Low stiff-branched shrub; spine-tipped evergreen needles in whorls of three along the branches; blue berrylike cones with a bloom.

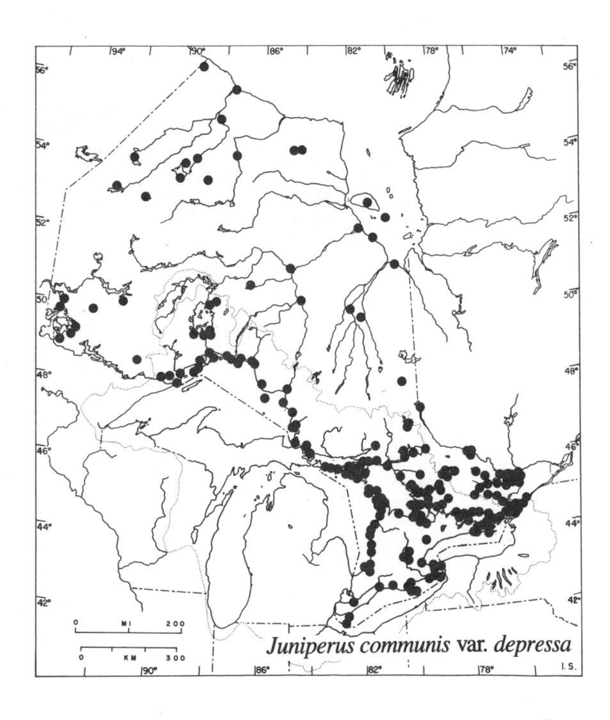

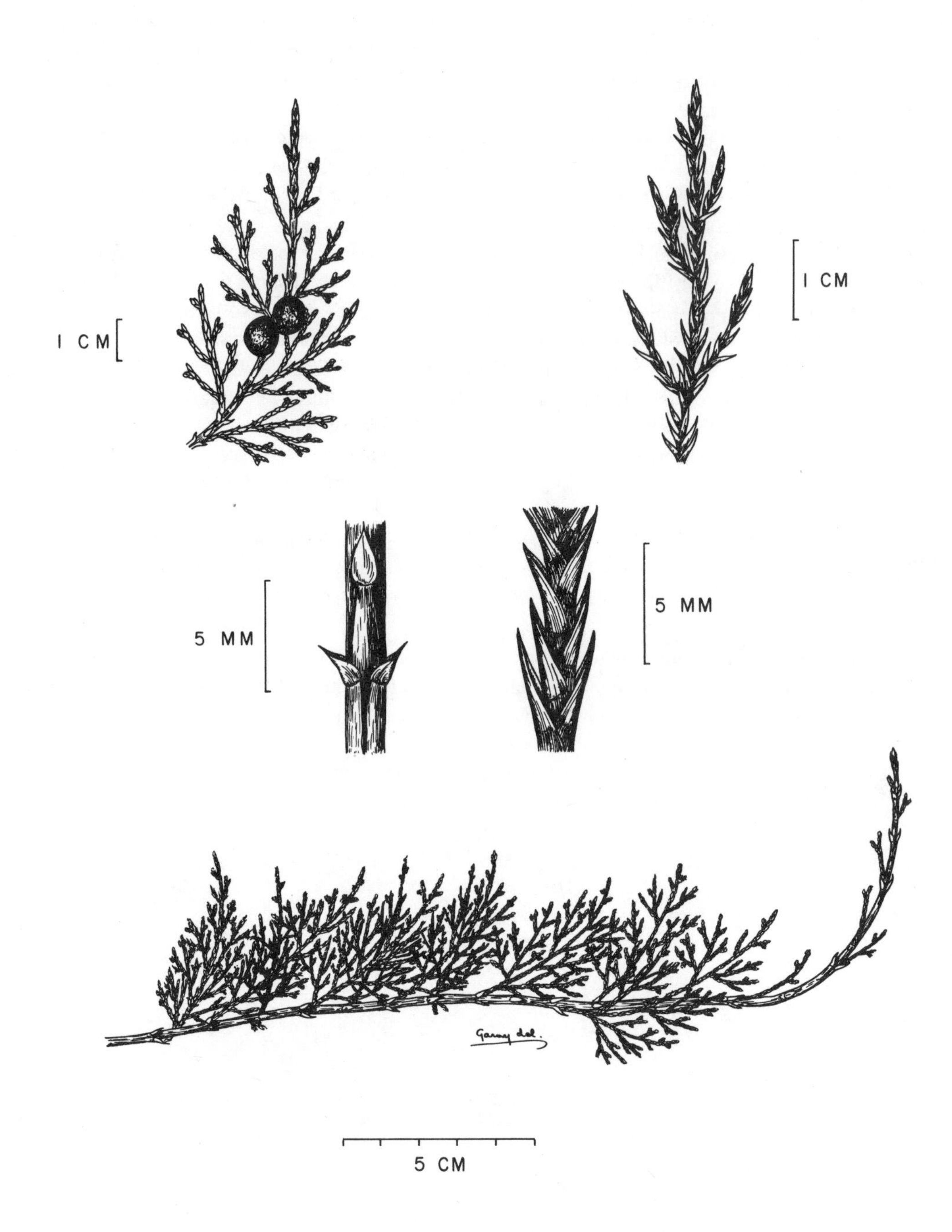

Juniperus horizontalis

Juniperus horizontalis Moench

Creeping juniper
Creeping savin

A prostrate shrub with a long trailing stem and many short erect or ascending branches, usually not more than 3 dm high. Branchlets greenish-brown and covered with numerous scalelike leaves; older stems becoming reddish-brown to blackish with scaly peeling bark.

Leaves (needles) evergreen, mostly small, scalelike, and decussate, closely appressed and overlapping, with their exposed triangular tips covering the slender branchlets and forming a diamond-shaped pattern; the leaves awl-shaped on young vigorous shoots.

Strobili minute and in dissimilar male and female cones borne at the ends of short branches, usually on separate plants; the male a tiny cluster of umbrella-shaped stamens; the female a tight round cluster of 3 – 8 small pointed scales, some or all with one or two ovules; May and June. The upper scales of the female cone become fleshy and coalesce to form a round or irregular-shaped berrylike mass enclosing the developing fertilized ovules; during the following season the berrylike cone matures, changing from green to blue and enlarging to 6 - 10 mm across on a short curved stalk. (*horizontalis*—lying flat, in reference to the growth habit)

On sand dunes, sandy or rocky shores, open rocky woods, slopes, and pastures.

Shores of Lake Huron and Lake Superior, north to James Bay and Hudson Bay and west to Lake-of-the-Woods; rare in the Deciduous Forest Region. (Nfld. to Alaska, south to Wyo. and N. Eng.)

Note Creeping juniper is a host for the fungus *Gymnosporangium nidus-avis* Thaxter. Infection causes a "witches'-broom" in which the branches are congested and the leaves awl-shaped like the juvenile leaves on vigorous new growth. (*gymnosporangium*—naked sporecases; *nidus-avis* —bird's-nest)

Field check Prostrate trailing shrub with many short branches; opposite, appressed, scalelike leaves; blue "berries" on short curved stalks.

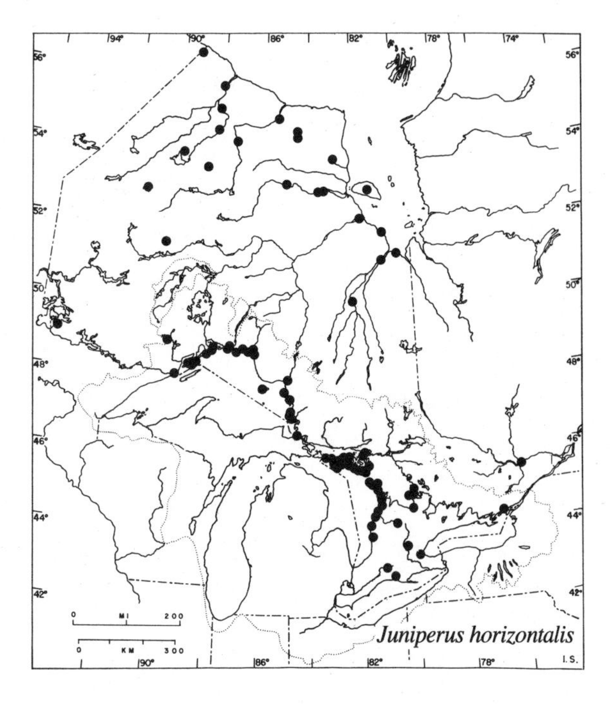

Smilax tamnoides var. *hispida*

LILIACEAE — LILY FAMILY

A large family of 250 genera and more than 3500 species, widely distributed. Mostly herbs and a few shrubs and trees. Leaves alternate, opposite or whorled; flowers 3-parted, usually bisexual and pollinated by insects; fruit a capsule or a berry. The family name is based on the genus *Lilium*, which includes the true lilies. (*Lilium*—the Latin name for the lily)

Smilax L. — Catbrier, Greenbrier

About 350 species, frequent in the tropics and subtropics but found occasionally in warm temperature regions. Both herbs and woody vines with alternate simple leaves, the broad blades net-veined between the longitudinal nerves, and the stipules prolonged as tendrils; dioecious. Flowers small, yellow or greenish-yellow, in umbel-like, stalked, axillary clusters; fruit a berry with up to six seeds. (*Smilax*—an ancient Greek name of uncertain application)

Key to *Smilax*

a. Stems usually round or slightly angular in cross-section, at least the two-year-old and older stems closely covered with slender bristlelike prickles; berries numerous, in an open umbel on a stalk more than twice as long as the petiole of the subtending leaf
S. tamnoides var. hispida

a. Stems frequently angular or four-sided, with well-spaced, flattened, black-tipped, stout prickles; berries few, in a compact cluster on a stalk about the same length as the petiole of the subtending leaf (found only in the western part of the Deciduous Forest Region)
S. rotundifolia

Smilax tamnoides L. var. *hispida* (Muhl.) Fern. (*S. hispida* Muhl.)

Bristly catbrier
Prickly greenbrier

A woody vine climbing over shrubs and trees sometimes to a height of 6 m or more; often hanging in tangled masses on the outer edges of the crown. Branchlets glabrous, often ridged, unarmed or with scattered brown to reddish prickles; older branches and lower woody stems usually beset with straight, blackish, bristly prickles.

Leaves alternate, simple, and deciduous; blades 6-13 cm long and 4-10 cm wide, ovate or elliptic to rounded, with 5-7 veins arising from the heart-shaped base and uniting at the abruptly long-pointed apex, thin and glabrous, dark green above, paler beneath, margins entire or minutely wavy; petioles 6-13 mm long, with a broad sheathing base bearing a pair of stipules in the form of slender curled tendrils.

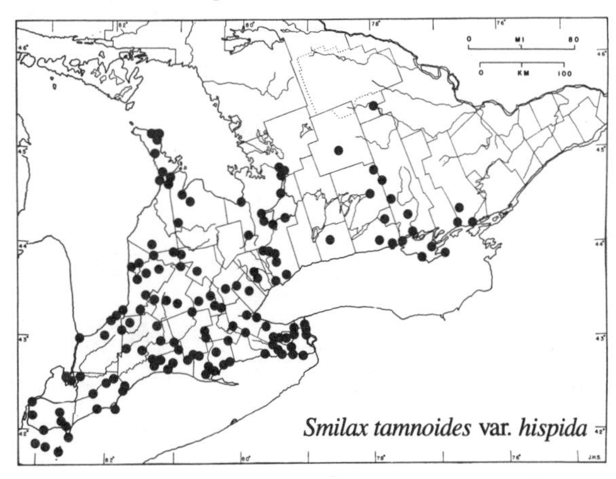
Smilax tamnoides var. *hispida*

Flowers small and unisexual, the male and female on separate plants, greenish-yellow, in clusters on slender stalks in the axils of the leaves of the current year's growth; June. Fruit a round, 1- or 2-seeded, blue-black berry about 6 mm across, in

Smilax rotundifolia

clusters of 5–20; Sept. and Oct. (*tamnoides* — like *Tamnus*, bryony; *hispida* — bristly)

In open thickets, at the edges of woods, along river banks and lake shores.

Common from the north shore of Lake Erie to the Bruce Peninsula and from Lake Ontario to the southern part of Nipissing District. (N.Y. to S.Dak., south to e. Tex. and Ga.)

Note In woodlands of southern Ontario a related species called carrion flower, *Smilax herbacea* L., is a slender herbaceous vine with long-stalked clusters of crowded blue-black berries.

Field check Woody vine with bristly stems; leaves with 5–7 prominent veins radiating from the base and coming together at the tip; berries blue-black, in long-stalked open clusters.

Smilax rotundifolia L. — Round-leaved greenbrier

A tough woody vine with numerous somewhat zigzag branches climbing over trees and shrubs. Branchlets glabrous, round or conspicuously four-sided, green or brownish, with scattered, stout, flattened, broad-based, black-tipped straight or slightly curved prickles.

Leaves much as in *S. tamnoides* var. *hispida* but somewhat smaller and more roundish, leathery in texture, frequently with minute spines on the lower surface near the petiole.

Flowers differ from those of *S. tamnoides* var. *hispida* mainly in being more bronze-coloured than yellow and having the stalks of the flower clusters short, flattened, and about as long as the subtending petioles; June. Fruit a small cluster of round, 2- or 3-seeded, blue-black berries with a bloom; Oct. or later. (*rotundifolia* — round-leaved)

In open woods and thickets.

Restricted in Ontario to the western part of the Deciduous Forest Region, but not recently collected there. (s. N.S. to Mo., south to Okla. and Fla.)

Field check Woody vine with scattered, stout, flattened, black-tipped prickles; berries few, blue-black, in short-stalked compact clusters; found only in Essex and Kent counties.

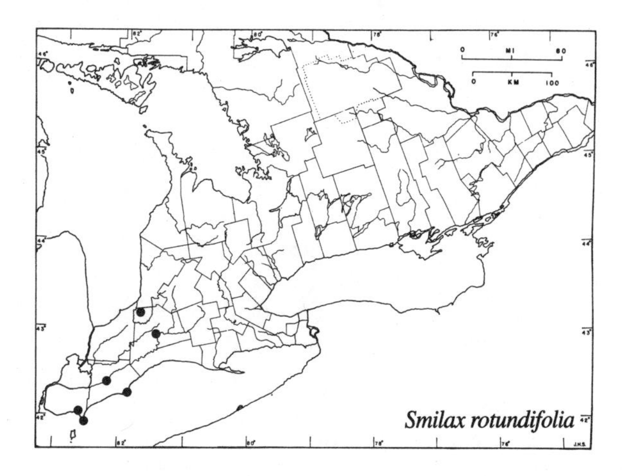

SALICACEAE — WILLOW FAMILY

Three genera and over 500 species, mostly of north temperate regions. Trees and shrubs with soft light wood and bitter-tasting bark. Leaves simple, mostly alternate, with stipules that may fall early or persist, then often leaflike. Flowers small, unisexual, borne in catkins (aments), the staminate and pistillate catkins on separate plants; each flower subtended by a small scale or bract, the true perianth lacking but represented by a cup-shaped disc or by one or more nectaries at the base of the flower; stamens 1–60, the slender filaments free or partly united below; pistil flask-shaped, composed of 2–4 fused carpels, but with a single locule containing several to many ovules. Fruit a capsule splitting open by valves; seeds provided with a tuft of silky hairs.

The name of the family is based on the genus *Salix*.

Salix L. — Willow

A large genus of about 500 species, primarily of north temperate and arctic regions but extending also into the tropics. Shrubs and trees varying in size from prostrate and dwarfed forms to large shrubs and medium-sized trees.

Buds not resinous, covered with a single hood-shaped scale, the margins of which may be fused or overlapping. Leaves simple, mostly alternate, narrowly linear to orbicular, stipules falling early or persistent. Catkins sessile, stalked or frequently on short lateral branches bearing reduced leaves, appearing before the leaves (precocious), at the same time as the leaves are unfolding (coetaneous), or after the leaves have expanded (serotinous). Staminate flowers typically with 2 stamens, rarely 1 or 3–8; pistils with a single style or the style lacking; stigmas 2, frequently lobed or branched into two parts; pollen sticky and adhering to catkin hairs. Bracts of the catkins persistent or shed early. Nectaries (glands) 1–4 per flower, basal. Fruit a broad to slender capsule which splits into two sections releasing the seeds with their tufts of silky hairs. Willows are usually insect-pollinated but wind may also be important, especially where insects are few. (*Salix* — willow)

Hybridization occurs frequently among species in certain groups of willows and even between apparently distantly related species. Thus, intergrading plants, which are difficult to assess, may be encountered in the field or herbarium. Further problems of identification arise because mature and sprout growths may differ markedly from each other and also because characters useful in species definition are present in both floral and vegetative parts. Often these are not available at the same time.

A number of willows are of economic importance: *S. alba* L., *S. babylonica* L., and *S. daphnoides* Vill., for example, are planted as ornamentals or wind-breaks; several are sources of flexible stems (osiers) for basketry or other weaving (*S. viminalis* L.); and the "pussywillows", easily forced to bloom early, are used in bouquets and spring flower arrangements (*S. discolor* Muhl.).

Special characters The following is a list of some of the characteristics that may be useful in the identification of certain species of *Salix*. Each characteristic is followed by the names of those species that exhibit this feature.

1. Buds and foliage with special fragrance when crushed — *pentandra* (bayleaf); *pyrifolia* (balsamic odour)
2. Leaves nearly round in outline — *lanata* var. *calcicola, reticulata, vestita* (all arctic)
3. Leaves rugose or impressed-veiny above — *reticulata, vestita* (both arctic)
4. Leaves shiny above — *arbusculoides, arctophila, lucida, pentandra, maccalliana, myrtillifolia, petiolaris, planifolia, serissima*
5. Leaves often drying blackish — *myricoides*

6. Young leaves often appearing reddish — *amygdaloides, eriocephala, lucida, maccalliana, nigra, pellita, planifolia, pseudomonticola*
7. Young leaves often with reddish or copper-coloured hairs — *discolor, humilis, myricoides, petiolaris*
8. Leaves densely pubescent — *candida, pellita, vestita, viminalis*
9. Leaves translucent when young — *pseudomonticola, pyrifolia*
10. Stipules prominent and persistent for more than one year — *lanata* ssp. *calcicola*
11. Stipules prominent — *cordata, eriocephala, nigra, pseudomonticola*
12. Precocious — *discolor, humilis, planifolia, lanata* ssp. *calcicola, pseudomonticola, viminalis*
13. Subprecocious to coetaneous — *arbusculoides, eriocephala, myricoides, pellita, purpurea*
14. Exceptions to stamens 2 (there are 2 in most Ontario species):
    2, but also 3 or 4 — *alba, fragilis*
    5 — *pentandra*
    3 - 6 (often 5) — *amygdaloides, lucida, nigra*
15. Capsules glabrous — *alba, amygdaloides, cordata, fragilis, lanata* ssp. *calcicola, eriocephala, lucida, myricoides, myrtillifolia, nigra, pedicellaris, pyrifolia, serissima*
16. Occurring typically (but not exclusively) in bogs — *bebbiana, glauca* ssp. *callicarpaea, pedicellaris, pyrifolia, serissima*
17. Occurring only in the tundra portion of Ontario — *arctica* var. *kophophylla, arctophila, lanata* ssp. *calcicola, reticulata*
18. Occurring in the tundra and nearby portions of the Hudson Bay and James Bay drainage basins — *arbusculoides, brachycarpa, glauca* ssp. *callicarpaea, vestita*
19. Occurring frequently in gardens, parks, or in street and highway plantings (introduced and sometimes escaping from cultivation) — *alba, babylonica, fragilis, pentandra, purpurea, viminalis.*

Species of *Salix* included in key: *alba, amygdaloides, arbusculoides, arctica* var. *kophophylla, arctophila, babylonica, bebbiana, brachycarpa, candida, cordata, discolor, eriocephala, exigua, fragilis, glauca* ssp. *callicarpaea, humilis, lanata* ssp. *calcicola, lucida, maccalliana, myricoides, myrtillifolia, nigra, pedicellaris, pellita, pentandra, petiolaris, planifolia, pseudomonticola, purpurea, pyrifolia, reticulata, serissima, vestita, viminalis.*

Key to *Salix*

A. Low shrubs (less than 0.5 m high) with prostrate or trailing branches, and dwarf shrubs with erect branches up to 1.5 m tall (or taller in *S. arbusculoides*), in Ontario chiefly restricted to the coastal tundra along the shores of James Bay and Hudson Bay
  B. Mature leaves glabrous or nearly so
    C. Leaves glaucous beneath
      D. Petioles more than half as long as the leaf blade; leaves with impressed veins above and prominently reticulate below; catkins developing after the leaves **S. *reticulata***
      D. Petioles short or not exceeding half the length of the leaf blade; leaves not prominently reticulate; catkins developing before or with the leaves
        E. Stipules prominent and persistent; catkins appearing before the leaves **S. *lanata* ssp. *calcicola***
        E. Stipules minute or absent; catkins appearing with the leaves
          F. Leaf blades narrowly oval to elliptic-oblong with nearly parallel sides; bog-inhabiting shrub of wide range **S. *pedicellaris***
          F. Leaf blades broadly elliptic, oblong, obovate, or suborbicular with obviously curved sides; tundra species of Hudson Bay drainage basin

G. Stems and leaves glabrous; margins of leaves glandular-serrulate **S. arctophila**

G. Stems and leaves pubescent or villous, at least when young; mature leaves usually pubescent along the veins; margins of leaves entire or with glands only near the base

H. Stems mostly trailing or with only the tips of branches erect, less than 3 dm high; mature stems usually smooth and glossy **S. *arctica* var. *kophophylla***

H. Stems mostly erect, up to 1.5 m tall, mature stems usually densely silky-villous **S. *glauca* ssp. *callicarpaea***

C. Leaves not glaucous, green and shiny on both surfaces **S. myrtillifolia**

B. Mature leaves prominently pubescent, at least beneath

I. Leaf blades densely tomentose below with a feltlike coating of tangled white hairs, the upper surface with tufts of hairs **S. candida**

I. Leaf blades not as above

J. Stipules prominent and persistent; catkins appearing before the leaves **S. *lanata* ssp. *calcicola***

J. Stipules minute or absent; catkins appearing with or after the leaves

K. Leaf blades dark green and rugose with impressed veins; the venation of the lower surface mostly obscured by a coating of long, shiny, silky, white hairs; catkins appearing after the leaves **S. vestita**

K. Leaf blade without sharp contrast between upper and lower surfaces; catkins appearing with the leaves

L. Margins entire or only remotely glandular-crenate

M. Petioles 1–3 mm long; catkins usually less than 2 cm long **S. brachycarpa**

M. Petioles 3–12 mm long; catkins 2–5 cm long

N. Stems mostly trailing or with only the tips of branches erect, less than 3 dm high; mature stems usually smooth and glossy **S. *arctica* var. *kophophylla***

N. Stems mostly erect, up to 1.5 m tall; mature stems usually densely silky-villous **S. *glauca* ssp. *callicarpaea***

L. Margins finely and evenly glandular-serrulate **S. arbusculoides**

A. Erect shrubs more than 1.5 m tall or trees; not restricted to the coastal tundra along the shores of James Bay and Hudson Bay

O. Leaves subopposite to opposite **S. purpurea**

O. Leaves alternate

P. Petiole glandular at or near its junction with the blade

Q. Leaves narrowly linear, linear-lanceolate, or elliptic-lanceolate; large trees, usually with a single stem

R. Leaves attenuate and falcate at the tip; stipules prominent on vigorous shoots; indigenous **S. nigra**

R. Leaves acute, acuminate or attenuate at the tip but not falcate; stipules small, lanceolate, usually falling early; introduced

S. Branchlets slender, long-drooping; pistillate catkins short, usually less than 4 cm long **S. babylonica**

S. Branchlets stout, sometimes pendent but not long-drooping; pistillate catkins usually more than 4 cm long

T. Leaves glabrous on both sides; margins closely serrate; twigs brittle at the base **S. fragilis**

T. Leaves usually silky-hairy beneath; margins finely serrate; twigs not brittle at the base **S. alba**

Q. Leaves broadly lanceolate to oblong-lanceolate or ovate-oblong
  U. Leaves glaucous beneath
    V. Leaves cordate to rounded at base, the apex blunt or rounded to acute or acuminate but not long-tapered; young leaves translucent; buds and leaves with resinous balsamlike fragrance *S. pyrifolia*
    V. Leaves acute to obtuse or rounded at the base, the apex acute, acuminate or long-tapered; young leaves not translucent; buds and leaves lacking balsam-like fragrance
      W. Young leaves glabrous; margins of mature blades glandular-serrate; fruiting in late summer or in autumn *S. serissima*
      W. Young leaves sparsely pubescent; margins of mature blades closely serrate; fruiting in early summer *S. amygdaloides*
  U. Leaves not glaucous beneath
    X. Apex of leaf blade long-attenuate with tail-like tip; branchlets and young leaves not aromatic; indigenous *S. lucida*
    X. Apex of leaf blade short-acuminate; branchlets and young leaves with resinous fragrance; introduced *S. pentandra*

P. Petiole not glandular at or near its junction with the blade
  Y. Mature leaves pubescent, at least beneath
    Z. Leaves linear to narrowly lanceolate
      a. Leaf blades densely tomentose below with a feltlike coating of tangled white hairs; young branchlets gray-white pubescent; shrub of calcareous bogs *S. candida*
      a. Leaf blades not as above; young branchlets glabrous, pruinose, or sparsely pubescent
        b. Margins of leaves entire, more or less revolute, or shallowly undulate, but lacking definite teeth; lower surface of leaf prominently hairy
          c. Branchlets numerous, brittle, usually glaucous, the leaves scattered or well-spaced; pubescence a dense layer of shiny white hairs; indigenous *S. pellita*
          c. Branchlets few, flexible, not glaucous, the numerous leaves closely crowded; pubescence of appressed, straightish, silky hairs; introduced *S. viminalis*
        b. Margins of leaves prominently or minutely denticulate (at least above the middle) with gland-tipped teeth; lower surface of leaf sparsely silky-hairy
          d. Margins of leaves remotely and often sharply toothed; petioles 0.5–5 mm long; colonial shrub of beaches and sand-bars *S. exigua*
          d. Margins of leaves minutely denticulate or irregularly serrate at least above the middle, sometimes entire near the base; petioles 3–10 mm long; forming clumps but not colonies *S. petiolaris*
    Z. Leaves broadly lanceolate to oblanceolate or oblong, elliptic or narrowly ovate
      e. Base of leaf usually rounded to cordate; margins regularly serrate, serrulate or denticulate; stipules frequently present
        f. Leaves oblong-lanceolate to oblanceolate, 1/8–1/3 as broad as long, the tips long-acuminate; young leaves reddish *S. eriocephala*
        f. Leaves oblong-ovate to broadly lance-oblong, 1/5–2/3 as broad as long, the tips acute to short-acuminate; young leaves not reddish *S. cordata*
      e. Base of leaf acute or tapered; margins entire, undulate or irregularly crenate to serrate or glandular-serrulate; stipules usually falling early except on vigorous shoots

g. Leaf blades linear-lanceolate to broadly lanceolate or oblanceolate, more than 5 times as long as wide, silky-velvety beneath with shiny white hairs **S. pellita**

g. Leaf blades obovate, oblanceolate, or elliptic, less than 5 times as long as wide, the lower surface with dull white or grayish pubescence

h. Branchlets divaricately spreading; pubescence of young leaves whitish; catkins appearing with the leaves; bracts yellowish to straw-coloured; capsules on pedicels 2–5 mm long **S. bebbiana**

h. Branchlets not divaricately spreading; pubescence of young leaves often partly of reddish to copper-coloured hairs; catkins appearing before the leaves; bracts dark brown to black; capsules on pedicels 1–2.5 mm long

i. Leaf blades smooth or the main veins slightly raised above; lower surface with sparse pubescence of short hairs; branchlets smooth, lustrous or pruinose **S. discolor**

i. Leaf blades often slightly rugose and with somewhat revolute margins, the veinlets impressed above; lower surface tomentose; branchlets pubescent or glabrous but dull, not shiny **S. humilis**

Y. Mature leaves glabrous or the midrib and petioles sometimes pubescent

j. Leaves glaucous or strongly whitened beneath

k. Margins entire to undulate-crenate, sometimes revolute, lacking definite teeth

l. Margins entire and more or less revolute

m. Leaves 4–12 cm long, tapered to an acute or acuminate tip **S. pellita** forma ***psila***

m. Leaves 2–7 cm long, the apex blunt, rounded or acute

n. Stems slender, creeping or stoloniferous, the lower branches often rooting in the moss; leaves with prominent reticulate venation, noticeably raised on the upper surface; petioles not yellow or red; catkins appearing with the leaves on leafy branches; capsules glabrous **S. pedicellaris**

n. Stems erect and spreading, not stoloniferous nor rooting; leaves not prominently reticulate-veined but veins impressed above; petioles yellowish to reddish; catkins appearing before the leaves, sessile with a few bracts at the base; capsules pubescent **S. planifolia**

l. Margins indistinctly toothed, undulate-crenate to irregularly serrate

o. Branchlets divaricately spreading; leaves dull green above, sometimes rugose beneath; catkins appearing with the leaves; bracts of pistillate catkins greenish-yellow to straw-coloured, sometimes reddish at the tip **S. bebbiana**

o. Branchlets not divaricately spreading; leaves dark green and shiny above; catkins appearing before the leaves; bracts of pistillate catkins dark brown to blackish

p. Stipules ovate and prominent on vigorous shoots; capsules borne on conspicuous pedicels 2–2.5 mm long **S. discolor**

p. Stipules lanceolate and soon deciduous, or minute or lacking; capsules sessile or on pedicels less than 2 mm long **S. planifolia**

k. Margins finely, distinctly or regularly serrate

q. Leaves linear-lanceolate, elliptic-lanceolate to ovate-lanceolate, long-attenuate or tapering to an acute or acuminate tip

r. Leaf blades more or less equally tapered at both ends

s. Leaves thin to membranaceous; petioles yellowish; stipules small or absent **S. petiolaris**

s. Leaves thick; petioles not yellowish; stipules prominent, at least on young shoots **S. myricoides**

r. Leaf blades unequally tapered at the two ends, the tip acuminate to long-attenuate, the base abruptly tapered, rounded or cordate

t. Young branchlets glabrous, stipules minute or absent; bracts of pistillate catkins pale yellow, usually falling early **S. amygdaloides**

t. Young branches gray-pubescent; stipules prominent; bracts of pistillate catkins dark brown to blackish, persistent **S. eriocephala**

q. Leaves broadly elliptic, ovate, oblong-lanceolate or obovate, acute to short-acuminate or blunt to rounded at the tip

u. Leaves and buds with balsamlike fragrance; petioles slender; stipules minute or absent; lower surface of leaf with reticulate veins; catkins appearing with the leaves on naked or leafy lateral branches **S. pyrifolia**

u. Leaves and buds without balsamic fragrance; stipules minute or prominent; catkins appearing before or with the unfolding leaves, sessile or on short leafy lateral branches

v. Young leaves translucent, sparsely pubescent with pale hairs; leaf base asymmetrical **S. pseudomonticola**

v. Young leaves not translucent, sometimes pubescent with rusty-red hairs; leaf base symmetrical **S. myricoides**

j. Leaves not glaucous beneath, green on both sides or only slightly paler beneath

w. Leaves linear to linear-lanceolate or linear-oblanceolate

x. Shrubs with many stems, spreading to form colonies; leaf tip acute to acuminate; margins remotely denticulate or sometimes entire; stipules minute or absent **S. exigua**

x. Trees or large shrubs with one or a few main stems, not forming colonies; leaf tip slender and prolonged, often falcate; margins finely serrulate; stipules prominent **S. nigra**

w. Leaves elliptic-lanceolate or oblong-ovate to obovate

y. Midrib prominently yellow to orange-coloured at least for half its length **S. maccalliana**

y. Midrib not as above

z. Leaf tip acute or abruptly acuminate **S. cordata**

z. Leaf tip rounded or blunt **S. myrtillifolia** var. **cordata**

Salix alba

Salix alba L. White willow

A large tree with ascending branches, growing to a height of 15 m or more. Branchlets greenish-brown or yellowish-brown, silky, flexible, and often pendent; bark of the older branches brownish-gray.

Leaves alternate, simple, and deciduous; blades lanceolate to elliptic-lanceolate with a close silky-white coating of hairs when young; blades of mature leaves 4-12 cm long, 0.5-2 cm wide, acuminate at the apex, acutish at base; margins finely serrate with minutely gland-tipped teeth; upper surface green and sparsely silky-hairy, lower surface glaucous and usually densely silky-hairy; petioles 2-10 mm long, pubescent and often with small glands at the junction with the blade; stipules small, lanceolate, and deciduous, often lacking.

Catkins slender, borne on leafy lateral branches 1-4 cm long, appearing with the leaves (May and June); staminate catkins 3-5 cm long, stamens 2 (occasionally 3), filaments pubescent at the base; pistillate catkins 4-6 cm long; nectary solitary. Capsules conic-ovoid, 3-5 mm long, glabrous, sessile or nearly so; styles short or absent, stigmas minute; bracts ovate-lanceolate, pale yellow, hairy near the base, falling early. (*alba*—white, referring to the silky coating of white hairs on the young leaves)

Along streams, roadsides, in waste ground, sand dunes, and wet bottomland.

Commonly planted and a frequent escape in southern Ontario south and west of the southern boundary of the Canadian Shield; rare in the eastern and northern parts of the province. (Introduced from Europe)

Note Much of the material collected in Ontario represents hybrids between *S. alba* and *S. fragilis* and shows various combinations of characters of the parental species. Hybrids with yellow non-brittle branchlets and with leaves and branches becoming glabrate are identified as *S. alba* var. *vitellina* Wimmer. (*vitellina*—of the colour of egg yolk) This hybrid derivate, and other hybrids of *S. alba* × *S. fragilis* found in Ontario, have been introduced from Europe and occur as escapes.

Field check Large coarse shrub or tree; leaves lanceolate, finely serrate and sericeous, appearing silvery-gray from a distance; capsules sessile or nearly so; bracts pale yellow, hairy near the base, soon falling off.

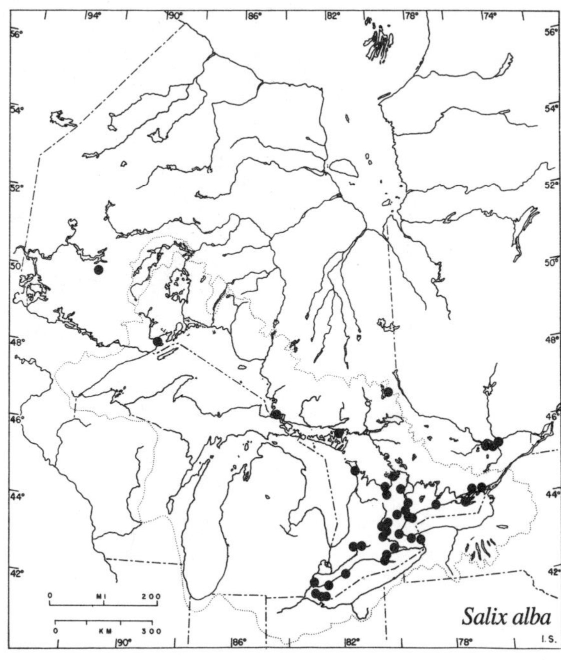

Salix amygdaloides

Salix amygdaloides Anderss. — Peach-leaved willow

A tall shrub or tree growing to a height of 10 - 20 m with several stems or trunks up to 4 dm in diameter and with fissured bark, the trunks often leaning away from the vertical. Branchlets slender, glabrous, yellowish-brown, tough and flexible (not brittle), somewhat drooping at the tip; older branches with dark brown to reddish-brown bark, becoming gray-brown in age.

Leaves alternate, simple, and deciduous; blades lanceolate to ovate-lanceolate, long attenuate at the apex, acute to obtuse or rounded at base; margins closely serrulate; young leaves sparsely pubescent, often reddish; mature blades 5 - 14 cm long, 1 - 3 cm wide, yellowish-green to dark green and glabrous above, glabrous and glaucous beneath, the pale midrib prominent on both surfaces; petioles 5 - 15 mm long, glabrous, sometimes with minute glands at the junction with the blade; stipules mostly absent except on vigorous vegetative shoots.

Catkins coetaneous, on lateral branches bearing short leaves which sometimes have entire margins (May and June); staminate catkins 3 - 6 cm long, slender and lax, the flowers appearing as if whorled on the axis, stamens 3 - 6, often 5, filaments pilose at the base; pistillate catkins 4 - 9 cm long, loosely flowered and lax, pistils ovoid, glabrous, short-beaked. Capsules 4 - 7 mm long on pedicels 1 - 2 mm long; style less than 0.5 mm long and stigmas shorter (June and July); bracts pale yellow, lanceolate to oblong, pubescent on the inner surface, usually deciduous in the pistillate catkins; nectary solitary, reddish. (*amygdaloides*—like *Amygdalus*, the old generic name for the peach, in reference to the resemblance between the leaves of this willow and those of the peach)

In low damp ground along river banks, lakeshores, in alluvial woods, margins of swamps and marshes, usually near water.

In southern Ontario from Lake Erie to Georgian Bay and eastward to the Ottawa Valley; also in the Lake-of-the-Woods region near the Manitoba border. (Vt. and s. Que. to s.-cent. B.C., south to Ariz., Tex., and Ky.)

Note This species is similar to *S. nigra* but may be distinguished by its broader leaves, not falcate at the tip, glaucous beneath, rarely as densely pubescent when young, and stipules mostly lacking.

Both *S. amygdaloides* and *S. lucida* have leaves with attenuate tips but they can be distinguished by the whitened lower surface of the leaves of the former in contrast to the shining surfaces of the latter.

Field check Tall shrub or tree with several stems; branchlets slender, tough but flexible; immature leaves usually reddish; mature blades lanceolate, long attenuate at the tip, and glaucous beneath; margins finely serrulate; stipules minute or absent.

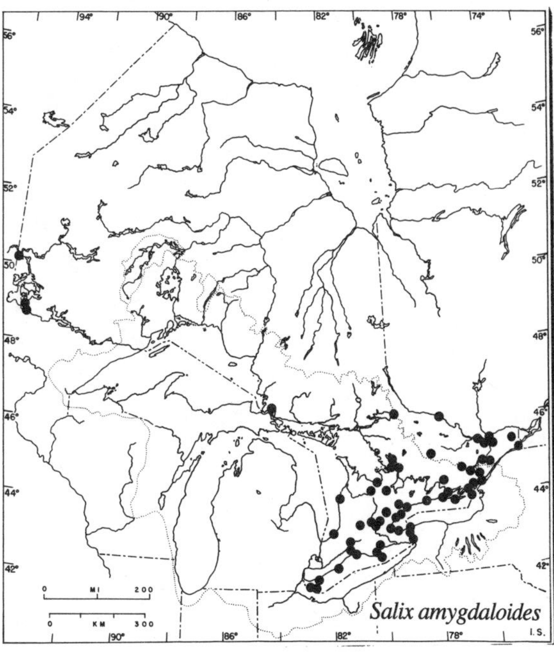

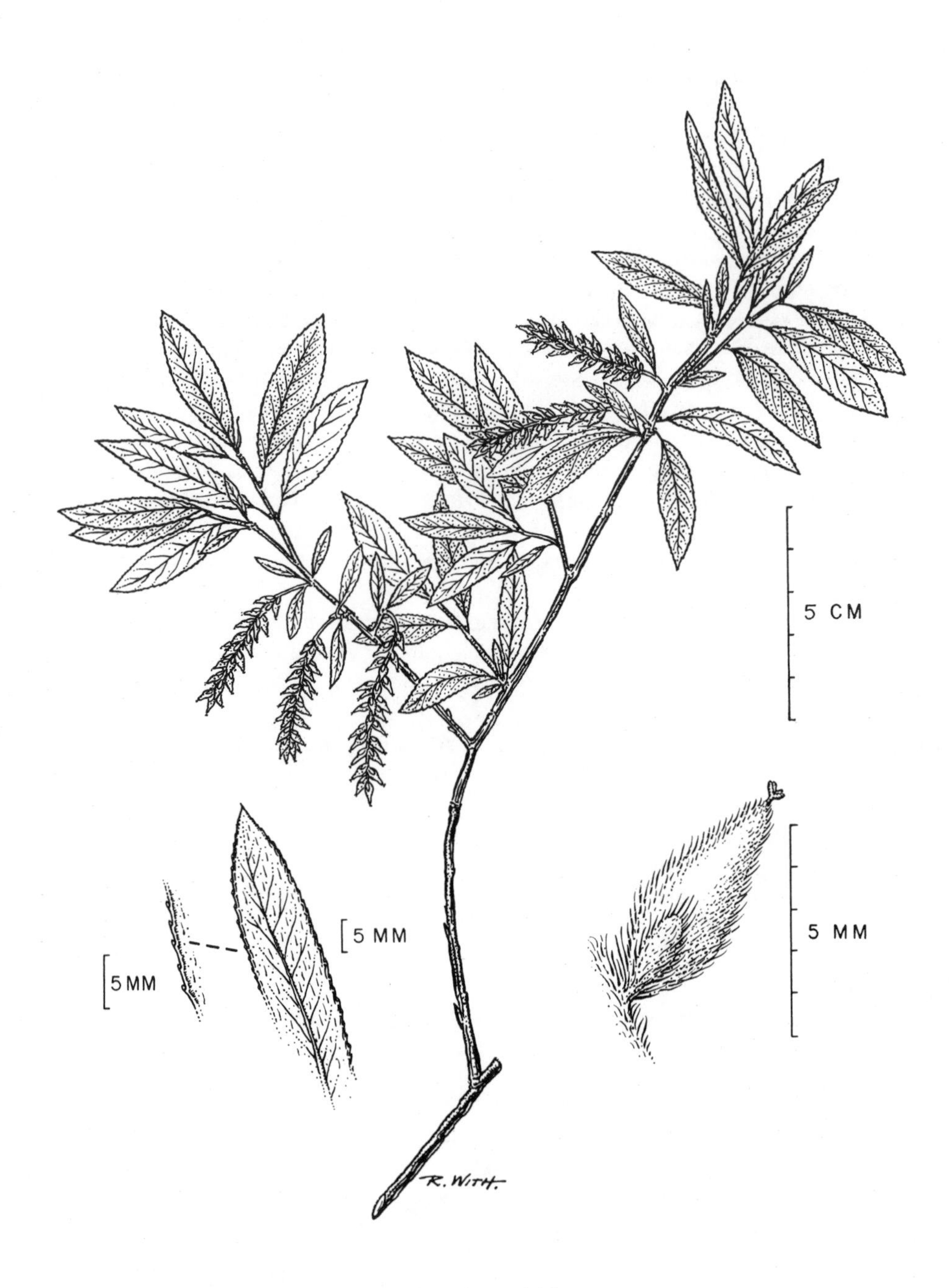

Salix arbusculoides

Salix arbusculoides Anderss. — Willow

A low to medium-sized shrub 1-4 m high. Branchlets slender, reddish-brown, shiny, and glabrous or nearly so, the pale epidermis sloughing off on older twigs.

Leaves alternate, simple, and deciduous; blades elliptic to narrowly ovate, 3-7 cm long, 0.5-1.5 cm wide, three to six times as long as wide, bright green, glabrous and shiny above, the lower surface paler and typically silky-hairy with short, appressed, whitish or rarely rusty-coloured hairs pointing towards the apex, or the lower surface rarely glabrescent; apex acute to obtuse, base tapered to a petiole 2-6 mm long; margins glandular-serrulate; stipules minute, linear-subulate, soon deciduous.

Catkins developing with the leaves or beginning slightly ahead of the leaves, nearly sessile or on short branchlets with several reduced leaves at the base; staminate catkins 1-2 cm long; stamens 2, filaments glabrous; pistillate catkins 2-6 cm long with loosely and irregularly spaced whorls of flowers; pistils 3-4.5 mm long; style short; stigma notched or 4-lobed. Capsules conical, 4-6 mm long, sparsely silky-hairy; pedicel less than 1 mm long; nectary solitary, barely exceeding the pedicel; bracts oblong to obovate, light to dark brown, hairy; June-August. (*arbusculoides*—resembling *S. arbuscula*, a Linnaean species native to northern Europe)

In littoral habitats; on calcareous clays and silts. In adjacent Manitoba this species has been collected in swampy areas, muskegs, black spruce-tamarack forest, on alluvial creek banks, gravel ridges, and rock outcrops.

In Ontario known only from the lower part of the Black Duck River near Hudson Bay and close to the eastern boundary of Manitoba, 56°41′ N, 89°14′ W. (n.w. Ont. to Alaska and B.C.; Lake Mistassini, Que.)

Note This is a distinctive species characteristic of the northern part of the Boreal Forest west of Ontario and having a disjunct occurrence in central Quebec. It is to be looked for in other parts of the Hudson Bay Lowlands in Ontario and Quebec.

Field check Leaves elliptic, glabrous and shiny above, silky-hairy beneath, the hairs pointing towards the apex; margins glandular-serrulate; catkins appearing with the leaves; capsules sparsely silky-hairy; extreme northwestern Ontario near Hudson Bay.

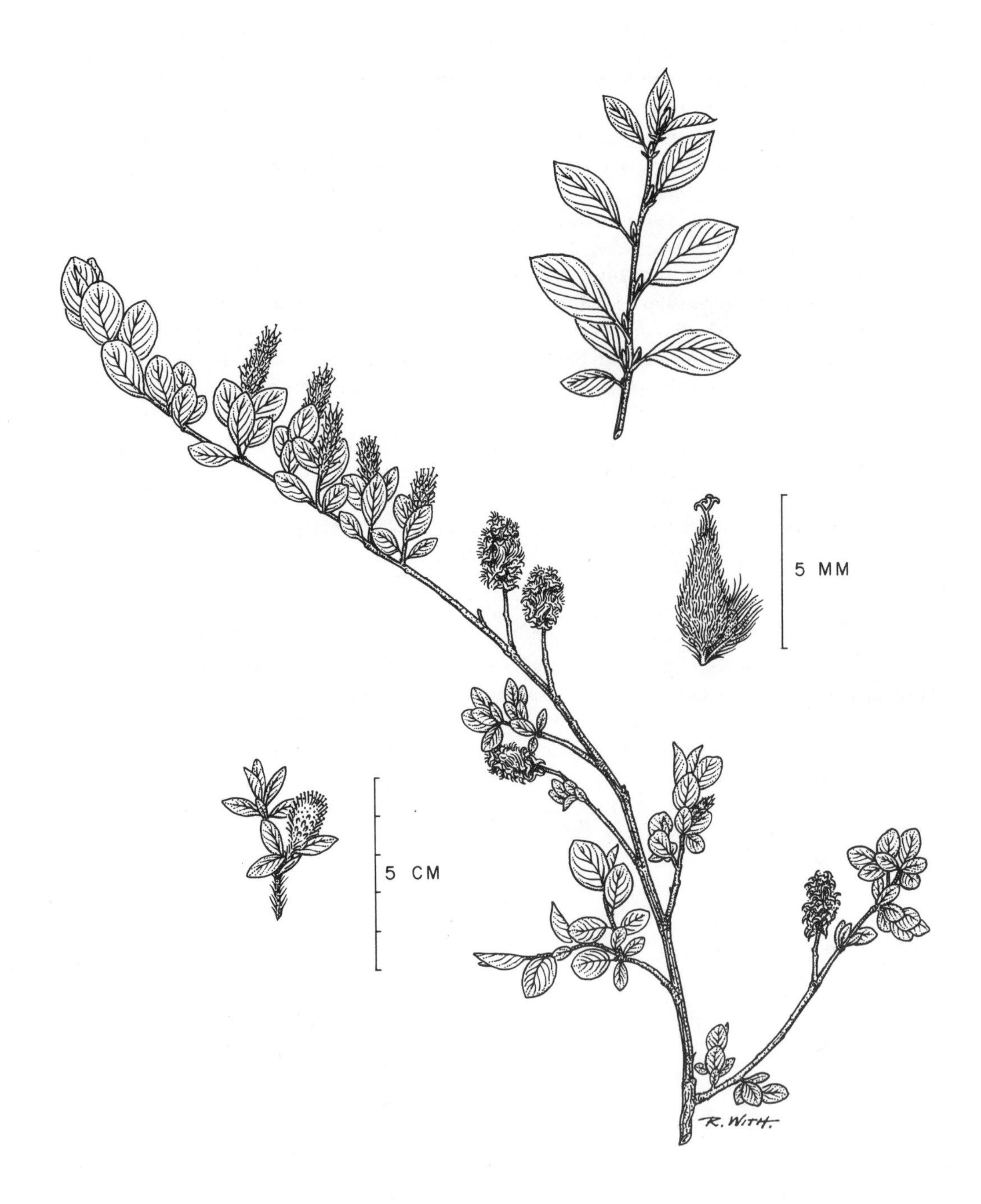

Salix arctica var. *kophophylla*

Salix arctica Pall. var. *kophophylla* (Schneid.) Polunin (*S. anglorum* Cham. var. *kophophylla* Schneid.) — Arctic willow

A dwarf shrub less than 3 dm high, straggling, creeping, and prostrate, the branches sometimes rooting. Branchlets flexible, greenish-brown, smooth and glossy or rarely sparsely pubescent at first, soon becoming dark brown to grayish and glabrous or occasionally purplish-gray and somewhat glaucous.

Leaves alternate, simple, and deciduous, the young leaves pubescent with long straight hairs; blades of mature leaves elliptic-ovate to obovate or rarely suborbicular, 1 - 3.5 cm long and 0.5 - 3 cm wide, firm or leathery, the apex acute, pointed or rounded, the base rounded or tapered; margins entire and slightly revolute; upper surface of blade dark green, dull, glabrous, lower surface much paler, glaucous, glabrous or with long white hairs persisting along the prominently raised veins and sometimes pubescent along the margins; petioles 3 - 12 mm long, somewhat channelled on the upper surface; stipules none or minute and glandlike.

Catkins erect on leafy lateral branches, expanding with the leaves (late June, early July); staminate catkins 1 - 1.5 cm long, stamens 2, filaments glabrous; pistillate catkins 1 - 3.5 cm long in fruit (late July and Aug.); pistils reddish, densely pubescent, style prominent with 2 or 4 linear stigmatic lobes. Capsules 5 - 7 mm long, reddish-purple, sparsely to densely pubescent; pedicels less than 2 mm long, pubescent, shorter than the slender nectary; bracts broadly oblong, dark brown, pubescent and long-bearded. (*arctica* — of the Arctic; *kophophylla* — blunt-leaved)

In tundra habitats along the Arctic coast; beach ridges, sandy barrens, and turfy slopes.

Shores of Hudson Bay. (Greenl. and Nfld. to Alaska, south to Calif. and N. Mex.)

Note Several varieties of Arctic willow have been described, but the Ontario material is generally referred to *S. arctica* var. *kophophylla*.

Some collections from the shores and river mouths around James Bay and Hudson Bay have been named *S. hudsonensis* Schneid., a species closely resembling *S. arctica* but supposedly differing in its shorter pedicels and more elongated nectaries; it is doubtful whether recognition of a distinct species is warranted in this case. (*hudsonensis* — of the Hudson Bay region)

Field check Dwarf shrub of coastal tundra habitats (Hudson Bay) with prostrate branches, elliptic to obovate or suborbicular leaves, usually glaucous and sparsely pubescent beneath; small erect fruiting catkins (1 - 3.5 cm long) on leafy lateral branches.

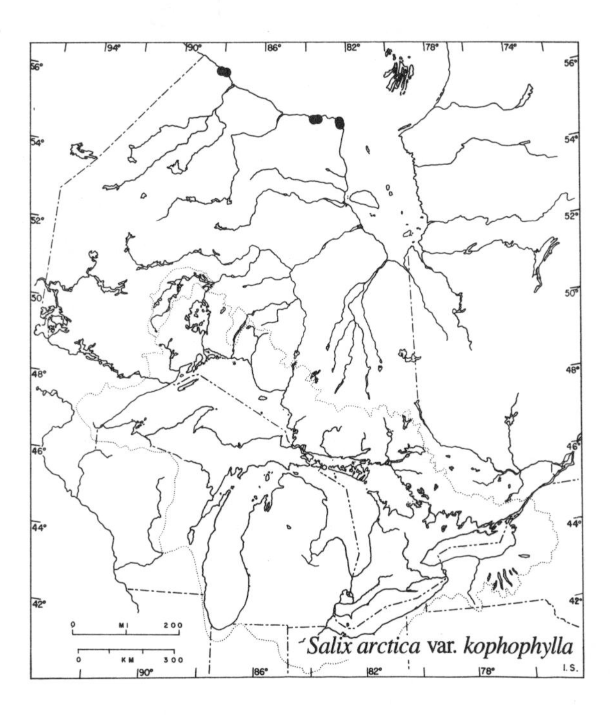

Salix arctica var. *kophophylla*

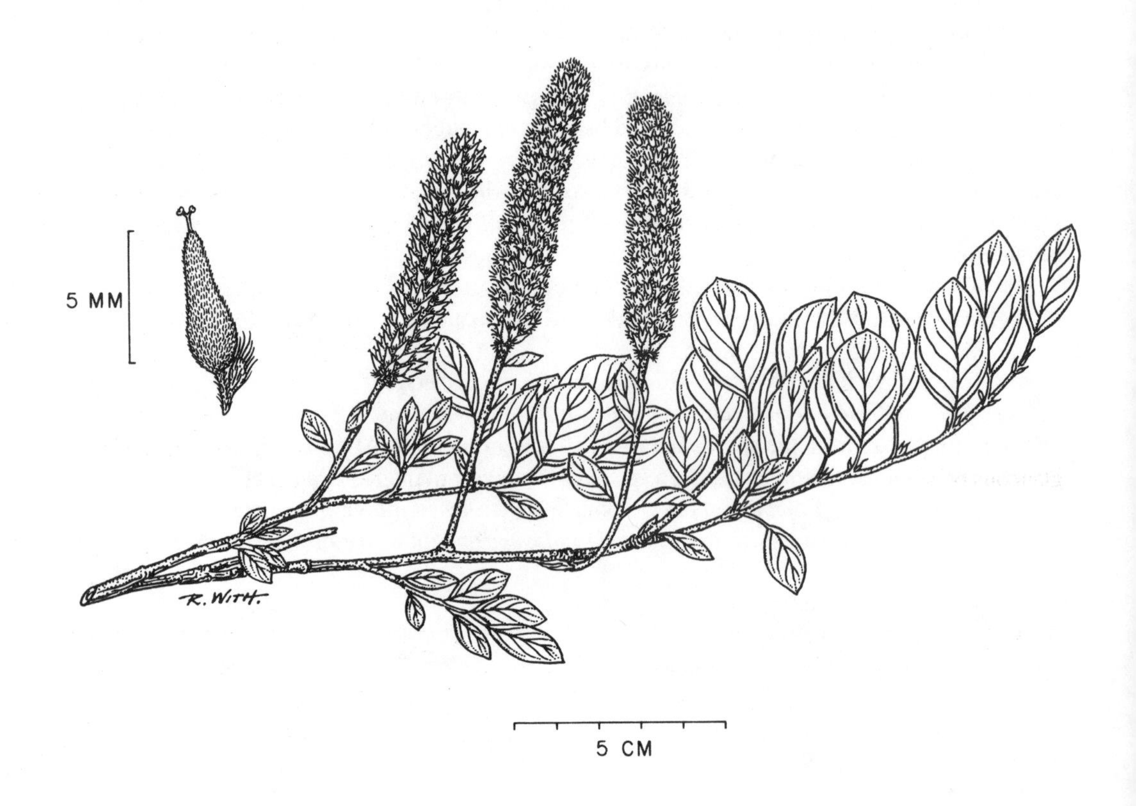

Salix arctophila

Salix arctophila Cock. ex Heller — Labrador willow

A dwarf shrub less than 3 dm high, usually depressed, decumbent or prostrate with trailing and sometimes rooting branches. Branchlets long and slender, glabrous, at first yellowish to greenish-brown, soon becoming chestnut-brown to dark brown.

Leaves alternate, simple, and deciduous; mature blades firm and leathery, elliptic to obovate, 2–4 cm long and 0.8–2.2 cm wide; apex acute, obtuse or rounded, base acute, cuneate or obtuse; upper surface dark green, glabrous, and lustrous, lower surface grayish to whitish, glabrous, and glaucous, the veins forming a slender, raised, netlike pattern; margins entire with minute glands or remotely and shallowly glandular-serrulate; petioles slender and conspicuous, 2–15 mm long; stipules narrowly ovate, 1–4 mm long with a few tiny marginal glands, or frequently absent.

Catkins erect on leafy lateral branches, opening with the unfolding leaves; staminate catkins 1–3 cm long, stamens 2, filaments purplish, pilose at the base; pistillate catkins 2–4 cm long, elongating to about twice their length in fruit; pistils 3–4 mm long, sparsely and evenly pubescent with short crinkled hairs that refract the light; stipe prominent, exceeding the single truncate nectary; bracts dark purplish-red to black, oblong with a rounded apex, and almost obscured by numerous long straight hairs; July and Aug. (*arctophila*—arctic-loving)

In wet tundra, turfy mats, on slopes and hummocks in the area of beach ridges and barrens near the Arctic coast; occasionally in open spruce muskeg or along clay banks of rivers a short distance inland.

Shores of Hudson Bay and James Bay and on islands in James Bay (N.W.T.); rarely inland in the Boreal Forest and Barren Region. (Greenl. to Yuk. and Alaska, south to n. Sask., Gaspé, Que., Nfld., and Maine; less common north of the Arctic Circle)

Field check Prostrate shrub of tundra habitats (Hudson Bay and James Bay) with trailing, sometimes rooting branches, elliptic to obovate glabrous leaves, dark green and shining above, glaucous beneath; large, erect, fruiting catkins (3–9 cm long) on leafy lateral branches.

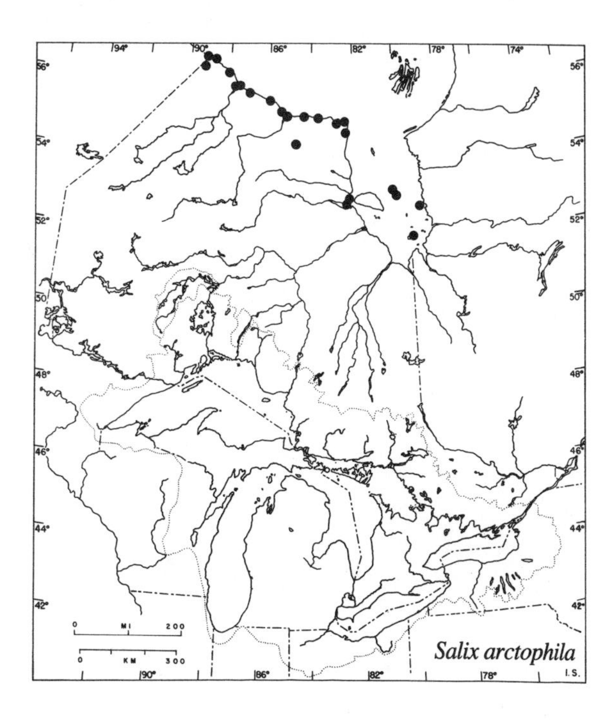

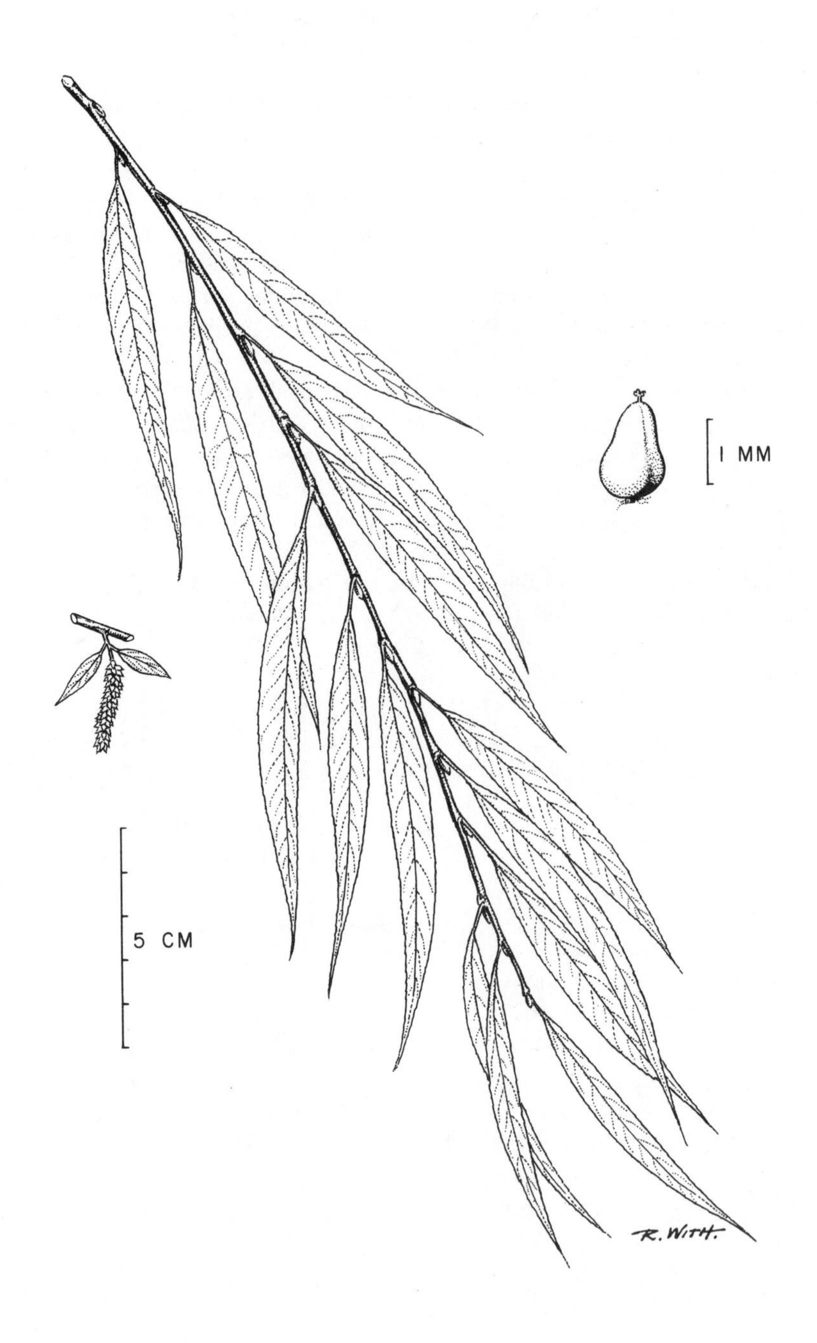

Salix babylonica

Salix babylonica L. **Weeping willow**

A tree up to 12 m tall. Branchlets yellowish to brown, smooth, slender, and pendulous.

Leaves alternate, simple, and deciduous, silky when young; blades of mature leaves linear-lanceolate, 5 - 12 cm long, 0.5 - 2 cm wide, long-acuminate at the tip, acute to tapered at base; margins finely and unevenly spinulose-serrulate; upper surface yellowish-green and glabrous, lower surface paler and also glabrous; petioles 2 - 10 mm long, often with small glands at the junction with the blade; stipules lanceolate, 2 - 7 mm long, usually absent.

Catkins slender, appearing with the leaves (April and May) on lateral branches, 0.5 - 1.5 cm long with a few reduced leaves at the base; staminate catkins 2 - 4 cm long, stamens 2 (occasionally 3 - 5 or more), filaments pubescent at base; pistillate catkins 2 - 3.5 cm long. Capsules ovoid-pyriform, 1 - 2.5 mm long, glabrous, sessile or nearly so, style and stigmas very short; bracts pale yellow, pubescent with crinkly hairs, deciduous. (*babylonica*—of Babylon)

A frequently planted ornamental tree which occasionally escapes to roadsides and stream banks.

An uncommon escape in the Deciduous Forest Region. (Native of northern China, probably introduced into North America from Europe)

Field check Tree with pendulous, slender, yellow-brown branchlets, linear-lanceolate spiny-serrulate leaves, and small catkins; bracts pale yellow, pubescent, deciduous.

Salix bebbiana Sarg. (*S. rostrata* Richards.) **Bebb's willow, Beaked willow Long-beaked willow**

A large coarse shrub with one or a few stems 1 - 6 m high or a small tree, rarely to 8 m, the branches ascending but somewhat divaricate. Branchlets reddish-brown and darkening in age, grayish-pubescent when young, becoming glabrous or, frequently, remaining thinly to densely hairy for several years; bark of older stems and trunks gray-brown, becoming fissured.

Leaves alternate, simple, and deciduous; blades varying from elliptic to obovate-oval to oblanceolate, 3 - 7 cm long, 1 - 3 cm wide, acute at both ends or the apex short-acuminate (rarely obtuse) and the base obtuse or rounded; relatively thin, more or less gray-pubescent to silky-hairy on both sides when young, and occasionally reddish-tinged; upper surface of mature leaves dull green, hairy or smooth, the lower surface glaucous and often with prominent rugose venation; margins nearly entire or undulate-crenate, sometimes irregularly gland-toothed; petioles 3 - 10 mm long, pubescent; stipules none or small (less than 2 mm long) and deciduous except on vigorous vegetative shoots where they are large and conspicuous, 5 - 15 mm long, semi-cordate to ovate-lanceolate with glandular serrate margins.

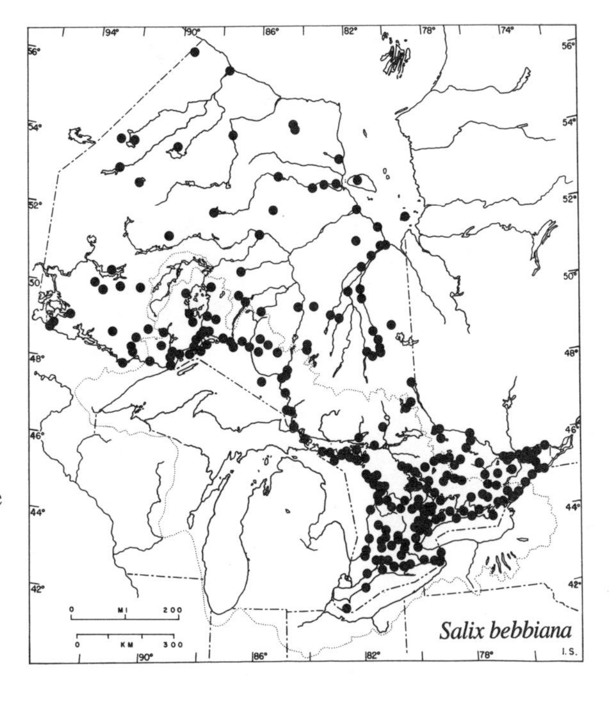

Catkins coetaneous, on leafy lateral branches; staminate catkins 1 - 3 cm long, stamens 2, filaments hairy at the base;

Salix bebbiana

pistillate catkins 2–7 cm long, loosely flowered, and frequently lax, pistils long-stalked, finely gray-pubescent, contracted above the base into a long slender neck or beak, style very short or obsolete, stigmas short, 2-parted or 4-parted, the lobes not widely spreading; bracts lanceolate, greenish-yellow to straw-coloured, sometimes reddish at the tip, pubescent; pedicels 2–5 mm long, at least twice as long as the bracts; nectary solitary, shorter than the bract. (*bebbiana*—named for M.S. Bebb, 1833–1895, a specialist in willows; *rostrata*—beaked)

In moist to wet habitats, including sedge meadows, swamps, bogs, lakeshores, river banks, alluvial flats, deciduous and coniferous forests; also on limestone flats and in sandy jack pine woods. One of the few forest species.

One of the commonest and most widely distributed of our willows, found from Lake Erie to Hudson Bay and from the Ottawa-St. Lawrence lowlands to the Manitoba border. (Nfld. to Alaska, south to Calif. and Md.)

Note Bebb's willow is very variable in leaf shape and leaf margin, and in the amount and distribution of pubescence on the leaves and shoots. Often the first leaves to mature on a particular shoot have entire margins and later leaves on the same shoot have crenate to serrate margins. Several varieties have been described and named, but because they are somewhat weakly defined, they have not been included.

Field check Leaves dull green above, rugose and gray-pubescent beneath, margins entire to crenate; capsules slender, long-beaked, on very long pedicels (2–5 mm long); bracts yellowish to straw-coloured.

Salix brachycarpa Nutt. — Willow

A dwarf shrub with ascending and erect stems usually less than 1 m high or sometimes prostrate and trailing. Branchlets gray-brown to blackish, with numerous short internodes and prominent leaf scars, the young branchlets yellowish-brown to reddish-brown under a heavy white to gray coating of woolly hairs; branches often short and twisted.

Leaves alternate, simple, and deciduous, characteristically crowded and overlapping; the blades elliptic to obovate or oblong, abruptly acute to obtuse at the apex, broadly tapered to rounded at base, densely tomentose when young; blades of mature leaves 1–4 cm long and 0.5–1.8 cm wide, dull green and sparsely hairy or nearly glabrous above, paler, glaucous, and usually more densely hairy below; margins entire to remotely glandular-crenate; petioles 1–3 mm long, about as long as the well-developed buds, pubescent, and often reddish; stipules none or minute and hidden in the pubescence of the stem.

Catkins short, ellipsoid to nearly globose, 0.5–1.5 cm long, the pistillate lengthening to 3 cm in fruit; on leafy lateral branches and opening fully as the leaves expand (July); stamens 2, filaments glabrous or pubescent near the base; pistils sessile or on very short pedicels, densely grayish-white pubescent, short-beaked, styles about 1 mm long, stigmas short, 2-lobed. Capsules light brown, lance-ovoid, 4–6 mm long; scales oblanceolate to oval, pale yellow to yellowish-brown, pubescent; nectaries usually 2 per flower; seeds released in Aug. and Sept. (*brachycarpa*—with short carpels, i.e., short-fruited)

On beach ridges and terraces of tundra along the Arctic shores and on gravel flats of rivers.

Confined chiefly to the shores of James Bay and Hudson Bay; also in the Severn River and Attawapiskat River drainage systems. (n. Que. to B.C., south to Oreg., Utah, and N.Mex. at high elevations)

Note *S. brachycarpa* and *S. glauca* ssp. *callicarpaea* are easily confused. They are alike in growth habit, may occur together in some parts of northern Ontario, have similar leaf shapes and about the same quantity and persistence of pubescence. The following list of characters may aid in their separation:

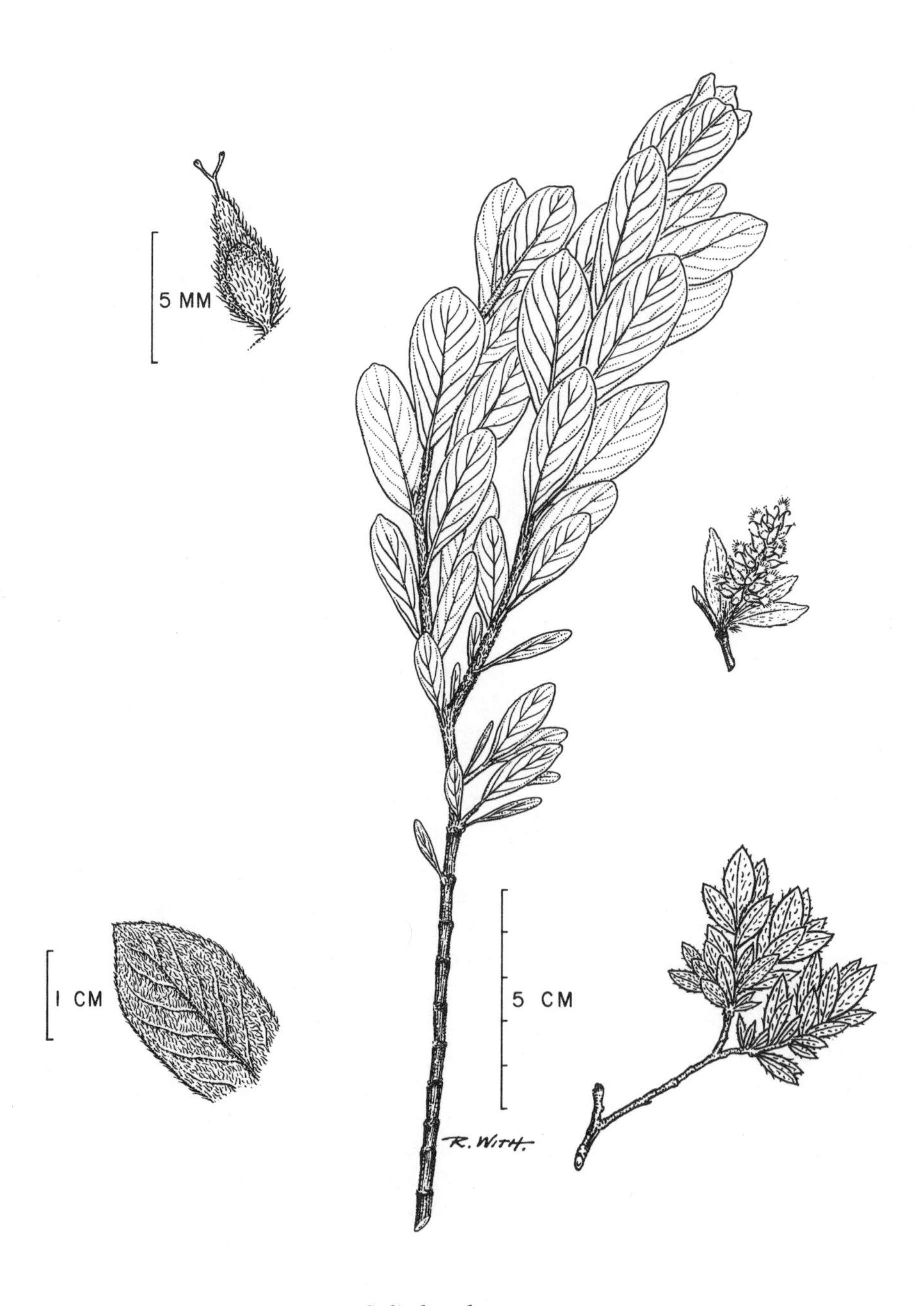

Salix brachycarpa

Character	*S. brachycarpa*	*S. glauca* ssp. *callicarpaea*
Maximum height	1 metre	1.5 metres
Stems	internodes short, the branches often twisted	branchlets flexible
Leaves	crowded and overlapping	more widely spaced
Pubescence	woolly	more silky
Petioles	1-3 mm long, frequently reddish	3-9 mm long, greenish
Capsules	4-6 mm long	6-10 mm long
Catkins	deciduous	often persisting until the following spring

Field check Dwarf shrub of tundra habitats along the shores of James Bay and Hudson Bay; leaves crowded and overlapping, gray-pubescent, glaucous beneath; petioles very short (1-3 mm long); catkins less than 2 cm long.

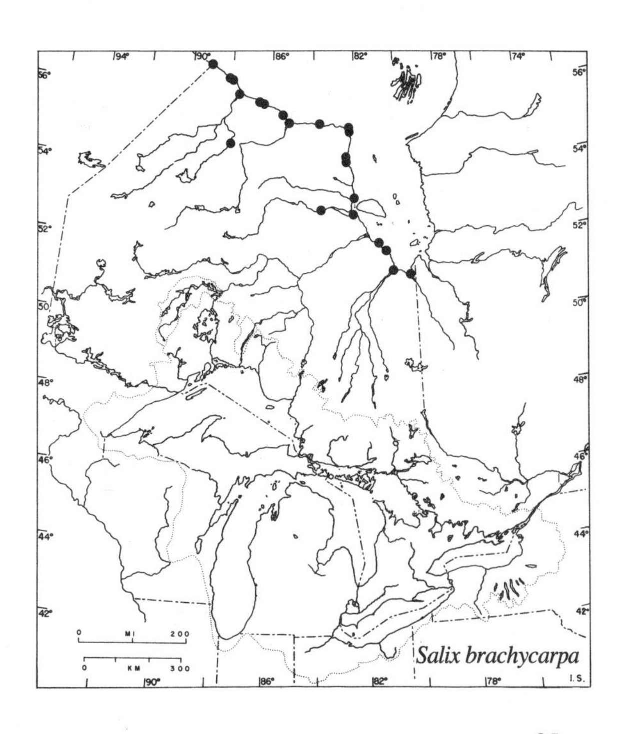

Salix candida

Salix candida Fluegge

Sage-leaved willow
Hoary willow

A shrub with erect or ascending stems and somewhat divaricate branches, mostly less than 2 or 3 m tall. Branchlets yellowish to brownish and white-tomentose at first, becoming smooth and reddish-brown to grayish on the older stems.

Leaves alternate, simple, and deciduous; blades linear-oblong to oblanceolate, 3 - 10 cm long, 0.5 - 2 cm wide, acute to acuminate at the apex, cuneate or abruptly narrowed at base; margins entire, revolute, sometimes undulate or glandular-crenate; upper surface of blade white-tomentose to floccose on unfolding but becoming glabrous, dark green, with impressed veins; lower surface tomentose with a feltlike covering of tangled white hairs, the prominent midrib often yellowish; petioles 3 - 10 mm long, tomentose; stipules small, lanceolate, glandular, tomentose, falling early or, on vigorous vegetative shoots, persistent.

Catkins coetaneous, on short branches, with several small leaves; staminate catkins 1 - 2.5 cm long, nearly sessile, stamens 2, filaments glabrous, the dark purple anthers forming a halolike fringe; pistillate catkins 2 - 5 cm long, cylindrical and rather densely flowered, pistil tomentose, style about 1 mm long, dark red at the scarcely branched tip which is divided into 4 subsessile stigmas; scales of the catkins pale to dark brown, heavily bearded with long white hairs; nectary solitary, dark. Capsules lanceolate, 5 - 8 mm long. (*candida* — white)

In moist sandy, peaty, or marshy ground, especially in calcareous areas; edges of sedge meadows, marshy lakeshores, shallow pools behind beach dunes, and on hummocks in spruce-larch bogs.

Widely distributed from Lake Erie and Lake Ontario northward to the shores of Hudson Bay, and from the lower Ottawa Valley to the Manitoba border; absent from the Canadian Shield portion of southern Ontario (Nfld. to Alaska, south to the mountains of Colo., to S. Dak. and n. N.J.)

Note A variant with narrower glabrous or glabrescent leaves, branchlets, and capsules (forma *denudata* (Anderss.) Rouleau) occurs scattered throughout the range of the species. (*denudata* — denuded, referring to specimens without the characteristic woolly hairs of the species)

Field check Leaves densely white-tomentose on lower surface, midrib usually yellowish; leaf margins revolute; capsules white-tomentose, anthers dark purple; styles and stigmas dark red; often in calcareous habitats.

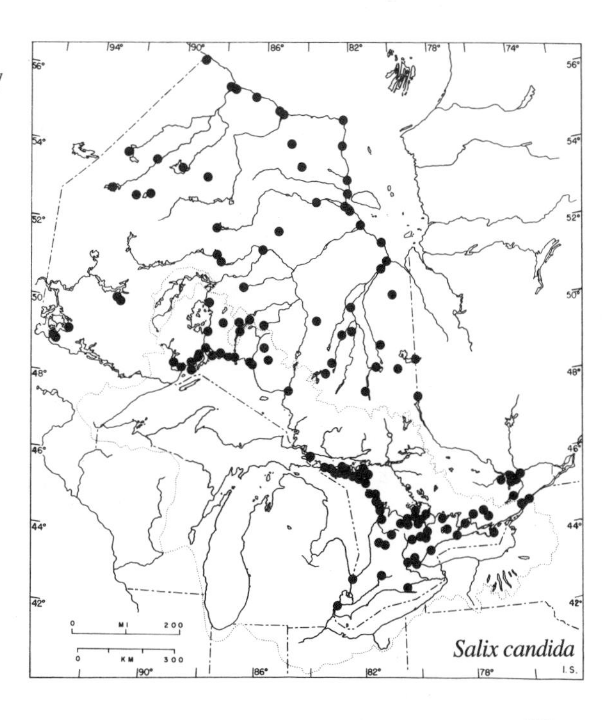

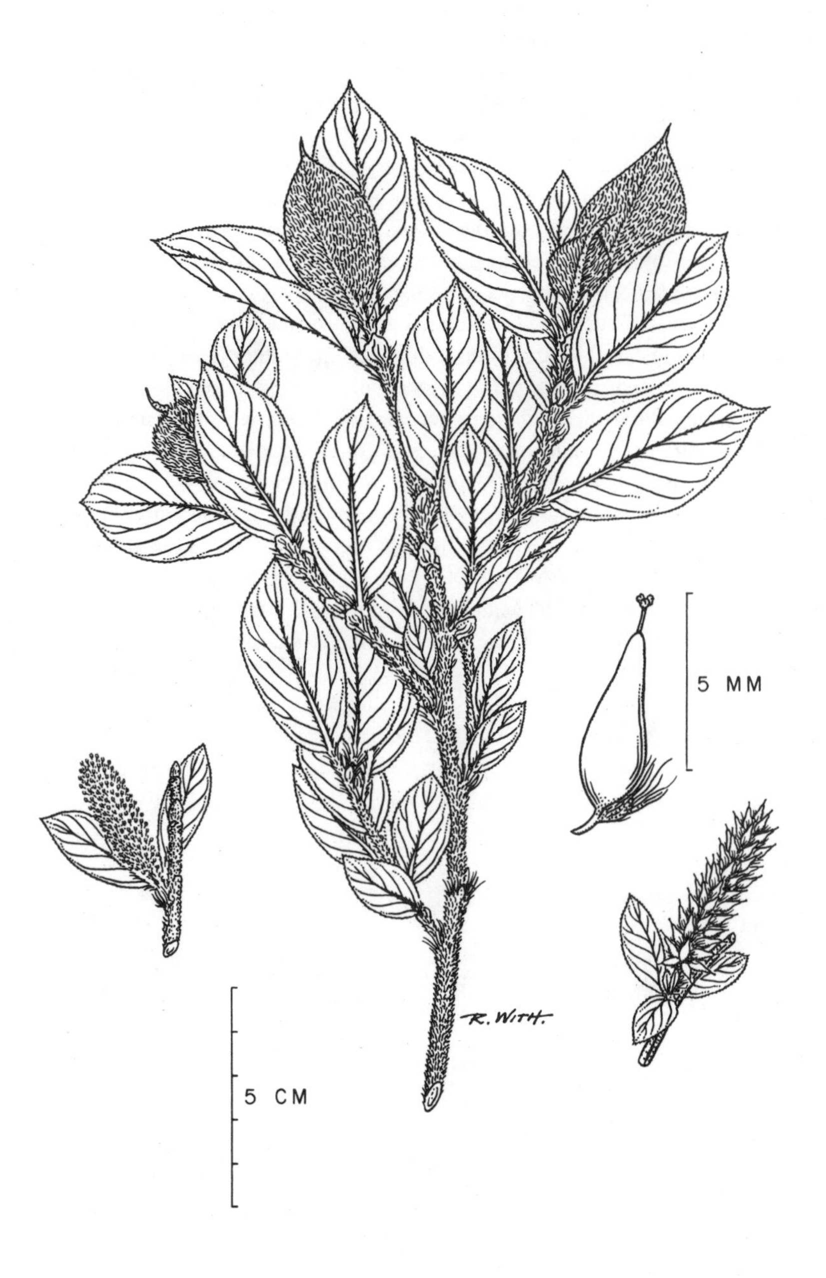

Salix cordata

Salix cordata Michx. **Heart-leaved willow**

A coarse, spreading, thicket-forming shrub with erect or ascending branches reaching a height of 1–3 m. Branchlets stout, reddish, and covered with a dense white to grayish tomentum when young; older branches becoming smooth.

Leaves alternate, simple, and deciduous, densely lustrous-hairy and sometimes reddish-tinged when young; blades of mature leaves 3–10 cm long, 1–5 cm wide, thickish, ovate-lanceolate, lance-oblong or oblong-ovate, the apex acute or abruptly acuminate, the base cordate or broadly rounded; green on both sides, often pubescent or lanate, at least below, or finally becoming smooth; margins glandular-serrate with numerous gland-tipped or prolonged teeth; veins prominent on the lower surface, the midvein often heavily pubescent at least near the petiole; petioles stout, 2–10 mm long, pubescent, somewhat dilated at the base; stipules 6–15 mm long, obliquely ovate or semi-cordate, glandular-dentate on the longer margin.

Catkins appearing with the leaves (May and June); staminate catkins 2–4.5 cm long, subsessile or on short lateral branches with several bracts at the base; stamens 2, filaments glabrous, the stamens barely exserted beyond the long-hairy, dark brown to blackish scales; pistillate catkins 3–8 cm long on leafy lateral branches 10–15 mm long. Capsules lance-ovoid, glabrous, 5–8 mm long, crowded, divergent or horizontally spreading when mature; pedicels 0.5–1 mm long, glabrous; bracts obovate to oblong-lanceolate, dark brown and densely villous; late June and July. (*cordata*—heart-shaped)

Banks and shores of rivers and lakes in sandy, silty, or gravelly soils; also on sand dunes.

Around the shores of the lower Great Lakes, in the Ottawa and St. Lawrence river valleys and along rivers in the drainage basin of James Bay and Hudson Bay. (Nfld. to n.e. Man., south to the lower Great Lakes, Ill., and n. Maine)

Note The present treatment includes plants formerly included in *S. adenophylla* Hook., and those referred to in *S. syrticola* Fern., described as an endemic of the Great Lakes region. (*adenophylla*—glandular-leaved; *syrticola*—sand-dwelling)

Field check Coarse shrub with pubescent young stems and leaves; leaf blades oblong-ovate to ovate-lanceolate, acute to abruptly acuminate at the apex; capsules green to brown, glabrous; scales longer than the pedicels and long-hairy; found chiefly on sand dunes, lake shores, and river banks.

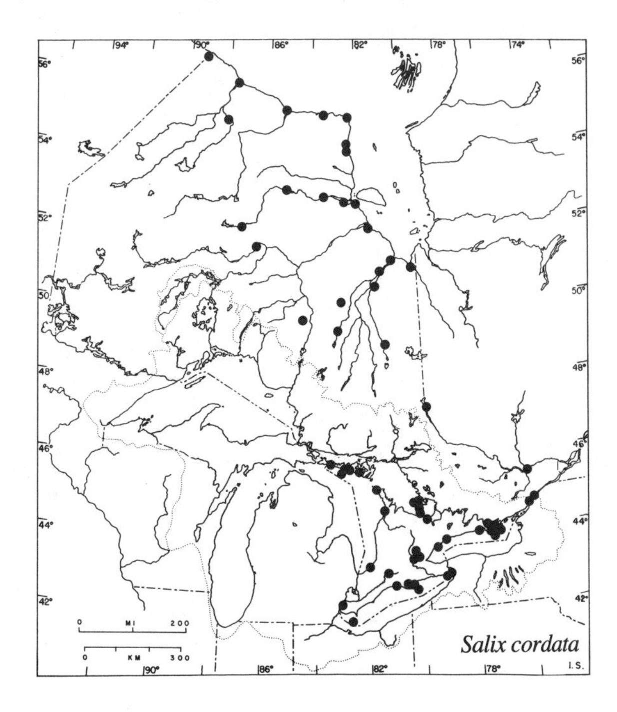

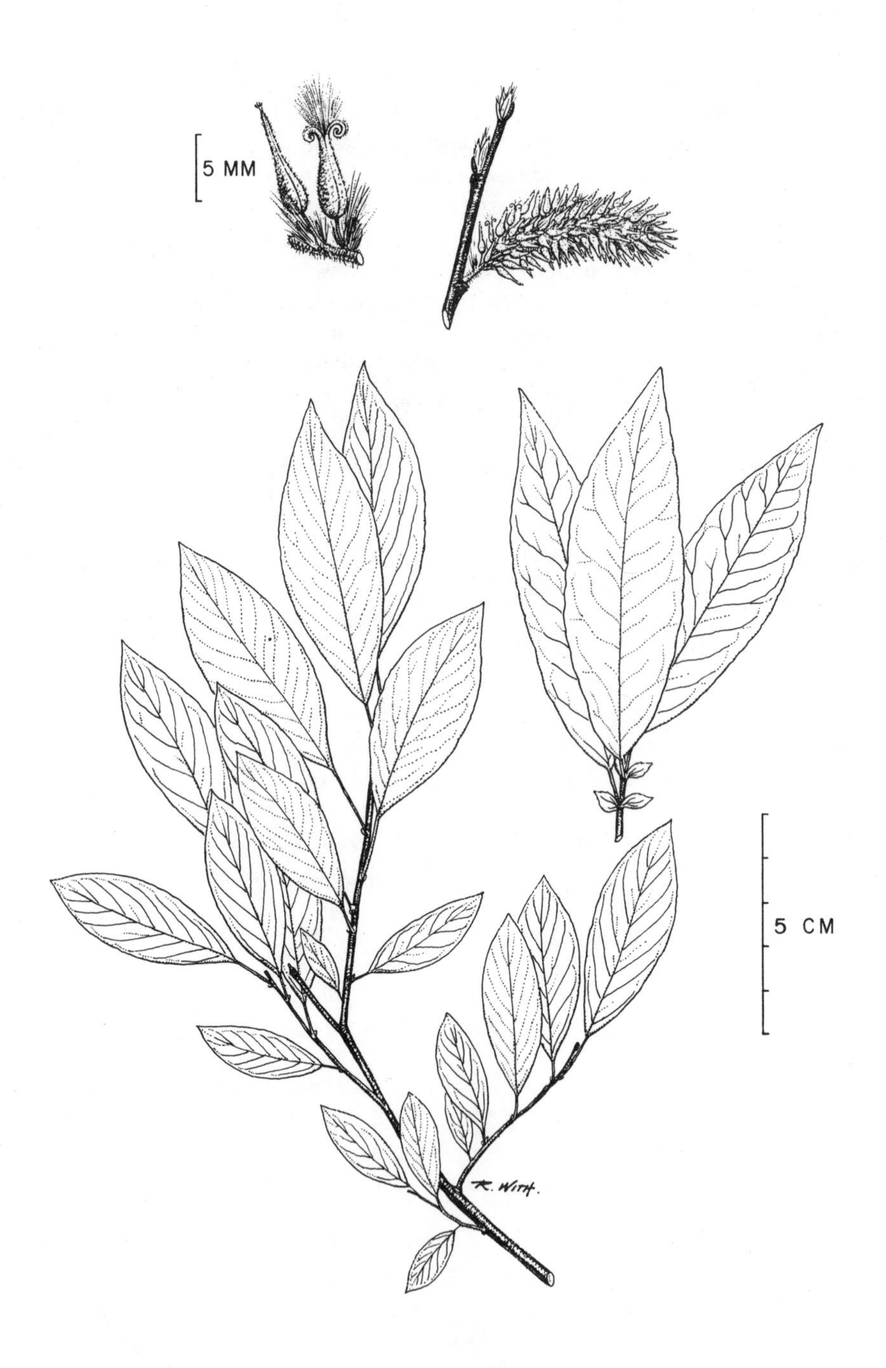

Salix discolor

Salix discolor Muhl. **Pussy willow**

A large, few-stemmed shrub 2 - 3 m tall or a small tree up to 6 m in height. Branchlets at first pubescent but usually becoming glabrous, the older branches rather stout, dark reddish-brown, smooth, lustrous or sometimes pruinose (with a waxy powdery bloom); bark of main stems and trunks grayish-brown; flower buds large, up to 1 cm long.

Leaves alternate, simple, and deciduous; blades oblong or elliptic to oblanceolate, the apex varying from acute or acuminate to blunt or rounded, the base usually acute or tapered, rarely obtuse; young leaves expanding after the catkins, often reddish, the blade thin and sparsely to densely pubescent with deciduous, rusty-coloured, curly hairs; mature blades firm, 3 - 10 cm long, 1 - 3 cm wide, sometimes with slightly raised venation, bright green and glabrous above, glaucous and glabrous or finely hairy beneath; margins irregularly crenate to serrate, especially near the middle; petioles 5 - 15 mm long, glabrous or minutely hairy; stipules minute or, on vigorous shoots and sprouts, ovate and prominent.

Catkins decidedly precocious (April and May in southern Ontario, May to June northward), fully developed before the leaves expand, sessile or on short stalks with several yellowish to greenish bracts at the base; staminate catkins 2 - 4 cm long; stamens 2, filaments glabrous; pistillate catkins densely flowered, 2 - 6 cm long, up to 9 cm long in fruit. Capsules 7 - 12 mm long, long-beaked, and minutely pubescent, on pubescent pedicels 2 - 2.5 mm long which exceed the dark brown to blackish and white-bearded scales; styles and stigmas about 1 mm long; nectary solitary; late May and June. (*discolor*—parti-coloured, in reference to the contrast between the dark green upper surface of the leaf and the pale lower surface)

In damp meadows, along shores of rivers and lakes, in alder swamps, cedar woods, wet thickets, and flooded ditches.

Common throughout southern and central Ontario and northward from Lake Superior to James Bay; no collections seen from north of 52° N. (Nfld. to B.C., south to Idaho and Ky.)

Note This species exhibits considerable variation in its vegetative characters. Plants with branchlets puberulent, bark dull, leaves becoming 2 - 5 cm wide and retaining the rusty pubescence on the lower surface have been named *S. discolor* var. *latifolia* Anderss. (*latifolia* —broad-leaved)

S. discolor and *S. planifolia* may be confused where they occur together north and east of Lake Superior. Both have precocious catkins with dark brown to blackish scales, leaves with bright green and shiny upper surface contrasting with the whitened lower surface, and often the young leaves reddish with rusty-coloured hairs which are soon shed.

They may be distinguished by their mature capsules: *S. discolor* has larger capsules (7 - 12 mm long) with shorter styles (less than 1 mm long), longer pedicels (more than 2 mm long), and a coating of minute, fine, soft hairs; *S. planifolia* has smaller capsules (7 mm or less in length), longer styles (exceeding 1 mm), shorter pedicels (less than 0.5 mm), and a coating of short, fine, silky hairs.

When capsules are not available, less

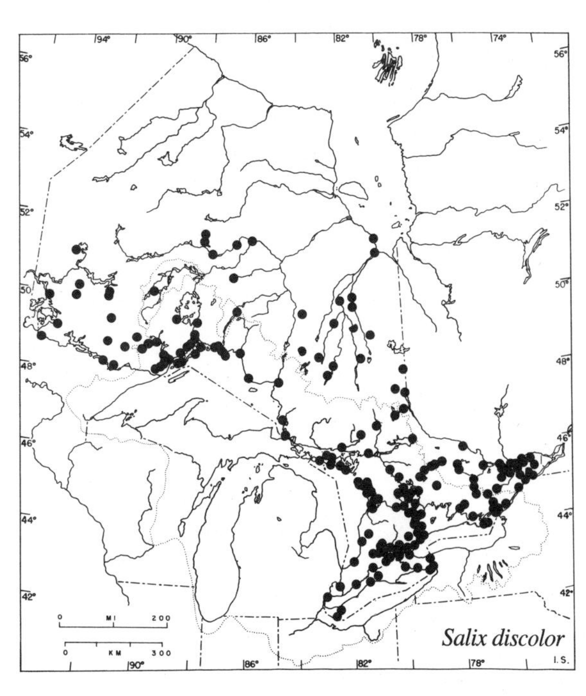

Salix eriocephala

conclusive differentiation can be made by the following comparisons: the petioles, which are yellowish to reddish and 10 mm or less in *S. planifolia* and green and a little longer (5 - 15 mm) in *S. discolor*; the stipules in *S. planifolia* are small and fall early, while in *S. discolor* they are small but more persistent, and usually larger on the vegetative shoots; and the leaves, which in *S. planifolia* are more regularly broadest about the middle and tapered to the apex and base and with a thick waxy bloom on the lower surface, and in *S. discolor*, more variable, broadest at or above the middle, with apex and base dissimilar, and a less prominent bloom beneath.

Field check Large shrub or small tree with conspicuously precocious catkins, the pistillate 4 - 9 cm long in fruit; capsules 7 - 12 mm long, long-beaked with a short style; young leaves with deciduous rusty-coloured hairs, older leaves whitened beneath, teeth irregular and mostly above the middle.

Salix eriocephala Michx. Willow
(*S. rigida* Muhl.; *S. cordata* Muhl., not Michx.)

A medium-sized to large shrub with erect or ascending branches reaching a height of 3 or 4 m. Branchlets reddish-brown to yellowish-green, often densely gray-pubescent when young, becoming smooth in age.

Leaves alternate, simple, and deciduous, the blades thin, conspicuously tinged with reddish-purple, and densely white-pubescent when young; blades of mature leaves oblong-lanceolate to slightly oblanceolate, 5 - 15 cm long, 1 - 4 cm wide, apex acuminate to long-attenuate, the base rounded to subcordate; margins glandular-serrulate, upper surface dark green and glabrous or finely pubescent, lower surface paler and somewhat glaucous; petioles 5 - 15 mm long, minutely pubescent; stipules lanceolate-ovate to semi-cordate, 5 - 10(- 20) mm long, remotely glandular-dentate; by autumn the leaves become rigid and veiny and often turn a deep reddish colour.

Catkins appearing before the leaves or as the leaves are unfolding (April and May), often forming long series on lateral branches with several basal leafy bracts; male catkins 1 - 3 cm long, stamens 2, filaments glabrous; pistillate catkins 2 - 8 cm long; pistils reddish- green, glabrous. Capsules lance-ovoid, 4 - 6 mm long, numerous, crowded and divergent, becoming brownish at maturity; pedicels slender, 1 - 2 mm long, exceeding the small, long-hairy, dark brown, oblanceolate bracts. (*eriocephala* — woolly-headed; *rigida* — stiff or rigid; *cordata* — heart-shaped)

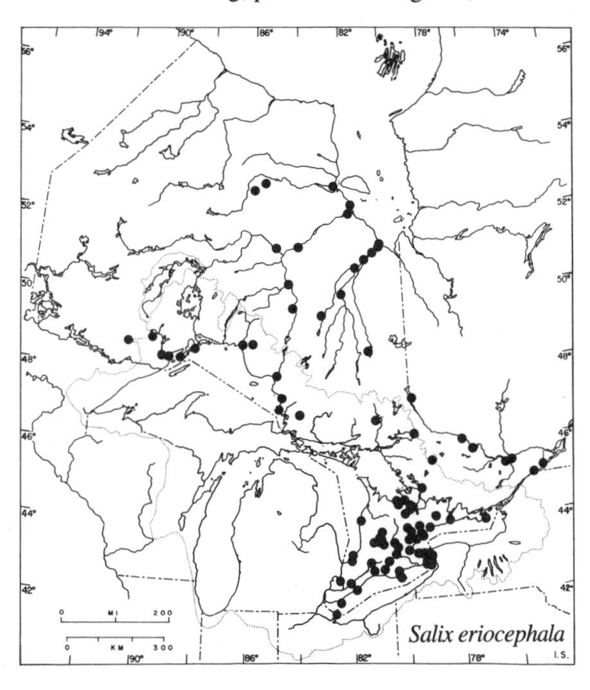

A shrub of river flats, banks of streams, edges of ponds, flooded ditches, willow swamps, bottomland forest, and wet roadsides.

From Lake Erie northward to James Bay and in the Lake Superior drainage basin. (N.S. to Yuk., south to Oreg., Mont., Neb., Mo., and Va.; also in Ark., Ala., and Ga.)

Note Argus (1980) has indicated that the correct name for this taxon is *S. eriocephala* Michx. Most manuals have

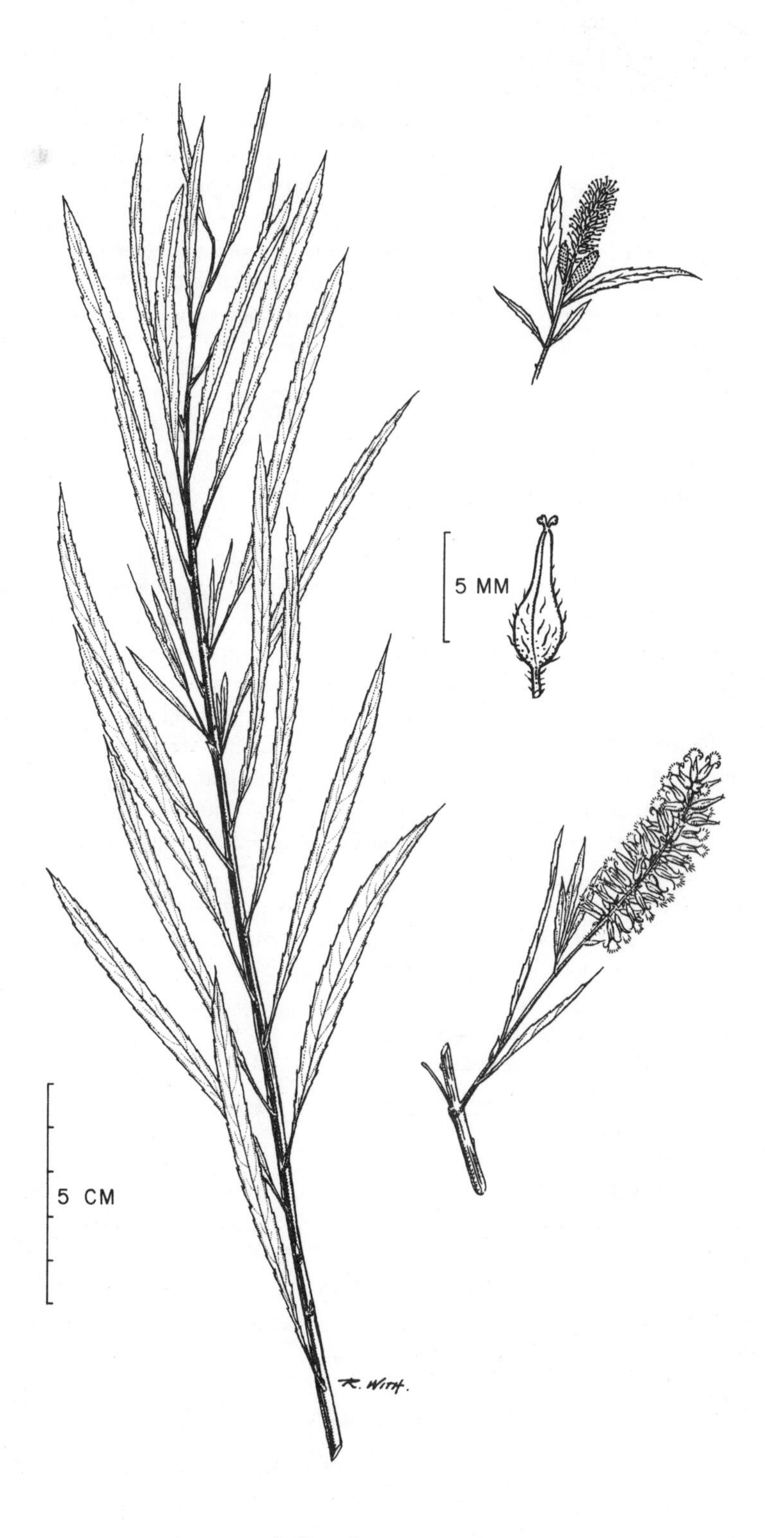

Salix exigua

been using the name *S. rigida* Muhl. and some botanists have interpreted it as a wide-ranging and highly variable species while others have recognized some of the variants in different parts of its vast range as separate species. One of the latter, *S. lutea* Nutt., has been reported in Ontario from three widely separated northern locations. The same entity has been reduced to a variety of *S. rigida* by Cronquist as *S. rigida* var. *watsonii* (Bebb) Cronq. In our treatment, this variant will key out to *S. eriocephala* and it seems best to retain it here until further study of the complex has resolved the relationships of the various geographical components. (*lutea* — yellow; *watsonii* — in honour of Sereno Watson (1826 - 1892), a Curator at the Gray Herbarium, Harvard University, and long-time student of the flora of North America)

Field check Medium to large shrub; young leaves reddish and pubescent; mature leaves oblong-lanceolate; young capsules reddish, glabrous, becoming divergent, scales shorter than the pedicels and long-hairy.

Salix exigua Nutt. (*S. interior* Rowlee) — Sandbar willow

A shrub up to 3 m or more in height, usually growing in colonies with many stems, spreading by underground stolons. Branchlets slender, reddish-brown, at first silky-hairy, soon becoming smooth; bark of older stems brown to grayish.

Leaves alternate, simple, and deciduous; blades linear to linear-oblanceolate, 5 - 12 cm long, 5 - 15 mm wide, acute to acuminate at the apex, narrowly cuneate at base or tapered gradually to a short petiole 0.5 - 5 mm long, green, and glabrous above, paler and sparsely silky-hairy or glabrous beneath; margins glandular-denticulate, the teeth irregularly and widely spaced, or the margins sometimes entire or nearly so; stipules lacking or present as minute glandlike appendages or as narrowly ovate to lanceolate scales up to 1.5 mm long.

Catkins on lateral leafy branches, often branched and opening as the shoots are developing; staminate catkins 2 - 5 cm long, stamens 2, filaments pubescent on their lower half; pistillate catkins 2 - 6 cm long, pistils densely to thinly silvery-villous, long-beaked with prominent 4-lobed sessile stigmas; bracts lanceolate, yellow-green to pale brown, deciduous before the capsules mature. Capsules 5 - 9 mm long, often glabrous. (*exigua* — small; *interior* — inland)

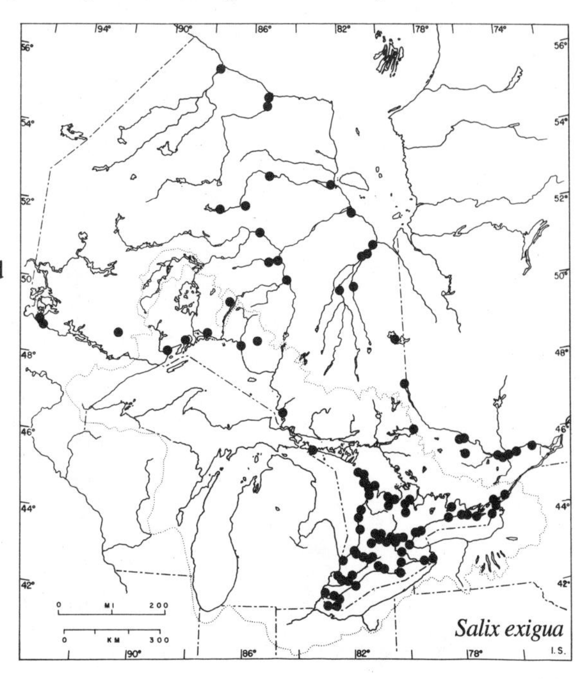

Salix exigua

Typical shrub of floodplains, alluvial flats, banks of rivers, lakeshores, and sandy beaches; also in shallow water around ponds, on the edges of swamps, and in wet lowland forests.

Widely distributed from Lake Erie northward to the shores of James Bay and Hudson Bay and from the lower Ottawa Valley westward to Lake-of-the-Woods. (N.B. to Alaska, south to Calif., Tex., and Va.)

Note This species is variable in leaf shape and in the amount and persistence of pubescence on the leaves. The name *S. interior* forma *wheeleri* (Rowlee) Rouleau

Salix fragilis

has been given to the variant in which the leaves are relatively shorter and broader with a dense and permanent coating of appressed hairs. (*wheeleri*—in honour of Charles F. Wheeler, 1842-1910, its discoverer)

Field check Leaves linear; margins remotely glandular-denticulate; bracts deciduous; colonial shrub of beaches and sandbars.

Salix fragilis L. — Crack willow, Brittle willow

A medium-sized or large tree up to 20-30 m high with divergent branches and a trunk up to 1 m in diameter. Branchlets slender, greenish-yellow to brown or dark red, smooth and shining; branchlets very brittle at the base and often shed in strong winds.

Leaves alternate, simple, and deciduous; blades lanceolate to elliptic-lanceolate (or oblong-ovate on sprouts), 7-15 cm long, 1-3 cm wide (or up to 20 cm long and 7 cm wide on sprouts); apex long-acuminate, base acute to tapered; margins coarsely serrate or undulate-serrate (with 4-7 large teeth per cm); mature leaves green and glabrous on both surfaces but usually paler or glaucous beneath; petioles 7-15 mm long, with prominent stalked glands near the junction with the blade; stipules small, semi-cordate, often lacking.

Catkins slender and lax, developing with the leaves (May and June), and borne on leafy lateral branches 1-5 cm long; staminate catkins 3-6 cm long, stamens 2 (occasionally 3 or 4), filaments pubescent at the base; pistillate catkins 5-8 cm long; nectaries 2 per flower. Capsules narrowly conic, 4-5 mm long, glabrous; styles less than 1 mm long; stigmas notched, short; pedicels 0.5-1 mm long; bracts oblong, pale yellow or greenish-yellow, crisply villous, deciduous. (*fragilis*—brittle or fragile)

In low areas along streams, rivers, edges of lakes; also in farmyards and pastures, and along roadsides.

Frequently planted and a common escape in southern Ontario, especially east and southwest of the Canadian Shield; less frequent northward to the upper portion of the Moose River drainage basin. (Introduced from Europe)

Field check Medium-sized tree with stiff yellowish-green twigs brittle at the base; leaves lanceolate, coarsely serrate or undulate-serrate (margins appearing "ragged"); slender lax catkins; glabrous capsules and deciduous yellowish bracts.

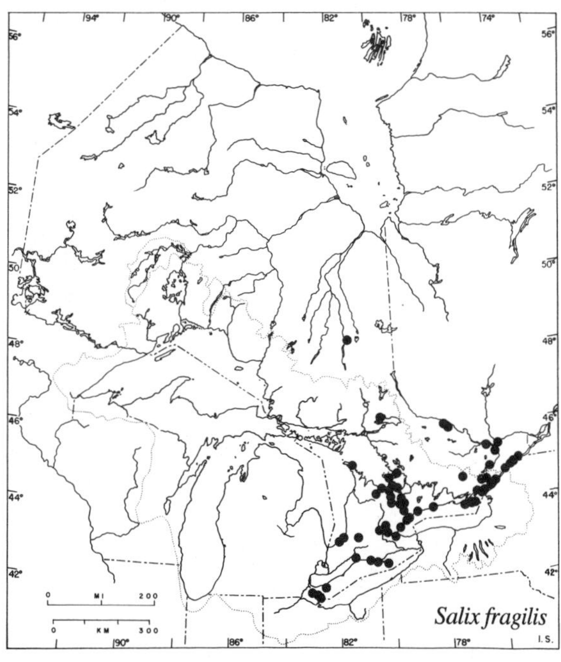

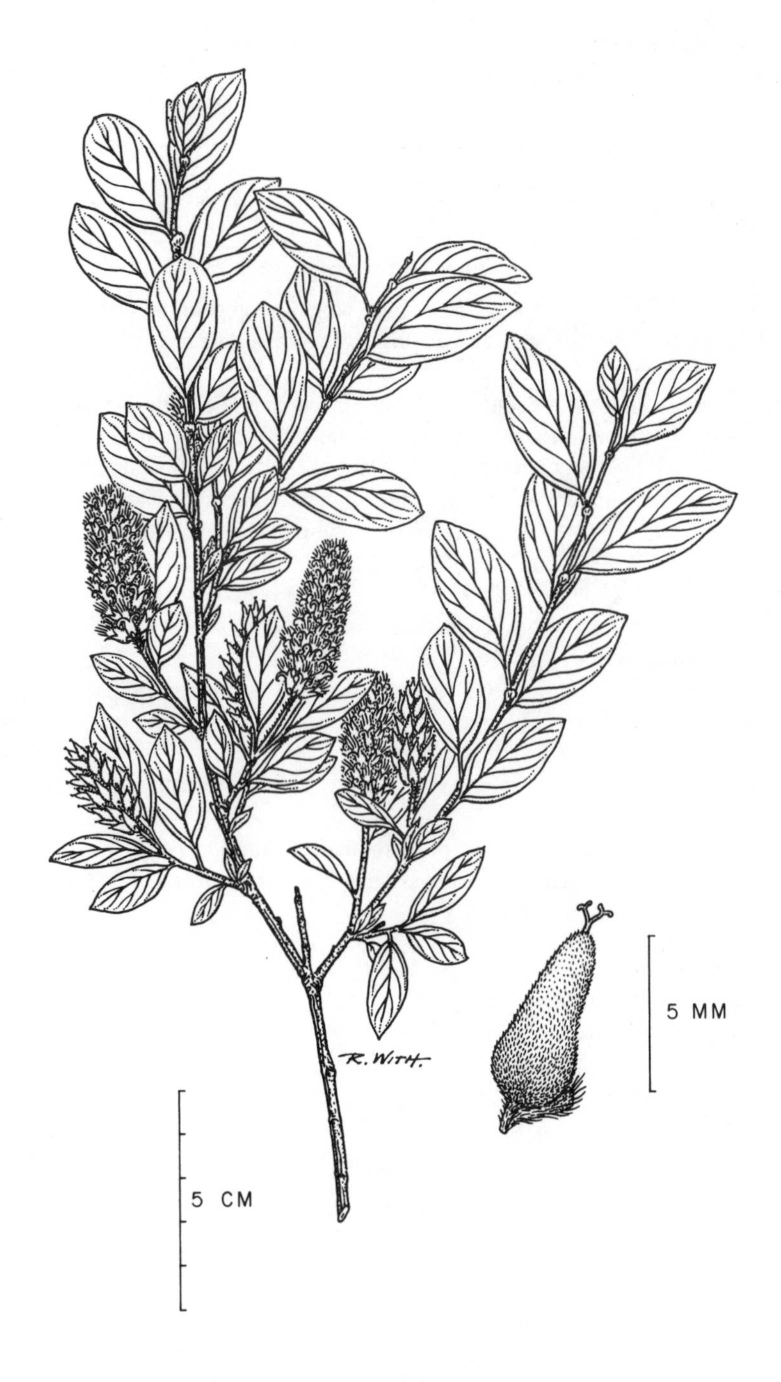

Salix glauca ssp. *callicarpaea*

Salix glauca L. ssp. *callicarpaea* (Trautv.) Böcher — Willow

An erect shrub with ascending branches and reaching a height of 1.5 m or depressed and trailing in exposed coastal sites. Branchlets flexible, at first densely silky-villous, the older stems becoming glabrous and reddish-brown to gray-brown.

Leaves alternate, simple, and deciduous; blades elliptic to oblong, obovate or oblanceolate, 2-6 cm long and 1-3 cm wide, acute to obtuse at the apex, the base tapered, rounded or, sometimes, cordate; margins entire or with glands near the base of the blade; young leaves silky-villous, becoming glabrous in age; leaves of the flowering and fruiting branches usually smaller and narrower than those of the vegetative shoots; mature leaves rather thin, mostly glabrate or sparsely villous along the veins on the lower surface; petioles slender, 3-9 mm long, longer than the well-developed buds, pubescent or glabrous; stipules minute (less than 1 mm long), often lacking.

Catkins appearing with the leaves at the ends of short, leafy, lateral branches; staminate catkins 1-4 cm long, stamens 2, filaments glabrous or hairy at the base; pistillate catkins 1.5-4.5 cm long, rather densely flowered; ovary and capsule pubescent; style about 1 mm long, stigmas forked; pedicel up to 1 mm long; bracts straw-coloured to dark brown, oblong, acute or rounded at the tip, with short or long hairs; nectaries 1 or 2 per flower; June and July. Capsules 6-10 mm long, pubescent; July to early Sept. The fruiting catkins occasionally persist into the next flowering season. (*glauca*—bluish-green; *callicarpaea*—with beautiful fruit)

On raised beaches, sand and gravel banks, and terraces along shores of rivers and on arctic coasts; also in marshes, bogs, and open fens.

Along the shores of Hudson Bay and James Bay and on islands in James Bay (N.W.T.); lower reaches of the Severn River and Attawapiskat River drainage systems. (Lab. and Baffin Is. to w. Hudson Bay, south to James Bay and Gaspé Pen.)

Note *Salix glauca* is a highly variable species ranging widely across arctic North America. The Ontario material falls in subspecies *callicarpaea* but two or three other variants have been recognized in the regions from Hudson Bay to Alaska, south through British Columbia and Alberta to Arizona, and in central Saskatchewan and Manitoba.

A comparison between this species and *S. brachycarpa* is given in the Note under *S. brachycarpa*.

Field check Prostrate or low erect shrub of tundra habitats (Hudson Bay and James Bay) with young leaves, branchlets, capsules, and bracts pubescent; petioles longer than the well-developed buds; catkins on short, lateral, leafy branches.

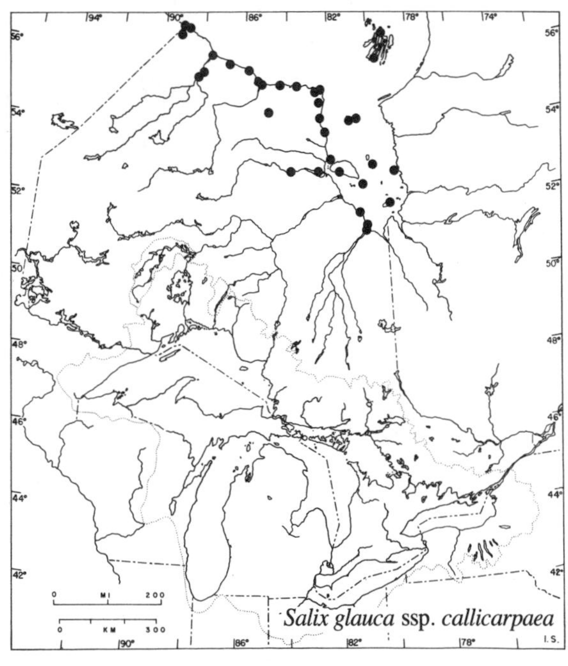

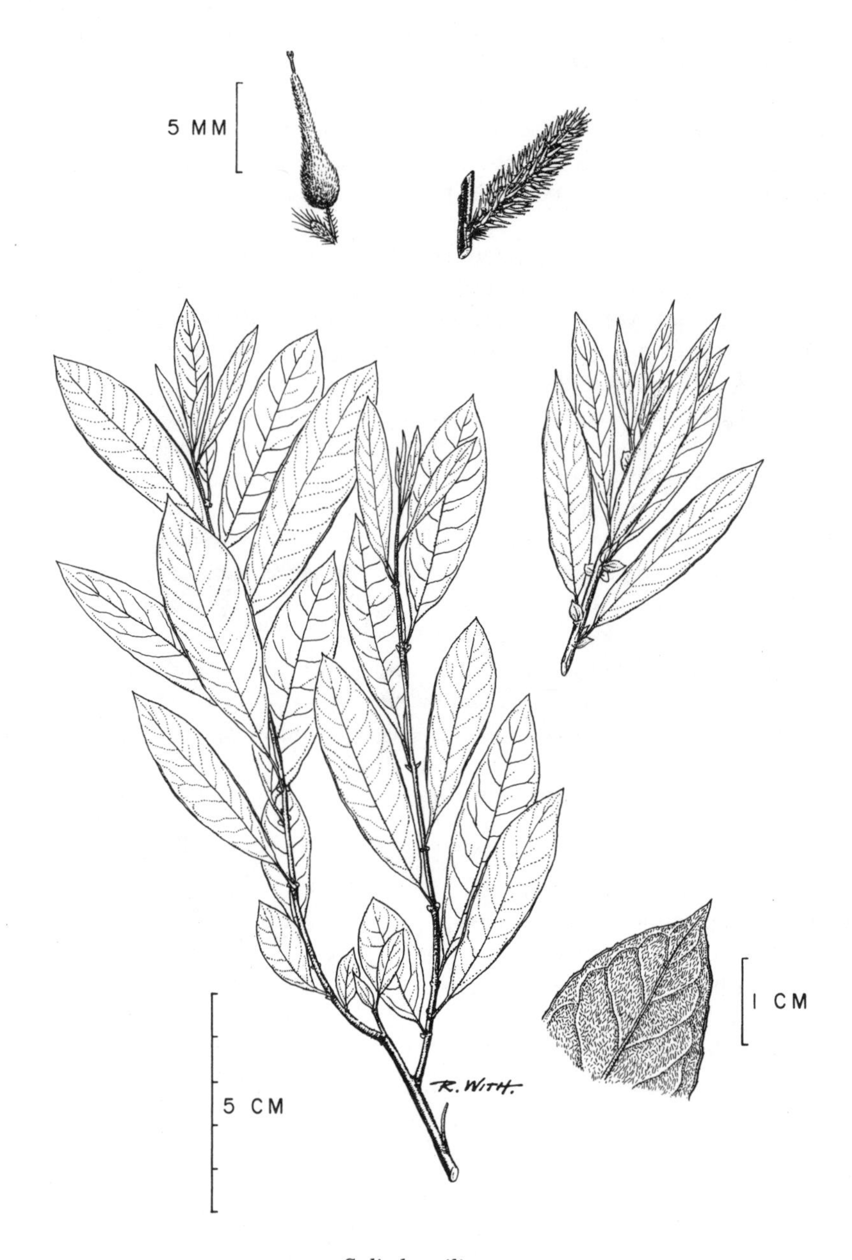

Salix humilis

Salix humilis Marsh.

Upland willow, Prairie willow Small pussy-willow

A low to medium-sized shrub, 1 – 3 m high. Branchlets yellow to brown, pubescent or becoming smooth, and dull, not shiny.

Leaves alternate, simple, and deciduous, expanding after the catkins have matured, the young leaves pubescent, often reddish-tinged with a coating of rusty-coloured hairs; blades of mature leaves obovate to nearly elliptic but usually oblanceolate and clearly broadest at or beyond the middle, 3 – 10 cm long, 1 – 3 cm wide, the apex acute to abruptly short-acuminate, the base tapered or acute; margins nearly entire or undulate-crenate and somewhat revolute; upper surface gray-green, puberulent to glabrate, and having somewhat impressed venation, the lower surface somewhat rugose, glaucous, and gray-pubescent to glabrate; petioles 2 – 10 mm long, pubescent; stipules lanceolate, 3 – 10 mm long, pubescent, usually deciduous.

Catkins precocious, appearing early in the season (late April to May in southern Ontario, May to early June northward), often numerous along the wandlike branches; staminate catkins 1 – 2.5 cm long, sessile or on short branchlets subtended by several pale scales; stamens 2, filaments glabrous, anthers red to purple; pistillate catkins 2 – 5 cm long, often recurved when young and elongating up to 8 cm in fruit, sessile or on short branchlets which may bear a few small green leaves or pale basal scales; pistils gray, pubescent, styles short, stigmas short and thick. Capsules slender and long-beaked, 6 – 9 mm long, gray-pubescent; pedicels 1 – 2 mm long, pubescent; nectary solitary; bracts 1.5 – 2 mm long, oblanceolate, dark brown to black, pubescent, and long-bearded; June and July. (*humilis* — low)

On dry sandy uplands in aspen, jack pine, or oak-pine woods; on outwash deltas and moraines; also on lakeshores and in boggy woods and alder swamps.

Common throughout southern Ontario, along the north shore of Lake Superior and northward in the Boreal Forest Region, becoming rare north of 54°N. (Lab. to n. Ont., south to Kans., La., and Fla.)

Note A number of intergrading varieties and forms have been described in this variable species.

Several collections from the north shore of Lake Superior (Pic Island and the Slate Islands) resemble closely the related western species *S. scouleriana* Barratt, and collections in the National Herbarium (CAN) from four widely separated locations, in the districts of Kenora, Rainy River, Sudbury, and Thunder Bay, have been identified as this species by G.W. Argus. (*scouleriana* — in honour of John Scouler, 1804 - 1871, Scottish botanist and collector in western North America)

Field check Catkins appearing very early, in series along the stems; leaves mostly oblanceolate, rugose, and gray-glaucous beneath.

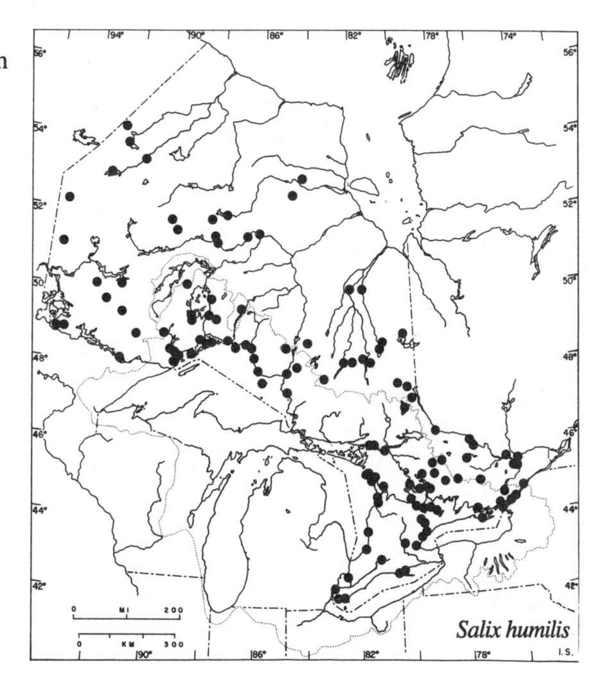

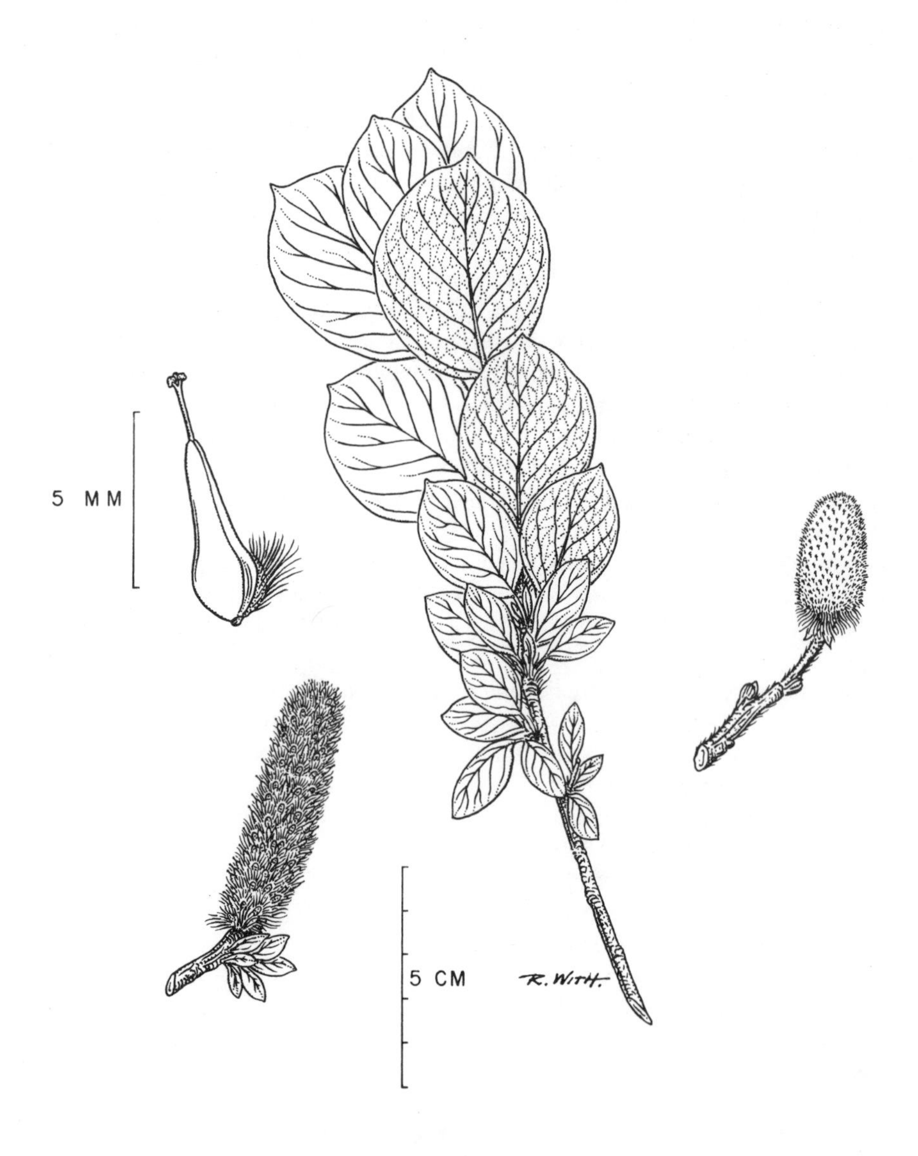

Salix lanata ssp. *calcicola*

Salix lanata L. ssp. *calcicola* (Fern. & Wieg.) Hult. Willow

A low, erect or prostrate shrub with spreading or matted, gnarled, stout, woody branches. Branchlets at first densely villous and grayish, the older stems dark brown and smooth with relatively short internodes and numerous bud scars; overwintering buds pubescent, stout, up to 1 cm long.

Leaves alternate, simple, and deciduous, with short petioles, somewhat crowded and overlapping at the ends of the branches due to the short internodes of the stems; blades of mature leaves thick and leathery, oblong-ovate to nearly orbicular, 1-5 cm long and 1-4 cm wide, acute to abruptly short-acuminate at the apex, the base tapered or rounded to subcordate; upper surface dull green and glabrous or with loose scattered pubescence, lower surface paler, glaucous, strongly veined, and pubescent, at least along the midvein; margins glandular-denticulate or entire, slightly revolute; petioles very short (1-4 mm long); stipules prominent, half-cordate to reniform, densely glandular-denticulate, often persisting even into the second year.

Catkins erect, stout, and dense, sessile at or near the ends of the branches, appearing before the leaves and developing as the leaves expand; bracts of the flowers blackish, villous, with long white to tawny hairs; staminate catkins 1-3.5 cm long, 1-1.5 cm thick, stamens 2, filaments glabrous; pistillate catkins 3-8 cm long, 1-2 cm thick; June and July. Capsules 6-9 mm long, glabrous, on very short, hairy pedicels; styles 1.5-2.5 mm long; nectary solitary. (*lanata*—woolly; *calcicola*—lime-inhabiting)

On barrens or in tundra.

Along or near the arctic shores of James Bay and Hudson Bay. (Baffin Is. to n.w. shore of Hudson Bay, south to the Gaspé Pen., Que., and Nfld.)

Note This willow was described in 1911 by Fernald and Wiegand as a distinct species (*S. calcicola* Fern. & Wieg.) but both it and the closely related *S. richardsonii* Hook. (known from Baffin Island and westward to Alaska and B.C.) have recently been grouped together as subspecies of the Eurasian *S. lanata*. (*richardsonii*—in honour of Sir John Richardson, 1787-1865, naturalist with Sir John Franklin's expedition to arctic North America)

Field check Prostrate or low, erect, dwarf shrubs of coastal tundra habitats (Hudson Bay and James Bay) with densely woolly young branchlets, thick roundish leaves, large persistent stipules, stout catkins, and glabrous capsules.

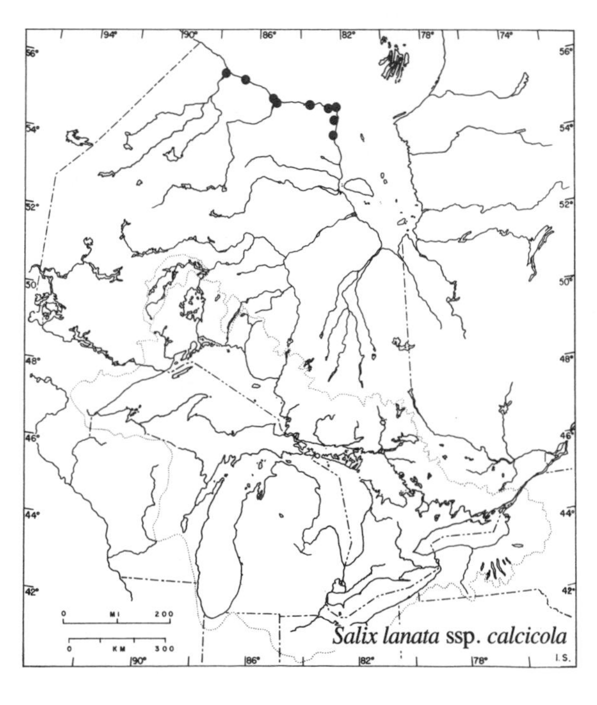

Salix lanata ssp. *calcicola*

Salix lucida

Salix lucida Muhl. — Shining willow

A shrub or small tree 3 - 6 m tall or occasionally a tree exceeding 10 m in height. Branchlets yellowish to reddish-brown, glabrous, and very shiny; young branchlets sometimes pubescent but soon becoming smooth (except in the var. *intonsa* Fern.); bark of older stems brown, glabrous.

Leaves alternate, simple, and deciduous; when young tinged with red, glabrous or pubescent, the unexpanded leaf blades and the stipules with conspicuously gland-dotted margins; blades of mature leaves lanceolate, ovate-lanceolate, or elliptic-ovate, 4 - 15 cm long, 1.5 - 4.5 cm wide, the apex acute to acuminate or, on the vegetative branches, long-attenuate, the base tapered or rounded to cordate, the upper surface dark green, glabrous, and glossy, the lower surface paler and glabrous (except in the variety); margins sharply serrate with gland-tipped teeth; petioles 5 - 15 mm long, glabrous or sparsely pubescent, glandular near the junction with the leaf blade; stipules 1 - 6 mm long, semi-circular to kidney-shaped with glandular margins.

Catkins coetaneous, on lateral leafy branchlets (May and June); staminate catkins 1.5 - 4 cm long, rather densely many-flowered, and with conspicuous, pubescent, pale yellow bracts; stamens 3 - 6, the filaments pilose near the base; glands 2, somewhat cup-shaped; pistillate catkins 1.5 - 5 cm long and 1 - 1.5 cm wide in fruit (June and July); pistils green to brownish, glabrous; styles less than 1 mm long, stigmas divided, thickish, somewhat club-shaped. Capsules slenderly ovoid-conic, 4 - 7 mm long, light brown, glabrous, bracts deciduous after flowering; nectaries minute, somewhat cup-shaped. (*lucida* — shining, in reference to the leaves and branchlets)

In wet places such as swamps, marshes, bogs, lakeshores, ditches, low meadows, mud flats, and sand bars along streams.

Throughout southern Ontario and northward beyond Lake Superior but becoming rare north of 51°N. (Nfld. to Sask., south to N.Dak., n. Iowa and Del.; also Neb. and Va.)

Note *S. lucida* var. *intonsa* is a variety in which the branchlets and the lower surface of the mature leaves remain somewhat pubescent with a mixture of rusty-red and colourless hairs. It has been collected at Lake Abitibi, along the east shore of Lake Superior, and on the Bruce Peninsula (*intonsa* — not shaved, in reference to the persistent pubescence)

S. lucida is similar in general aspect to *S. serissima* and both resemble an introduced European species, *S. pentandra* L. (bay-leaved willow). The latter is a shrub or small tree up to 7 m high with acute to short-acuminate, glabrous, shiny leaves, green on both sides or a little paler beneath. The staminate flowers have five stamens and the pistillate catkins are yellow-green. It spreads from cultivation occasionally and has been found growing wild in a few widely separated parts of Ontario. The common name, bay-leaved willow, refers to the resinous fragrance produced by its branchlets and gummy young leaves. (*pentandra* — with five stamens)

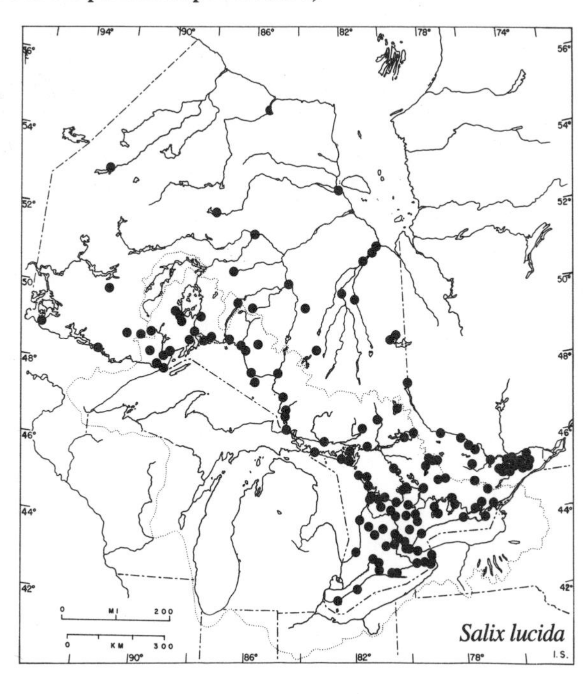

The above trio of similar species can be distinguished as follows: leaf tips of *S. pentandra* and *S. serissima* are merely short-acuminate, those of *S. lucida*

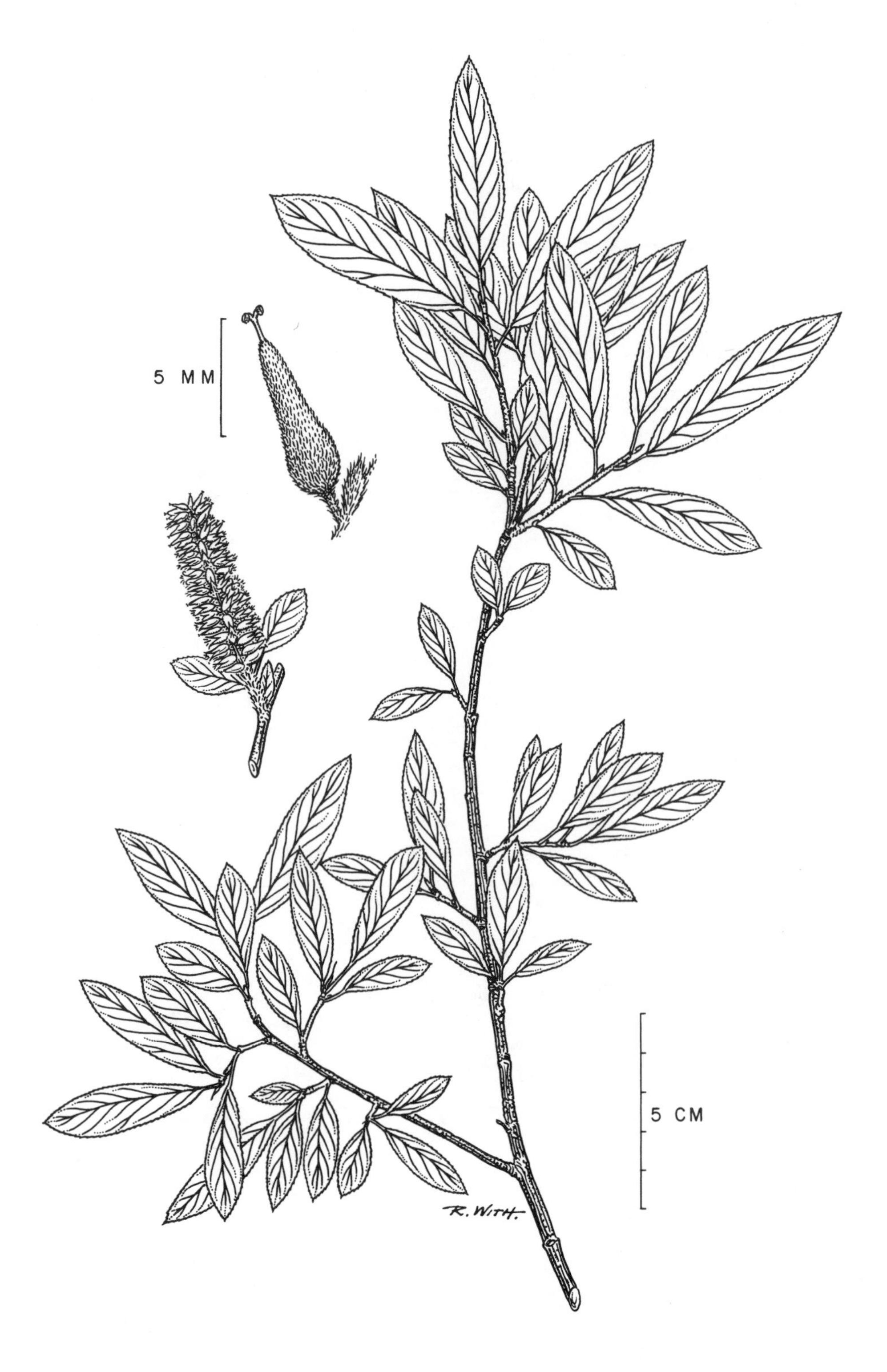

Salix maccalliana

long-attenuate. Leaves of *S. serissima* are whitened beneath, those of *S. pentandra* and *S. lucida* shiny and the lower surface only a little paler than the upper surface.

Field check Young branchlets and immature leaves with rusty-coloured pubescence; leaves shiny on both surfaces; apex of leaf long-attenuate; stipules 1 - 6 mm long with glandular margins; pistillate catkins 1.5 - 5 cm long and 1 - 1.5 cm wide; capsules splitting in June and July.

Salix maccalliana Rowlee — McCalla's willow

An upright shrub 1 - 3 m high. Branchlets yellowish and minutely hairy at first, becoming reddish-brown or purplish-red, glossy, and smooth.

Leaves alternate, simple, and deciduous, reddish-tinged when young, becoming leathery, green, and glossy on both sides, the midrib prominently yellow to orange-coloured for at least half of its length; blades of mature leaves elliptic-lanceolate to oblong, 3 - 8 cm long, 0.5 - 2.5 cm wide, the apex acute or somewhat acuminate, the base tapered or rounded; margins serrulate to crenate with gland-tipped teeth, rarely almost entire; lower surface of blade paler than the upper but not glaucous, sparsely pubescent with short white and rusty-coloured hairs; petioles 3 - 12 mm long, yellow-brown, slightly grooved, and minutely hairy on the upper side; stipules absent or reduced to minute glandular lobes.

Catkins coetaneous, on leafy lateral branches (June); staminate catkins 2 - 5 cm long, stamens 2, filaments distinct, pubescent near the base; pistillate catkins 2 - 5 cm long, densely flowered and up to 2.5 cm wide at maturity (late June and July), the gray to greenish-brown bracts contrasting conspicuously with the paler silky-hairy capsules; styles about 1 mm long, stigmas 2-lobed; bracts of the catkins pubescent near the base and glabrous but long-ciliate near the apex, persistent in the pistillate catkins. (*maccalliana* — named for William Copeland McCalla, 1872 - 1962, a Canadian school teacher, amateur naturalist, skilled photographer, and author of a book on the wild flowers of western Canada)

Chiefly in muskegs and wet black spruce forests.

In the James Bay and Husdon Bay drainage basins. (Que. to B.C. and s.e. Yuk.)

Note Somewhat similar in habit and characters to *S. serissima* but differing in its leaves being only somewhat paler beneath (not glaucous), in its dark persistent bracts, pubescent capsules, and earlier flowering period.

Field check Upright shrub of the James Bay drainage basin; leaves glossy green on both sides with yellow to orange midrib; capsules pubescent and much paler than the persistent darkish bracts.

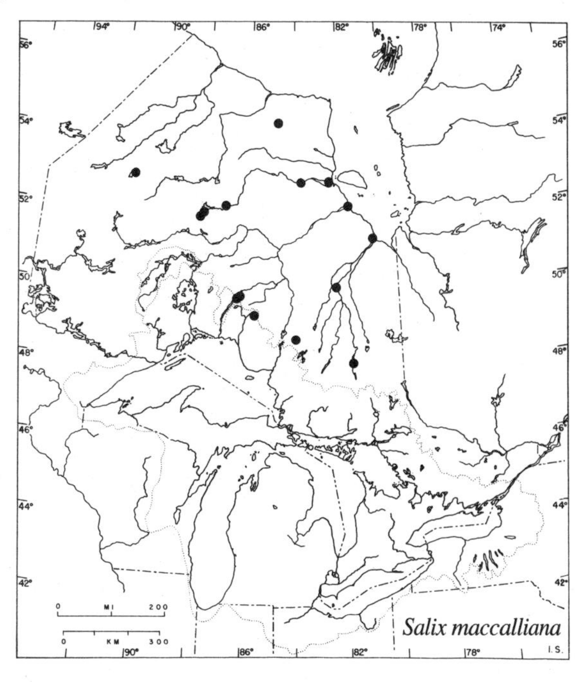

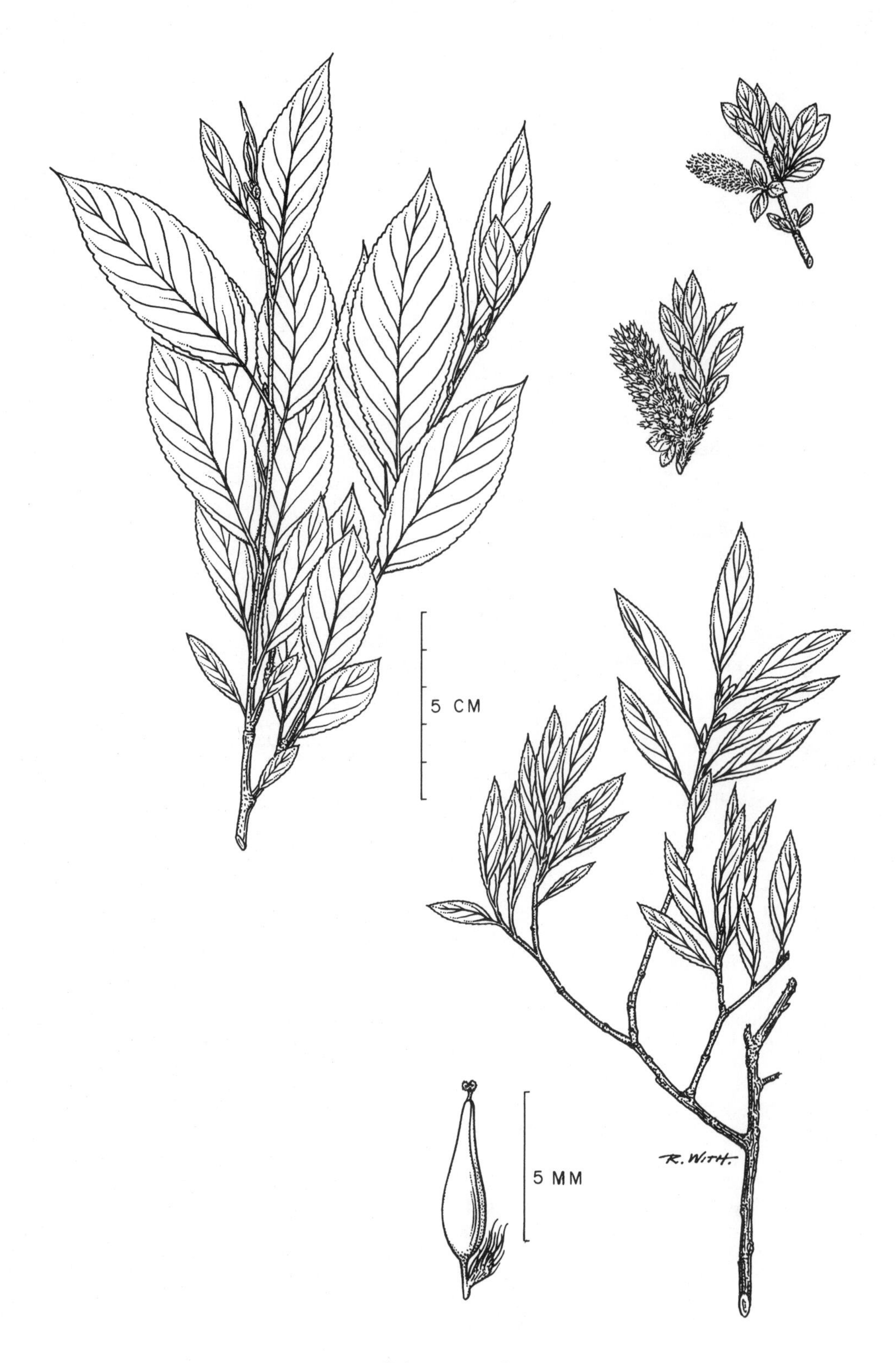

Salix myricoides

Salix myricoides Muhl. (*S. glaucophylloides* Fern.) — Blue-leaved willow

A coarse shrub with spreading, ascending, or erect branches up to 3 m in height, rarely a small tree up to 5 m tall. Branchlets brown to yellowish, usually pubescent when young but soon becoming smooth and lustrous, rarely permanently gray-velvety.

Leaves alternate, simple, and deciduous, often reddish-tinged when young, glabrous or with a temporary coating of rusty-coloured hairs; blades of mature leaves rather thickish, ovate to obovate but mostly lanceolate or elliptic, 4-12 cm long, 1-6 cm wide, apex acute or abruptly short-accuminate, base obtuse to rounded or narrowed, rarely cordate; upper surface usually dark green and glabrous, lower surface glaucous, often strongly so, with a whitish to bluish coating of wax, usually drying black; margins glandular-serrate or serrate-crenate with gland-tipped teeth; petioles 3-12 mm long, pubescent, the base somewhat dilated, the pubescence often extending along the midrib of the blade; stipules prominent (at least on vigorous vegetative shoots), semi-ovate, 5-10 mm long, the margins glandular-toothed.

Catkins appearing before or with the unfolding leaves (May and June), usually in flower before the leaves are fully grown; staminate catkins 2-4 cm long, subsessile or on short, leafy-bracted, lateral branches 2-6 mm long; stamens 2, filaments glabrous; pistillate catkins 3-6 cm long, on leafy lateral branches 4-14 mm long; capsules narrowly lance-ovoid, 4-8 mm long, glabrous; pedicels 1-4 mm long; flowers numerous and crowded at first but catkins more open and lax when the capsules are mature; bracts dark brown to blackish, obovate to oblanceolate, 1-2 mm long, densely villous. (*myricoides*—resembling *Myrica*, in reference to the similarity of leaf shape; *glaucophylloides*—with glaucous leaves)

Primarily on sand dunes, sand flats, or gravelly shores and in alluvial thickets, especially on calcareous soils.

Around the shores of the lower Great Lakes and less commonly along the north shore of Lake Superior and rivers of the James Bay drainage basin. (Nfld. to Hudson Bay, south to Ill., Pa., and Maine)

Note The leaves show considerable variation in the amount of pubescence and type of margin. Plants with the young leaves (and branchlets) densely white-tomentose and leaf margins more or less entire have been named *S. myricoides* var. *albovestita* (Ball) Dorn. The older leaves may retain some of the pubescence, especially along the midvein. (*albovestita*—clothed in white)

Some collections from the east shore of Lake Huron have rather small leaves and belong to the small-leaved variant that has been recognized by some authors as *S. glaucophylloides* var. *brevifolia* (Bebb) Ball. (*brevifolia*—with short leaves)

Field check Leaves elliptic-lanceolate to ovate or obovate, pubescent at least when young, usually strongly glaucous (whitened or bluish-white) beneath, drying blackish; bracts of catkins dark brown, long-hairy.

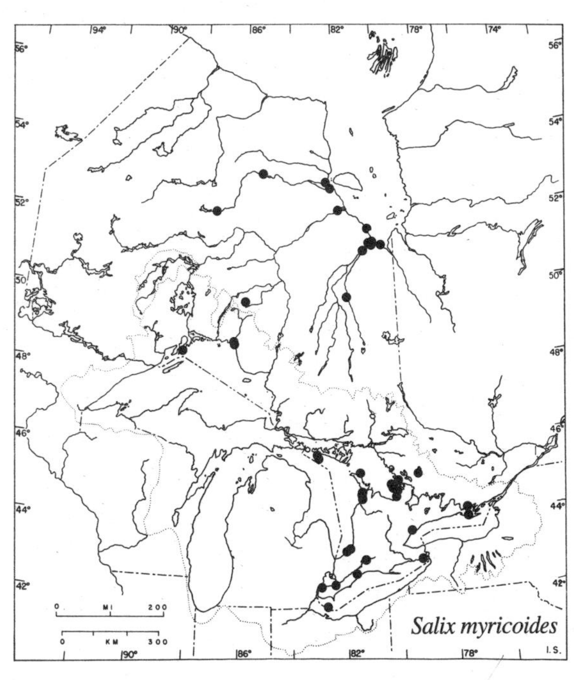

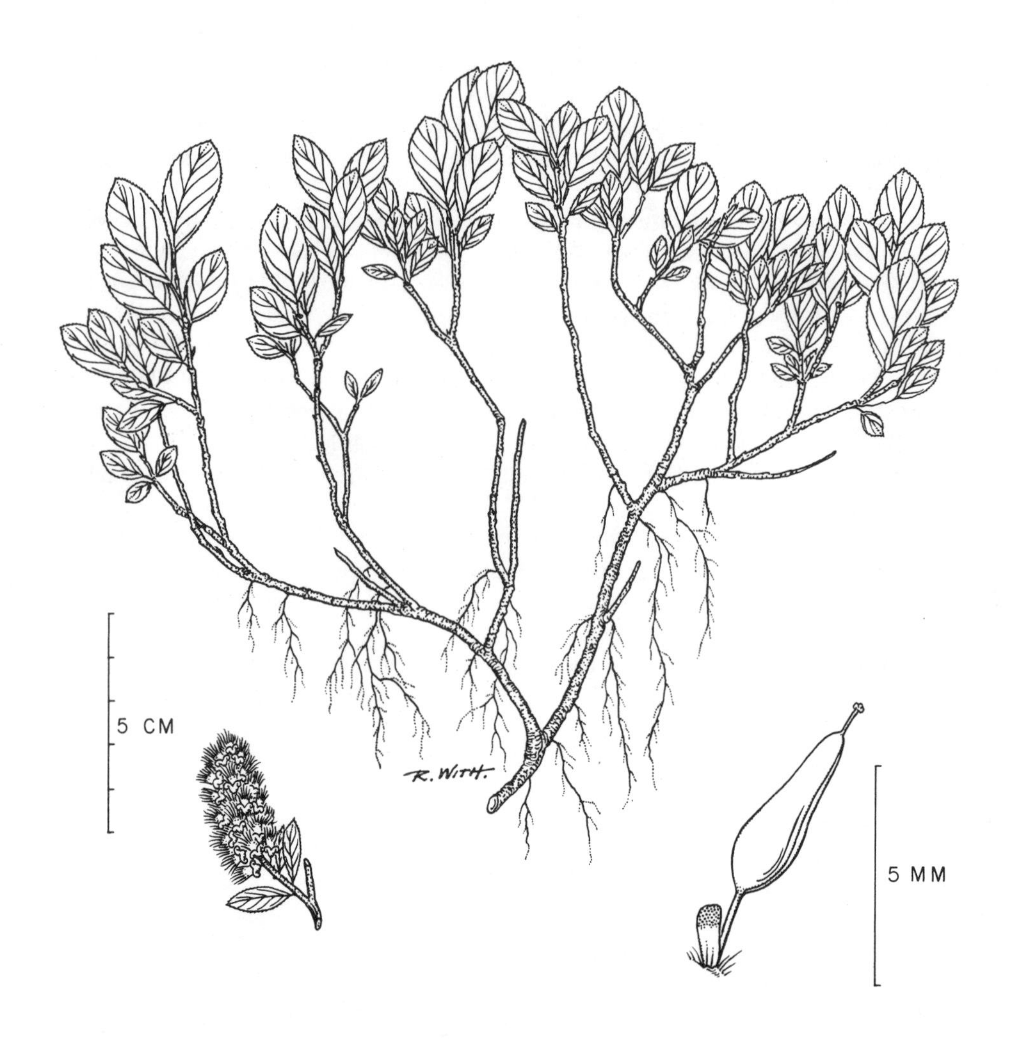

Salix myrtillifolia

A dwarf shrub usually with decumbent or trailing habit, the lower branches rooting in the moss; in some habitats it develops erect and ascending branches and forms a bush or develops thickets 0.5 - 1.5 m high. Branchlets greenish-brown to dark reddish-brown, glossy and sparsely pubescent, becoming brown to gray-brown and smooth.

Leaves alternate, simple, and deciduous; blades obovate to elliptic or lanceolate, relatively small, 1 - 5 cm long and 0.5 - 2 cm wide (or larger, up to 9 cm long, on vigorous vegetative shoots); apex rounded and blunt, but the sides often forming an acute angle, rounded or gradually tapered at base; uniformly green and shiny on both surfaces, glabrous or sometimes pubescent when young; margins glandular-crenulate or serrulate; petioles 2 - 10 mm long, glabrous; stipules narrowly elliptic to semi-ovate, generally small (1 - 3 mm long) and soon falling, or absent except on vigorous vegetative shoots where they may be up to 12 mm long and conspicuous.

Catkins appearing with the leaves, erect or ascending, on short, leafy, lateral branches; staminate catkins 1 - 3 cm long; stamens 2, filaments glabrous; pistillate catkins 2 - 5 cm long; June and July. Capsules thickly conical, greenish or brown, glabrous, 4 - 6 mm long; pedicels 0.5 - 1.5 mm long; bracts narrowly oblong, brown to blackish or pale near the base and darker at the rounded tip, sparsely to heavily silky-hairy. (*myrtillifolia* — with leaves like *Myrtillus*, referring to *Vaccinium myrtillus*, a European blueberry)

On moss-covered boulders and hummocks in open woods, ravine bottoms, on beach ridges, river banks, and alluvial terraces.

Restricted to northern Ontario where it is common along the shores of James Bay and Hudson Bay; also inland along the main river systems of the Hudson Bay drainage basin, rare on and near the north shore of Lake Superior and at Lake-of-the-Woods. (Nfld. to cent. Alaska, south to B.C., Ont., and N.B.)

Note Plants 1 - 3 m high with leaves about 5 cm long, occurring in well-drained areas within the range of the species may be referred to the variant *S. myrtillifolia* var. *cordata* (Anderss.) Dorn. A second variety, *S. myrtillifolia* var. *brachypoda* Fern., has recently been treated at the species level as *S. ballii* Dorn based on collections studied from northern Ontario, Gaspé, Que., and Newfoundland. We are including the single Ontario record of the latter species under *S. myrtillifolia* to which it is closely related. (*cordata* — heart-shaped; *brachypoda* — short-stalked; *ballii* — named for Carleton Roy Ball, 1873 - 1958, American student of willows)

Field check Low erect shrub or decumbent and trailing in moss; leaves obovate to elliptic, green on both sides; capsules brownish, glabrous; bracts darkish or two-toned, long-hairy; northern habitats.

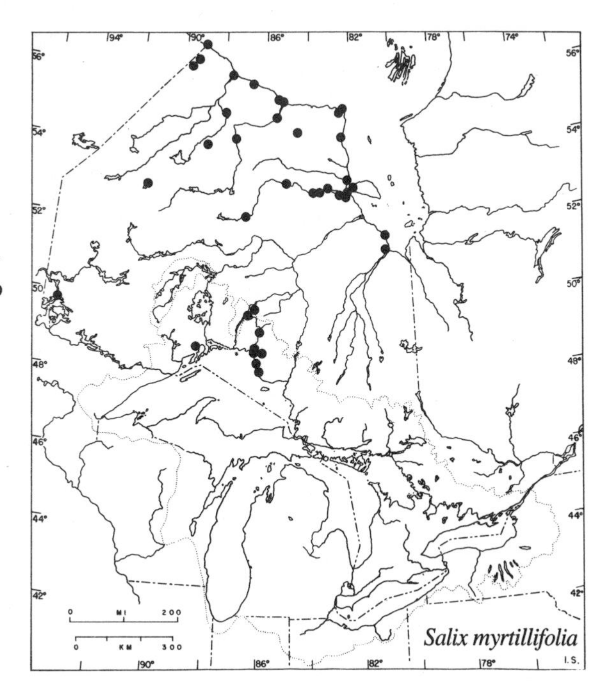

Salix nigra

Salix nigra Marsh. **Black willow**

A tall shrub or tree 3 - 20 m high, often with several stems or trunks up to 5 dm in diameter; trunk covered with dark brown to blackish flaky bark and sometimes leaning away from the vertical position. Branchlets slender, brittle at the base but tough and flexible above, yellowish to brownish or purplish red, pubescent when young but soon glabrous, with longitudinal ridges and furrows.

Leaves alternate, simple, and deciduous, often reddish and densely pubescent when young, becoming glabrous or nearly so; blades of mature leaves deep green above, the lower surface only a little paler and non-glaucous, the midrib minutely hairy, linear to linear-lanceolate, 5 - 12 cm long and 0.5 - 2 cm wide, the apex narrowed to a prolonged slender tip, often falcate, the base acute to rounded; margins finely serrulate; petioles 3 - 10 mm long, usually pubescent, and with glands near the junction with the blade; stipules prominent on vegetative branches, up to 10 mm long, glandular-serrulate.

Catkins coetaneous, on leafy lateral branches, the flowers appearing whorled on the axis (May and June); staminate catkins slender, lax, 3 - 10 cm long, stamens 3 - 6, filaments long-hairy near the base, nectaries 2 or more; pistillate catkins 4 - 8 cm long, slender, and loosely flowered. Capsules conic-ovoid, brown, glabrous, 3 - 5 mm long, on short pedicels 0.5 - 1 mm long; style and stigmas very short; nectary solitary, shorter than the pedicel; scales oblong to obovate, pale yellow, 2 - 3 mm long, pubescent, those of the pistillate catkins usually falling early; June and July. (*nigra* — black)

In low ground, often pioneering on alluvial flats, sand bars, shores, and disturbed areas; along streams, river banks, flood plains and in swamps, lowland pastures, and forests.

Restricted to southern Ontario where it is common from Lake Erie to the Bruce Peninsula and in the Ottawa-St. Lawrence lowlands. (N.B. and s. Que. to cent. Minn., south to w. Tex. and Fla.)

Note Vegetative specimens may be confused with those of *S. eriocephala* but in the latter species the leaves are broader, less attenuate, and mostly glaucous beneath, the petiole is not glandular, and the habit is shrubby and never treelike.

Field check Tree or tall shrub with several stems; branchlets slender, brittle at the base; leaves lanceolate, long-attenuate, often falcate, non-glaucous beneath; stipules prominent on vegetative branches.

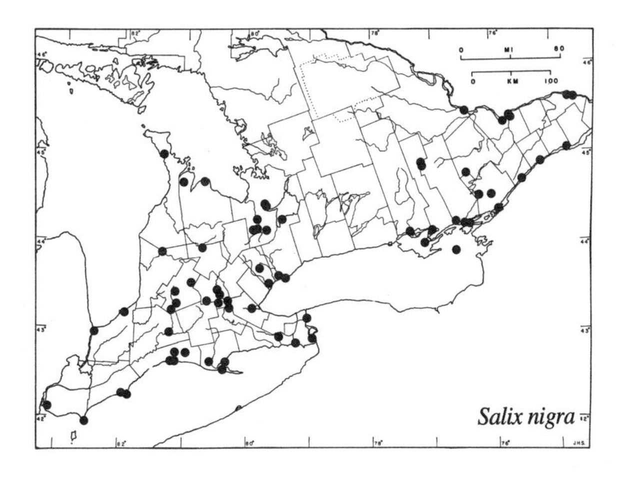

Salix pedicellaris

Salix pedicellaris Pursh **Bog willow**

A low shrub with erect or ascending branches usually about 1 m or less in height but occasionally a little taller; often decumbent near the base and rooting along the lower branches or, in peat bogs, forming adventitious roots along the lower branches as they become buried in the moss. Branchlets glabrous, yellowish to olive-green, becoming reddish to purplish-brown and grayish on the older stems.

Leaves alternate, simple, and deciduous; blades narrowly oval to elliptic-oblong or oblanceolate, rarely broadly obovate, 2-5 cm long, 0.5-2 cm wide, obtuse, rounded or acute at the apex, the base rounded or tapered, firm to leathery, glabrous, dark green above, paler and glaucous on the lower surface; both surfaces with reticulate venation, prominent pale to reddish-brown midrib and often with a waxy coating; margins entire, usually somewhat revolute; petioles 2-6 mm long; stipules absent.

Catkins coetaneous, terminating leafy branches 1-5 cm long; staminate catkins 0.5-2 cm long, stamens 2, filaments glabrous; pistillate catkins 1-3 cm long and up to 2 cm wide in fruit; pistils narrowly lanceolate, yellowish to dark red-brown. Capsules yellow to brown, glabrous, 4-7 mm long, capped by short, divided, sessile stigmas; bracts oblong, yellowish-brown, glabrous on the oustide, and sparsely pubescent on the inner surface; nectary very short, much exceeded by the slender pedicel which is 2-4 mm long. (*pedicellaris*—having pedicels, in reference to the conspicuous stalks of the capsules)

Chiefly in sphagnum bogs and muskegs and around the edges of swamps, marshes, and beaver meadows; associated with tamarack, black spruce, dwarf birch, leatherleaf, and species of *Carex*.

Throughout Ontario but apparently rare or absent in the Algonquin highlands. (Nfld. to s.w. Mack. Dist., N.W.T., and s.e. Yuk., south to Oreg., Iowa, and N. Eng.)

Note The lower leaf surface in this species is commonly glaucous. However, as pointed out by Argus (1964), the waxy bloom can be destroyed by heat. Thus, some herbarium specimens that have been dried over excessive heat do not show the glaucescence. Also, some herbarium specimens have leaves partly and irregularly glaucous, indicating a non-uniform destruction of the waxy coating. Therefore, there is no reason to take up the name proposed by Fernald in 1909, *S. pedicellaris* var. *hypoglauca* Fern. (*hypoglauca*—glaucous below)

Field check Low shrub with leathery glaucous leaves and prominent reticulate venation; pistils usually reddish, on long pedicels; mainly in bogs and other wet habitats.

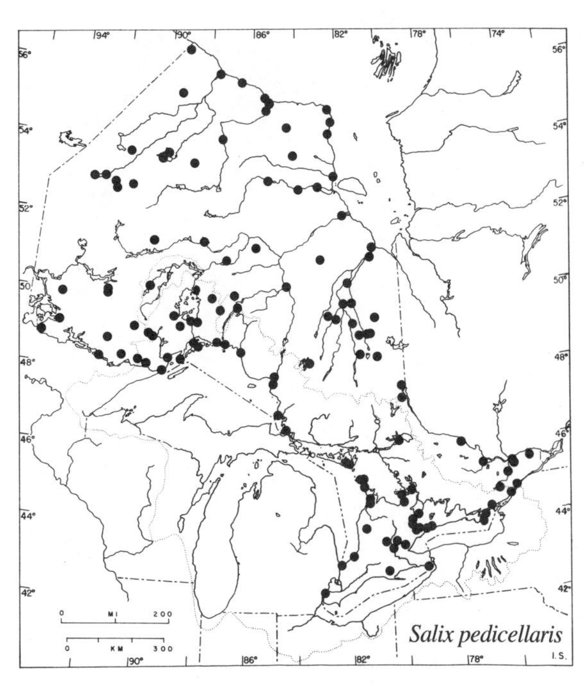

Salix pellita

Salix pellita Anderss. **Willow**

A medium-sized shrub up to 4 m or occasionally a small tree. Branchlets glabrous or thinly pubescent, brittle, yellowish to dark reddish-brown, often with a bluish-white waxy bloom.

Leaves alternate, simple, and deciduous, often reddish-tinged when young; blades of mature leaves thick and firm, dark green and glabrous above, glaucous and with dense shiny white pubescence beneath or rarely becoming smooth (forma *psila* Schneid.), linear-lanceolate to broadly lanceolate or oblanceolate, 4-12 cm long, 0.5-2.5 cm wide, acute, acuminate or attenuate at the tip, the base tapered and mostly acute; primary veins impressed above, numerous, and closely parallel, the midvein prominent on the lower surface; margins entire and revolute or very shallowly undulate-crenulate; petioles 2-10 mm long, glabrous; stipules none or minute and deciduous.

Catkins appearing before or with the developing leaves (May and June), sessile or on short bracted peduncles; scales dark brown to blackish, long-villous; staminate catkins 2-3 cm long, stamens 2, filaments glabrous; pistillate catkins 3-5(-8) cm long. Mature capsules sessile or nearly so, 4-6 mm long, conic-ovoid, densely pubescent with silky white hairs; styles slender, 0.8-1.2 mm long; nectary solitary, elongate; June and July. (*pellita*—clad in skins; *psila*—smooth)

In clay, sand, or gravel or among boulders on river banks and lakeshores; damp depressions of fens and spruce woods.

Throughout northern Ontario from the shores of Hudson Bay and James Bay southward to the north shore of Lake Superior and the upper part of the Ottawa Valley. (Nfld. to Sask., south to Mich., N.H., Vt., and Maine)

Note Related to *S. drummondiana* Barratt of western Canada but usually treated as distinct from that species. (*drummondiana*—named for Thomas Drummond, 1780-1853, Scottish botanical collector)

Field check Linear-lanceolate to oblanceolate leaves, green and smooth above, densely silky-hairy and lustrous beneath, the margins revolute and essentially entire; catkins with dark silky-villous bracts; capsules densely pubescent.

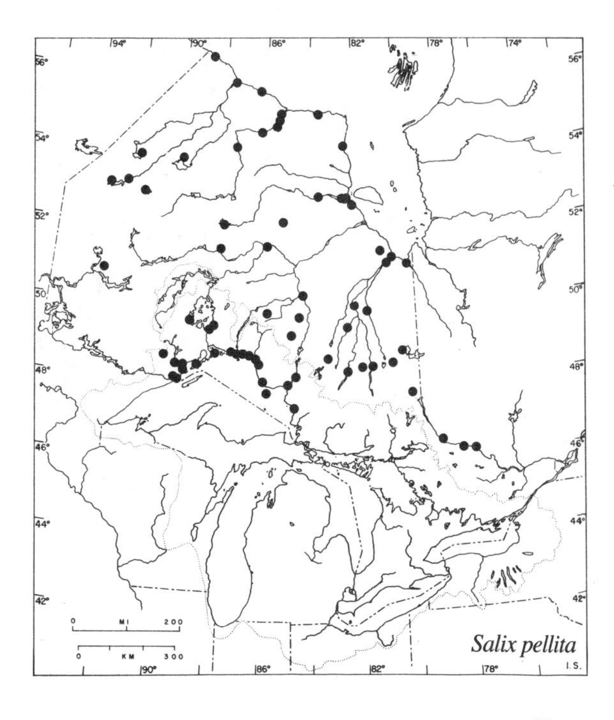

Salix petiolaris

Salix petiolaris J.E.Sm. (*S. gracilis* Anderss.) — Slender willow

A low to medium-sized shrub 1–3 m high, the branches erect or ascending, often clumped. Branchlets slender, yellow-green to olive-brown, pubescent or glaucous but becoming glabrous and dark brown to nearly black.

Leaves alternate, simple, and deciduous, numerous, ascending, and often overlapping on the stem, the young leaves silky-hairy and often reddish-tinged from rusty-coloured hairs; blades of mature leaves usually glabrous, green, and lustrous above, glaucous and glabrous or thinly silky-hairy beneath; blades linear to lanceolate, thin to membranaceous, 2–7 cm long, 0.5–1.5 cm wide (or up to 12 cm long and 2 cm wide on vigorous vegetative branches); apex acuminate and base acute; margins minutely denticulate or irregularly serrate with gland-tipped or prolonged teeth, at least above the middle, sometimes entire or nearly so towards the base; petioles 3–10 mm long, yellowish; stipules minute and deciduous, usually absent.

Catkins appearing with the leaves (May and June) on short leafy branchlets and often forming long series on the numerous ascending branches; staminate catkins ellipsoid-obovoid, 1–2 cm long; stamens 2, filaments glabrous or pubescent at the base; pistillate catkins rather loosely flowered, 1.5–2.5 cm long, in fruit usually more lax and up to 4 cm long; pistils densely silky-hairy. Capsules elongate, slender-beaked, 5–7 mm long, styles obsolete, stigmas forked with 2 diverging linear lobes; pedicels slender, 1–3 mm long, pubescent; nectary solitary, very short; bracts oblanceolate, 1–2 mm long, pubescent; late May and June. (*petiolaris*—pertaining to the petiole; *gracilis*—slender)

Chiefly in low damp or wet habitats; in meadows, boggy or grassy flats; along the edges of ponds, swamps, rivers, and lakes; in ditches and ravines; also in dry upland sites such as jack pine woods.

Common throughout southern Ontario and westward along the north shore of Lake Superior but becoming rare north of 50°N. (N.B. to B.C., south to Colo., Okla., Mass., and Va.)

Field check Leaves linear-lanceolate, glaucous beneath; margins finely serrate; catkins appearing with the leaves on short lateral branchlets with reduced leaves, often forming long series.

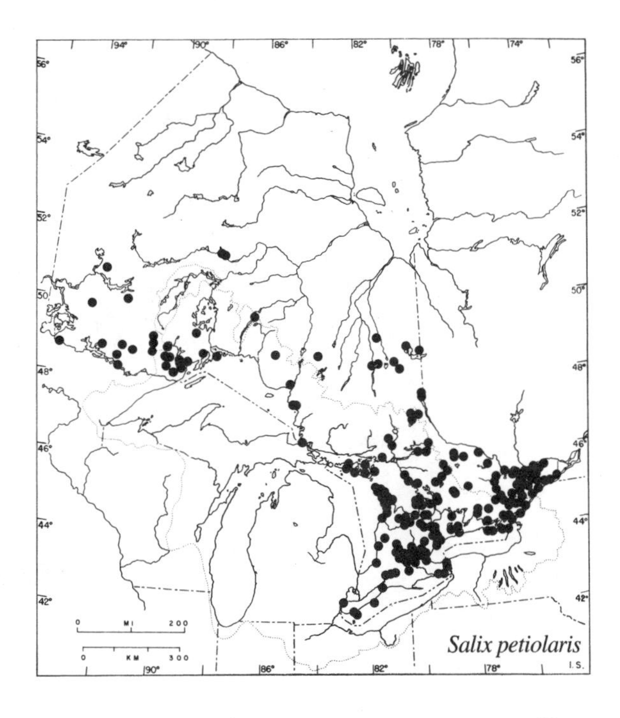

Salix planifolia

Salix planifolia Pursh **Willow**

An erect shrub 1–4 m high with stout divaricate branches. Branchlets chestnut-brown, minutely pubescent when young, becoming purplish-brown, smooth, and shiny or sometimes with a waxy bloom.

Leaves alternate, simple, and deciduous, sometimes reddish-tinged when unfolding; blades of mature leaves glabrous, somewhat leathery, dark green, and shiny above, with impressed veins, glaucous beneath, the bloom often thick and waxy, the midrib pale and prominent for most of its length, elliptic-lanceolate to rhombic or obovate, 2–7 cm long, 1–2.5 cm wide, acute to obtuse at the apex, tapered or (the larger ones) obtuse at the base; margins entire and slightly revolute, indistinctly toothed; lateral veins rather numerous and closely spaced; petioles yellowish to reddish, 2–10 mm long, glabrous; stipules minute, narrowly elliptic, glabrous with glandular margins, early deciduous, thus usually absent.

Catkins precocious (May and June) and sessile, with a few green to brown bracts at the base, the scales dark brown to blackish and with long silky hairs; staminate catkins 2–4 cm long, stamens 2, filaments glabrous; pistillate catkins 2–6.5 cm long, rather compact. Mature capsules (June and July) conic-ovoid, 5–7 mm long, covered with short, white, silky hairs; styles prominent, yellowish-brown, slender, each with two branches; pedicels very short (about 0.5 mm long); nectary solitary. (*planifolia*—flat-leaved)

In cool, moist habitats; along rocky lakeshores, in cedar swamps and black spruce bogs, at the edges of creeks and sedge marshes, and on moss-covered boulders in canyons.

Throughout northern Ontario from the shores of Hudson Bay and James Bay southward to the north shore of Lake Superior; rare south of 48°N. (Nfld. to N.W.T. and s. Yuk., south to Calif., N. Mex., Minn., N.H., Vt. and Maine)

Field check Leaves dark green and glossy above, glaucous beneath, entire to shallowly crenate-serrulate, and slightly revolute; catkins precocious; stamens glabrous; capsules silky-hairy; scales darkish and long-villous.

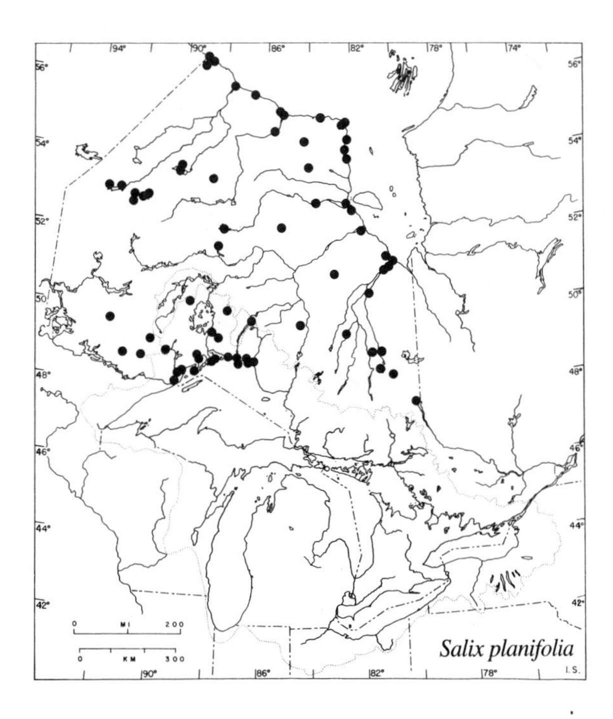

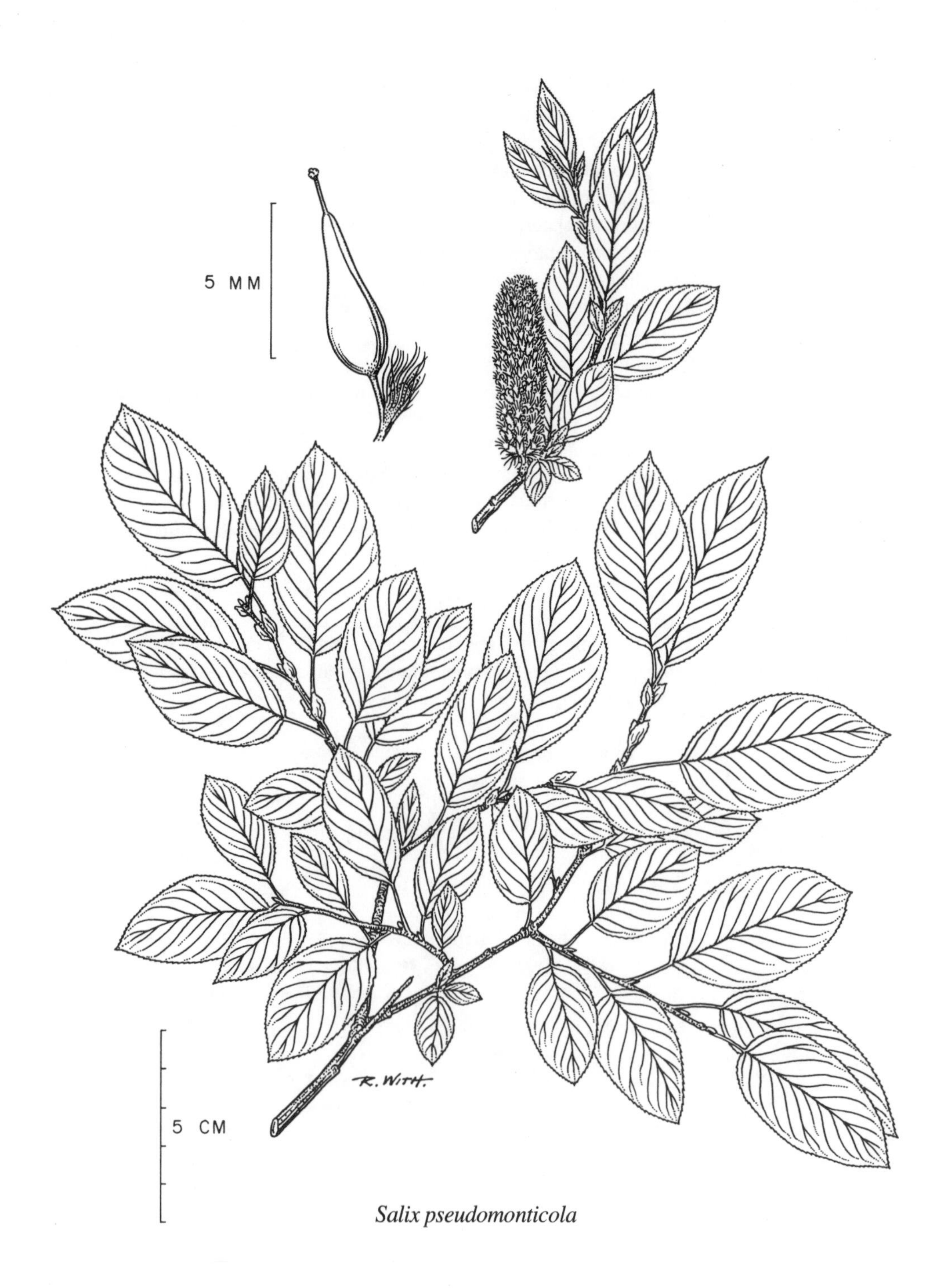

Salix pseudomonticola

Salix pseudomonticola Ball **Willow**

A large erect shrub or small tree up to 6 m high. Branchlets yellowish-green, glabrous, and glossy or sparsely pubescent; older branches yellow-brown to dark reddish-brown, smooth, and shiny.

Leaves alternate, simple, and deciduous, translucent when young, sparsely pubescent, and usually reddish-tinged; blades of mature leaves elliptic to ovate or obovate, 3–10 cm long, 1.5–4.5 cm wide, acute to acuminate at the often broad apex, rounded to cordate at the base which is frequently asymmetrical (oblique); upper surface dark green and glabrous or nearly so, lower surface glaucous and becoming glabrous; margins glandular-crenate or glandular-serrulate; petioles 0.6–2 cm long, glabrous or pubescent, often reddish; stipules ovate, up to 10 mm long, their margins more or less toothed; the portion of the midrib nearest the petiole is often reddish and conspicuous.

Catkins appearing before the leaves (May) and usually mature before the leaves have fully developed; staminate catkins 1.5–4 cm long, sessile; stamens 2, filaments glabrous; pistillate catkins 2–9 cm long, sessile or on a leafy lateral branch less than 5 mm long; capsules glabrous, brownish, 4–6 mm long; pedicels 0.5–2.5 mm long, about as long as the narrowly oblong, brown to blackish, long-hairy bracts. (*pseudomonticola*—false *monticola*, in reference to the similarity and confusion between this species and *S. monticola* Bebb; *monticola*—mountain-dwelling)

On river banks and lakeshores, in ravine bottoms, aspen woods, and gravelly clearings.

From Lake Superior and the Clay Belt of eastern Ontario northward to James Bay; rare south of 48°N; to be expected also in the western half of northern Ontario and to the limit of trees near Hudson Bay. (Que. to Alaska, south to Idaho, Wyo., S.Dak., and Ont.)

Note In a recent study, this species has been clearly distinguished from *S. monticola* Bebb (see Dorn, 1975) and reports of the latter from Ontario should be referred here. The true *S. monticola* occurs in southern Wyoming, Colorado, Utah, New Mexico, and Arizona.

Field check Coarse erect shrub or small tree; leaves broad, reddish, and translucent when young; catkins sessile or subsessile, maturing before the leaves; petioles and adjacent portion of the midrib reddish; stipules prominent; capsules glabrous and bracts long-hairy.

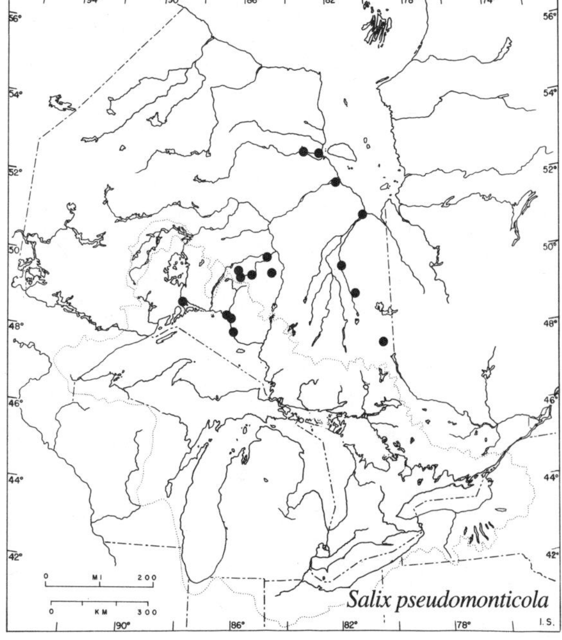

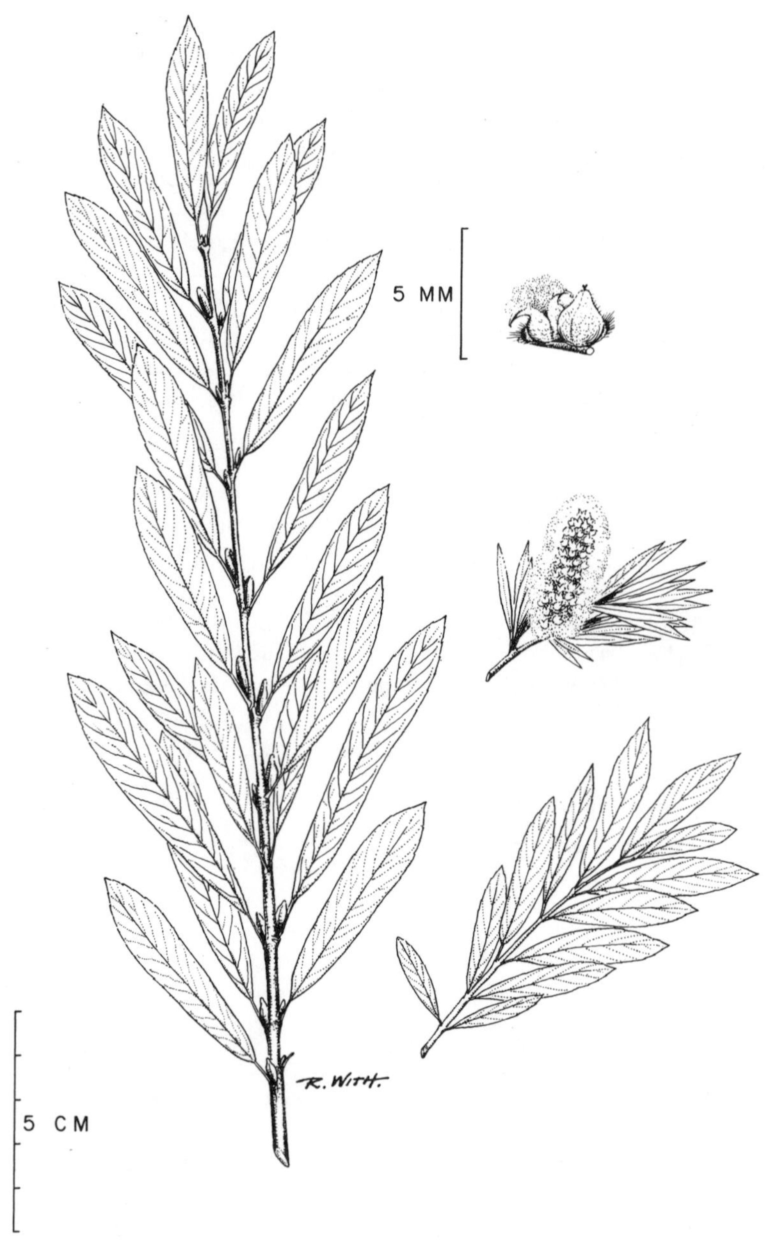

Salix purpurea

Salix purpurea L.

Basket willow
Purple osier

A medium-sized shrub with numerous stems and ascending branches reaching a height of 3 m or more. Branchlets glabrous, yellowish-green to purplish, slender, flexible, and often dark purple, especially on young vegetative growth.

Leaves opposite or nearly so, simple, and deciduous; blades oblanceolate, spatulate or narrowly elliptic to linear-lanceolate, 2–6 cm long and 0.5–1.5 cm wide, apex acute to acuminate, base obtuse or abruptly rounded; margins entire or irregularly serrulate towards the apex with low gland-tipped teeth, glabrous on both surfaces, and glaucous beneath, with raised venation; petioles 1–4 mm long; stipules absent.

Catkins appearing before the leaves, often in pairs (subopposite), slender, cylindrical, sessile or on short branches with a few small leaves or bracts at the base; staminate catkins 2–3 cm long, stamens 2, filaments united to each other and pubescent below, sometimes the anthers also fused; pistillate catkins 2–3 cm long, pistils densely pubescent, styles and stigmas minute. Capsules white-pubescent, plump-ovoid, 2–3 mm long, sessile; bracts obovate, pubescent, blackish or with paler central portion. (*purpurea*—purplish)

Escaping from cultivation and persisting in low ground along streams, edges of pools, and sandy beaches; possibly spreading.

Found in widely separated localities from Lake Erie and the western end of Lake Ontario to Lake Huron and near Georgian Bay; also in the valleys of the Ottawa and St. Lawrence rivers. (Introduced from Europe)

Field check Our only willow with leaves, buds, and catkins opposite to subopposite; immature branches usually dark purple; the spatulate to elliptic leaves purple-tinged; plump, sessile, pubescent capsules in short catkins (2–3 cm long); introduced and found occasionally as an escape.

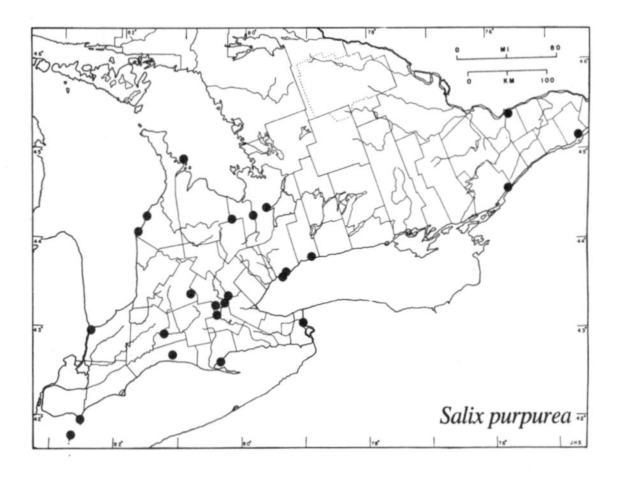

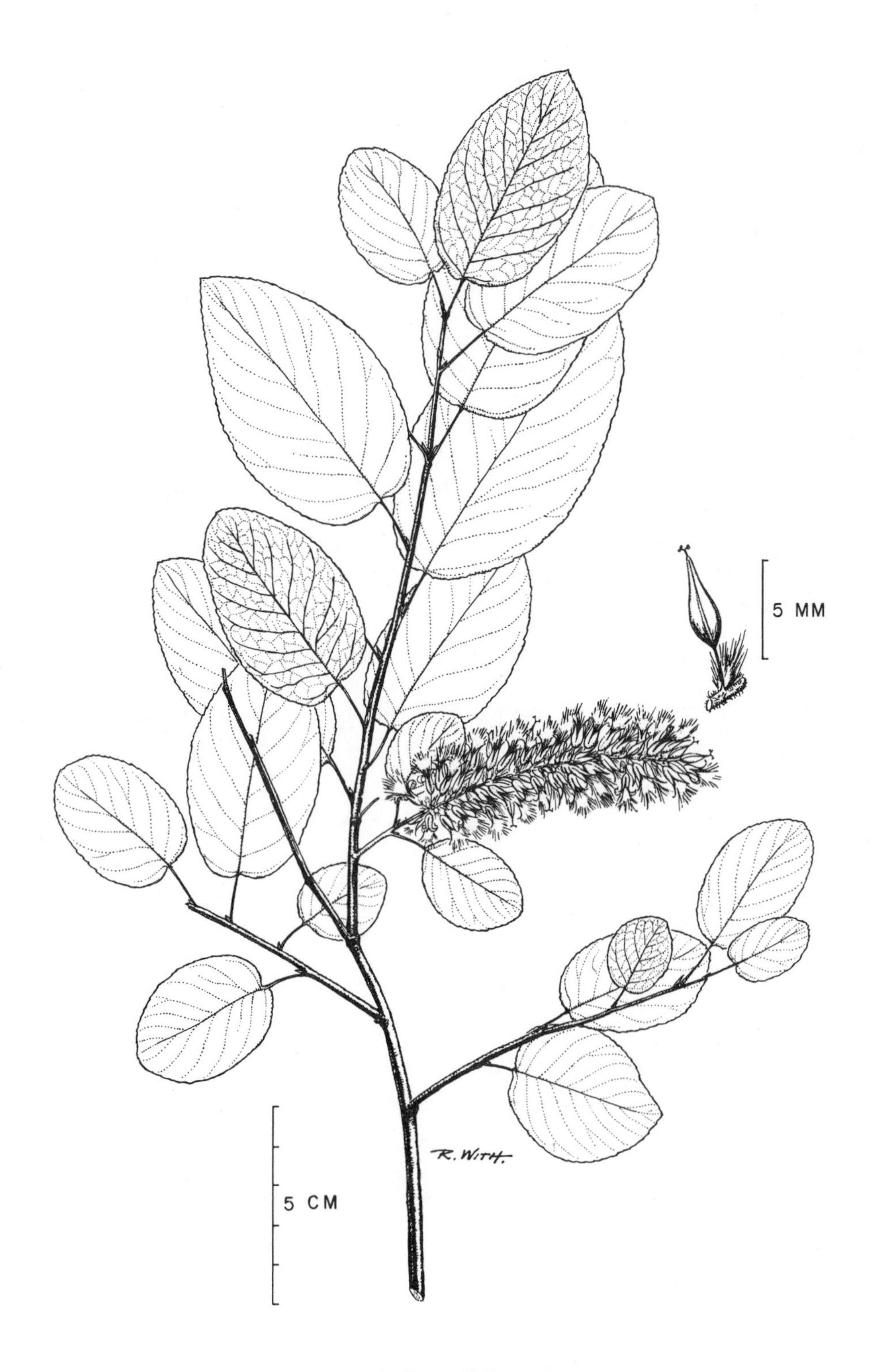

Salix pyrifolia

Salix pyrifolia Anderss. (*S. balsamifera* Barratt) — Balsam willow

A medium-sized shrub growing to a height of about 2 or 3 m. Branchlets glabrous, yellowish or olivaceous at first, becoming reddish-brown and shiny; older stems dark purplish-red to gray-brown.

Leaves alternate, simple, and deciduous, often purplish-red when unfolding; young leaves thin and translucent; blades of mature leaves firm to coriaceous, oblong-lanceolate to ovate, 3-8 cm long and 2-4 cm wide, dark green and glabrous above, paler to strongly glaucous beneath with a rather evident reticulate venation, the apex acute or acuminate, the base cordate to rounded; margins shallowly crenate-serrulate with short, ascending or incurved, gland-tipped teeth; petioles 0.5-2.0 cm long, sometimes glandular at or near the base of the leaf; stipules minute or wanting.

Catkins appearing with the leaves on naked or leafy lateral branches; male catkins 2-5 cm long; stamens 2, filaments glabrous or pubescent at the base; pistillate catkins 2-6 cm long. Capsules glabrous, 5-8 mm long; style 0.5-1 mm long; stigmas short; bracts reddish-brown and with long white hairs. (*pyrifolia*—with pear-shaped leaves; *balsamifera*—balsam-bearing, in reference to the balsamlike fragrance)

In wet places, particularly black spruce-tamarack bogs, alder swamps, and sandy shores of ponds, creeks, and lakes; also in damp sandy ditches.

From the Ottawa Valley westward to Georgian Bay and northward to Lake Superior and James Bay; apparently absent south of 44°N and rare north of 52°N. (Nfld. to Yuk., south to cent. B.C., n. Wis., and n. N.Eng.)

Note The buds and foliage have a balsamlike fragrance which is the origin of the common name. The odour may be detected even on herbarium specimens.

Field check Erect shrub primarily of bog habitats; young leaves thin and translucent, often tinged purplish-red; mature leaves with obvious reticulate venation on the lower surface; catkins developing with the leaves on naked or leafy lateral branches; buds and foliage with balsamlike fragrance.

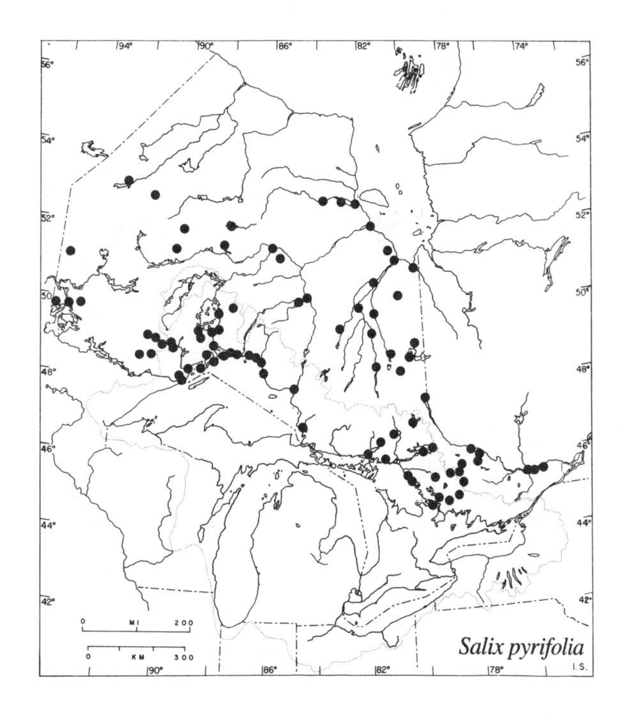

Salix pyrifolia

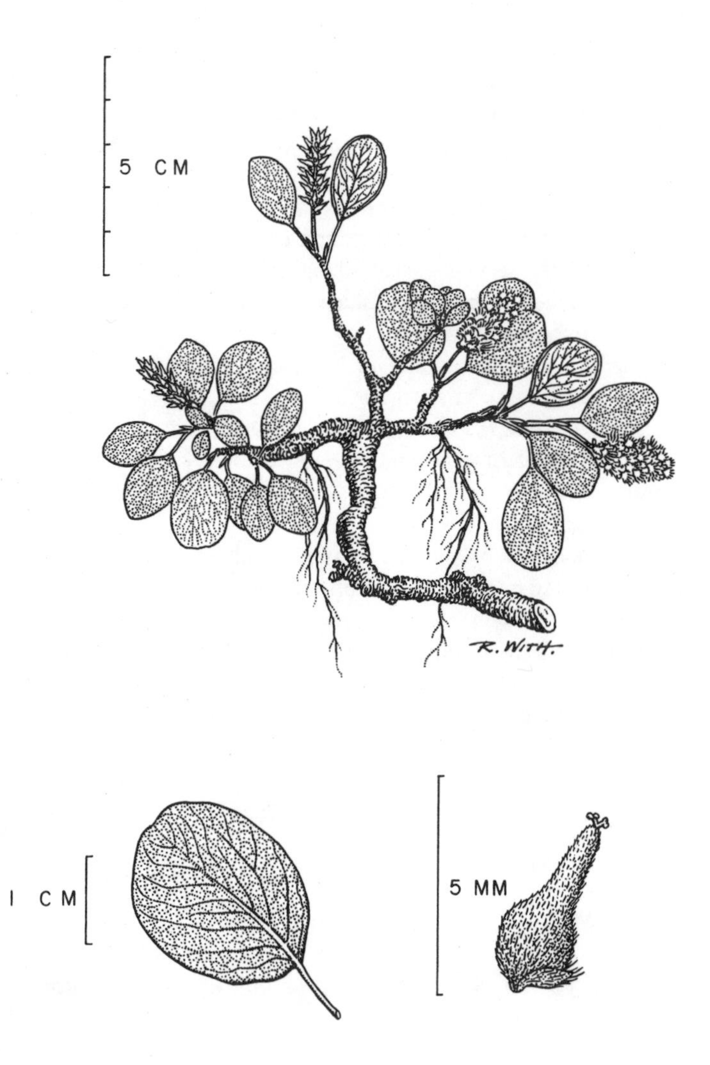

Salix reticulata

Salix reticulata L. **Net-veined willow**

A dwarf prostrate shrub with short thick stems rooting at the nodes; branches usually less than 2 dm long, their ascending tips rarely more than 1 dm above the ground. Branchlets light brown, glabrous, with short internodes.

Leaves alternate, simple, and deciduous; blades oval to oblong or suborbicular, 1 - 4 cm long and 0.5 - 3.5 cm wide, firm or leathery, the apex obtuse, rounded or notched, the base obtuse, rounded or cordate; upper surface dark green and with a distinctly wrinkled appearance due to the impressed veins, lower surface pale green to grayish, glabrous or with a few, scattered, silky hairs, the reticulate venation prominent and raised, the veins somewhat reddish; margins entire to shallowly glandular-crenate, revolute; petioles thin, 3 - 20 mm long, yellow to reddish, glabrous; stipules none or minute glandular lobes less than 1 mm long, soon deciduous.

Catkins serotinous, slender, solitary, on subterminal, slender, leafless but usually villous peduncles; staminate catkins 1 - 4 cm long; stamens 2, filaments pubescent near the base; nectaries 2 or 3, more or less surrounding the stamens; pistillate catkins 1 - 5 cm long; pistils subsessile, brown to reddish, 3 - 5 mm long, thinly to densely pubescent, style very short, the four lobes of the stigmas divaricate; nectaries 2, sometimes divided; July. Capsules splitting in late July and early August. Bracts of the staminate and pistillate flowers oblong to obovate or obcordate, yellowish to reddish-brown, at least the inner surface densely hairy. (*reticulata* — netted or reticulate)

Chiefly in tundra; also on turfy slopes of river banks and beach ridges.

An arctic species characteristic of the coastal tundra of Ontario along the shores of Hudson Bay and the upper part of James Bay, rarely a short distance inland. (Nfld. to Alaska, south to cent. B.C. and the James Bay region)

Note A comparison between this species and the similar northern dwarf species *S. vestita* is given in the Note under *S. vestita*.

Field check Prostrate dwarf shrub of tundra habitats along the coasts of Hudson Bay and James Bay; stems creeping and rooting; leaves leathery, oval to roundish, dark green and with impressed veins above, pale and with conspicuous raised net-venation below; catkins solitary, on leafless peduncles opposite the terminal leaf.

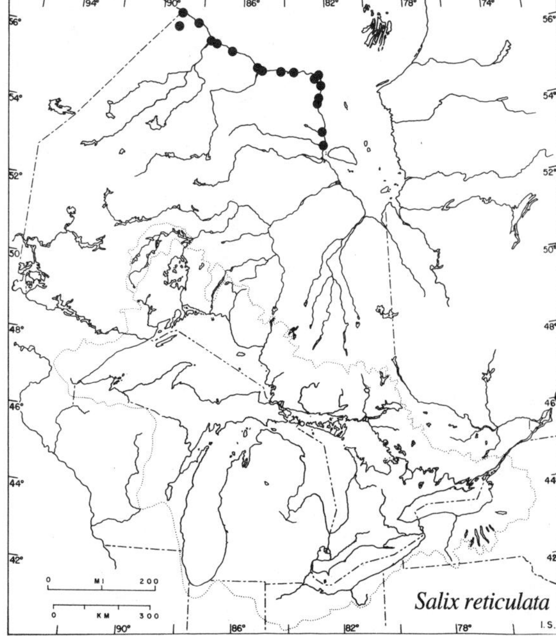

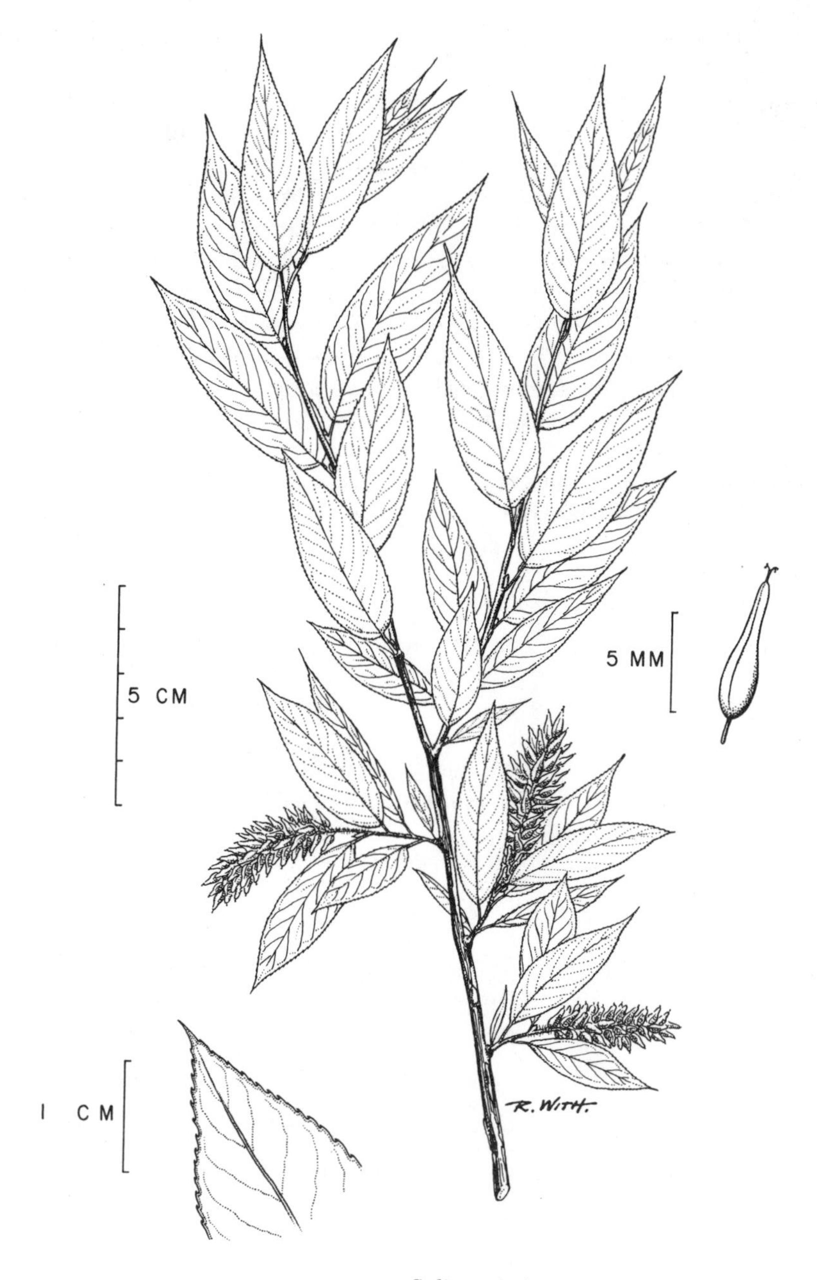

Salix serissima

Salix serissima (Bailey) Fern. — Autumn willow

A medium-sized shrub growing to a height of 2-4 m. Branchlets yellowish to olive-brown or reddish-brown, glabrous, and shiny; bark of older stems gray-brown.

Leaves alternate, simple, and deciduous, glabrous and reddish when unfolding, becoming somewhat firm and leathery; blades of mature leaves lanceolate to elliptic-lanceolate or oblong-lanceolate, 5-10 cm long, 1-3 cm wide, the apex acute to acuminate or long-tapered, the base acute to obtuse or rounded; dark green, glabrous, and shiny above, paler and slightly glaucous beneath with the venation evident as a fine network; margins finely glandular-serrate; petioles 4-10 mm long with several small glands near the junction with the blade; stipules absent or minute, often reduced to a small gland at the node.

Catkins serotinous, on lateral leafy branchlets (late May and early June to mid-July); staminate catkins 1-3.5 cm long, stamens 3-5(-7), filaments pilose on the lower portion; nectaries 2 per flower; pistillate catkins 1.5-4.5 cm long and up to 2 cm wide in fruit (from mid-July to Sept.); pistils reddish, glabrous, style short, stigmas divided, very short. Capsules narrowly conic, light brown, thick-walled, glabrous, 7-10 mm long, deciduous after dehiscence; pedicels 0.5-2 mm long, about twice as long as the nectaries; bracts pale yellow, oblong to obovate, 2-3 mm long, pubescent, deciduous in the pistillate catkins. (*serissima*—very late)

Primarily in bogs and marshes; on sedge mats and around the edges of lakes and ponds; also on seepage areas, marly meadows, limestone pavements, and damp sandy shores.

Throughout Ontario and common in calcareous areas; apparently absent in the Algonquin highlands. (Nfld. to Alta., south to cent. Minn. and N.J.; also in S.Dak. and Colo.)

Note Of all our willows, this is the last in the season to flower, and its capsules, which mature in late summer, often persist well into the fall, hence the common name autumn willow.

Field check Late-blooming shrub of marshes and bogs; branchlets shiny; leaves dark green and glossy above, whitened beneath and with fine reticulate pattern of veins; margins finely glandular-serrulate.

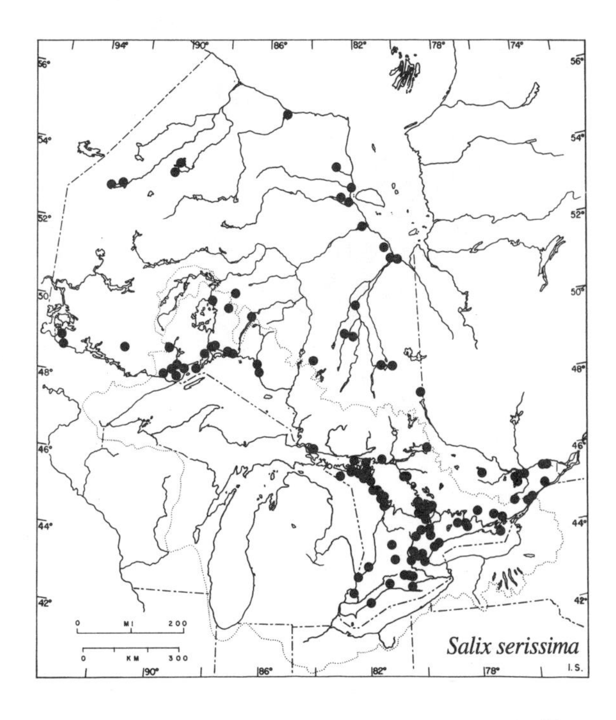

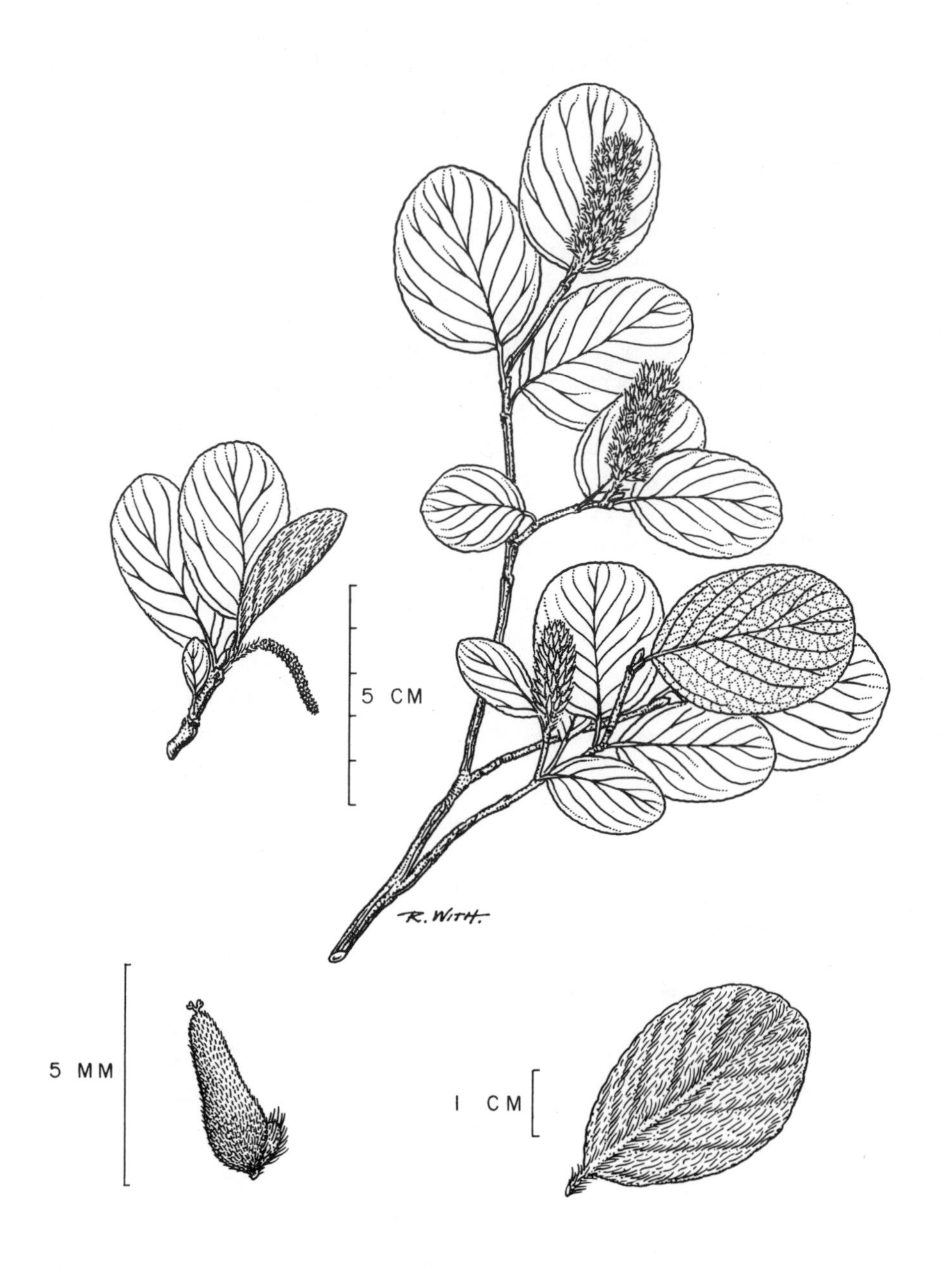

Salix vestita

Salix vestita Pursh **Hairy willow**

A depressed to ascending dwarf shrub 1-6 dm in height. Branchlets stout, somewhat keeled or angled, not normally rooting, light to dark brown or grayish, glabrous, the short internodes often thickened upwards.

Leaves alternate, simple, and deciduous; blades oval to broadly oblong, obovate or nearly orbicular, 1-5 cm long and 0.5-4 cm wide, obtuse, rounded or retuse at the apex, broadly cuneate to obtuse or rounded at base, thick, leathery, dark green and rugose above with impressed veins, glabrous or soon becoming so, the lower surface whitened, the conspicuous venation mostly obscured by an appressed covering of long, shining, white hairs, these often disappearing except along the veins; margins closely and shallowly crenate or scalloped; petioles pubescent, 2-10 mm long, often shorter than the well-developed axillary buds; stipules none.

Catkins serotinous, 2-4.5 cm long, solitary, opposite the terminal leaf at the end of a leafy branch; floral bracts obovate, about 1 mm long, silky-hairy; staminate catkins slender, stamens 2, filaments glabrous; pistillate catkins at first slender but up to 1 cm wide in fruit. Capsules subsessile, brownish, pubescent, ovoid-conic, 4-6 mm long, style very short or obsolete, stigmas forked at the tip.(*vestita*—clothed, in reference to the coating of silky white hairs on the leaves)

On dry or well-drained ridges, shoulders, and crests of river banks, cliffs, and occasionally in open muskegs of the interior.

In the Hudson Bay lowlands from about 51°N latitude to the shores of Hudson Bay. (Nfld. to n. Man.; also in the Rocky Mts. of Alta. and B.C., south to n.e. Oreg. and Mont.)

Note The two arctic willows *S. reticulata* and *S. vestita* are similar in many respects. Both are dwarf shrubs with dark green roundish leaves, impressed veins on the upper surface, and serotinous, pseudoterminal catkins. However, *S. reticulata* regularly roots at the nodes while *S. vestita* seldom does. In addition, the leaves of *S. reticulata* are strongly revolute and glabrous or with a few silky hairs beneath; those of *S. vestita* are glabrous on the upper surface only, the lower surface being coated with shining white hairs at least along the veins and on the petiole.

Field check Dwarf shrub with short, erect or ascending stout branches and internodes somewhat thickened upwards; leaves ovate to elliptic and short-stalked, thick, leathery, dark green and rugose above and conspicuously white, silky-pubescent beneath; pseudoterminal pistillate caktins with pubescent capsules and silky-hairy bracts.

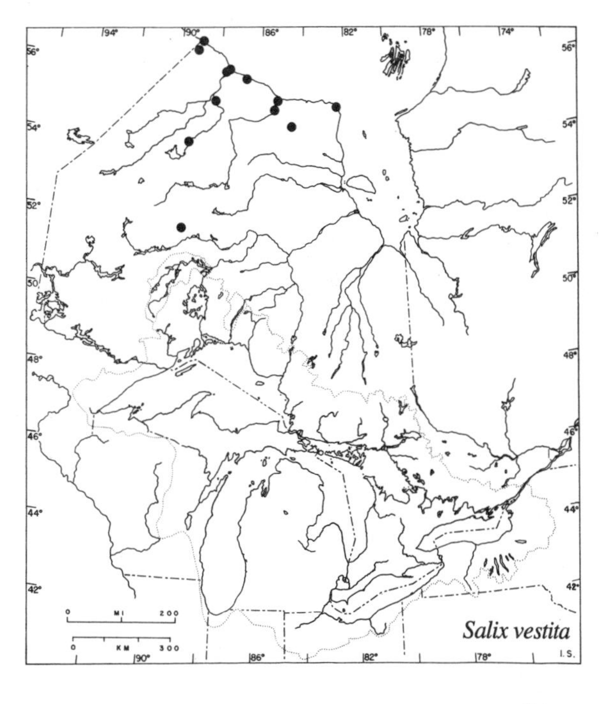

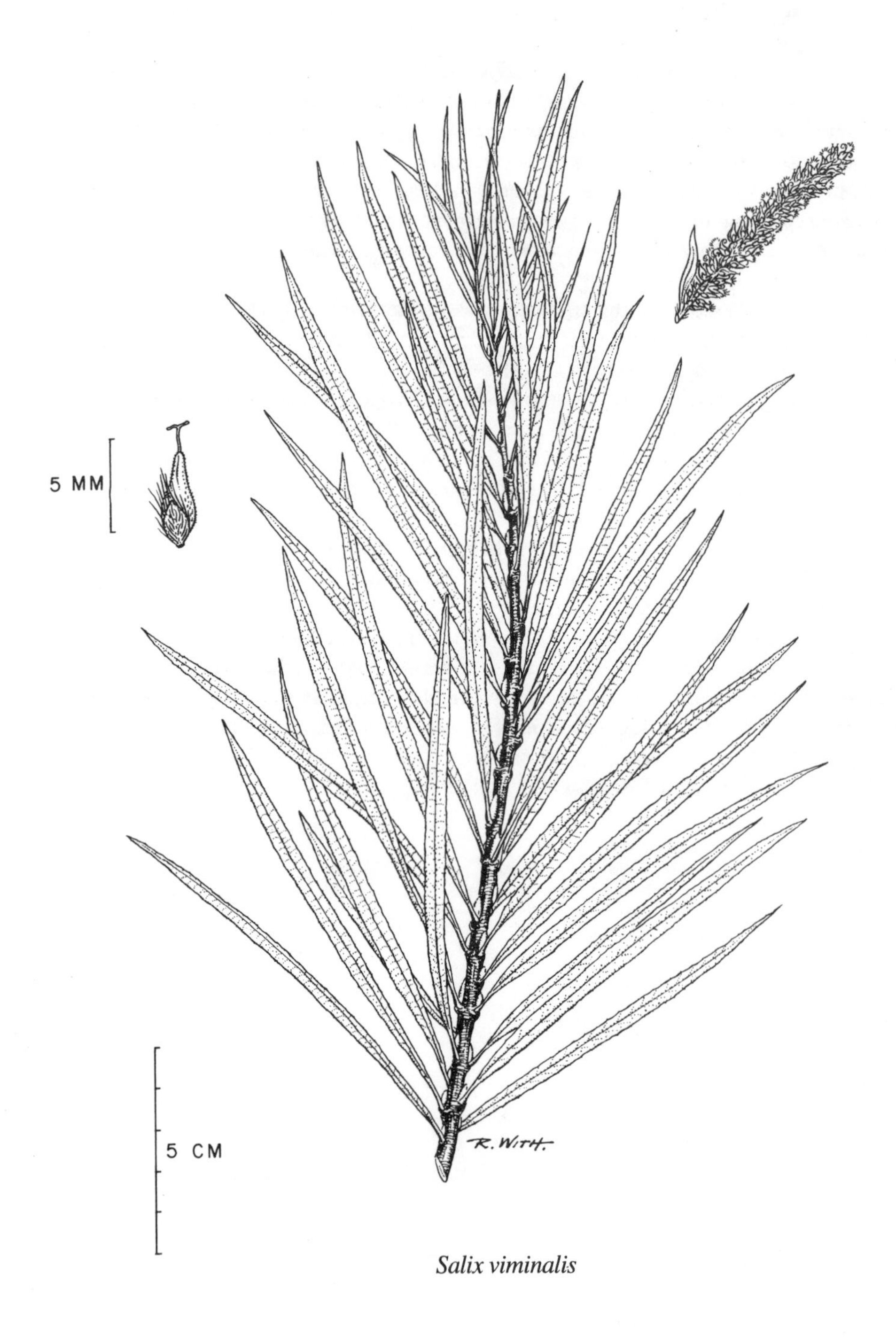

Salix viminalis

Salix viminalis L. — Basket willow

A large, few-stemmed shrub or small tree up to 8 m or more in height. Branchlets slender, greenish-yellow to reddish-brown, minutely silky at first but soon becoming smooth and shiny; branches long and flexible.

Leaves alternate, simple, and deciduous; blades linear to linear-lanceolate, elongate, the apex acuminate with a slender tip, the base acute or tapered, 5-20 cm long, 5-15 mm wide, dull green and puberulent above, densely silky-hairy below, with prominent yellowish midrib; margins entire or shallowly undulate, revolute; petioles slender, 2-10 mm long, dilated at the base; stipules narrowly lanceolate, glandular-dentate, deciduous.

Catkins developing before the leaves have expanded (April and May), sessile or on short lateral branches with a few reduced leaves at the base; staminate catkins slenderly cylindric, 2-3 cm long, stamens 2, filaments glabrous; pistillate catkins 4-6 cm long, dense, and rather crowded in a series along the stem; pistils covered with short silky hairs, the styles conspicuously branched and clearly exserted beyond the fringe formed by the long silky hairs of the broadly oval to elliptic, blackish bracts. Capsules 4-7 mm long, densely sericeous, subsessile or on stout pubescent pedicels less than 1 mm long. (*viminalis*—bearing withes, in reference to the slender flexible branches that can be woven together)

Edges of streams and along river banks, an escape from cultivation.

Reported from the vicinity of several cities in southern Ontario (Ottawa, Belleville, Toronto, Hamilton); also on Manitoulin Island, Lake Huron. (Introduced from Europe)

Note The basket willow has been introduced in various parts of eastern Canada as an ornamental and for the production of the long slender branches used in basketry.

Field check Large few-stemmed shrub or small tree with long, flexible, yellowish branches; leaves elongate, linear-lanceolate, dull green above, and silky-satiny beneath; pistillate catkins dense and rather crowded; bracts blackish and long-hairy, contrasting with the silvery, short-hairy capsules.

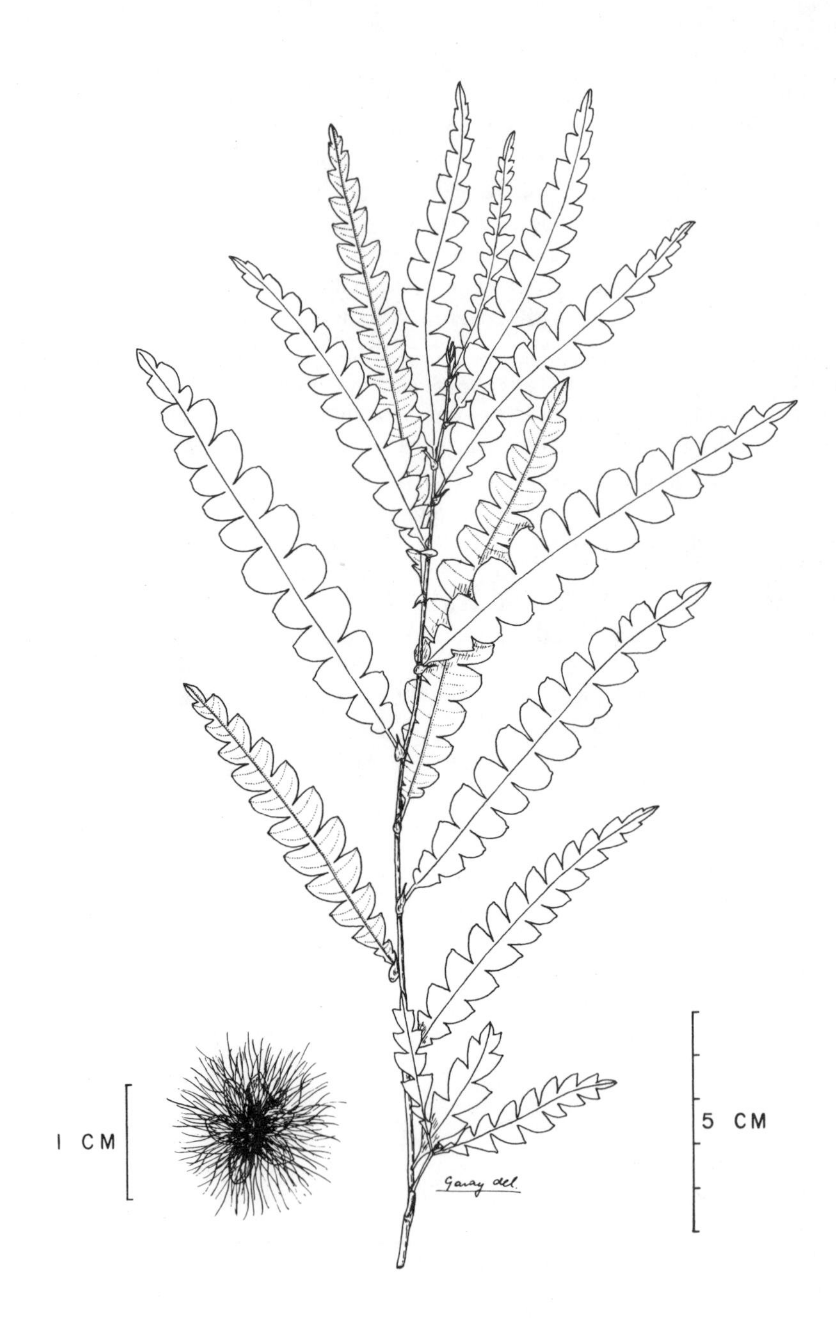

Comptonia peregrina

MYRICACEAE — BAYBERRY FAMILY

A small family of aromatic trees and shrubs comprising 3 genera and about 50 species of wide distribution. Leaves simple, alternate or subopposite, entire or dentate to pinnatifid, often resin-dotted, and fragrant when bruised or crushed. Flowers small, in short cylindric to globose axillary clusters. Fruit a small drupe, often waxy-coated. These plants are frequently found in swampy or in sandy (acidic) soils and in some species the roots develop nodules with nitrogen-fixing bacteria (see Morris et al., 1974; Rodriguez-Barrueco, 1969; Schramm, 1966). The name of the family is based on the genus *Myrica*.

Comptonia L'Hér. — Sweet-fern

A single species, native to eastern North America. A low, freely branching, aromatic shrub with soft pinnatifid leaves and prominent semi-cordate stipules. Fruit burlike. Plants of dry sandy, gravelly, or rocky habitats. (*Comptonia* — named after Henry Compton, 1632–1713, Bishop of London, an amateur botanist, gardener, and patron of botany)

Comptonia peregrina (L.) Coult. — Sweet-fern
(*Myrica asplenifolia* of various manuals, not L.)

A low, much-branched, erect shrub of dry habitats, spreading freely by long underground stems and forming dense thickets usually less than 1 m high. Branchlets hairy and gland-dotted, fragrant when bruised; bark varying from reddish-brown to gray or nearly black.

Leaves alternate, simple, and deciduous; blades 4–10 cm long and 0.5–2.5 cm wide, dark green above and paler beneath, long and narrow, tapered at both ends, and lobed in a pinnatifid or fernlike manner, the deep indentations between the rounded or pointed segments reaching almost to the midrib, gland-dotted, especially above, and very fragrant when crushed; petioles up to 6 mm long, with a pair of semi-cordate long-pointed stipules at the base.

Flowers unisexual, either on the same or on separate plants; male flowers in slender flexuous catkins 2.5 cm or more in length, in clusters near the ends of the branches; female flowers in small, dense, bristly clusters usually at the ends of short side branches; late April and May. Fruit a bristly burlike cluster about 2.5 cm in diameter, consisting of a number of smooth, bony, conical or barrel-shaped nutlets surrounded by numerous slender, glandular, and ciliate scales; June and July. (*peregrina* — foreign, i.e., not known to Linnaeus who was the first to describe the plant; *asplenifolia* — with leaves like *Asplenium*, a genus of ferns)

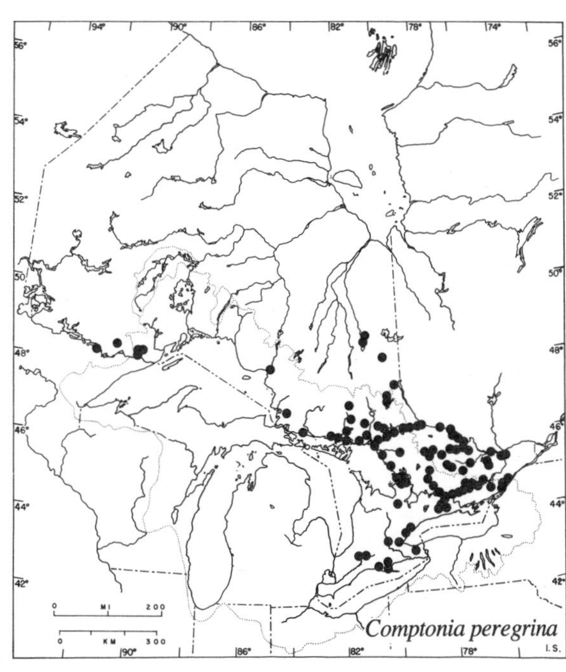

In dry sandy, gravelly, or rocky soil, often in association with jack pine in the north.

Common on the Canadian Shield and north to the latitude of Lake Abitibi (about

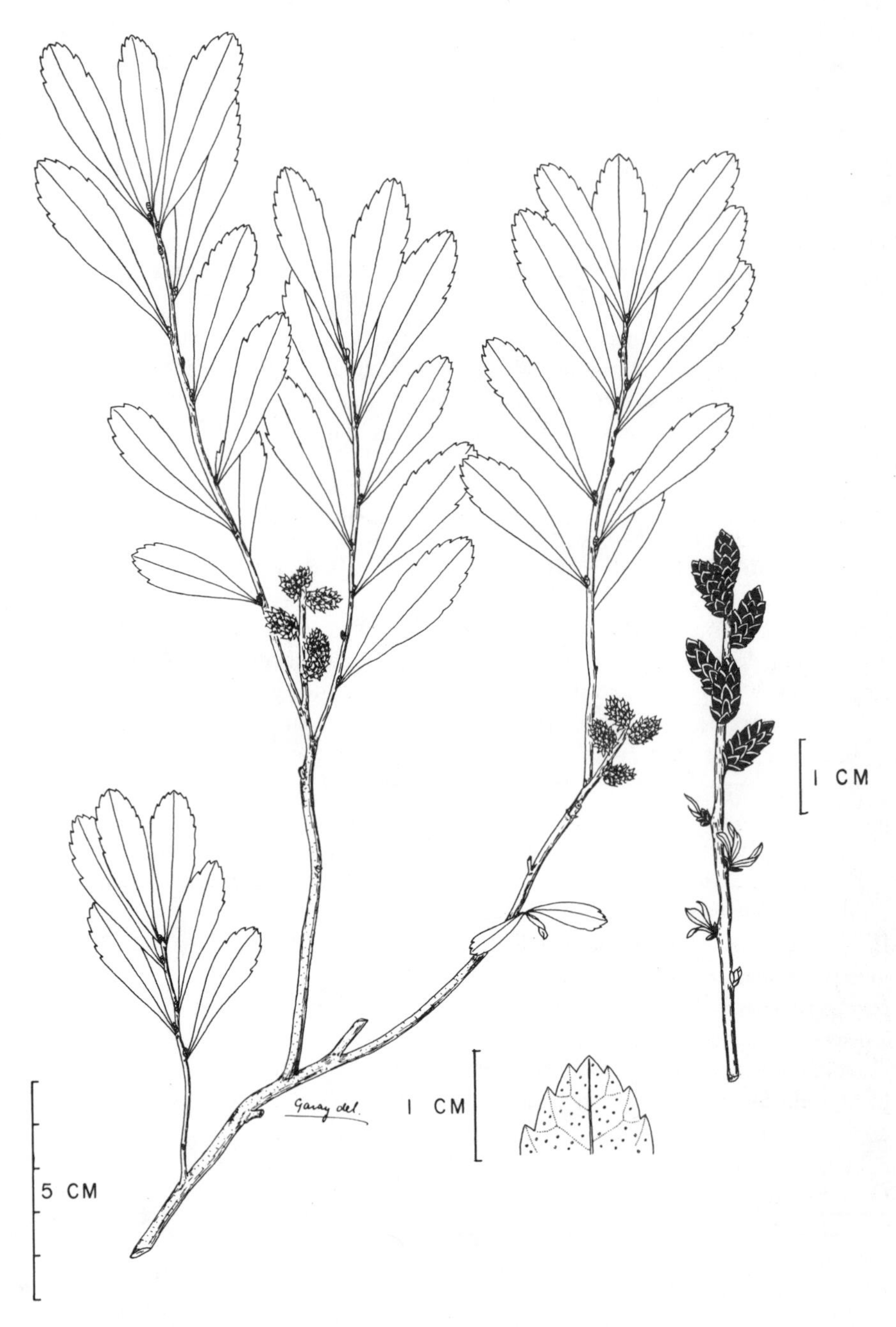

Myrica gale

49° N); also in the southern part of Thunder Bay district adjacent to the Minnesota border; rare south of the Canadian Shield and apparently absent on limestone or dolomite outcrops. (Cape Breton to w. Ont., south to Minn., Ill., Tenn., and Ga.)

Note Sweet-fern is the alternate host for a blister rust that occurs on jack pine. The root nodules of sweet-fern can fix atmospheric nitrogen which may account for its ability to colonize such impoverished habitats as coal waste dumps in Pennsylvania (see Schramm, 1966). Retention of viability for a minimum of 70 years has been reported (Del Tredici, 1977) for seeds of the sweet-fern found in forest soils.

Field check Low shrub of dry sandy or rocky ground; gland-dotted and fernlike leaves and young stems fragrant when crushed.

Myrica L.

About 35 species of shrubs and trees with wide distribution in tropical and temperate regions. Leaves entire or serrate to dentate or slightly incised, frequently gland-dotted, and aromatic; stipules absent. Drupes resin-dotted or coated with minute waxy particles, those of *M. cerifera* L. being the source of the wax used in making bayberry candles. The wax is removed by placing the fruits in boiling water. (*Myrica*—Greek *myrike*, the tamarisk or some other fragrant plant; *cerifera*—bearing wax)

Key to *Myrica*

a. Nutlets brownish, ovoid, in scaly clusters dotted with golden resinous glands; found north of a line joining the west end of Lake Ontario and the south end of Lake Huron **_M. gale_**

a. Nutlets bluish-white, globular, and encrusted with pale waxy particles; found only in the Regional Municipality of Haldimand-Norfolk, midway along the north shore of Lake Erie **_M. pensylvanica_**

Myrica gale L. — Sweet gale

A much-branched upright shrub growing to a height of 6–15 dm. Branchlets hairy, gland-dotted, and fragrant when bruised; bark dark gray to reddish-brown with small pale lenticels.

Leaves alternate, simple, and deciduous; blades rather firm in texture, dark green above and paler beneath, dotted on both surfaces with shiny yellow glands, and fragrant when crushed, up to 6 cm long and 2 cm wide, narrowly elliptic to oblanceolate, usually broadest above the middle, the rounded tip conspicuously toothed, the long basal portion wedge-shaped and entire; petioles 1–3 mm long.

Flowers small and unisexual, usually in scaly catkins or conelike clusters at the tips of the branches, the two kinds on separate plants; opening in early spring before or with the unfolding leaves; male catkins with dark brown, shiny, triangular, pale-margined scales which become loose and open when the pollen is shed; female flowers inconspicuous when young, maturing into ovoid to globular brownish catkins of gland-dotted nutlets. (*gale*—from *Gale*, the former generic name still recognized by some botanists who separate *Gale* and *Myrica* by the deciduous catkins and winged seeds found in *M. gale*)

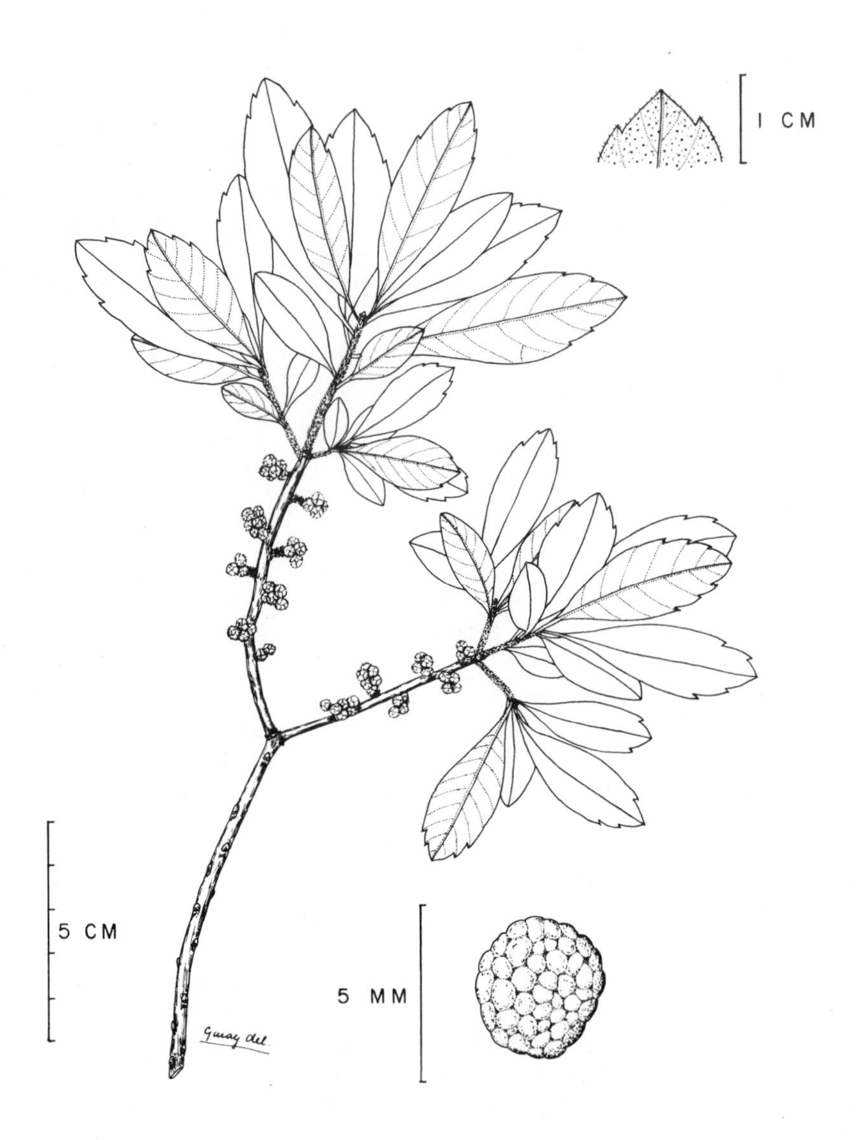

Myrica pensylvanica

In damp soil and shallow water along the shores of lakes and streams and in swamps and bogs.

Common throughout central Ontario and northward to Hudson Bay and James Bay; Manitoulin Island and the Bruce Peninsula; rare south of 44° N. (Nfld. to Alaska, south to Oreg., Wis., Tenn., and N.C.; Eurasia)

Note The fruits have two corky or spongy bractlets which cause them to float on water, thus assisting in their dispersal to other suitable wet habitats.

The roots have nodules containing nitrogen-fixing bacteria living in a symbiotic relationship with the shrub.

In the early days, leaves of sweet gale were used to scent linen and to flavour beer before the introduction of hops.

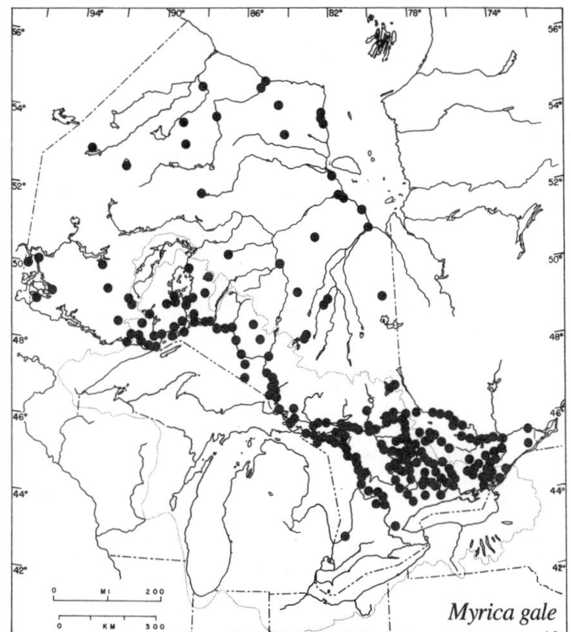

Field check Gland-dotted leaves and young branchlets are fragrant when crushed; leaves wedge-shaped at base, entire except for toothing on the rounded tips; in wet habitats north of a line joining the west end of Lake Ontario and the south end of Lake Huron.

Myrica pensylvanica Loisel. — Bayberry

Erect and rather stiffly-branched shrub reaching a height of 1.5 – 2 m or occasionally higher. Branchlets brown and hairy, gland-dotted, and fragrant when bruised; bark dark gray to reddish-brown.

Leaves simple and arranged alternately along the stem but crowded towards the ends of the branches, and although deciduous, often remaining attached to the stems late into the fall; blades up to 8 cm long and 2.5 cm wide, thin, bright green and sparsely gland-dotted above, paler and copiously gland-dotted beneath, fragrant when crushed, elliptic or oblong-oval in shape, rounded or obtuse at the tip and somewhat wedge-shaped or tapered at the base; margins often entire in the smaller leaves but the larger ones coarsely toothed from the tip about halfway down each side; petioles up to 4 mm long.

Flowers unisexual, in small catkins on the old wood below the current leaf-bearing tips of the branches; female catkins maturing into a cluster of globular bony nutlets encrusted with grayish-white to pale bluish, waxy particles. (The original spelling of the specific name, although not current usage for the name of the state, should be retained, according to the International Rules of Botanical Nomenclature. *pensylvanica*—Pennsylvanian)

In low, often swampy woods and occasionally in dry sandy soil.

The only known stations in Ontario are at and near Turkey Point in Charlotteville Township, Regional Municipality of Haldimand-Norfolk. (Atlantic Coastal Plain from s. Nfld. to N.C. and locally inland to the Lake Erie region)

Field check Gland-dotted leaves and young twigs, fragrant when crushed; leaves toothed only near the apex; fruit a cluster of small rounded nutlets encrusted with bluish-white wax; restricted to the Turkey Point area, Regional Municipality of Haldimand-Norfolk.

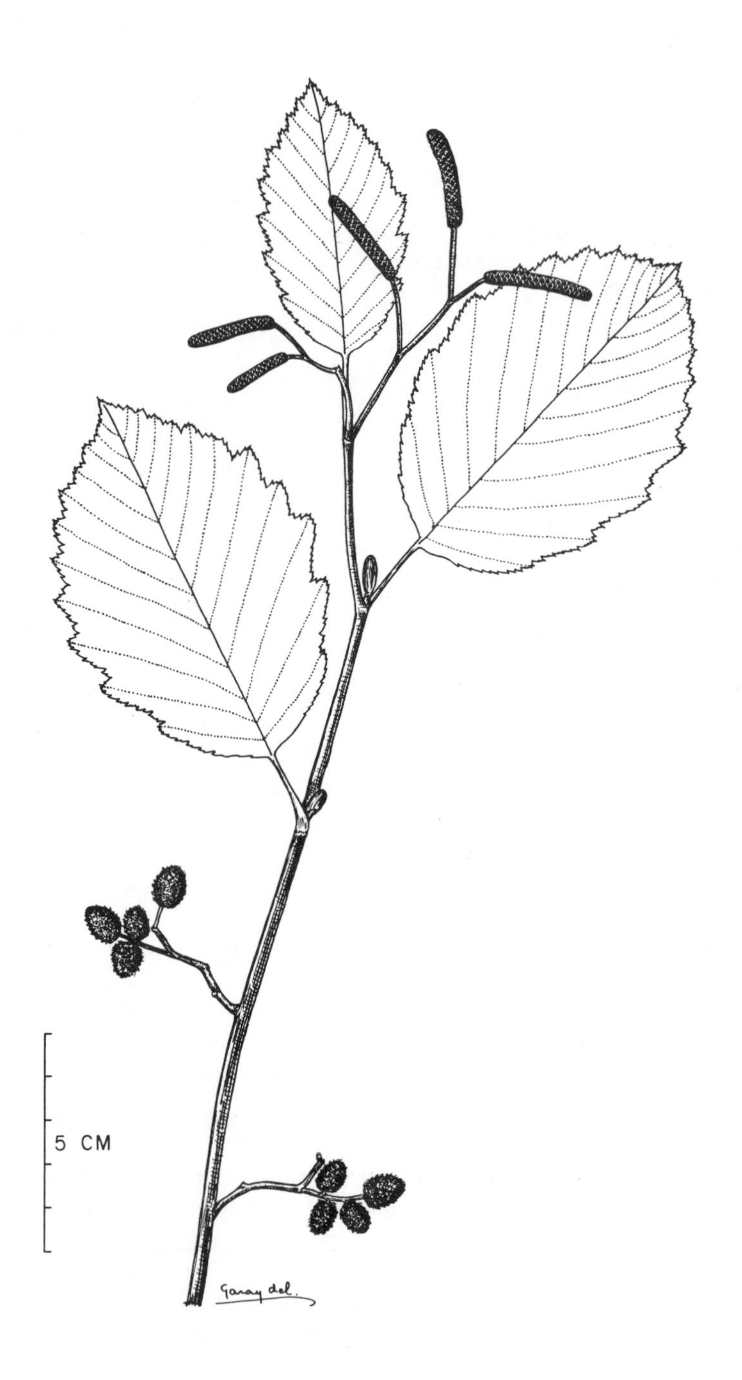

Alnus incana ssp. *rugosa*

BETULACEAE — BIRCH FAMILY

Trees and shrubs with alternate, simple, serrate, straight-veined leaves and deciduous stipules. Flowers small, wind-pollinated; the staminate borne in slender cylindrical catkins, the pistillate clustered, in spikes or in scaly catkins; perianth none or represented by tiny bracts and bractlets considered to be the remains of a reduced inflorescence. Fruit a 1-seeded nutlet with or without a foliaceous involucre. Six genera and about 150 species, primarily of north temperate regions. The family name is derived from the genus *Betula*—the birch.

Alnus B. Ehrh. — Alder

Shrubs or trees with alternate, simple, and deciduous leaves, their blades broadly oval to obovate and conspicuously veiny; pith 3-angled; fruit a winged nutlet borne on a 5-lobed scale in short, somewhat woody catkins resembling small cones. The roots often produce nodules that contain a micro-organism capable of fixing free nitrogen from the atmosphere. This capacity has been demonstrated for both species of *Alnus* which are native to Ontario and may have been an important factor in post-glacial invasion by these plants (see Dalton & Naylor, 1975; Daly, 1966). Our nomenclature follows that used by J. J. Furlow (1979).

About 35 species in the northern hemisphere and western South America. (*Alnus*—the Latin name for the European alder)

Key to *Alnus*

a. Branchlets and young leaves sticky; winter bud sessile, with sharp-pointed curved tip; leaves finely sharp-toothed; catkins long-stalked — ***A. viridis* ssp. *crispa***

a. Branchlets and young leaves not sticky; winter bud stalked, blunt at the tip; leaves unevenly doubly serrate; catkins sessile or short-stalked — ***A. incana* ssp. *rugosa***

Alnus incana (L.) Moench ssp. *rugosa* (DuRoi) Clausen (*A. rugosa* (DuRoi) Spreng.) — Speckled alder

A coarse shrub up to 4 m tall. Branchlets light reddish-brown and hairy, older stems glabrous, dark brown to purple-black, with scattered pale lenticels, pith three-sided in section; winter buds stalked, slenderly oval, and blunt at the tip.

Leaves alternate, simple, and deciduous; blades ovate to broadly oval, 6–10 cm long and 3–6 cm wide, rounded or slightly heart-shaped at the base, pointed at the tip, dark green and glabrous above, with the lower surface green or tawny to whitened and more or less pubescent along the veins, sometimes densely hairy throughout; margins undulate or wavy and doubly serrate; petioles 1–2 cm long.

Flowers small and in catkins; male flowers appearing in late summer as a group of stalked catkins 1–2.5 cm long, elongating the following spring as drooping tails 5–8 cm long; female catkins much smaller, appearing also in late summer at the tips of some of the branches in unexpanded clusters, becoming functional the following spring; April and May. After fertilization, the female catkins mature into sessile or short-stalked, blunt, nearly globose "cones" about 1 cm long with persistent woody scales which bear tiny wingless nutlets that are shed in the fall, the empty "cones" remaining on the branches for a year of more. (*incana*—grayish, hoary; *rugosa*—wrinkled or creased, in reference to the wrinkled appearance of the leaves)

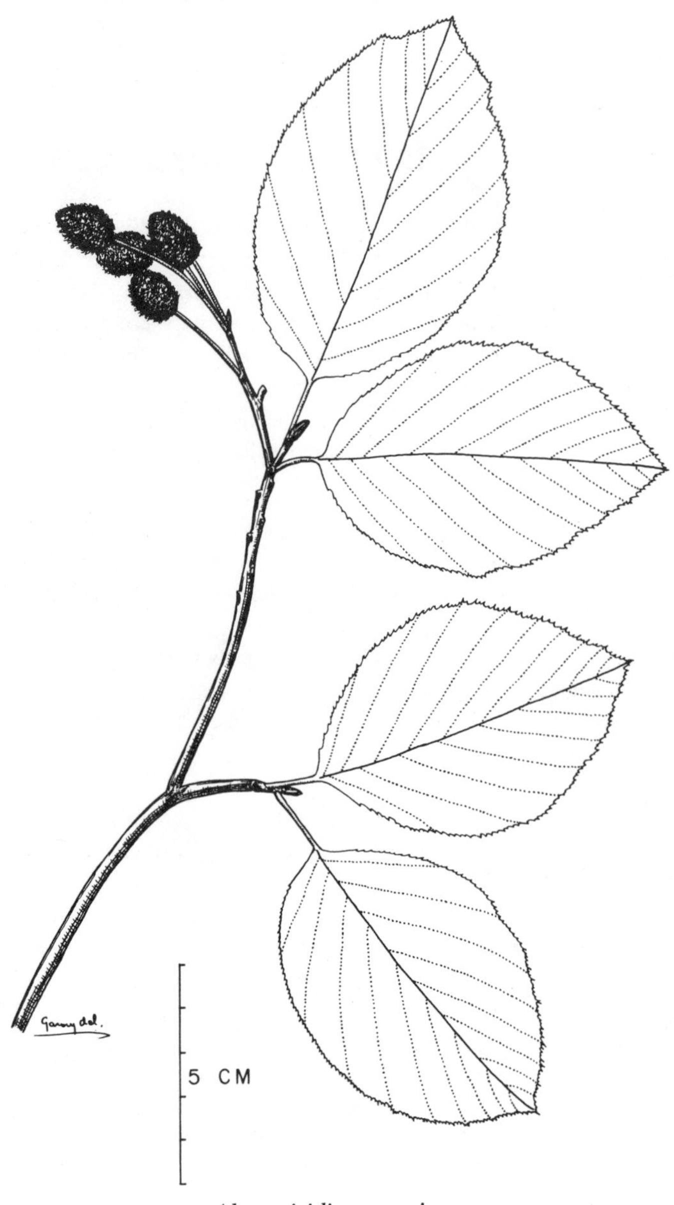

Alnus viridis ssp. *crispa*

Along the edges of streams, rivers, swamps, and lakes, or in wet depressions or open woods.

Throughout Ontario except in the counties near the west end of Lake Erie. (Nfld. and Lab. to Sask., south to Iowa, W. Va., and N.Y.; Eurasia)

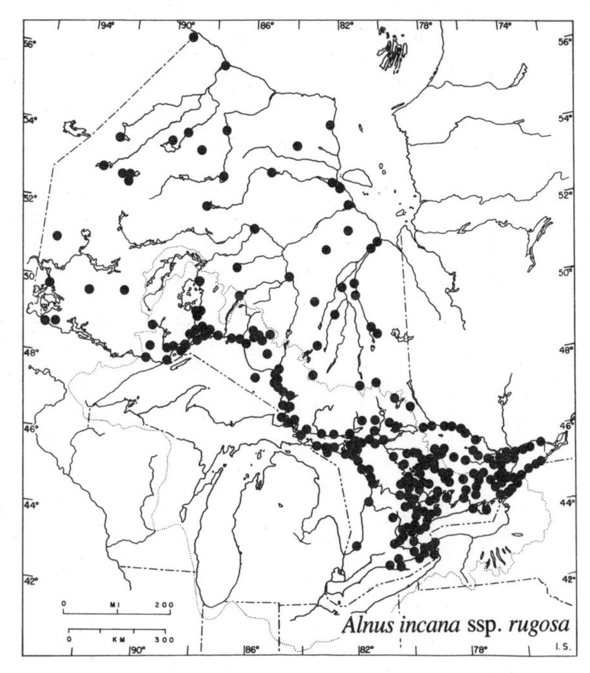

Note Specimens in which the leaves are glaucous or whitened beneath are treated in some manuals as *A. rugosa* var. *americana* (Regel) Fern. (*A. incana β americana* Regel), in contrast to the typical variety with the leaves green or tawny but not whitened on the lower surface. The eastern North American taxon is considered to be a subspecies of the circumpolar species *A. incana*. (*americana* — American)

The common name "speckled alder" refers to the conspicuous lenticels of the stems and branches.

The common European or black alder (*A. glutinosa* (L.) Gaertn.) is sometimes planted in Ontario, as at Hanlan's Point, Toronto Island. When mature, it is a broadly pyramidal tree up to 20 metres or more in height, usually with one or several trunks. It differs from our two native shrubby species not only in its growth habit but also in its obovate to suborbicular leaves (4 – 7.5 cm long, 3 – 7 cm wide), bluntly rounded to retuse at the apex, and its flat, doubly serrate to denticulate leaf margins. (*glutinosa* — sticky)

A. glutinosa has been collected in the Lake Erie region at Port Dover and near St. Williams (in the Regional Municipality of Haldimand-Norfolk), at Port Burwell (Elgin County), and north of Canning (Oxford County). (Native of western and central Europe; planted and occasionally escaped and naturalized in eastern North America: Mass. to Ill., south to N.J.)

Field check Leaves broadly ovate with heart-shaped base, coarsely and doubly serrate; pith 3-sided in section; pistillate catkins woody and conelike, sessile or short-stalked in clusters; cones long-persisting after the seeds are shed.

Alnus viridis (Villars) Lam. & DC. ssp. *crispa* (Ait.) Turrill (*A. crispa* (Ait.) Pursh)

Green alder
Mountain alder

A tall coarse shrub reaching a height of 3 m or more. Branchlets somewhat hairy, brownish, and sticky, with a few, scattered, pale lenticels; older stems glabrous, reddish-brown to gray; pith 3-sided in section; winter buds sessile with sharp-pointed curved tips.

Leaves alternate, simple, and deciduous; blades ovate, round-oval, or slightly heart-shaped, 4 – 9 cm long and 2.5 – 5 cm wide, rounded or heart-shaped at the base, blunt or pointed at the apex, thin, bright green above, slightly paler beneath, and sticky when young; margins somewhat crispy-wavy with many fine sharp-pointed teeth; petioles 6 – 12 mm long.

Flowers small and borne in catkins; male catkins in long-stalked clusters which develop in late summer, each catkin slender, scaly, conelike, and about 1 cm long, elongating the following spring to a length of 5 – 8 cm and shedding its pollen as the leaves expand; May and June; the female catkins develop within the scaly winter buds and appear in spring in small clusters; after fertilization, catkins becoming long-stalked, blunt, and conelike, 1-2 cm long, the woody persistent scales bearing tiny winged nutlets. After the seeds are shed (late summer to fall), the empty woody catkins remain on the branches for a year or more. (*viridis*—green; *crispa*—curly, in reference to the margins of the leaves)

In wet depressions, on damp sandy or gravelly shores; also on rocky crests, slopes, and cliffs.

Throughout northern Ontario but rare south of 46° N and notably absent in the area south of the Canadian Shield. (Greenl. and Lab. to Alaska, south to n. Calif., Minn., and N.Eng., also disjunct in Wyo., Pa., and N.C.; Eurasia)

Note This taxon was formerly described as a North American species, *A. crispa*, but is now regarded as part of a circumboreal complex, *A. viridis*.

Field check Branchlets and young leaves sticky; pith 3-sided in section; leaves round-oval, finely sharp-toothed, and wavy-margined; woody conelike catkins in long-stalked clusters, persistent.

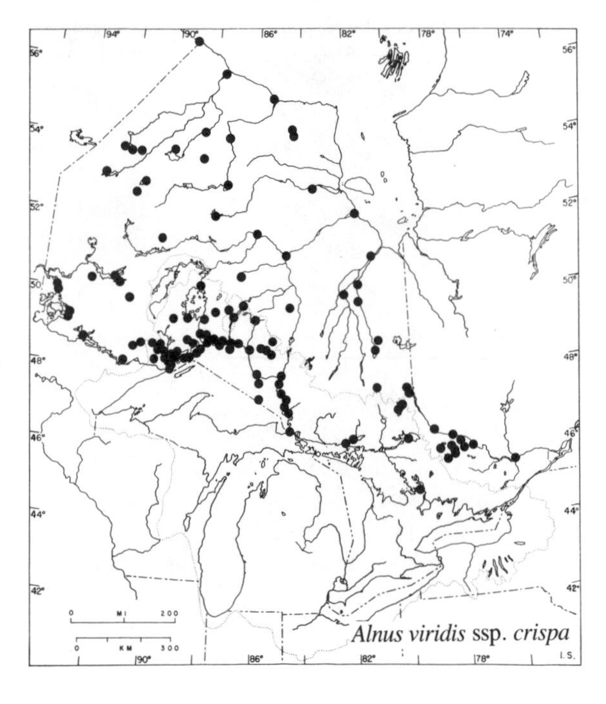

Betula L. – Birch

Trees and shrubs with alternate, simple, deciduous, and serrate leaves; bark with prominent lenticels, that of older stems and trunks often separable in sheets. Staminate catkins terminal and lateral, sessile, developing in the fall and persisting through the winter to open the following spring. Pistillate catkins ovoid to cylindrical, usually borne at the ends of the 2-leaved short shoots. Fruit a nutlet with 2 lateral membranaceous wings. The bark of the eastern American white birch (*B. papyrifera* Marsh.), also called paper birch and canoe birch, was used by the Indians for the building of their canoes. About 60 species, several occurring in the arctic regions of both hemispheres. (*Betula*—the Latin name for the birch; *papyrifera*—paper-bearing)

Key to *Betula*

a. Low shrubs, rarely exceeding 2 m in height; bark dull; branchlets with whitish or yellowish glands; leaves short-stalked; margins of leaves crenate to crenate-serrate; body of nutlet wider than the wing
  - b. Branchlets densely white-glandular; leaves green on both sides, not conspicuously reticulate-veiny beneath; scales of the catkins resinous; chiefly northern (James Bay and Hudson Bay) ***B. glandulosa***
  - b. Branchlets sparsely yellow-glandular; leaves paler or whitened beneath and with obvious reticulate venation; scales of the catkins not resinous; widely distributed from Lake Ontario to Hudson Bay ***B. pumila* var. *glandulifera***

a. Tall shrubs (3–5 m) with stout stems, often thicket-forming, branchlets with reddish glands; leaves with conspicuous petioles (1–2 cm long); margins of leaves sharply serrate to doubly serrate; body of nutlet narrower than the wing ***B. occidentalis***

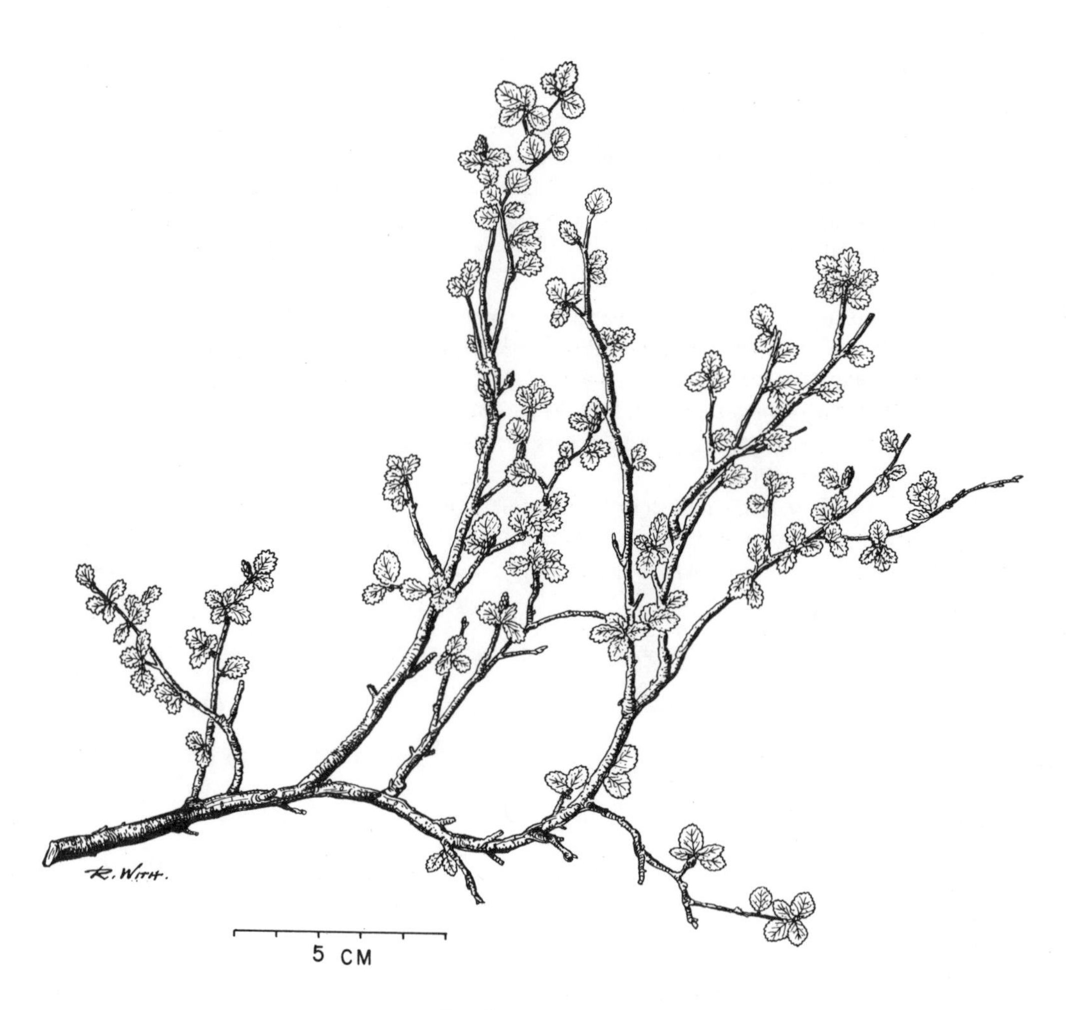

Betula glandulosa

Betula glandulosa Michx. — Dwarf birch

A low, depressed or mat-forming shrub with few gnarled or twisted stems, usually less than 1 m high. Branchlets glabrous or very finely hairy but conspicuously dotted with resinous warty glands and covered with a grayish waxy layer.

Leaves alternate, simple, and deciduous; blades obovate to suborbicular, 0.5 - 3 cm long and nearly as broad, cuneate to rounded at the base, rounded or blunt at the apex and with 3 or 4 pairs of veins, not strongly reticulate beneath; firm and leathery in texture and somewhat glutinous, green on both sides, glabrous or minutely glandular-puberulent; margins crenate-dentate with low simple teeth; petioles short, stout, pubescent.

Flowers small and unisexual, in scaly catkins; mature female catkins up to 2.5 cm long, the scales with a resinous hump on the back; nutlets with a narrow wing on each side which is usually broader towards the apex. (*glandulosa* — glandular)

On dry sandy hillsides, rocky ridges, cliffs, and stream banks; also in bogs and muskegs.

Common along and near the shores of James Bay and Hudson Bay and at a few sites inland; islands in James Bay (N.W.T.); not reported south of 53° N. (Greenl. and Nfld. to Alaska, south to n. Calif. and Maine)

Note *B. nana* L. ssp. *exilis* (Sukatch.) Hultén, another closely related and widespread arctic species, is similar to *B. glandulosa* but differs with respect to the reticulate venation which is well marked on the lower surface of the leaves and to the scales of the catkins which are not resinous. Although it is said to extend from the Alaska-Yukon region eastward to Greenland, its occurrence in Ontario has not yet been substantiated. (*nana* — dwarf; *exilis* — slender)

Field check Low, depressed or matted shrub; branchlets waxy-coated and with warty glands; scales of catkins resinous; northern (James Bay and Hudson Bay shores and lowlands north of 53° N).

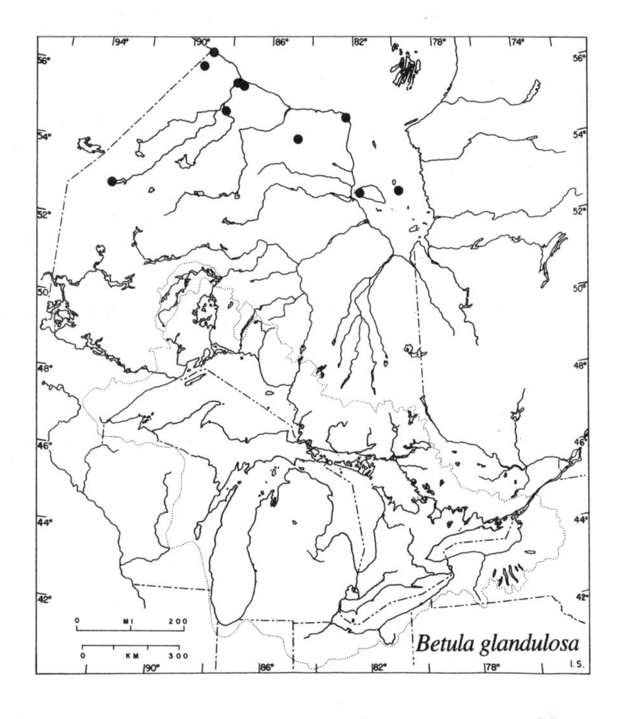

Betula occidentalis

Betula occidentalis Hook. (*B. fontinalis* Sarg.) — River birch

A large shrub or small tree up to 5 m or more in height, usually with several stems and often forming clumps or thickets, the branches ascending and spreading to form a broad open head; branchlets glabrous or finely pubescent, covered with shiny, reddish, resinous glands; bark reddish to purplish-brown, the papery layers not peeling off easily even on the older branches; trunk or main stems 1–3 dm in diameter at the base.

Leaves alternate, simple, and deciduous; blades broadly ovate, 2–5 cm long and 1–4 cm wide, thin and firm, rather shiny and gland-dotted, at least when young, acute to obtuse at the apex, cuneate to truncate or rounded at the base; the margins sharply and often doubly serrate except at the base; veins 4–6 pairs; petioles stout, 1–2 cm long, usually glandular.

Flowers small and unisexual, borne in catkins which open in early spring before the leaves unfold; male catkins clustered, 1–3 cm long, lengthening to 4–5 cm; female catkins 2–3 cm long in fruit, erect or pendulous on short glandular stalks. Fruit a winged nutlet 3–6 mm wide, the wings wider than the nutlet or about the same width. (*fontinalis*—growing in or by springs; *occidentalis*—western)

On river banks.

A predominantly western species found in a few localities in northern Ontario, not east of 82° W. (Ont. to Alaska, south to n. Calif., Ariz., and S.Dak.)

Field check Coarse thicket-forming shrub or small tree with reddish to purplish-brown bark not peeling in layers; branchlets with persistent reddish glands; leaves sharply toothed and with 4–6 pairs of veins.

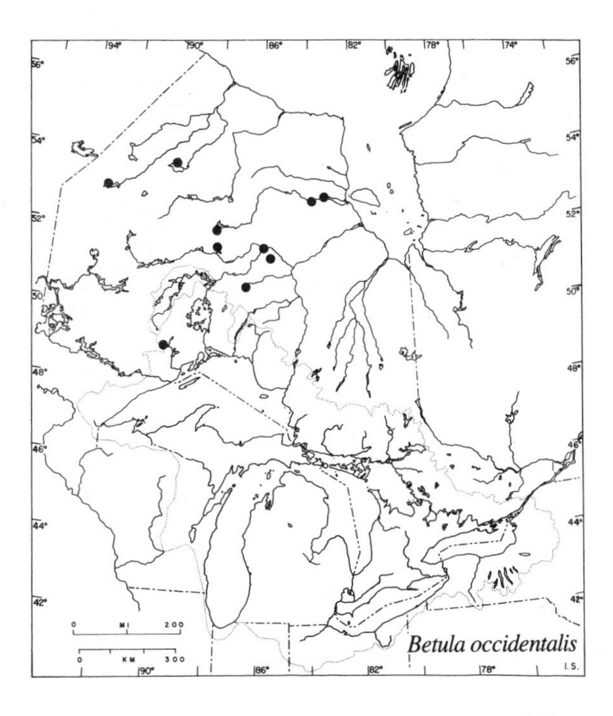

Betula pumila var. *glandulifera*

Betula pumila L. var. *glandulifera* Regel (*B. glandulosa* Michx. var. *glandulifera* (Regel) Gleason) — Dwarf birch, Swamp birch

A slender much-branched shrub with erect or ascending stems seldom exceeding 2 m. Branchlets finely hairy at first, becoming glabrous, usually gland-dotted; bark of older stems smooth, dark gray to reddish-brown, with numerous scattered light-coloured lenticels.

Leaves alternate, simple, and deciduous; blades obovate, orbicular or reniform, 1-4 cm long and 1-2 cm wide (but often larger on vigorous vegetative shoots), rounded or wedge-shaped at the base, rounded or blunt at the apex, firm and leathery in texture, dark green above, paler beneath, glabrous to yellow gland-dotted on both surfaces and with 4 or 5 pairs of veins, the reticulate venation obvious on the lower surface; margins coarsely toothed with simple and rounded, blunt, or pointed teeth; petioles 3-6 mm long.

Flowers small and unisexual, in catkins which develop in late summer and open the following spring (May and early June) as the leaves are unfolding; male catkins erect, slender, 12-20 mm long, and deciduous; female catkins 12-25 mm long and about 6 mm thick, with numerous overlapping dry papery scales on which develop small flattish nutlets; when ripe, the nutlets are shed together with the scales of the catkins; Aug. and Sept. (*pumila* — dwarf; *glandulifera* — gland-bearing; *glandulosa* — glandular)

In sphagnum bogs, swamps, and moist depressions and along shores of lakes and rivers.

Widely distributed from Lake Ontario to Hudson Bay but absent in the extreme southwestern part of Ontario near Lake Erie. (Nfld. to cent. Yukon, south to Oreg. and N.J.)

Note E. Lepage has described sixteen hybrids from Canada and Alaska (Lepage, 1976), at least two of which have been collected or reported from Ontario: *B. sandbergii* Britt. (*B. papyrifera* × *pumila* var. *glandulifera*) has been collected frequently in northern Ontario and in a few sites on the Bruce Peninsula; *B. purpusii* Schneid. (*B. lutea* Michx. f. × *pumila* var. *glandulifera*) has been reported from southern Ontario. (*sandbergii* — named for John Herman Sandberg, 1848-1917; *purpusii* — named for Joseph Anton Purpus, 1860-1933, of the Darmstadt Botanical Garden; *lutea* — yellow)

Field check Upright much-branched shrub of wet places, leaves short-stalked, small, rounded, coarsely toothed, and gland-dotted; fruit a conelike catkin of papery scales bearing small nutlets.

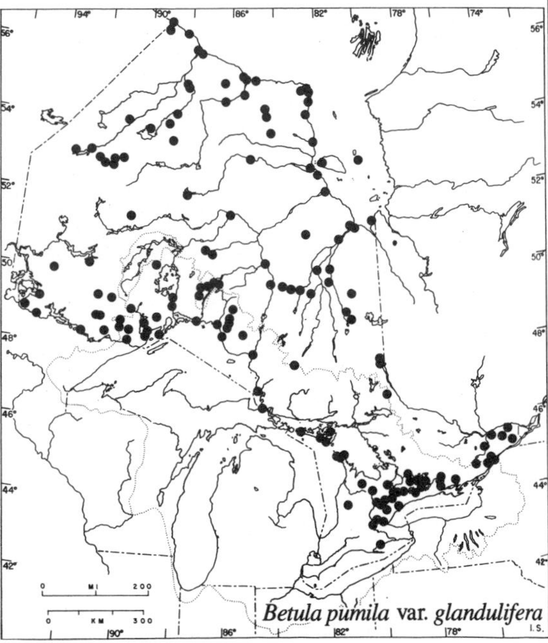

Betula pumila var. *glandulifera*

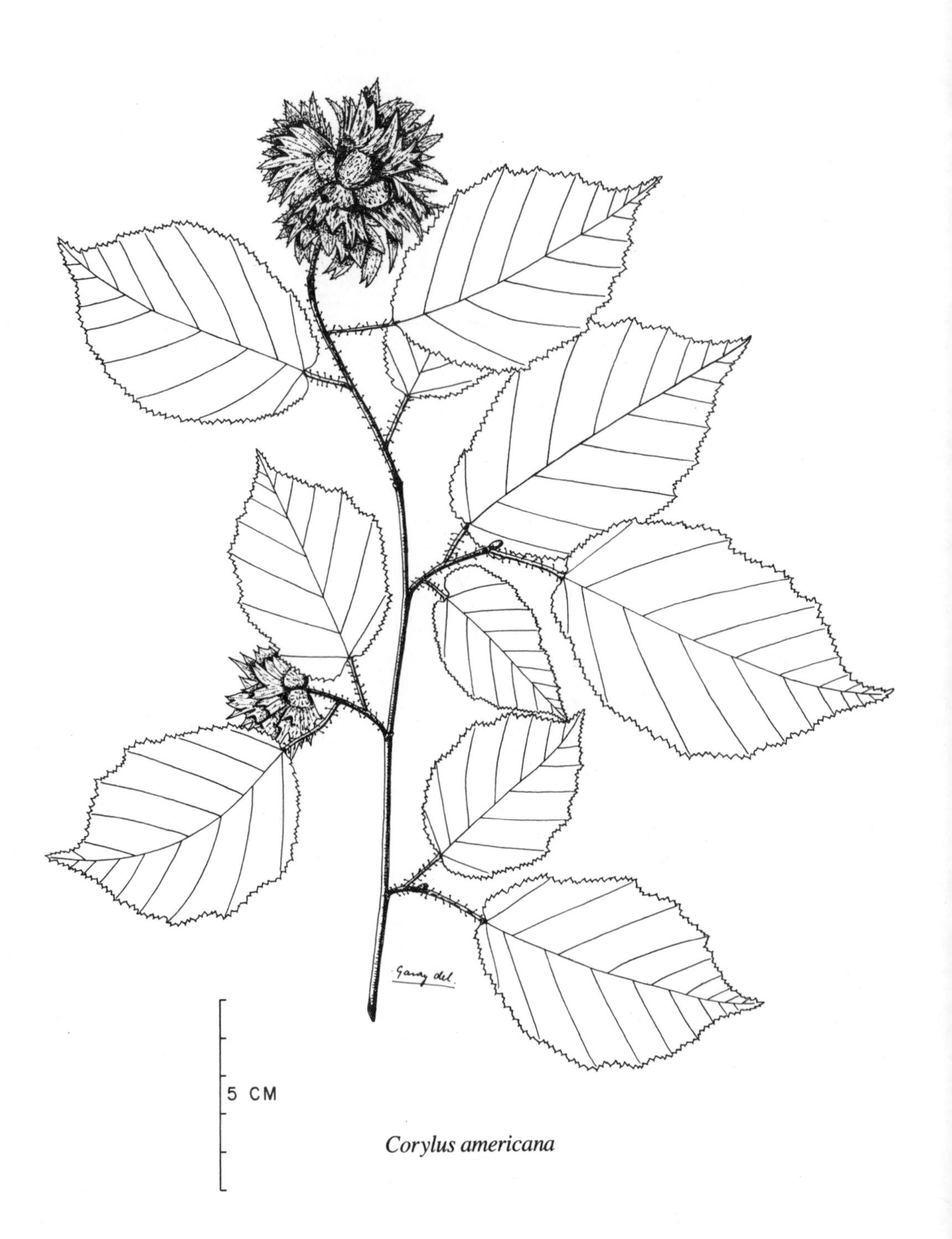

Corylus americana

Corylus L. — Hazel

Shrubs or small trees with alternate, simple, deciduous leaves which are folded lengthwise in the bud, the blades thinnish, doubly serrate. Flowers small and unisexual, developed in the fall and emerging early the following spring. Male catkins solitary or in clusters, from scaly buds; female catkins at the ends of the current leafy shoots. The nuts are edible, those of our native shrubs usually taken by squirrels. The hazel nut of commerce is the fruit of the European hazel *C. avellana* L. About 15 species. (*Corylus*—the Latin name for the hazel, thought to have been derived from the Greek *korys*, a helmet, in reference to the involucre which subtends or encloses the fruit; *avellana*—an early name used for a section in the genus *Corylus*, derived from Abella, a city of Campania, Italy)

Key to *Corylus*

a. Branchlets and petioles glandular-bristly; nuts several, borne in a tight cluster but visible among the numerous fringed bracts of the involucre ***C. americana***

a. Branchlets and petioles smooth or with a few long hairs but lacking glandular hairs; nuts solitary or a few together, each nut hidden in the base of a densely bristly involucre with a contracted tubular beak which is lacerate at the end ***C. cornuta***

Corylus americana Walt. — Hazelnut, American hazel

A low or medium-sized shrub growing to a height of 2–3 m and often forming thickets. Branchlets somewhat hairy with small, reddish to dark-coloured, glandular bristles; bark smooth and gray on the older stems.

Leaves alternate, simple, and deciduous; blades ovate to broadly oval, 5–12 cm long and 2.5–7 cm wide, bright green above, paler beneath, rounded or heart-shaped at the base and taper-pointed at the apex; margins sharply and irregularly toothed to coarsely dentate; petioles 8–18 mm long, glandular-bristly, with a pair of small scalelike stipules at the base which fall off early in the season.

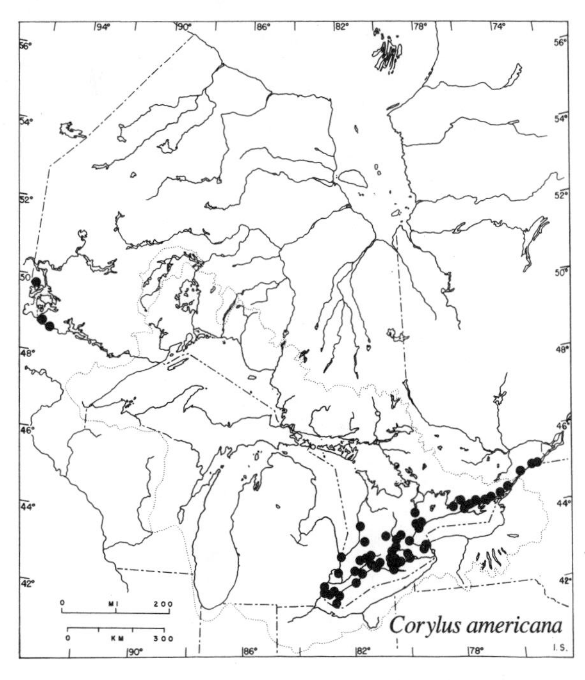

Male flowers in catkins up to 7 cm long, the female in tiny clusters with only the red stigmas protruding from the bud, both kinds opening in the early spring; April and May. Fruit a group of 2–6 round, hard-shelled nuts, each cupped by two, broadly enlarged, leafy bracts which are deeply cut or lacerate along their margins; nuts light brown, striate, about 12 mm long; seeds edible. (*americana*—American)

In thickets, at the edges of woods, in old pastures, and on the slopes of ravines, especially on well-drained sandy, gravelly, or rocky soils.

From the St. Lawrence River along Lake

Corylus cornuta

Ontario and westward to Lake Huron; also in the Lake-of-the-Woods region near the Ontario-Minnesota boundary. (Maine and s.w. Que. to Sask., south to Okla. and Ga.)

Field check Stems and petioles glandular-bristly; leaves broadly oval, coarsely or doubly serrate; nuts in clusters and visible among the deeply cut or fringed bracts.

Corylus cornuta Marsh. (*C. rostrata* Ait.) — Beaked hazel

A coarse shrub up to 3–4 m in height with smooth gray bark, often forming thickets in the open; branchlets smooth or with a few long hairs but usually without glandular hairs.

Leaves alternate, simple, and deciduous; blades ovate to broadly oval, 5–12 cm long and 2.5–7 cm wide, bright green above, paler beneath, rounded or heart-shaped at the base and taper-pointed at the apex; margins sharply and irregularly toothed to coarsely dentate; petioles 8–18 mm long, minutely hairy but not glandular-bristly, and in the early part of the season bearing a pair of small scalelike stipules.

Male flowers in catkins up to 5 cm long, the female in tiny clusters with the ovaries concealed by the bud scales from which only the vivid crimson styles emerge; both male and female flowers open before or with the unfolding leaves; April and May. Fruit solitary or in groups of 2–6 round hard-shelled nuts, each hidden in the base of a densely bristly flask-shaped involucre which is prolonged into a lacerate beak open at the end; nuts light brown, striate, about 12 mm long; seeds edible; Aug. and Sept. (*cornuta*—horned, in reference to the prolonged beak; *rostrata*—beaked)

In thickets, clearings, borders of woods; a typical understorey plant of mixed conifer-hardwood forests.

From Lake Erie northward through most of southern Ontario and to about 50° N. (Nfld. to s. B.C., south to cent. Calif. and Ga.)

Note The nuts are edible but the bristly hairs on the involucre make it difficult to extract them with impunity.

Field check Thicket-forming shrub with sharply and doubly serrate, broadly oval leaves; nuts enclosed in long-beaked bristly involucres.

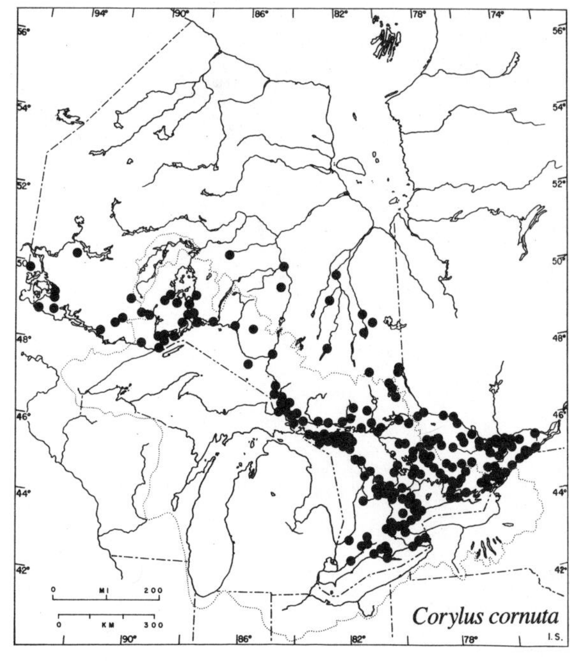

Corylus cornuta

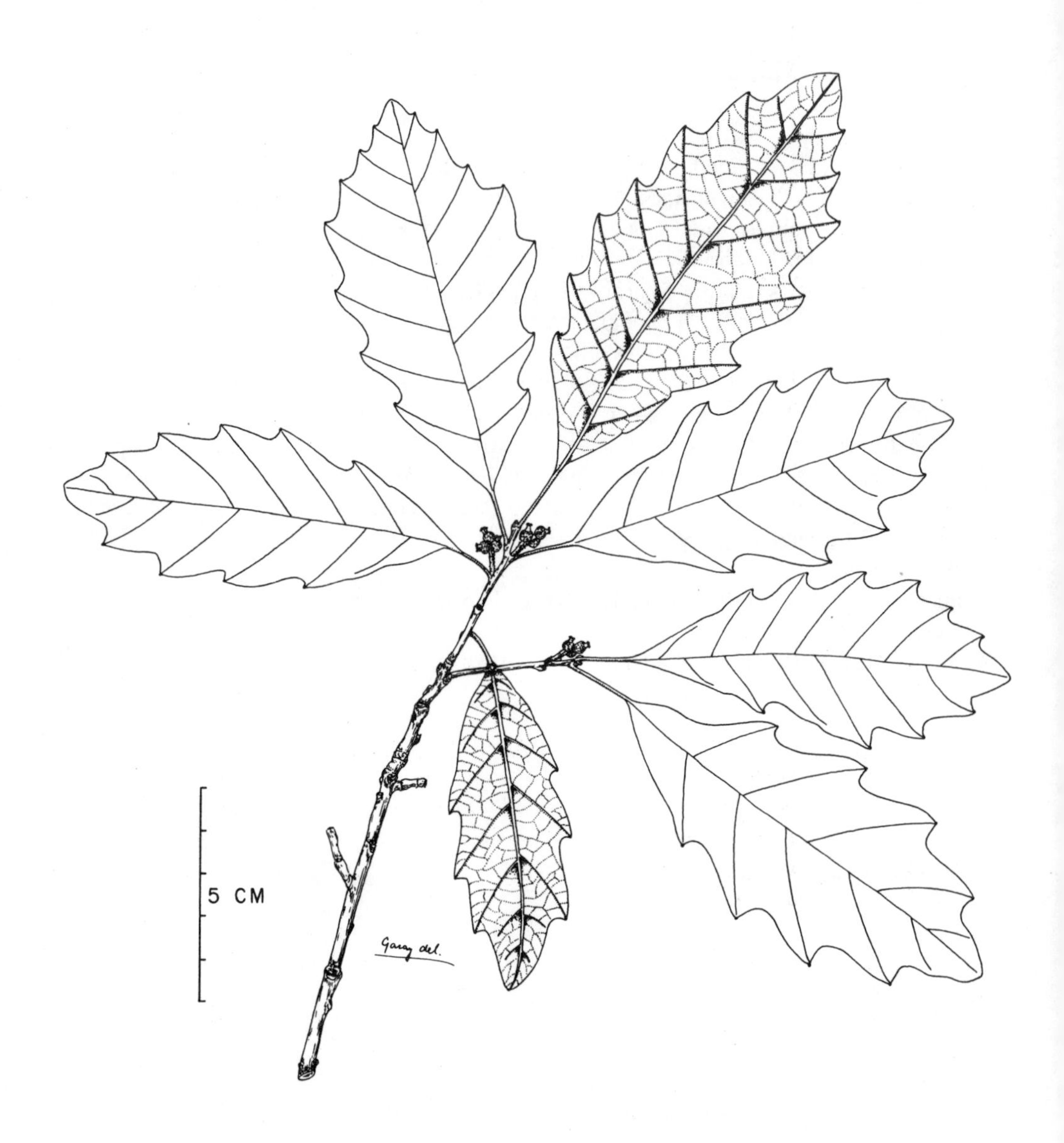

Quercus prinoides

FAGACEAE — BEECH FAMILY

Eight genera and nearly 900 species widely distributed in temperate and tropical regions. Deciduous and evergreen trees and shrubs with simple, alternate (or rarely whorled), straight-veined leaves and deciduous stipules. Flowers small, unisexual, often borne in aments (pendent catkins). The only shrub in our area is a dwarf species of oak. The family name is taken from the genus *Fagus*, the beech. (*Fagus* — from the Greek *phagein*, to eat, in reference to the edible beech nuts)

Quercus L. — Oak

A large genus of 500 species or more, including trees and shrubs, some deciduous and others evergreen. Leaves alternate, simple, entire or toothed, often lobed. Staminate flowers solitary and arranged in pendent catkins appearing with the leaves, the pistillate solitary or slightly clustered; each flower surrounded by a scaly involucre; pollinated by wind and gravity. Fruit a 1-seeded nut (acorn) partly enclosed at the base by a cup (cupule) formed by the fusion and hardening of the numerous enlarged scales of the floral involucre.

A commercially important genus which furnishes valuable timber. The bark of certain species is used in tanning and to make a yellow dye. The inner bark of the cork oak, *Q. suber* L., is the source of cork for stoppers and cork insulation. (*Quercus* — the Latin name for the oak; *suber* — cork)

Quercus prinoides Willd. — Dwarf chestnut oak, Chinquapin oak

A medium-sized shrub or small tree usually with several stems and reaching a height of about 3 m. Branchlets slender, smooth, and brittle, the bark pale gray to reddish-brown.

Leaves alternate, simple, and deciduous; blades bright green and shiny above, pale beneath with a dense flat coating of white stellate hairs, elliptic to oblanceolate, with acute or broadly pointed tip and wedge-shaped base, 5 - 10 cm long and 2 - 5 cm wide; margins wavy-toothed with 3 - 7 teeth on each side, the teeth blunt but each with a definite dark tip; petioles 3 - 10 mm long.

Flowers appearing about the end of May with the unfolding leaves; male flowers in slender catkins up to 2.5 cm long clustered at the ends of the branches; female flowers sessile or short-stalked in small clusters in the axils of developing leaves. Fruit an acorn 1 - 2 cm long and up to 1 cm wide, the nut covered at the base and for about one-third its length by the hairy cup; acorns solitary or in groups of two or three, maturing in a single season. (*prinoides* — like *Prinus*, i.e., the leaves similar to those of the chestnut oak, *Q. prinus*; *prinus* — Greek *prinos*, an oak tree)

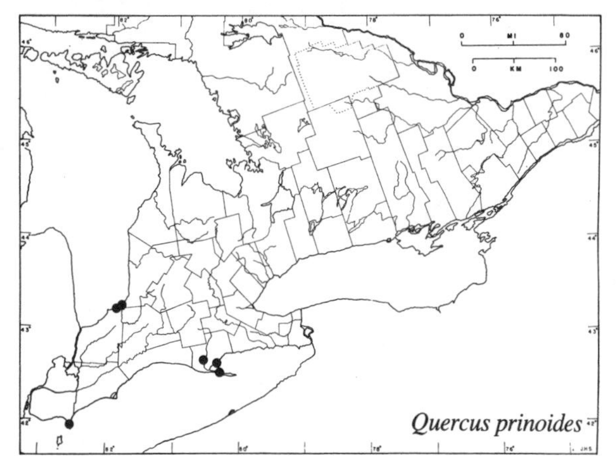

Quercus prinoides

In sandy soil; sand plains, open or wooded sand dunes.

Indigenous stands of this southern shrub are known in three areas of southern

Celtis tenuifolia

Ontario: the Grand Bend Pinery on Lake Huron; Point Pelee on Lake Erie; and the sand plains of Walsingham Township, Regional Municipality of Haldimand-Norfolk, near Lake Erie.
(s. Maine to Neb., south to Tex. and n. Fla.)

Note This species is truly a dwarf as it produces acorns when barely one metre high.

Field check Leaves chestnutlike (wavy-toothed wih 3 - 7 teeth on each side); producing acorns on low shrubby plants (less than 3 m tall); found only along Lake Erie and the southeastern shore of Lake Huron.

ULMACEAE — ELM FAMILY

A small family of about 15 genera and 150 - 200 species found in temperate and tropical regions. Trees and shrubs with mostly a 2-ranked arrangement of leaves along the branches. Leaves simple, pinnately veined, and asymmetrical (oblique) at the base; stipules falling early. Flowers small, solitary or in fascicles or cymose clusters, perfect or unisexual; perianth with 4 - 8 lobes, petals none, stamens 4 - 8, and pistil of two fused carpels but usually with a single locule and two styles with inner stigmatic surfaces. Fruit a samara, nutlet, or drupe.

Several trees provide valuable timber or wood useful in turnery.

Our only native shrub is in the largest genus *Celtis*. The name of the family is based on the genus *Ulmus*, the Latin name for the elm.

Celtis L. — Hackberry

Some 80 species are known from the northern hemisphere and South Africa. Shrubs or medium-sized to large trees. Leaves alternate, simple, entire or serrate, 3-nerved from the usually asymmetrical base. Flowers tiny, unisexual, appearing in early spring. Fruit a drupe with a thin layer of sweet pulpy flesh surrounding the thick-walled hard central stone. (*Celtis*— a Latin name applied by Pliny to the lotus and transferred to this genus for reasons which are obscure)

Celtis tenuifolia Nutt. (*C. occidentalis* L. var. *pumila* Gray) — Dwarf hackberry

A scraggly shrub or small tree growing to a height of 1 - 4 m, the stems rigid with numerous divergent short branches, often intertwined and with spinelike tips. Branchlets finely pubescent, at least when young, reddish at first but the bark finally turning gray on older stems.

Leaves alternate, simple, and deciduous; blades firm and leathery, ovate, 3 - 8 cm long and 1 - 4 cm wide, with 3 main veins from the base, the tip blunt to acuminate, and the base rounded and somewhat lopsided, dark green, and slightly scabrous above and paler beneath; margins entire or shallowly but sharply dentate, mostly beyond the middle; petioles 2 - 8 mm long.

Flowers small and unisexual, the male and female on the same plant, solitary or in groups of two or three on short pedicels along the basal portion of the current season's growth, opening as the leaves begin to expand; late May and early June. Fruit a dry, spherical, prominently stalked berrylike drupe 5 - 9 mm in diameter, a distinctive orange-brown to brick red colour, shrinking little in drying. (*tenuifolia*—with thin leaves)

In dry, well-drained, sandy soils of open woods and on slopes and crests of sand dunes.

Known only from the Grand Bend Pinery on Lake Huron and from Pelee Island and Point Pelee in Lake Erie. (Pa. to Ind. and Mo., south to La. and Fla.)

Note The Ontario populations are the northernmost indigenous stands of this southern species and are disjunct from the main area which is south of the limits of Wisconsin glaciation. The shrubs in our region often exhibit a witches'-broom effect, the result of an infection which causes deformation of the trunk and branches and a characteristic flattening of some of the branches.

The thin fleshy layer surrounding the large central pit of the fruit is edible and has a sugary flavour.

As indicated by the synonym, this species has also been treated as a variety of *Celtis occidentalis* (hackberry, bastard elm, or nettle tree), a large tree of eastern North America. W. H. Wagner Jr. (1974) concluded that the two constitute distinct species. (*occidentalis* — western, in contrast to the Old World; *pumila* — dwarf)

Field check Scrubby much-branched shrub; leaves alternate, entire to dentate, ovate with 3 main veins from the usually lopsided base; drupes dry, orange-brown; restricted in Ontario to the Pelee and Grand Bend regions.

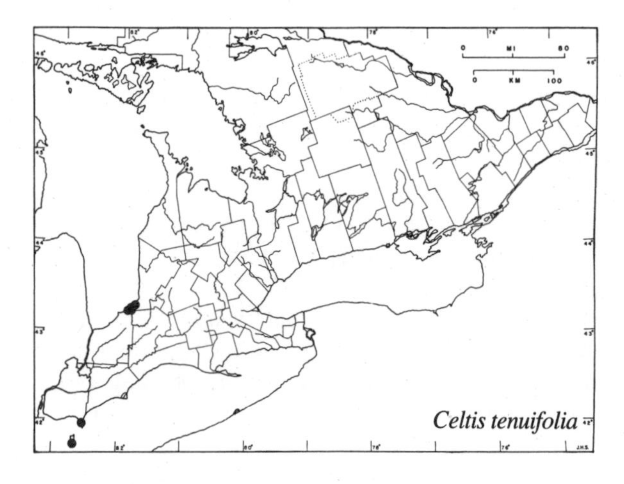

RANUNCULACEAE — BUTTERCUP FAMILY

A family of 50 - 70 genera and 2000 - 3000 species chiefly in the northern hemisphere, many in arctic and alpine regions; represented sparingly in the southern hemisphere. Mostly perennial herbs but with a few woody representatives. Leaves usually alternate (opposite in *Clematis*), with a sheathing base, the blade simple, lobed, or compound; stipules lacking. Members of this family exhibit such great diversity in flower form that it is often difficult to recognize their affinities. Nevertheless, the sepals, petals, stamens, and pistils are all separate; stamens usually numerous; pistils one to many, ovary superior. Pollination mainly by insects, also by wind and by hummingbirds. Fruit a capsule, achene, or berry.

A number of genera which are showy as wildflowers have been introduced into cultivation, e.g., columbine (*Aquilegia*), larkspur (*Delphinium*), monkshood (*Aconitum*), anemone (*Anemone*), buttercup (*Ranunculus*), winter aconite (*Eranthis*), globe-flower (*Trollius*), and Christmas rose (*Helleborus*).

Clematis L. — Clematis

A large genus of more than 200 species widely distributed but chiefly in the temperate regions of both hemispheres. Herbaceous perennials to woody vines with opposite, simple or compound leaves, the petioles twisting around available supports in the manner of tendrils. Flowers unisexual or perfect, solitary or in clusters, sepals white or coloured, petals lacking; pistils numerous, style persistent and hairy, effective as an aid to dispersal of the achenes by wind. Many handsome varieties, some of hybrid derivation, are in cultivation and offer a variety of colours. (*Clematis* —from the Green *klema*, a shoot; hence the name *klematis* assigned to some climbing plant with long shoots or stems)

Key to *Clematis*

a. Flowers opening as the leaves unfold (May - June); sepals 3 - 5.5 cm long, bluish to pink-purple; fruiting heads solitary or 2 or 3 in a group, each on a long peduncle **C. occidentalis**

a. Flowers opening after the leaves have unfolded (July - August); sepals less than 1.5 cm long, creamy-white; fruiting heads three or more (usually numerous) in open-branched stalked clusters **C. virginiana**

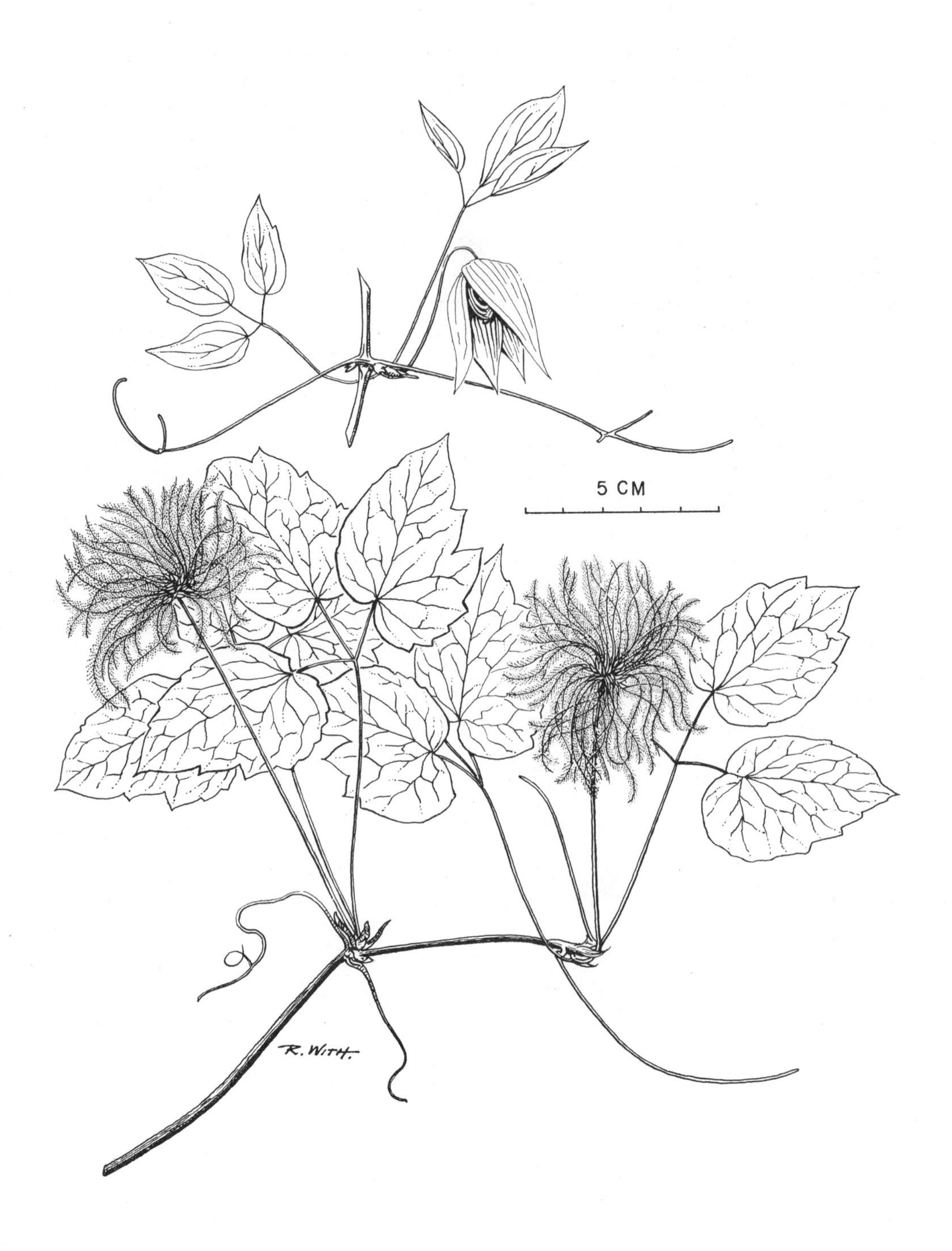

Clematis occidentalis

Clematis occidentalis (Hornem.) DC. (*C. verticillaris* DC. of eastern N. Amer.; *C. verticillaris* var. *grandiflora* Boivin)

Purple clematis

A slender woody vine climbing on trees and shrubs or trailing along the ground and over rocks to a length of 5 m or more. Branchlets smooth, mostly light brown at first, becoming reddish-brown to purplish in age.

Leaves opposite or occasionally in whorls of 3, ternately compound, and deciduous; each of the 3 leaflets long-stalked, ovate or heart-shaped with a pointed tip, 4–7.5 cm long and 2.5–5 cm wide; margins almost entire or coarsely and sharply toothed, sometimes with 1 or 2 lobes; petioles often twining around branches of shrubs, trees, fences, or other support.

Flowers perfect, with 4 bluish to pinkish-purple sepals 3–5.5 cm long, usually solitary but sometimes in twos or threes in the axils of the leaves; petals lacking; peduncles stout and as long as, or longer, than the subtending petiole; outer stamens modified into staminodes; May and early June. Fruit a fluffy head of hairy brown achenes each with a feathery persistent style 3.5–5 cm long; July to Sept. (*occidentalis* — western; *verticillaris* — in a whorl; *grandiflora* — with large flowers)

In clearings, on talus slopes, banks, and cliffs.

Widely scattered between 43° N and 51° N: on the Niagara Escarpment at the western end of Lake Ontario, on the Canadian Shield in eastern Ontario, and on Manitoulin Island; common in the western part of Thunder Bay district. (e. Que. to w. Ont., south to La. and N.C.)

Note The spring-flowering purple clematis attracts early emerging insects by its bluish-purple colour which is visible to them and by its copious nectar.

According to Pringle (1971) the eastern purple clematis may be differentiated as the typical variety and the widespread western North American plants as *C. occidentalis* var. *grosseserrata* (Rydb.) Pringle. (*grosseserrata* — with large teeth)

Field check Vine with trifoliolate opposite leaves, climbing by long twining petioles; bluish to pinkish-purple flowers and fluffy fruiting heads solitary on long stalks.

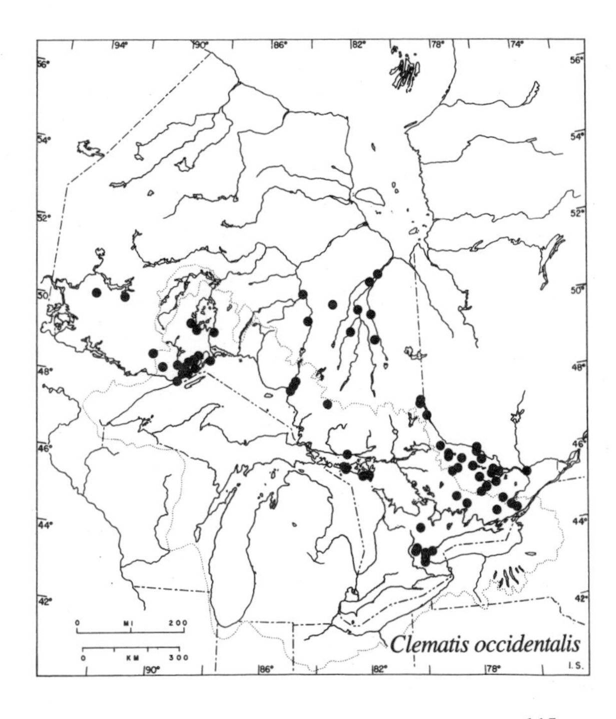

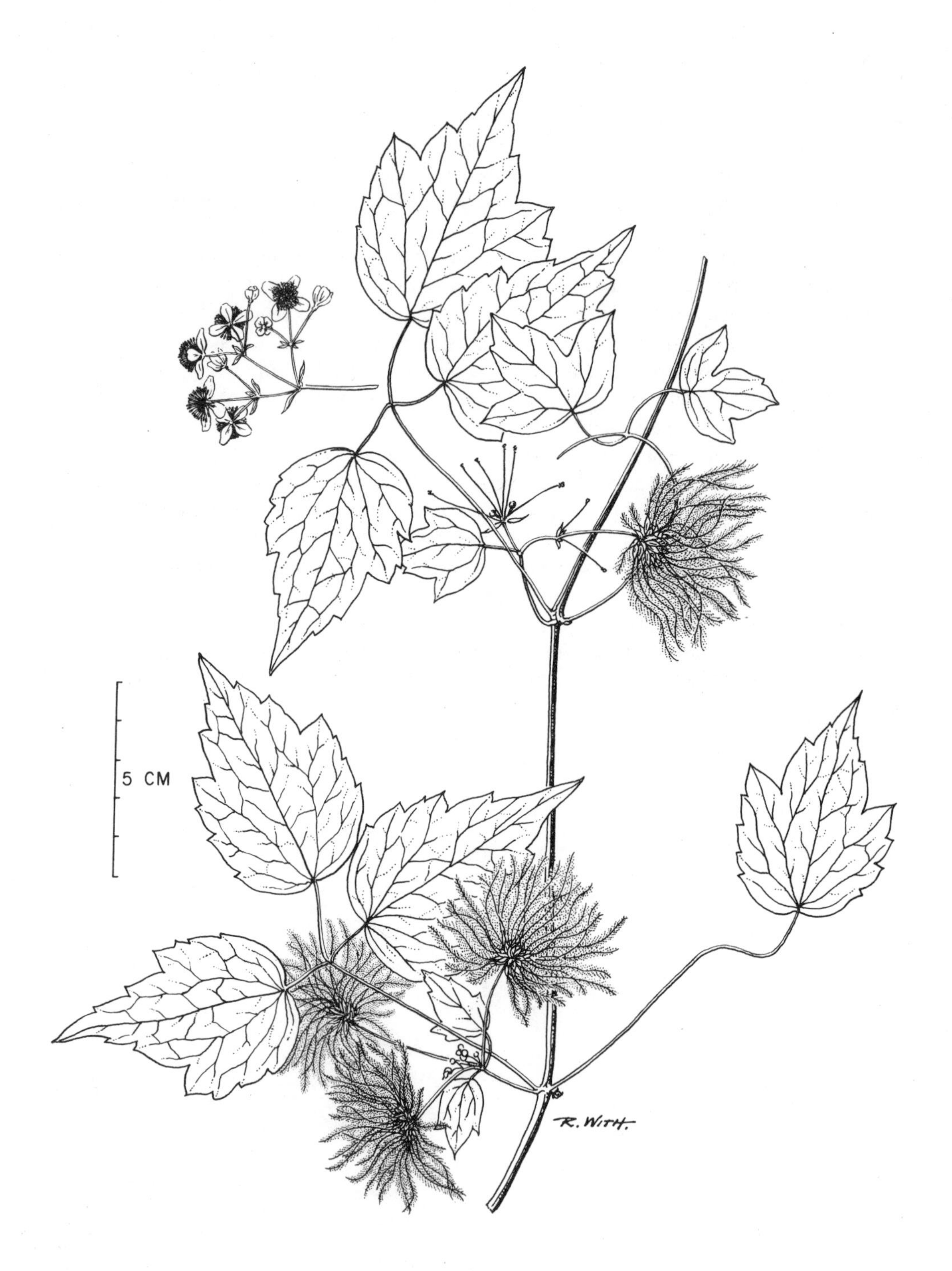

Clematis virginiana

Clematis virginiana L. **Virgin's bower**

A slender woody vine trailing on the ground and rampant over trees, shrubs, and fences, climbing to a height of 5 m or more; dioecious. Stems smooth, light brown to reddish-purple, round, and somewhat ridged in section.

Leaves opposite, ternately compound, and deciduous; the leaflets glabrous, broadly ovate or slightly heart-shaped, sharp-pointed, 5 - 9 cm long and 2.5 - 6 cm wide, frequently deeply lobed or incised; margins entire or with irregularly spaced, rounded, and abruptly pointed teeth; petioles 5 - 9 cm long, twining around any available support.

Flowers white or cream-coloured, unisexual, the pistillate flowers containing sterile stamens, the staminate with stamens only, both types numerous, in large, open, leafy clusters on a main flower stalk 1 - 8 cm long which is usually shorter than the subtending petiole; sepals 6 - 10 mm long; petals lacking; July and early Aug. Fruit a large cluster of fluffy heads of hairy brown achenes, each achene with a feathery persistent style 2.5 - 3.8 cm long; Sept. and Oct. (*virginiana* — Virginian)

In clearings, on borders of woods or thickets, and trailing over rocks.

Common throughout southern Ontario and northward to about 48° 30′N on the shores of Lake Superior. (N.S. to s.e. Man., south to Kans. and Ga.)

Note Although virgin's bower has small unisexual flowers in clusters as do many wind-pollinated plants, it is nevertheless insect-pollinated. The flies, bees, and beetles which are most abundant at the peak of flowering in midsummer are attracted by the masses of white flowers visible to them by ultra-violet rays and by fragrance and are rewarded by an abundance of pollen.

Field check Vine with trifoliolate opposite leaves, climbing by means of long twining petioles; flowers white to cream-coloured, in profuse clusters in midsummer followed by fluffy fruiting heads.

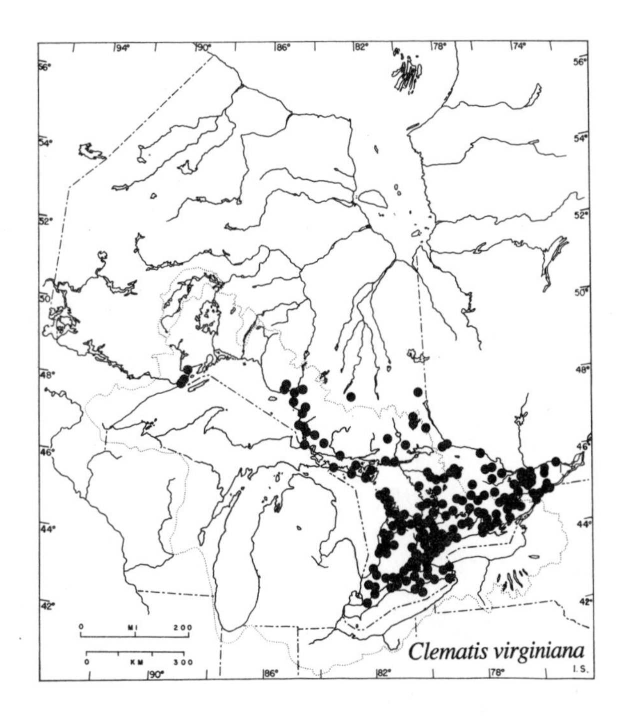

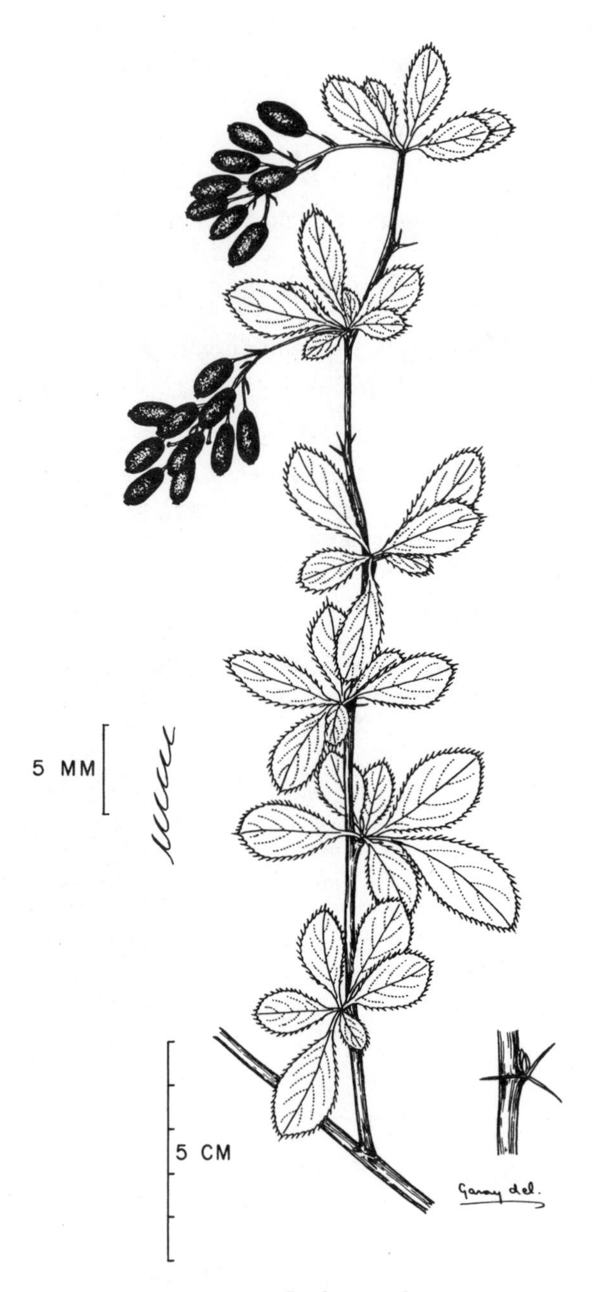

Berberis vulgaris

BERBERIDACEAE – BARBERRY FAMILY

Shrubs, small trees, and herbaceous perennials, with alternate, often spiny, simple to ternate or pinnately compound leaves, 3-parted flowers, and the fruit a berry. Ten to 20 genera with about half of the 500–600 described species contained in the genus *Berberis*. Primarily in north temperate regions, also in South America and on mountains of the tropics.

Berberis L. – Barberry

Shrubs with both long and short shoots, the simple leaf probably a reduced form of a compound leaf as indicated by the presence of a joint at the junction of petiole and blade; spines 3-pronged; flowers 3-parted, the calyx, corolla, and stamens each of two series of three; calyx and corolla both yellow. The flowers are insect-pollinated. When the flowers open the stamens lie close to the surface of the petals. Their filaments are sensitive to contact so that, when touched by an insect in search of nectar in the glands on either side of the filament at the base of a petal, a mature stamen springs upwards, showering the visitor with pollen which may be deposited on the stigma of the next flower visited. The berries are edible and the seeds are distributed by birds.

The 200–300 species are mainly Asian and South American in distribution but some also occur in Europe, Africa, and North America; introduced into Ontario. (*Berberis*—from *berberys*, an Arabic name for the plant)

Berberis vulgaris L. — Barberry

An upright many-stemmed shrub with numerous spiny branches, 2–3 m in height. Stems smooth with gray bark and yellow wood, the branchlets bearing strong, slender, sharp, 3-pronged spines with short leafy shoots in their axils.

Leaves alternate, simple, and deciduous, borne singly along the shoots of the current season and often replaced by slender single or 3-pronged spines 9–18 mm long; during the second year short leafy shoots arising in the axils of the spines; leaf blades obovate to oblong, broader at or above the middle, 2.5–5.5 cm long and 1–2.5 cm wide, glabrous on both surfaces, obtuse or abruptly pointed at the tip, tapered at base; margins prominently spiny-toothed; petioles winged and jointed near the base.

Flowers small, yellow, numerous, in drooping clusters; late May and early June. Fruit a bright scarlet, few-seeded, oblong berry, in drooping clusters; Sept. and Oct. (*vulgaris*—common)

An escape along roadsides, in fields, old pastures, clearings, thickets, and open woods.

Almost entirely restricted to calcareous regions in southern Ontario south and east of the Canadian Shield. (Native to Europe; a horticultural introduction in North America now naturalized from Nfld. to s. B.C., south to Mo., La., and Del.)

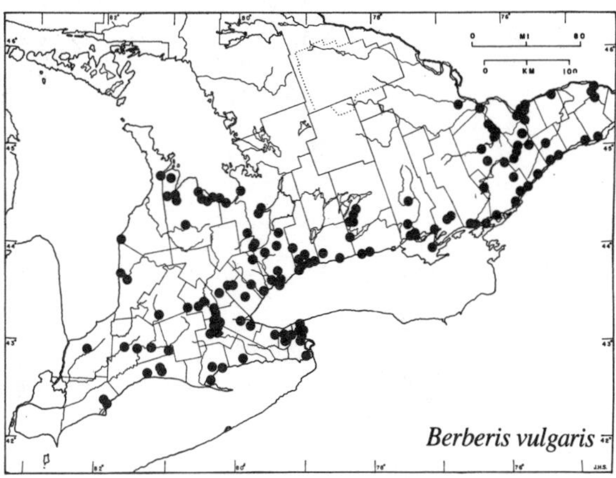

Berberis vulgaris

Note Common barberry is an attractive shrub and its berries can be used to make jellies and jams. However, it should be

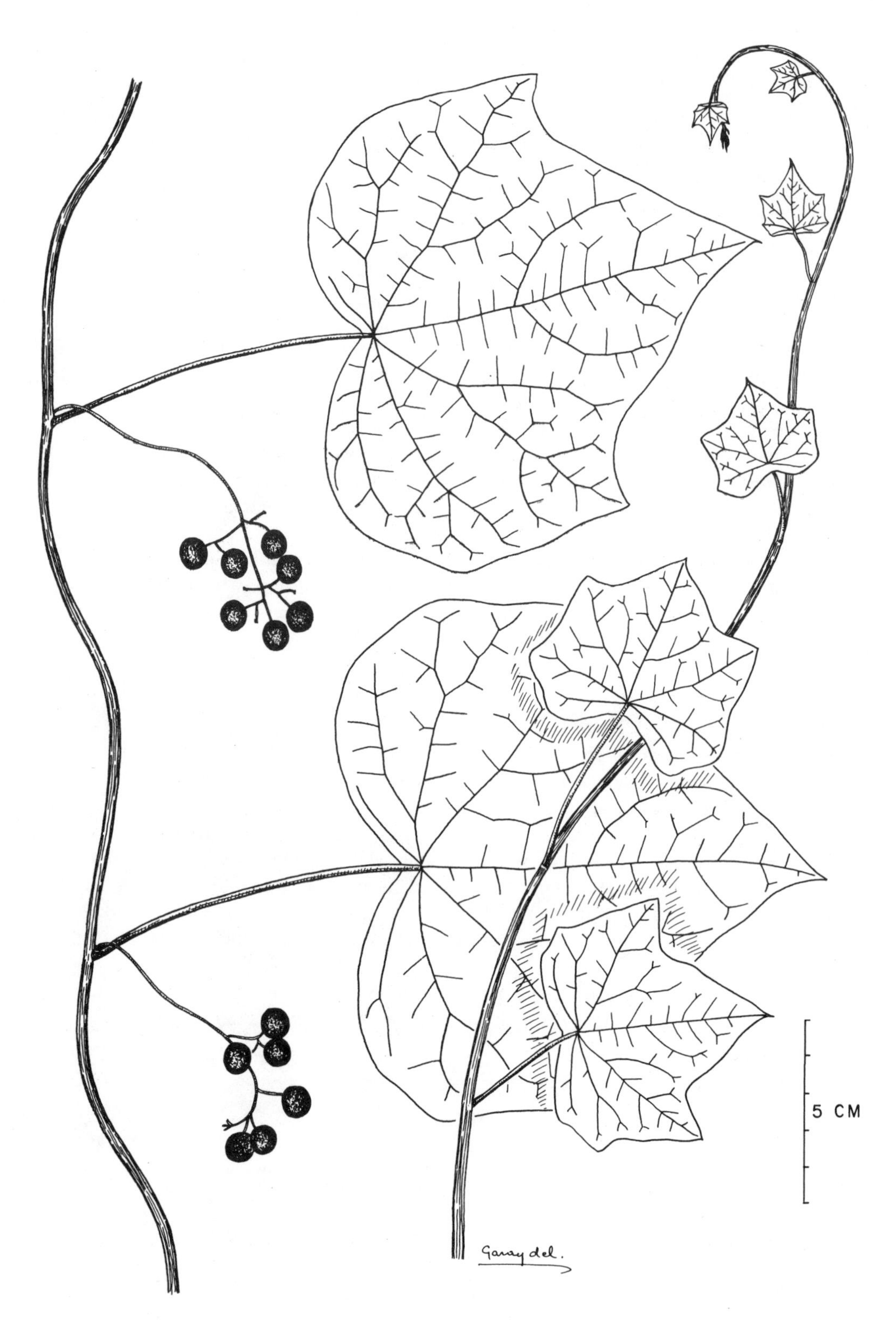

Menispermum canadense

eradicated wherever found as it is the alternate host of the stem rust of wheat. The Japanese barberry, *B. thunbergii* DC., is not involved in the life cycle of the stem rust and is widely used in hedges and other shrub plantings. It is a compact thorny shrub with fine autumn coloration of foliage and red berries that persist during the winter. (*thunbergii* — in honour of Carl Peter Thunberg, 1743 - 1828, author of *Icones Plantarum Japonicarum*, a first flora of Japan)

Field check Stems with single and 3-pronged spines at the nodes and yellow wood; leaves spiny-toothed, mostly in dense clusters; yellow flowers and scarlet berries in drooping clusters.

MENISPERMACEAE – MOONSEED FAMILY

Primarily a tropical family with 65 genera and 300 - 400 species, some extending into warm temperate regions. Mostly twining shrubs or vines, but some are herbs and others are trees; dioecious. Leaves alternate, usually simply, palmately veined, and lacking stipules. Flowers small, white or greenish-white, unisexual, mostly 3-parted. Fruit a drupe. Some South American species are poisonous and are used as sources for arrow poisons by the natives. The only genus which occurs as far north as Ontario is *Menispermum*, from which the family name is derived.

Menispermum L. – Moonseed

There is a single species in eastern North America and a second one in eastern Asia. (*Menispermum* — from the Greek *mene*, moon, and *sperma*, seed, in reference to the crescentic, laterally flattened stone in the globular drupe)

Menispermum canadense L. — Moonseed vine

A slender woody vine trailing along the ground or climbing on shrubs and trees. Stems somewhat hairy, light to dark brown, with fine longitudinal ridges and furrows.

Leaves alternate, simple, and deciduous; blades reniform, orbicular, or broadly ovate, often with 3, 5, or 7 shallow lobes, 5 - 11 cm long and 5 - 15 cm wide, apex rounded or pointed, dark green and thinly pubescent to glabrous above, much paler and pubescent beneath; petiole 5 - 12 cm long, attached to the lower surface near, but not exactly at, the heart-shaped base; main veins radiating from the point of attachment in a palmate fashion.

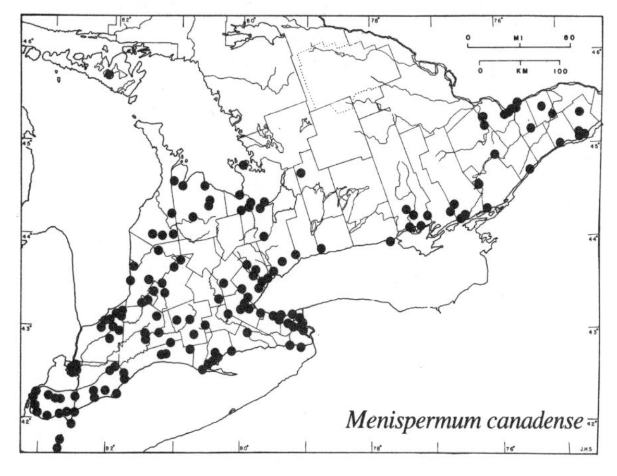
Menispermum canadense

Flowers small, creamy-white, in slender-stalked, bracted, open clusters from the axils of the leaves; June and early July. Fruit a round, dark blue, berrylike drupe somewhat resembling a small grape,

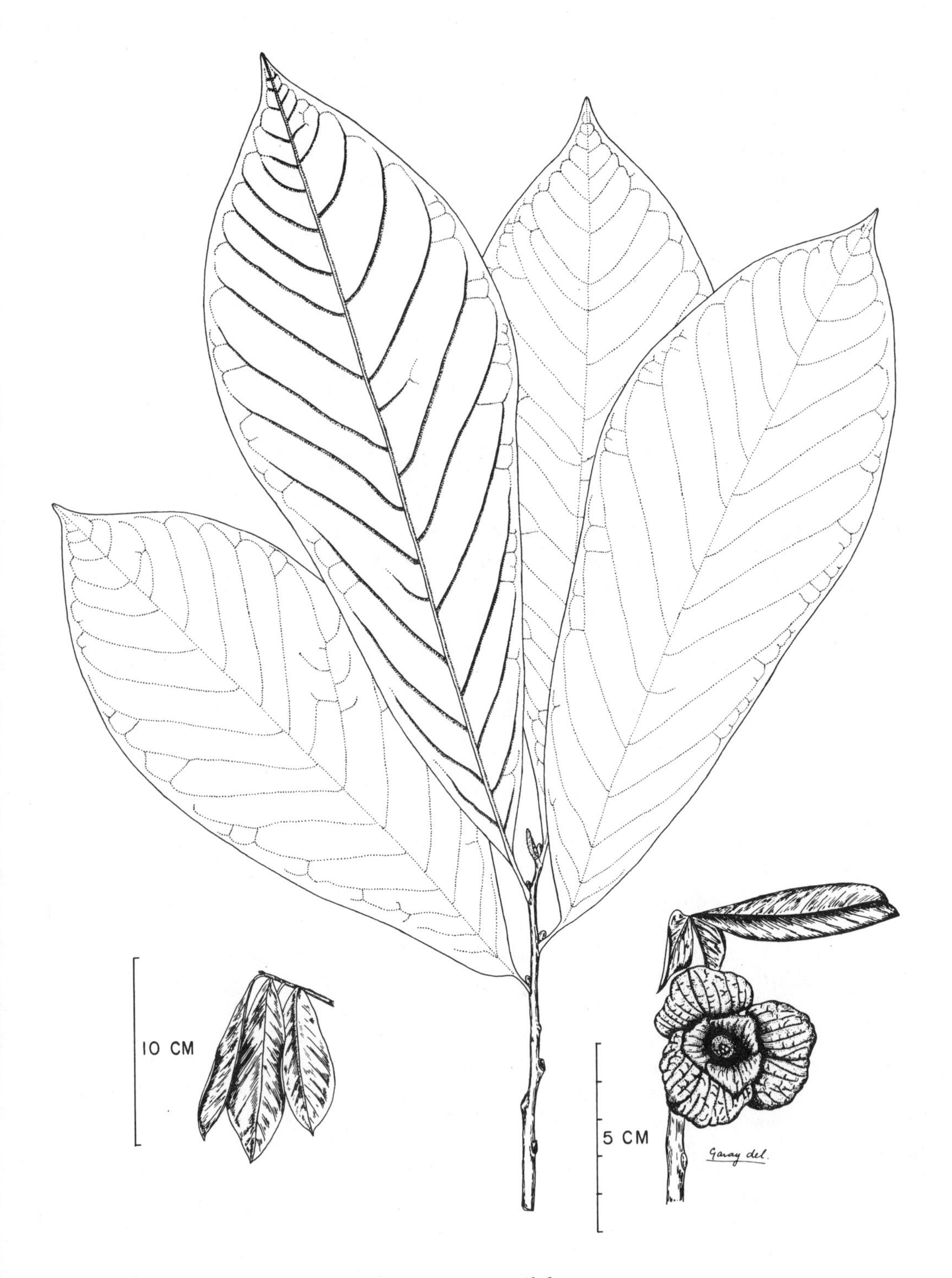

Asimina triloba

6–10 mm in diameter, with a bloom, containing a crescent-shaped pit; Sept. (*canadense*—Canadian)

In moist thickets, cedar woods, maple woods, on shaded banks of streams and rivers, and on talus slopes.

Chiefly in the areas east and southwest of the Canadian Shield in southern Ontario; also on Manitoulin Island. (s.w. Que. to s.e. Man., south to Okla. and Ga.)

Note The distribution suggests that this vine is probably a calciphile since it skirts the Canadian Shield in southern Ontario.

Field check Slender twining vine with broad rounded or shallowly lobed leaves attached to the petiole near the base; fruit dark blue, grapelike, in open stalked clusters.

ANNONACEAE – CUSTARD-APPLE FAMILY

Trees and shrubs with alternate entire pinnately-veined leaves lacking stipules, flowers with a 3-parted perianth in 3 whorls, and fleshy fruits with large seeds. More than 100 genera, chiefly in the tropics and including such edible and flavourful American fruits as custard-apple, sweetsop, soursop, and cherimoya (*Annona* spp.) as well as being the source of ylang-ylang (*Cananga*), an important and valuable oil from Asia used in perfumery. The name of the family is based on the genus *Annona*. (*Annona*—year's harvest; *Cananga*—from *kenanga*, the native Malayan name for the tree yielding the perfume)

Asimina Adans. – Pawpaw

The only genus represented as far north as southern Ontario is *Asimina*, with about 8 species in eastern North America. (*Asimina*—from *assimin*, the Indian name for the pawpaw, converted to *asimminier* by settlers in the French colonies)

Asimina triloba (L.) Dunal — Pawpaw

A tall shrub or small tree up to 7.5 m in height, commonly spreading by root suckers to form thickets, less frequently found as individual specimens. Branchlets reddish-brown and hairy at first, later becoming smooth, gray to brown, ridged and somewhat scaly in age; buds naked (without bud scales) but covered by a coating of rusty-brown hairs, the leaf buds flattened, the flower buds smaller and rounded.

Leaves alternate, simple, and deciduous, characteristically in a partially or completely drooping position at the ends of the branches; blades obovate to oblanceolate, usually broadest beyond the middle, and often very large, up to 30 cm long and 13 cm wide, smooth or slightly veiny, acute to sharp-pointed at the tip, wedge-shaped at the base, gradually tapering to a grooved petiole 6–10 mm long; margins entire; venation pinnate and conspicuous on the lower surface.

Flowers solitary on the wood of the previous season, opening as the leaves unfold, at first whitish or greenish-yellow but soon becoming reddish-purple to maroon in colour as they expand; 3-parted, reticulate-veiny, and fleshy in texture; late May. Fruit a large berry which resembles an irregular-shaped pear, roughly oblong-cylindric, 7 - 13 cm long, the skin green at first, then turning yellow to brown as the fruit matures; flesh soft and yellow to orange with one or two rows of flat brown beanlike seeds; Oct. (*triloba*—three-lobed, probably in reference to the 3-parted flowers)

In damp sandy or clayey soils, chiefly in alluvial flats, on stream banks and the slopes of sheltered ravines.

Restricted to the Deciduous Forest Region; most abundant in the counties of Essex, Kent, and Lambton and in the Regional Municipality of Niagara. (N.J. and w. N.Y. to s. Mich. and s.e. Neb., south to Fla.)

Note In Ontario, where this species is at the northern limit of its natural range, the fruit is not always as flavoursome as it is in the southern states. Some pawpaws have yellowish flesh and although edible do not compare in quality with others in which the flesh is orange in colour. The distinctive flavour of the pawpaw is difficult to describe and various attempts include references to such elemental flavours or aromas as those of banana, custard, apple, pineapple, eau-de-cologne, and turpentine.

As the stamens and pistils mature at different times, the perfect flowers of pawpaw are functionally unisexual, thus preventing self-pollination. When the petals have spread enough to admit insects, mostly flies attracted by the foetid odour of the flower, the stigmas of the pistils are receptive but the stamens are immature. In older flowers, the petals are spread exposing mature stamens shedding their pollen but the stigmas are no longer receptive. Thus, the first flowers to open may not be pollinated.

Field check Large, drooping, entire leaves; reddish-brown flowers and pearlike fruit with yellow-orange edible flesh and flat beanlike seeds; in Ontario, restricted to the Deciduous Forest Region.

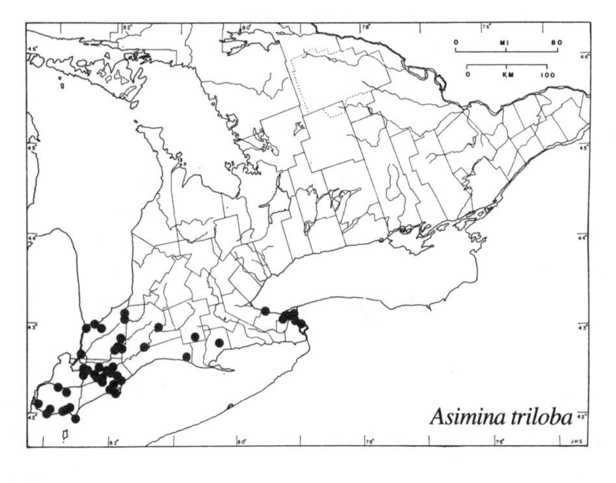

LAURACEAE – LAUREL FAMILY

About 40 genera and over 2000 species of tropical to warm temperate regions, especially well represented in southeastern Asia and in South America. Aromatic trees and shrubs, usually with numerous oil canals or glands in the tissues; leaves often leathery, simple or lobed, alternate or opposite, and lacking stipules; flowers 3-parted, usually yellow; fruit a 1-seeded berry or drupe.

Several genera are of economic importance as sources of oils, spices, edible fruit, and timber, e.g., *Cinnamomum* (camphor, cassia bark, and cinnamon), *Persea* (avocado), and *Ocotea* (greenheart, an oil used in perfume). The family name is based on the genus *Laurus* which contains the laurel or sweet bay (*Laurus nobilis* L.). (*Cinnamomum*—Greek *kinnamomon*, cinnamon; *Persea*—the Greek name for an Egyptian tree bearing sweet fruit; *Ocotea*—from a native name for the tree; *Laurus*—the Latin name for the laurel; *nobilis*—noble)

Lindera Thunb. – Spicebush

A genus of 100 species primarily of the Himalayas to eastern Asia and Malaysia. Aromatic shrubs with deciduous leaves; small yellow flowers in almost sessile lateral clusters; fruit (drupe) borne singly, or paired, or in small clusters along the stems. A single species occurs in Ontario. (*Lindera*—in honour of the Swedish botanist Johann Linder, 1676-1723)

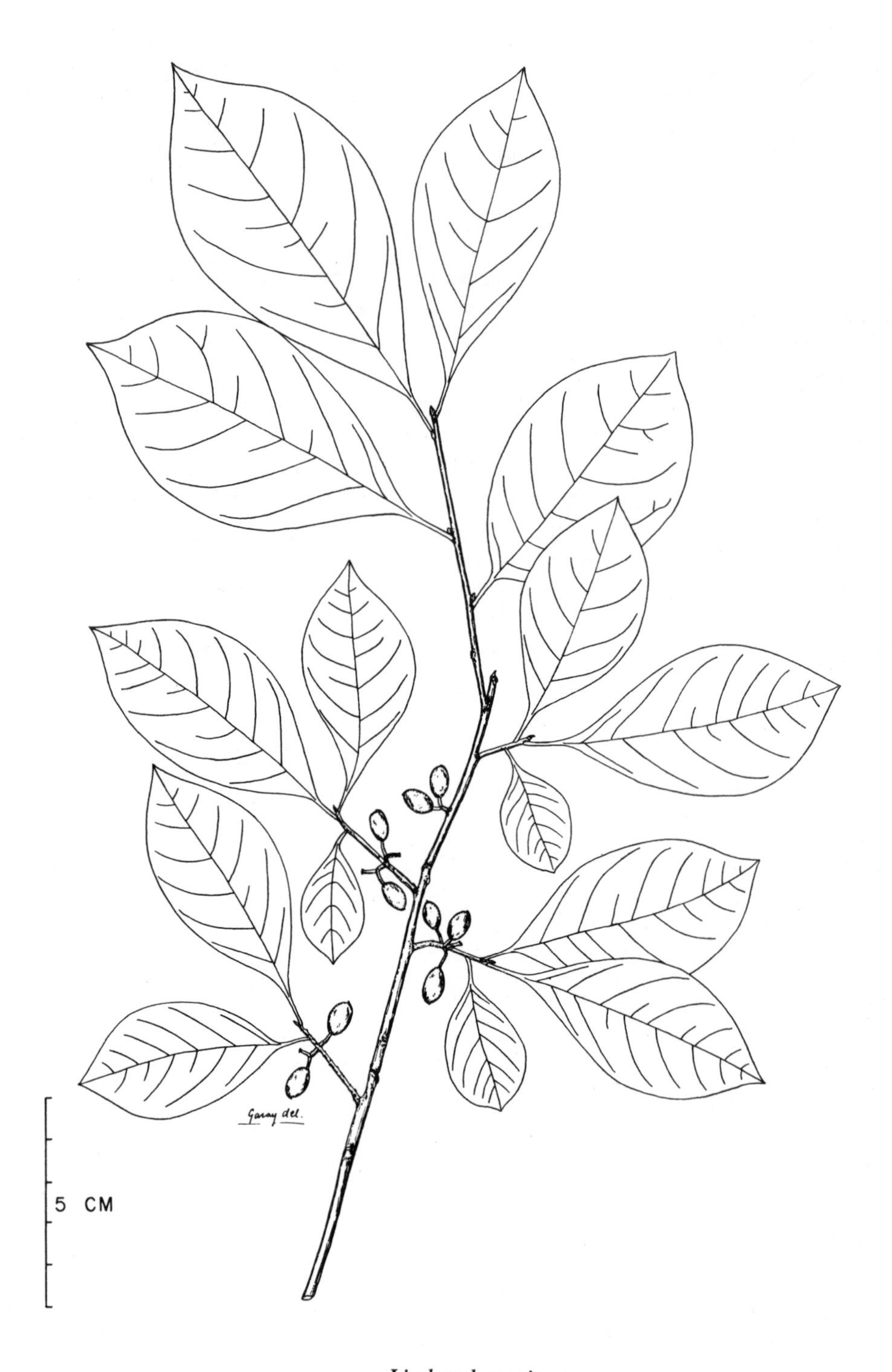

Lindera benzoin

Lindera benzoin (L.) Blume

Spicebush
Allspice

A medium-sized spreading shrub up to 3 m tall; dioecious. Branchlets slender, spicy-aromatic when bruised, slightly hairy at first but later becoming smooth; bark gray-brown and roughened in age.

Leaves alternate, simple, and deciduous; blades thin, aromatic when crushed, oval, or oblong-ovate, broadest at or beyond the middle, the larger ones reaching 15 cm in length and 6 cm in width, acute or acuminate at the apex, the smaller leaves usually rounded or blunt at the tip; margins entire; petioles grooved, 7 – 10 mm long.

Flowers small, yellow, and unisexual; in dense clusters of 4 – 6 surrounded by 4 deciduous scales and appearing before the leaves in early spring; April and early May. Fruit a smooth, bright red, glossy, berrylike drupe about 1 cm long in clusters of 2 – 6 on short stalks, spicy-aromatic when bruised or chewed. (*benzoin* — a name once used for a member of the laurel family)

In low moist thickets or woods along or near water courses or rarely on higher well-drained sites such as river banks, open meadows, and sand dunes.

Common in the Deciduous Forest Region and extending sporadically northward to Georgian Bay and eastward along the north shore of Lake Ontario in Hastings, Northumberland, and Prince Edward counties. (s.w. Maine to e. Kans., south to Tex. and Fla.)

Field check Aromatic; early yellow flowers in dense sessile clusters; leaves oval or obovate, broadest beyond the middle, entire; fruit short-stalked, berrylike, glossy red.

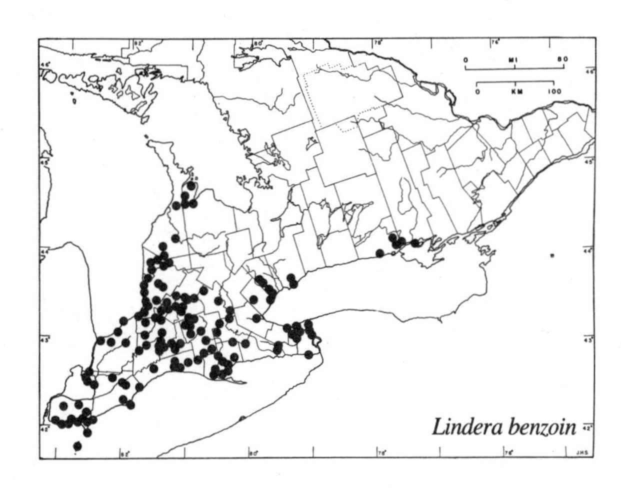

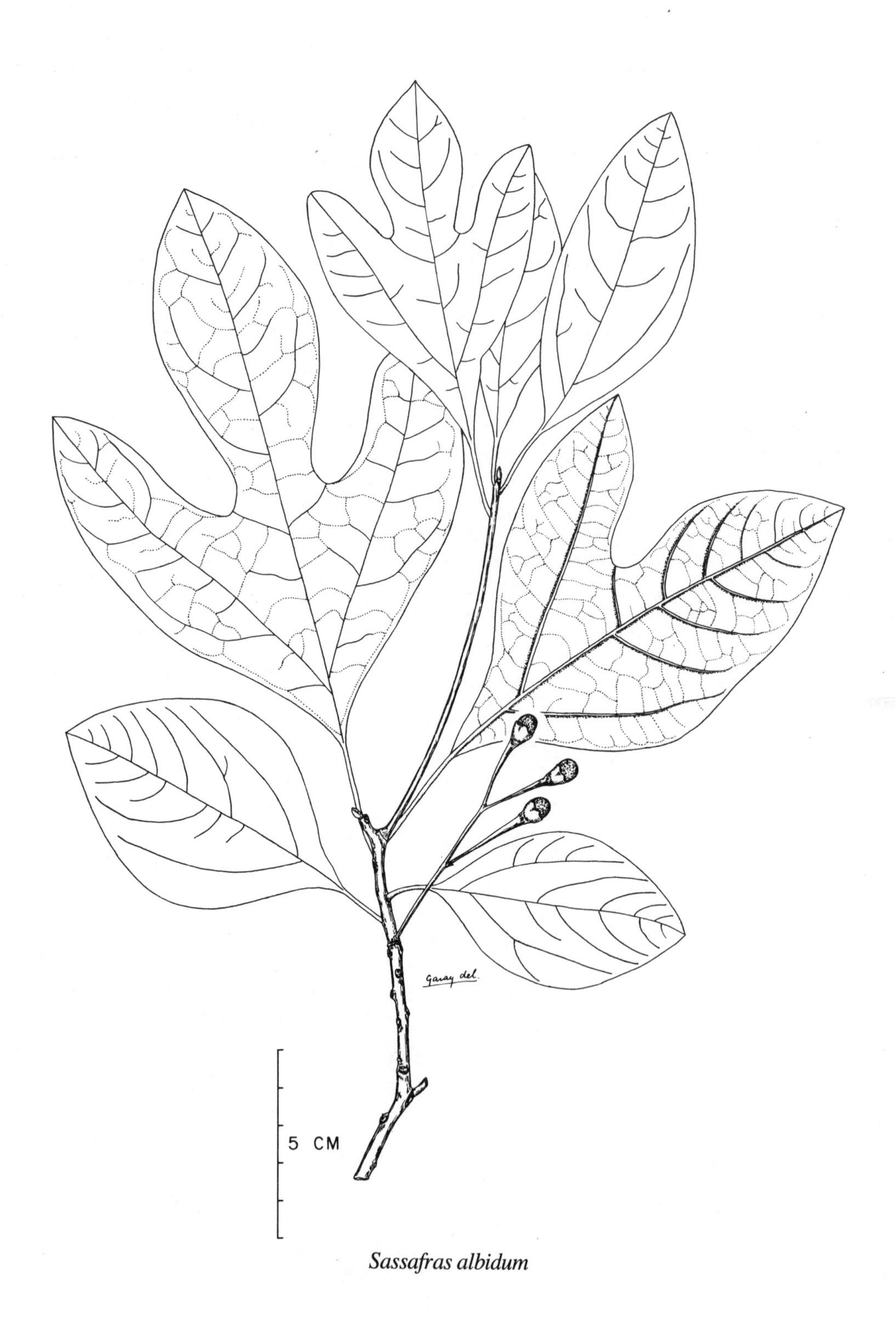

Sassafras albidum

Sassafras Trew — Sassafras

Only three species are known, two in eastern Asia (China and Taiwan) and one in eastern North America. Trees with aromatic bark and deciduous leaves; first leaves opening before the flowers, not lobed, later leaves lobed or variously asymmetrical in shape; flowers greenish-yellow, borne on long stalks in clusters, with conspicuous scaly bracts at the base; fruit a drupe. (*Sassafras*—the popular name used by early French settlers in Florida; adopted by Trew as the botanical name)

Sassafras albidum (Nutt.) Nees

Sassafras
Mitten tree

A large shrub spreading by root suckers to form thickets but under good growing conditions developing into a medium-sized or large forest tree; primary branches horizontally spreading with upcurved tips; dioecious. Branchlets dark green to purplish at first; smooth, shiny, brittle, and spicy-aromatic when bruised; soon becoming rough, mottled, and reddish-brown with conspicuous, flat, corky ridges.

Leaves alternate and deciduous with characteristic aromatic fragrance when crushed; the blade simple, mitten-shaped, or 3-lobed, elliptic to broadly ovate in general shape, 20 - 25 cm long and almost as wide when tri-lobed; margins entire; petioles firm, 3 - 10 mm or more in length.

Flowers unisexual, greenish-yellow, in stalked racemose clusters subtended by several scaly bracts and appearing with the developing leaves; late May. Fruit a dark blue, ovoid, berrylike drupe with a large brown stone, the pedicel red and thickened at the end into a club-shaped base or collar supporting the fruit; July - Aug. (*albidum*—white)

Chiefly in sandy soil and rich loam of semi-open deciduous woods.

Restricted to the Deciduous Forest Region and not found north of the Toronto region. (s.w. Maine to Kans., south to Tex. and Fla.)

Field check Aromatic; leaves simple, mitten-shaped, or tri-lobed, their margins entire; flowers yellow; fruit blue and berrylike on club-shaped red stalks.

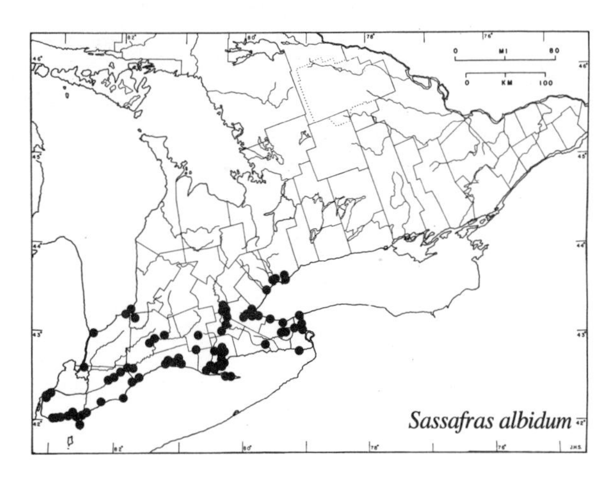

GROSSULARIACEAE — GOOSEBERRY FAMILY

About 150 species of shrubs in the genus *Ribes* (including *Grossularia*, which is maintained as a separate genus by some botanists). Widely distributed in Eurasia, North America, Central America, and southward in the Andes; also northwestern Africa.

Low prostrate to erect or upwardly-arching shrubs; the stems often armed with prickles or nodal spines or both. Leaves alternate, palmately 3- or 5-lobed, deciduous, sometimes with resinous dots. Flowers in small clusters or in racemes, usually perfect, 5-parted, white, greenish-white pink, red, purple, or yellow; hypanthium saucer-shaped, rotate, or tubular with 5 free calyx lobes always longer than the petals which are separate and inserted on the calyx; stamens alternating with the petals; styles 2, united below; ovary inferior (in the Ontario species) and maturing into a berry capped by the persistent but withering calyx.

Members of the *Grossulariaceae* are also treated as comprising a subfamily of the *Saxifragaceae*. The family name is derived from *Grossularia*, the classical generic name for a European gooseberry, probably based on *groseille*, the French name for a gooseberry.

Ribes L. — Currants and Gooseberries

The species of *Ribes* fall into two groups: (a) the currants, characterized by stems lacking spines and prickles (except *R. lacustre*) and by the pedicels being jointed at the summit so that the berries fall from the stalk; (b) the gooseberries, in which the stems have spines and prickles and the pedicels are not jointed at the summit.

The flowers display considerable variation in the shape and length of the hypanthium and in the relative length of stamens and styles, apparent adaptations for insect pollination in some and for self-pollination in others.

Dispersal of seed is by birds, the seeds passing through their digestive tract undamaged and possibly dropped in a better state of preparation for speedy germination.

The cultivated currants and gooseberries of the garden are European species or their derivatives. *Ribes* is the alternate host of the fungus that causes white pine blister rust. (*Ribes*—from the Danish *ribs*, currants)

Note The chief difficulties in identifying our native species of *Ribes* lie in distinguishing between the closely related and frequently confused *Ribes hirtellum* and *R. oxyacanthoides*. In southern Ontario *R. hirtellum* can be recognized by the characters given in the field check for that species. In the north where its range overlaps that of *R. oxyacanthoides*, particularly along the north shore of Lake Superior and in the James Bay area, the distinctions between these two species are often obscure. For example, plants are found which are *R. hirtellum* in most respects but with some characters of *R. oxyacanthoides*, such as glands on the lower leaf surface and in the sinuses of the leaf margin, and with shorter stamens. *R. hirtellum* has been found in the Clay Belt, but some plants in that area resemble *R. oxyacanthoides* in having glands on the lower leaf surfaces and on the floral bracts. Also, plants of *R. oxyacanthoides* occur in the Thunder Bay district that lack glands on the upper leaf surface, have wedge-shaped leaf bases, and weak nodal spines and prickles, all of which is characteristic of *R. hirtellum*. It is highly probable that these species are hybridizing where their ranges overlap.

Key to *Ribes*

a. Stems unarmed
 b. Blade of leaf mostly more than 3.8 cm long (3–9.5 cm), broad or roundish but not fan-shaped, with many coarse teeth; flowers less than 12 mm long, not spicy-fragrant
  c. Stems erect, leaves resin-dotted, at least beneath; berries black
   d. Leaves copiously resin-dotted on both surfaces, flowers and fruit in drooping clusters; floral bracts persistent, longer than the flower stalk ***R. americanum***
   d. Leaves mostly resin-dotted only on the lower surface; flowers and fruit in erect or ascending clusters; floral bracts short and deciduous ***R. hudsonianum***
  c. Stems reclining or straggling; leaves not resin-dotted; berries red
   e. Stems, leaves, and berries with skunklike odour when bruised; berries glandular ***R. glandulosum***
   e. Stems, leaves, and berries without skunklike odour when bruised; berries smooth ***R. triste***
 b. Blade of leaf mostly less than 3.8 cm long (1.2–3.8 cm), fan-shaped, entire or with a few teeth at the ends of the lobes; flowers more than 12 mm long, bright golden-yellow, spicy fragrant ***R. odoratum***
a. Stems armed with spines or bristles, sometimes only at the nodes or on new wood
 f. Spines and prickles persistent; fruit with glands or prickles
  g. Stems densely prickly between the nodes; leaves ill-scented when bruised; fruit purplish-black, glandular-hairy ***R. lacustre***
  g. Stems usually without prickles between the nodes; leaves not ill-scented; fruit wine red, with prickles ***R. cynosbati***
 f. Spines and prickles shed with peeling bark; flowers borne close to the stem; fruit without glands or prickles
  h. Spines and prickles weak, soon shed with outer bark; leaves without glands beneath; floral bracts with long hairs ***R. hirtellum***
  h. Spines and prickles firm, tardily shed with outer bark; leaves with scattered glands beneath, at least on the veins; floral bracts glandular-hairy ***R. oxyacanthoides***

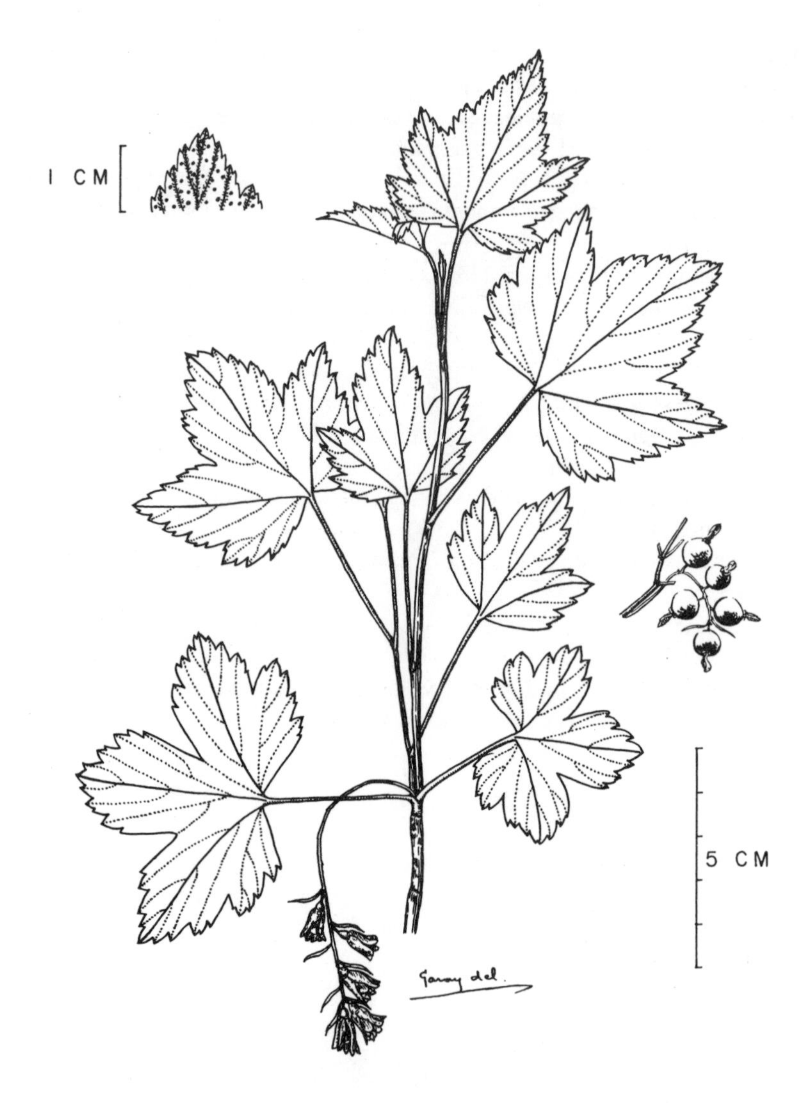

Ribes americanum

Ribes americanum Mill. — Wild black currant

A small unarmed shrub with upright or ascending branches growing to a height of 9 - 12 dm. Branchlets grayish, finely hairy, and more or less angled, with the outer layer of bark peeling off and exposing an inner reddish layer; older branches glabrous and reddish to black.

Leaves alternate, simple, and deciduous; blades rounded to suborbicular, 3 - 7.5 cm long and 3.5 - 10 cm wide, 3 - 5-lobed, dark green above with scattered resinous dots and sparsely hairy at first but becoming glabrous, paler beneath with a thin covering of fine hairs and copiously sprinkled with golden resinous dots; the lobes sharply pointed, base heart-shaped, truncate, or rarely cuneate; margins coarsely and doubly serrate; petioles 3 - 5 cm long, resin-dotted, finely hairy, and with a few large branched hairs, especially near the base.

Flowers in drooping clusters 2.5 - 7.5 cm long, creamy-white to yellowish, bell-shaped, about 9 mm long, each flower with a persistent bract longer than the pedicel; May and early June. Fruit an edible black berry, 6 - 9 mm in diameter, in drooping clusters; July and Aug. (*americanum*—American)

In damp soil along streams, on wooded slopes, in open meadows, low wet woods, and on rocky ground.

Throughout southern Ontario south and east of the Canadian Shield; widely scattered in the eastern and western parts of northern Ontario at least as far north as 50° N; bordering the Boreal Forest and Barren Region only in the Moose River drainage basin. (N.B. to Alta., south to N. Mex. and Del.)

Note The black currant cultivated in gardens is a related species, *R. nigrum* L., which differs in its greenish-white to purplish flowers and a floral bract shorter than the pedicel. It is occasionally found as an escape from cultivation along roadsides and in thickets. (*nigrum*—black)

Field check Leaves roundish, 3 - 5-lobed, gland-dotted on both surfaces; floral bract longer than the flower stalk and often present in the fruiting stage; berries black, edible, without glands or prickles.

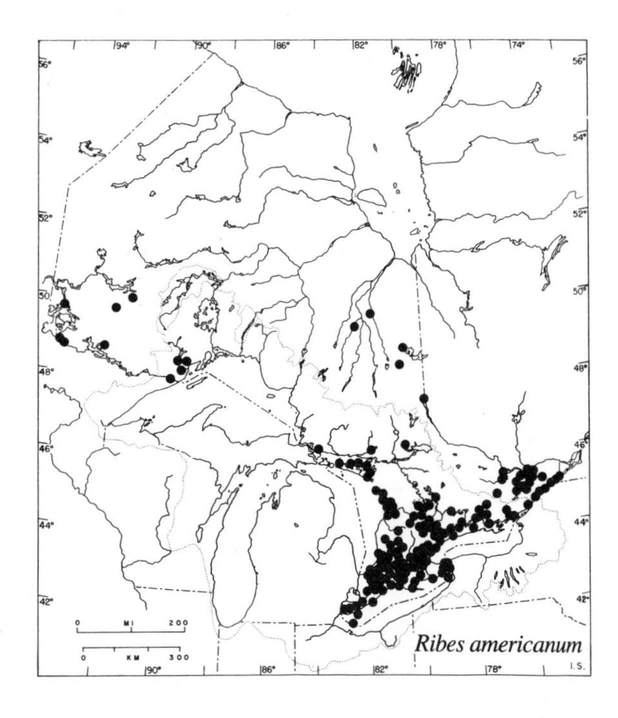

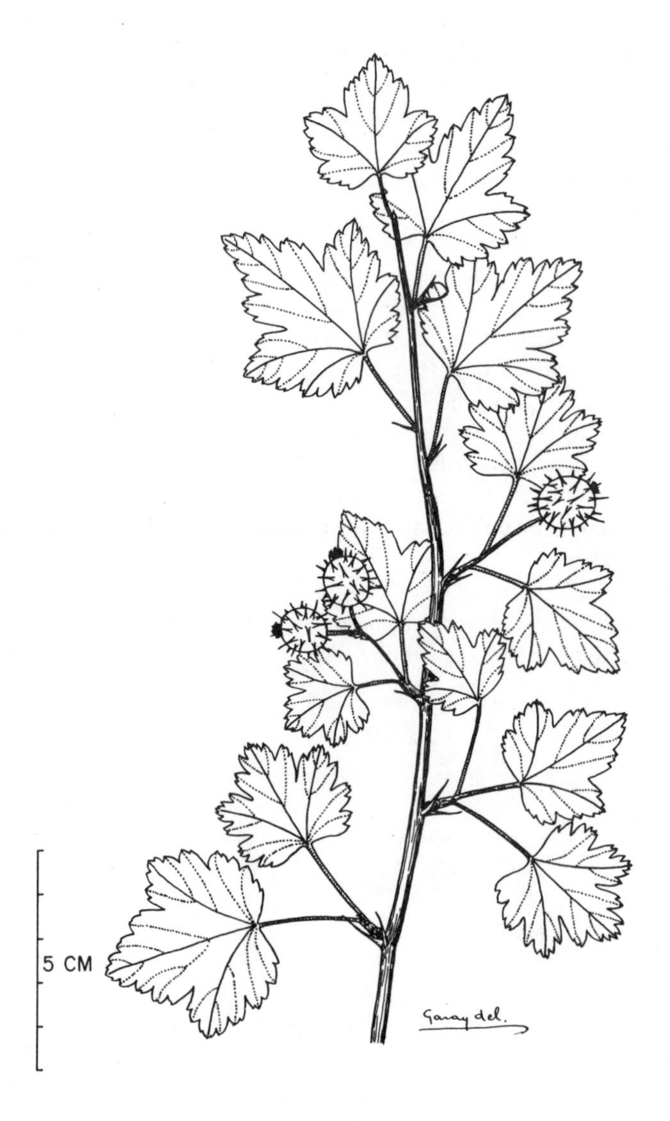

Ribes cynosbati

Ribes cynosbati L. — Prickly gooseberry

A low shrub with upright or spreading branches from 6–9 dm high. Branchlets brownish-gray, finely hairy, and with a scattering of gland-tipped hairs or weak prickles and with 1–3 slender spines at the nodes; on older stems the nodal spines become firm, about 1 cm long, and the outer bark peels off, exposing a brownish-purple to blackish inner layer.

Leaves alternate, simple, and deciduous; blades 3–7.5 cm long and about as wide, upper surface dark green and sparsely hairy, the lower surface paler and finely hairy, often with gland-tipped hairs along the veins, prominently 3- or 5-lobed, the lobes blunt or rounded at the tip, the base shallowly to deeply heart-shaped; margins with coarse rounded teeth; petioles 2.5–4 cm long, finely hairy, with scattered gland-tipped hairs, especially near the base.

Flowers bell-shaped, 6–9 mm long, greenish-yellow, in groups of 2 or 3 on slender stalks, chiefly on spur shoots of the old wood; pedicels and tiny floral bracts with gland-tipped hairs; May and early June. Fruit a reddish-purple or wine-coloured edible berry, 8–12 mm in diameter, covered with stiff, pale brown prickles or rarely smooth; July and Aug. (*cynosbati*—dog-berry)

In dry or moist woods and on rocky ground.

Common in southern Ontario but not reported from north of 47° N. (N.B. to Minn., south to Mo. and Ala.)

Note A variety with densely prickly stems, *R. cynosbati* var. *atrox* Fern., has been collected near Lisle in Simcoe Co., and in the area south of Little Current on Manitoulin Is. (*atrox*—cruel)

Field check Stems with 1–3 sharp spines at the nodes; berries large, edible, dull wine-red, with numerous, pale brown prickles.

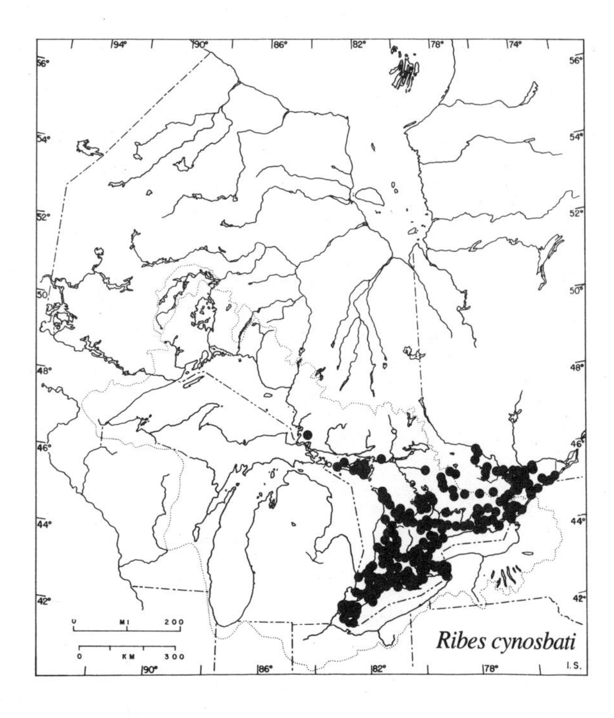

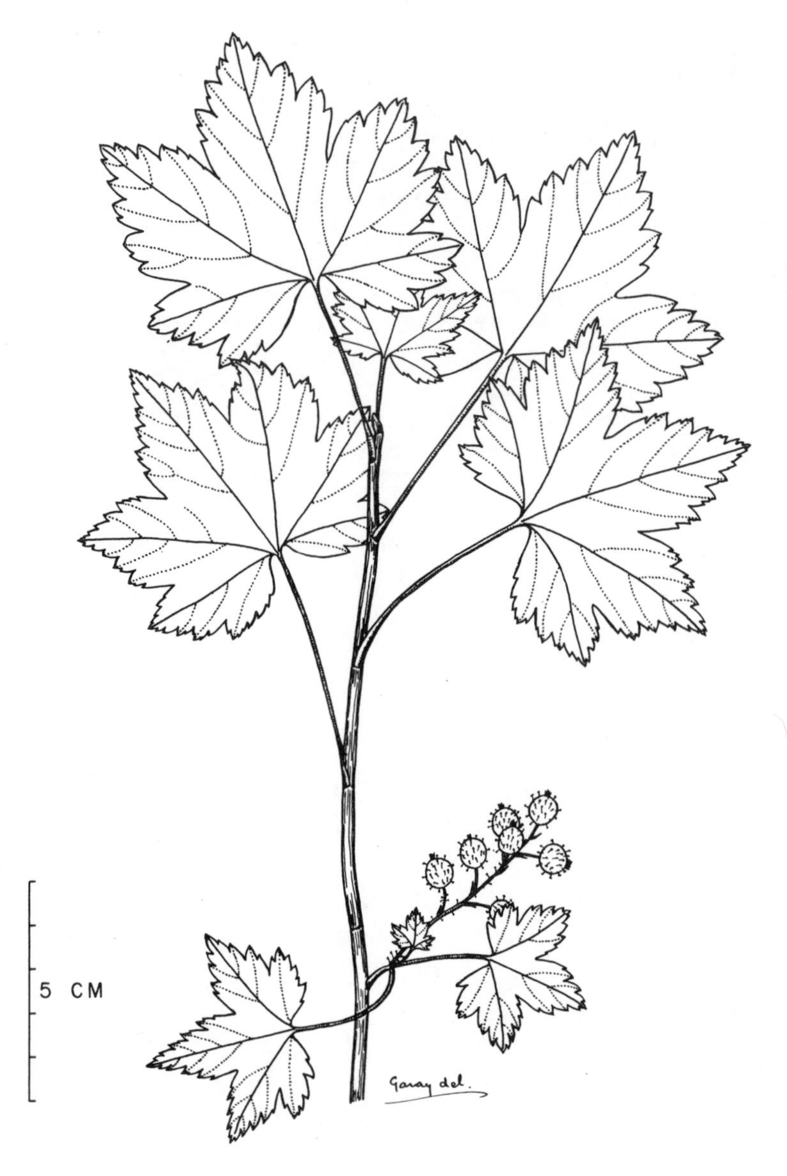

Ribes glandulosum

Ribes glandulosum Grauer **Skunk currant**

A low trailing shrub with reclining or ascending unarmed stems usually less than 1 m tall. Branchlets minutely hairy to glabrous, brown to purplish-gray, and somewhat ridged or angled, becoming smooth and blackish as the outer bark peels off; stems and foliage with a skunklike odour when bruised.

Leaves alternate, simple, and deciduous; blades 2 - 8 cm long and 4 - 8 cm wide, with 3 or 5 pointed lobes, the base deeply heart-shaped, dark green and glabrous above, paler and finely hairy beneath, at least along the veins; margins doubly serrate with rounded teeth; petioles 3 - 5.5 cm long, finely hairy, with a few gland-tipped hairs on the expanded sheathlike base.

Flowers less than 6 mm long, yellow-green to purplish, and deeply saucer-shaped in loose erect or ascending clusters 2.5 - 6 cm long, the slender flower stalks and tiny floral bracts with gland-tipped hairs; May and June. Fruit a red, glandular-bristly berry about 6 mm in diameter, disagreeable to the taste; July and Aug. (*glandulosum* — glandular)

In cool moist woods, rocky ground, along streams and shores, and in open, black spruce woods.

From Lake Ontario north to James Bay, Hudson Bay, and Lake-of-the-Woods; absent from the Deciduous Forest Region. (Nfld. to Alaska, south to n. B.C., Minn., and the mountains of N.C.)

Field check Stems reclining or ascending; all parts of the plant with skunklike odour when bruised; leaves roundish, 3 - 5-lobed; flowers and fruits in erect or ascending clusters; berries red and glandular-bristly.

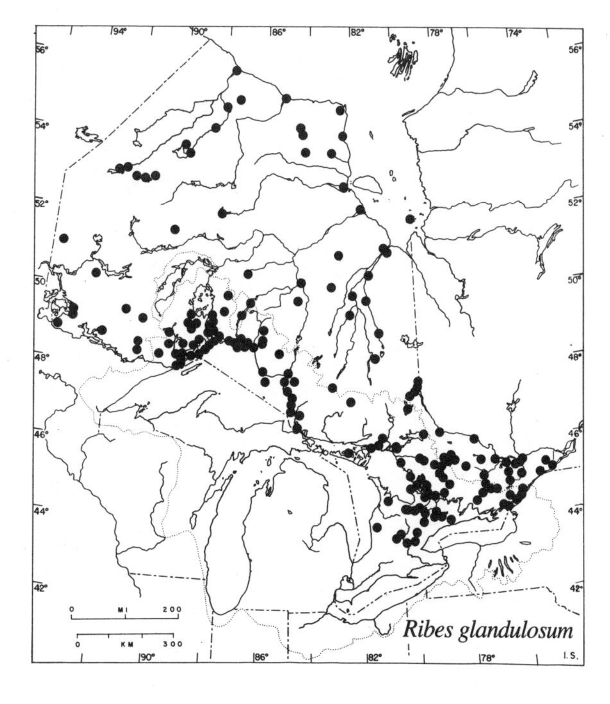

Ribes hirtellum

A low shrub with erect or ascending branches up to 9 dm high. Branchlets grayish and smooth or weakly armed with scattered bristlelike prickles and usually with 1–3 sharp slender spines 3–8 mm long at the nodes; outer pale bark soon peeling from the older stems and exposing the dark reddish-brown to blackish, smooth, inner layer.

Leaves alternate, simple, and deciduous; blades 2.5–6 cm long and about as wide, with 3 or 5 pointed lobes, the base broadly cuneate or occasionally truncate to rounded, upper surface dark green, glabrous to sparsely hairy, and without glands, the lower surface paler and hairy, at least along the veins, or in the Bruce Peninsula and locally elsewhere, densely hairy throughout; generally without glands in the southern part of Ontario; margins coarsely toothed and hairy; petioles finely hairy, 1–3 cm long, and with a broad sheathlike base, with scattered feathery hairs and often a mixture of sessile and stalked glands. In northern Ontario, plants may have stalked glands on the leaf margins near the base of the leaf and sometimes in the sinuses between the main lobes.

Flowers narrowly bell-shaped, 6–9 mm long, greenish-yellow to purplish, in groups of 2 or 3 on short, usually glabrous and glandless pedicels, the bracts with a fringe of long hairs on the margin (or in northern plants, a mixture of hairs and a few sessile or short-stalked glands), and often scattered long hairs on the outer surface of the floral tube and sepals; stamens usually well exserted (longer than the sepals) in the south; but in the north they may be slightly shorter than, or equal to, the sepals; late May and June. Fruit an edible, smooth, dark blue-black berry, 8–12 mm in diameter; July and Aug. (*hirtellum*—bristly)

In bogs and wet woods, along rocky shores and river banks, on hillsides, and in clearings.

Widely distributed from the St. Lawrence River west to Rainy River and from Lake Erie north to Hudson Bay. (Nfld. to Man., south to Minn. and W. Va.)

Note The densely hairy variant may be identified as *R. hirtellum* var. *calcicola* Fern. (*calcicola*—growing on lime)

The wild gooseberry is a source of cultivated gooseberries through selection and through hybridization with a cultivated European species, *R. uva-crispa* L. (*R. grossularia* in some manuals). (*uva-crispa*—a compound of *uva*, grape, and *crispa*, curly)

Field check Stems smooth or prickly and with spines at the nodes but bark soon peeling and exposing dark reddish-brown to blackish smooth inner bark; leaves roundish, 3–5-lobed, floral bracts with fringe of long hairs; stamens exserted (exceptions may be found in northern Ontario); berries edible, blue-black, short-stalked, without glands or prickles.

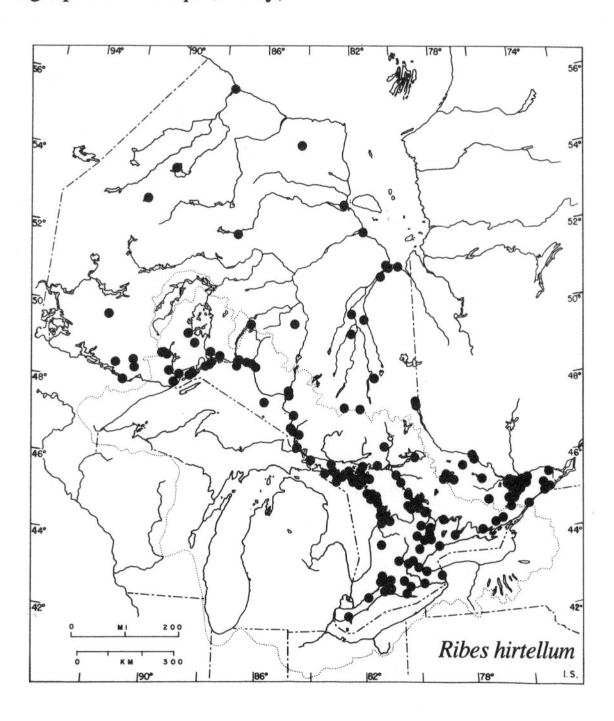

Ribes hirtellum

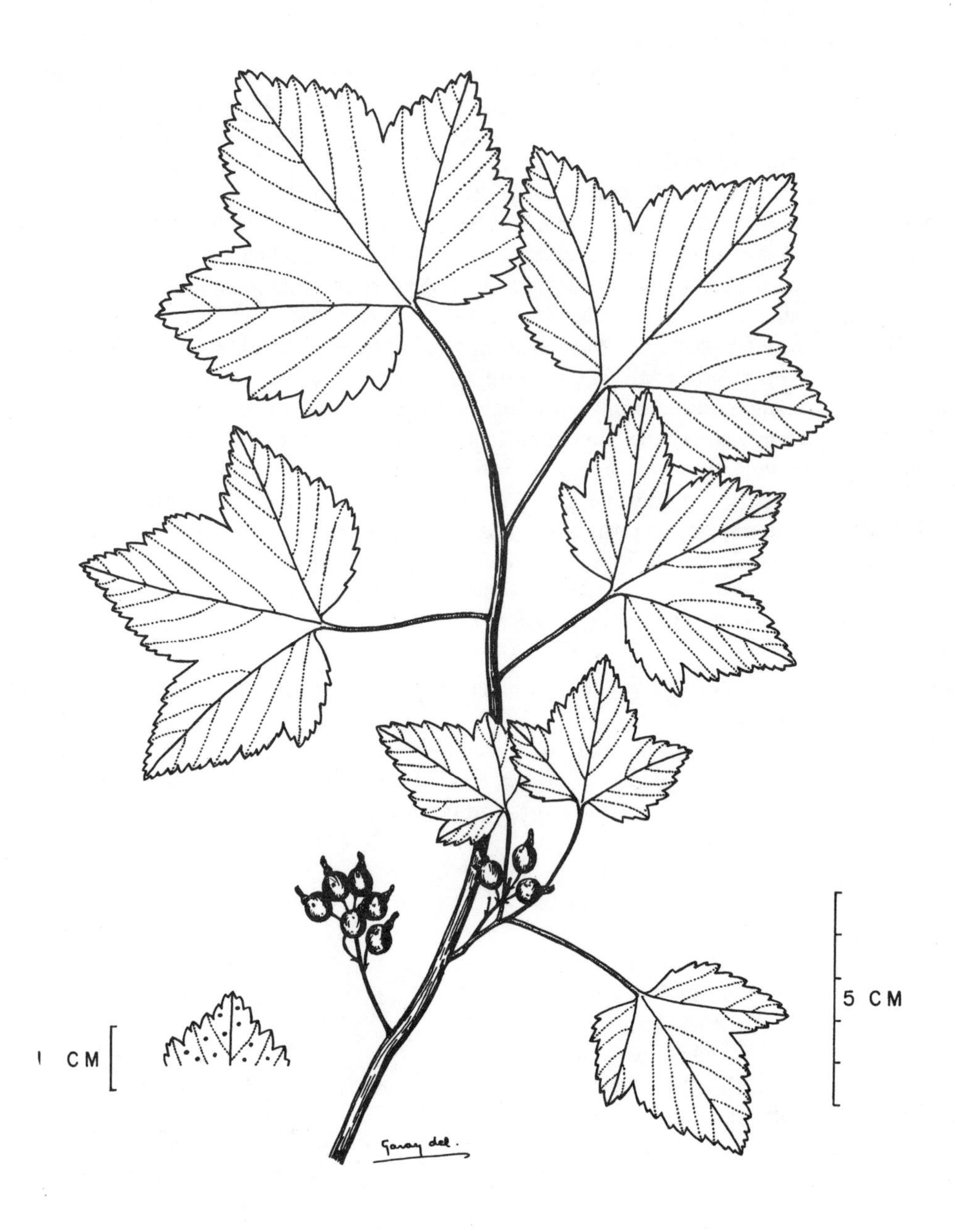

Ribes hudsonianum

Ribes hudsonianum Richards.

Northern wild black currant
Hudson Bay currant

A low shrub with stiffly ascending or upright unarmed stems reaching a height of 6 - 9 dm. Branchlets somewhat angled, minutely hairy, with scattered, yellowish, resinous dots; outer gray bark with its resinous dots peeling from the older stems and exposing a smooth, dark purplish to blackish layer.

Leaves alternate, simple, deciduous; blades dark green and glabrous or with a few scattered small hairs above, grayish to whitish and glabrous to hairy beneath, with scattered yellowish resin dots, 5 - 9 cm long and 6 - 13 cm wide, rather firm, with an unpleasant odour when crushed, 3 - 5-lobed, the lobes pointed, the base deeply heart-shaped; margins coarsely toothed with callus-tipped teeth; petioles 2.5 - 7.5 cm long, finely hairy, resin-dotted, and with a few long feathery hairs on the slightly widened base.

Flowers less than 6 mm long, bell-shaped, and whitish, on threadlike stalks in upright or ascending clusters 2.5 - 4.5 cm long; late May and early June. Fruit a blue-black berry about 9 mm in diameter, in erect or ascending clusters, scarcely edible; late July and Aug. (*hudsonianum* — of the Hudson Bay region)

In damp soil of swamps, thickets, and low woods, especially along streams and lake shores.

Rare across the central part of southern Ontario, but becoming common northward from Lake Superior to Hudson Bay. (Que. to Alaska, south to n. Calif., Oreg., and Mich.)

Field check Erect shrub; leaves broadly roundish, 3 - 5-lobed, gland-dotted on the lower surface; flowers and smooth black berries in erect clusters.

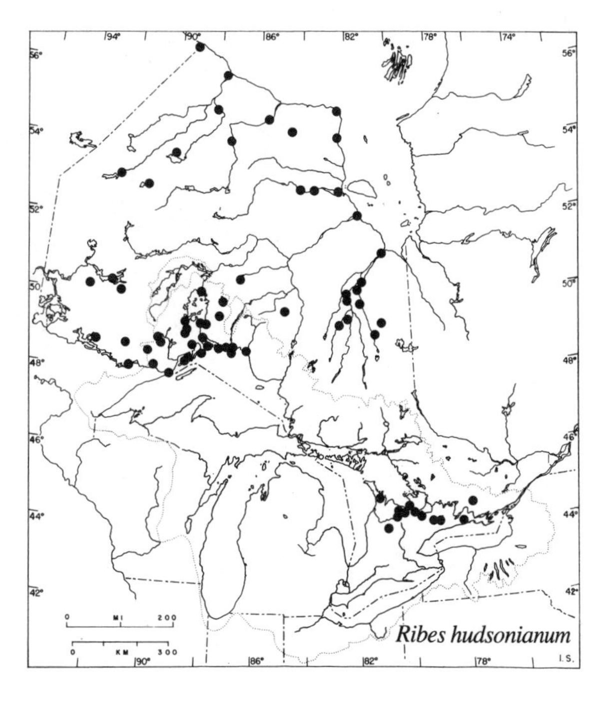

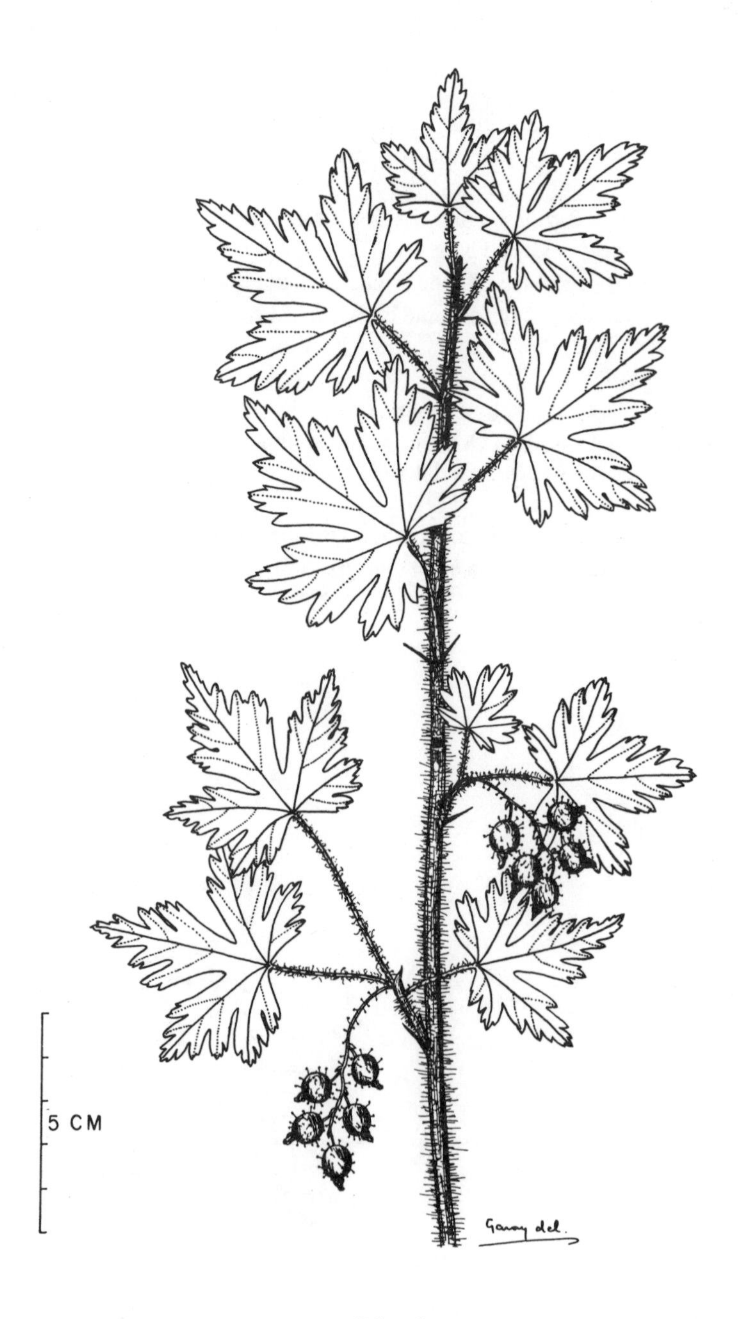

Ribes lacustre

Ribes lacustre (Pers.) Poir. — Bristly black currant

A low shrub up to 1 m in height with spreading or ascending, strongly armed branches. Branchlets with an unpleasant odour when bruised, pale brown, minutely hairy, somewhat ridged, and densely covered with sharp slender prickles, with longer spines at the nodes; bark of older stems grayish and peeling, then exposing the inner blackish layer.

Leaves alternate, simple, and deciduous; blades 4 - 8 cm long and about as wide, with 3 or 5 deeply incised and pointed lobes which are further irregularly cut into lobes with rounded teeth; upper surface dark green and nearly glabrous, the lower surface paler with scattered gland-tipped hairs, the base truncate or heart-shaped; petioles 2.5 - 4 cm long, glandular-hairy.

Flowers yellowish-green to pinkish, deeply saucer-shaped, and less than 6 mm across, in slender arching or drooping clusters; pedicels and tiny floral bracts with scattered, dark, gland-tipped hairs; May and June. Fruit a purple-black berry covered with gland-tipped hairs and disagreeable to the taste, 9 - 12 mm in diameter; late July and Aug. (*lacustre*—of lakes)

In thickets, moist woods, and swamps, on slopes, in rock crevices, and along the edges of woods.

From the St. Lawrence and Ottawa rivers across the central part of southern Ontario to Manitoulin Is., northward to Lake Superior and the shores of James Bay and Hudson Bay; absent from the Deciduous Forest Region; characteristic of the Boreal Forest and Barren Region, particularly in the Moose River drainage basin and northward along the shores of James Bay and Hudson Bay. (Nfld. to Alaska, south to Calif. and the mountains of Tenn.)

Field check Stems bristly with longer spines at the nodes; leaves roundish, with 3 or 5 deeply cut, pointed lobes and glandular petioles; berries purplish-black, glandular-hairy, and disagreeable in taste.

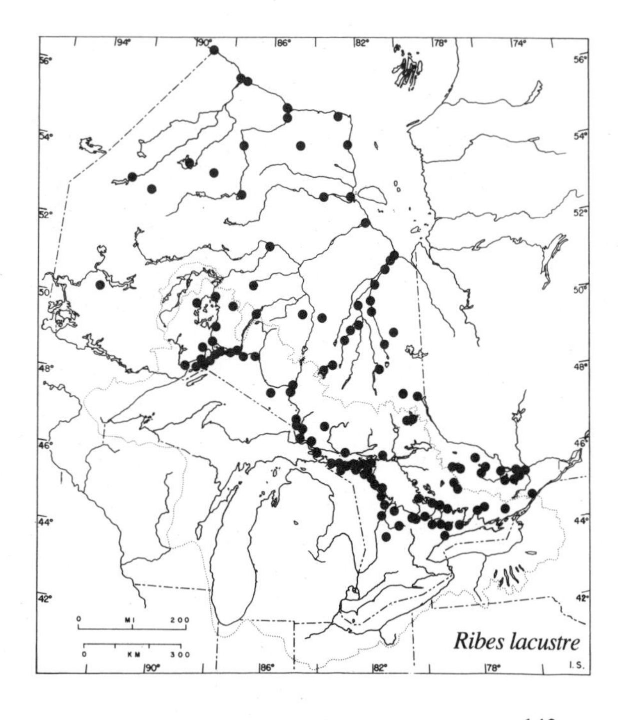

Ribes odoratum

Ribes odoratum Wendl. fil. — Golden currant, Buffalo currant, Missouri currant

A tall unarmed shrub with erect stems reaching a height of 1.5 - 2 m. Branchlets yellowish-brown to brownish-gray and minutely hairy; older twigs purplish-gray to blackish, glabrous, somewhat ridged, and becoming roughened with small sections of the outer bark peeling off in shreds.

Leaves alternate, simple, and deciduous; blades broadly fan-shaped, usually deeply 3-lobed, 1 - 4 cm long and about as wide, firm, dull above and paler beneath, at first minutely hairy, later becoming glabrous, both surfaces copiously resin-dotted in some specimens but in others lacking the resin dots, lobes entire or with a few coarse teeth at their tips, base wedge-shaped; margins with a fringe of fine hairs; petioles about as long as the blades, minutely hairy.

Flowers in leafy clusters on spur shoots on the old wood, golden-yellow, and spicy-fragrant, trumpet-shaped, about 15 mm long, the tube of the hypanthium two to three times as long as the widely spreading oval lobes of the calyx; floral bracts elliptic to ovate with hairy margins, often persisting after flowering; early June. Fruit a smooth blackish berry 6 - 9 mm in diameter; July and Aug. (*odoratum*—fragrant, in reference to the fine spicy fragrance)

In fields, along roadsides and fencerows.

Local from Lake Erie to the Bruce Peninsula and eastward to the Ottawa district. (Introduced from midwestern U.S.A. and escaped from cultivation)

Field check Unarmed upright shrub, leaves small, fan-shaped, 3-lobed, entire or with a few teeth at the ends of the lobes; flowers golden-yellow, spicy-fragrant; berries black.

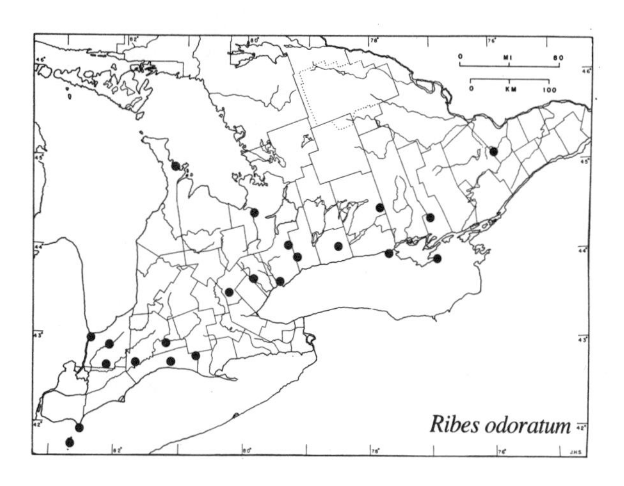

Ribes oxyacanthoides

Ribes oxyacanthoides L. — Bristly wild gooseberry

A low shrub with erect or stiffly ascending and usally armed branches up to 1 m in height. Branchlets brownish to grayish, finely hairy, and sometimes glandular, most frequently with scattered prickles, and with 1-3 stout spines up to 1 cm long at the nodes; outer bark close and not peeling for several years, the stems thus remaining prickly longer than in *R. hirtellum*.

Leaves alternate, simple, and deciduous; blades 2.5-5 cm long and about as wide, with 3 or 5 lobes, the lobes blunt or rounded, the base rounded, truncate, or heart-shaped, upper surface glabrous to sparsely hairy and usually with scattered stalked glands, the lower surface sparsely to densely hairy, with stalked glands, at least along the veins, resin-dotted between the veins; margins coarsely toothed and hairy, often with scattered glands; petioles 0.5-3 cm long, finely hairy, with a broad sheathlike base, scattered feathery hairs, and sometimes also a mixture of sessile and stalked glands.

Flowers greenish-yellow, bell-shaped, 6-9 mm long in groups of 2 or 3 on short stalks barely exceeding the bud scales; pedicels usually hairy or glandular, the floral bracts glandular-hairy or with a fringe of glands on the margin; ovary, floral tube, and sepals without hairs or glands; stamens not exserted, being shorter than the sepals but mostly longer than the petals; June. Fruit an edible, smooth, blue-black berry, 9-12 mm in diameter; late July and Aug. (*oxyacanthoides* —like *oxyacantha*, which is the name of a European hawthorn with leaves rather like those of *Crataegus monogyna*. The suitability of the name can be checked by comparison of leaves of the bristly wild gooseberry and those of *Crataegus monogyna*, which is found in Ontario.)

On rocky and sandy shores, talus slopes, and stony banks and in clearings.

From the north shore of Lake Superior to Lake-of-the-Woods and northward; also around James Bay and in the Severn River drainage basin. (James Bay to Yuk., south to Mont. and n. Mich.)

Field check Stems with close prickly bark and stout spines at the nodes; bark peeling tardily; leaves broadly roundish, 3-5-lobed, with stalked glands at least along the veins on lower surface; floral bracts with glands along the margins; stamens shorter than the perianth; berries on very short stalks, edible, and without glands or prickles.

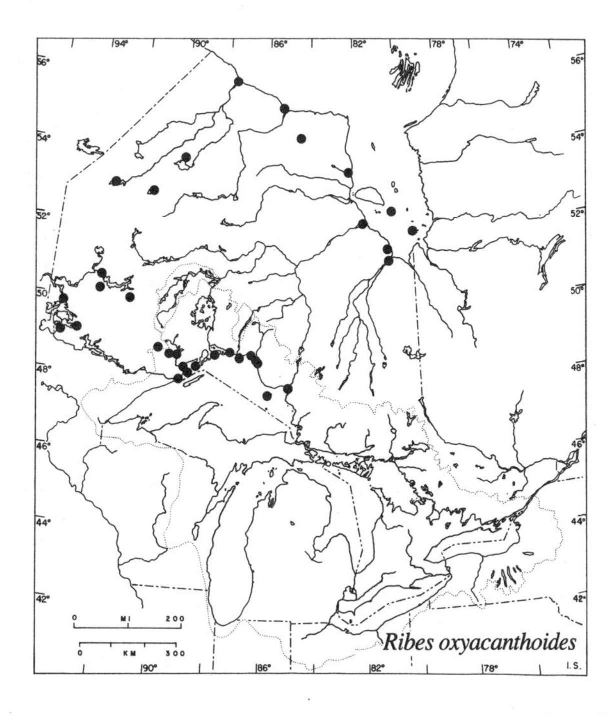

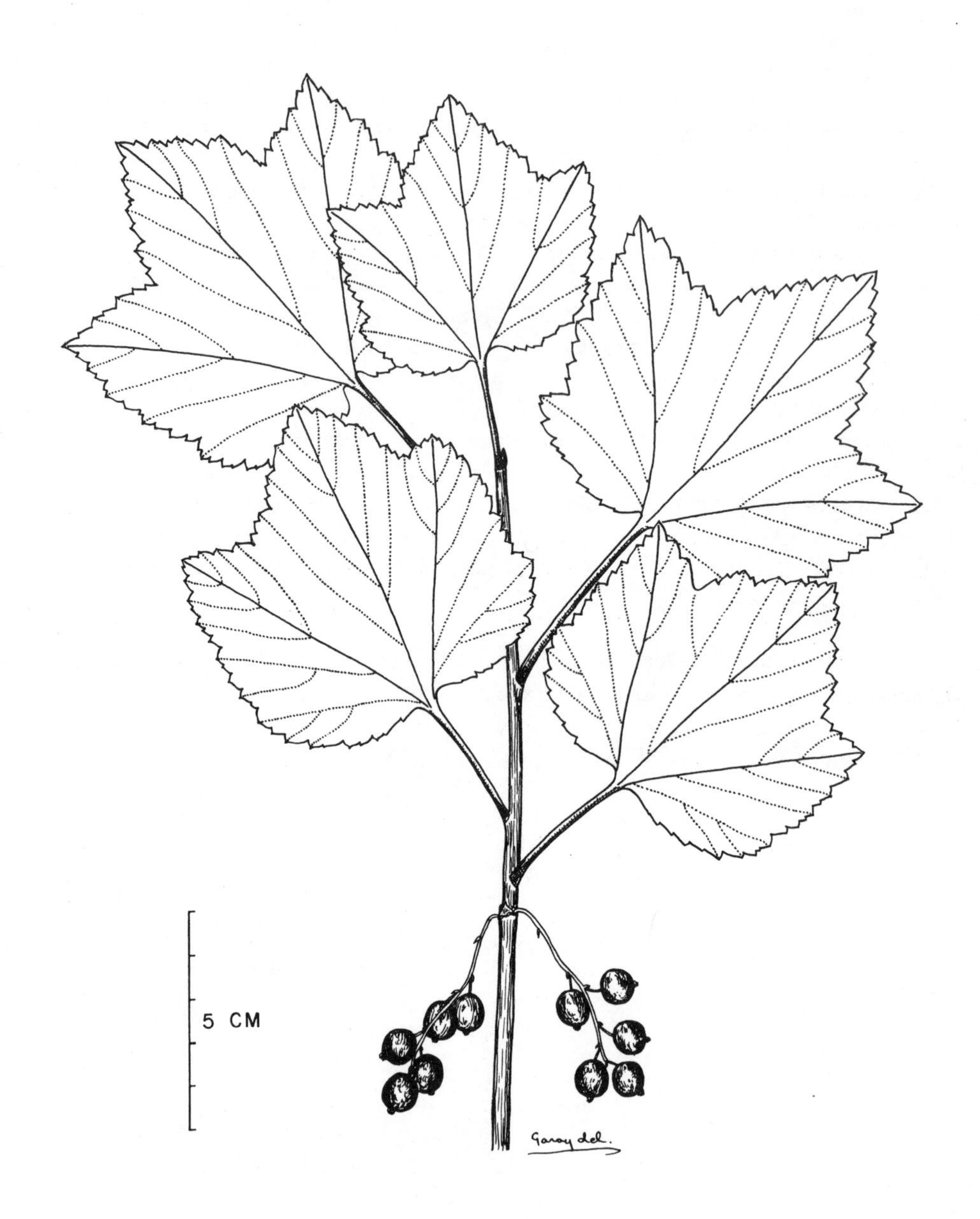

Ribes triste

Ribes triste Pall.

Wild red currant
Swamp red currant

A low, spreading or reclining, unarmed shrub seldom more than 1 m high, with stems often rooting at the nodes. Branchlets somewhat ridged, minutely hairy, gray to brownish; older stems becoming glabrous and reddish-purple to blackish after the outer bark peels off.

Leaves alternate, simple, and deciduous; blades 4 - 10 cm long and 5 - 10 cm wide, with 3 or 5 broad lobes, the lobes pointed or rounded, the base heart-shaped or truncate, dark green and nearly glabrous above, paler and usually hairy beneath, at least along the veins; margins with numerous rounded and abruptly pointed callus-tipped teeth; petioles 2.5 - 6 cm long, sparsely hairy, and with a few gland-tipped hairs.

Flowers less than 6 mm in diameter, greenish-purple, on slender pedicels bearing scattered gland-tipped hairs, in arching or drooping clusters 2.5 - 7.5 cm long, from scaly lateral buds, at first partially hidden under the tufts of expanding leaves; late May and early June. Fruit a smooth, bright red berry 6 - 9 mm in diameter, in drooping clusters; July and Aug. (*triste*—sad, possibly in reference to the drooping flower clusters)

In deciduous, mixed, or coniferous woods, alder and cedar swamps, spruce bogs, and on damp soil along stream banks.

Throughout Ontario from Lake Erie to Hudson Bay and from the Ottawa River to Lake-of-the-Woods. (Nfld. to Alaska, south to Oreg. and W. Va.)

Note The red currant cultivated in gardens is *R. rubrum* L., a species of European origin, which differs in its upright stems, yellowish or greenish flowers, absence of gland-tipped hairs on the flower stalks, and larger, juicier, red berries. It escapes infrequently to open woods and thickets. (*rubrum*—red)

Field check Leaves broadly roundish with 3 or 5 broad lobes; stems reclining and rooting at the nodes; berries red and smooth, tart.

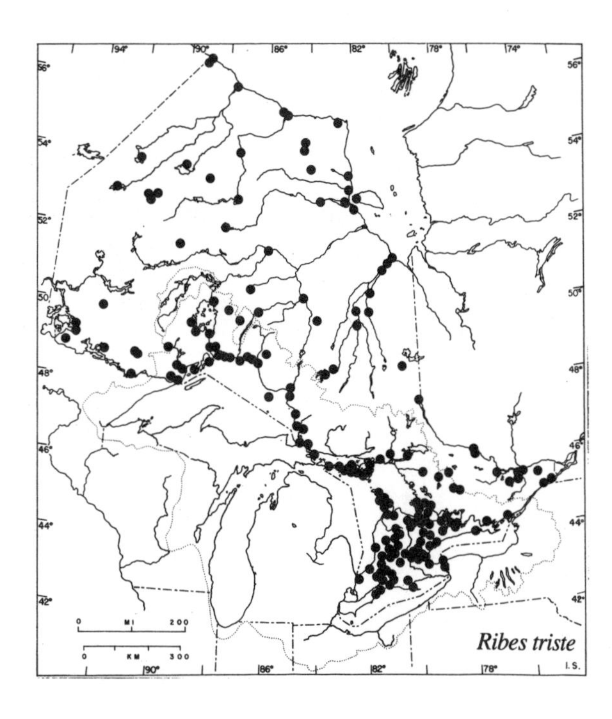

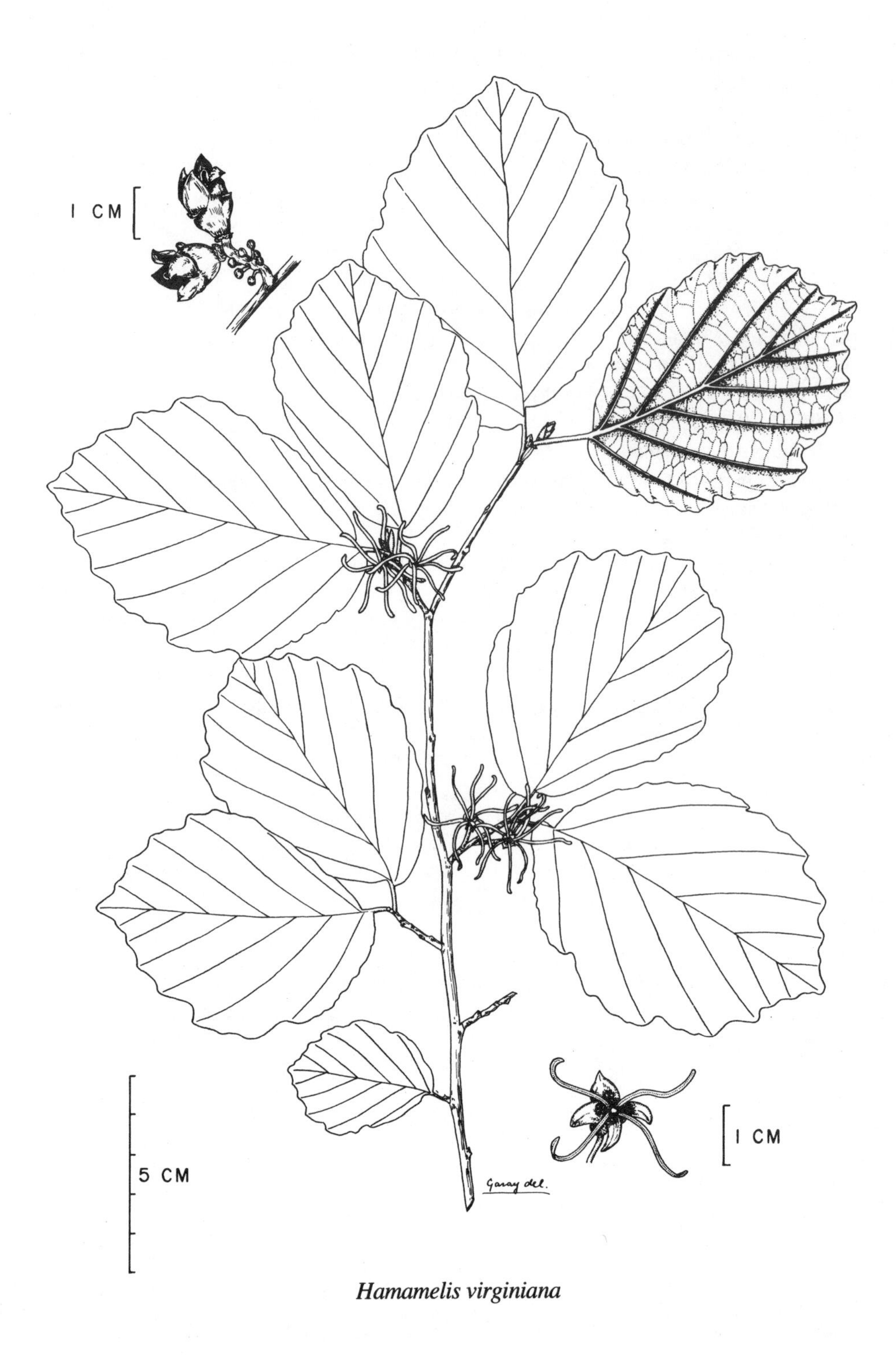

Hamamelis virginiana

HAMAMELIDACEAE — WITCH-HAZEL FAMILY

A small family of about 20 genera and fewer than 100 species which are chiefly subtropical or warm temperate with their best development in China and neighbouring parts of eastern Asia. Trees and shrubs with alternate simple leaves, deciduous stipules, and flowers arranged in heads or spikes. A single genus is native in Ontario.

Hamamelis L. — Witch-hazel

This genus has 5 or 6 species which are native in eastern North America and eastern Asia. They are tall shrubs or small trees with straight-veined alternate leaves and mostly yellow flowers. (*Hamamelis* — an ancient Greek name for the medlar or some similar tree with applelike fruit)

Hamamelis virginiana L. — Witch-hazel

A large spreading shrub or small tree 4.5 - 6 m in height. Branchlets slender with more or less rusty-stellate pubescence; buds naked and stalked, the terminal one flattish, slightly curved, and covered with light brown hairs; bark gray-brown.

Leaves alternate, simple, and deciduous; blade broad and rounded, oval, or obovate, 7 - 15 cm long and 3 - 10 cm wide, the base obviously asymmetrical, i.e., with the leaf tissue joining the petiole farther up one side than on the other; margins irregularly wavy; petioles short but distinct.

Flowers perfect and usually in groups of 3, opening in the autumn about the time the leaves are falling, the 4 very narrow, pale yellow, ribbonlike petals of each flower about 1 - 2 cm long and crumpled or twisted so that the flower appears ragged; Sept. and Oct. Fruit a pale brown, woody capsule about 1 cm long, urn-shaped and two-beaked with a prominent ring around the middle formed by the adherent calyx; capsule maturing a year after fertilization of the flower and when fully ripe splitting open at the top. Two shiny, black, slippery seeds are shot out from the capsule for a distance of several meters, as a fresh apple seed may be shot from between the finger and thumb, the empty capsules remaining on the shrub for another season. (*virginiana* — Virginian)

In dry, well-drained, sandy situations of open woods, edges of woods, and slopes of ravines.

Common in the Deciduous Forest Region from Grand Bend to Toronto and also along the St. Lawrence River; less frequent northwards to about 45°30′ N; along both sides of the Ottawa River within the Ottawa district. (N.S. to Minn., south to Mo., Tenn., and Ga.)

Note The yellow flowers appearing in autumn with the yellowing leaves and maturing capsules make witch-hazel an attractive ornamental for large gardens.

The bark, leaves, and branches yield a volatile oil which is used in pharmaceuticals.

Field check Alternate broadly roundish leaves with irregular wavy margins and an asymmetrical base; yellow flowers with 4 slender, ribbonlike petals, in the fall; woody and persistent capsules ejecting shiny black seeds.

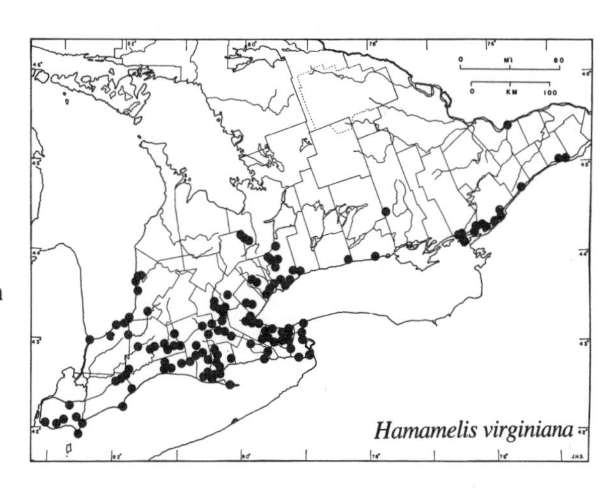

ROSACEAE — ROSE FAMILY

About 100 genera and 3000 species of wide distribution, mainly in north temperate regions. Trees, shrubs, and perennial herbs with leaves alternate, simple or compound, and usually bearing stipules. Thorns and prickles are sometimes present and some groups have vegetative reproduction by runners, stolons, or underground suckers. The flowers are mostly 5-parted, perfect, and regular, but they display great diversity in structure; borne singly or in various types of inflorescences, often in terminal, racemose or cymose clusters. The calyx may have an additional outer set of bractlets (e.g., *Potentilla*) and varies from saucer-shaped to cup-shaped or urn-shaped; the petals and stamens are inserted on the edge of the hypanthium; the stamens are usually numerous, often as a multiple of the number of petals; the pistil may consist of one, or of several to many, carpels and the fruit may be a dry and dehiscent follicle or capsule, an indehiscent achene, a fleshy pome or drupe, or an aggregation of achenes or drupelets often accompanied by, or imbedded in, fleshy tissues developed from the hypanthium or receptacle.

Because it is such a large family and exhibits great diversity in the structure of flower and, especially, of fruit, the Rosaceae is usually divided into tribes or subfamilies and, by some botanists, these are even regrouped into three separate families. The family is noted for the perplexing problems encountered in attempts to identify species in a number of genera including, among our Ontario shrubs, *Amelanchier*, *Crataegus*, *Rosa*, and *Rubus* in particular. Such problems are by no means confined to the rose family, and though in this case they are now known to be due to hybridization, polyploidy, and apomixis, their solution remains elusive.

A large number of representatives of this family are familiar as the source of our most prized ornamentals and favourite fruits, including various Ontario shrubs such as roses, meadowsweets, chokeberries, cherries, raspberries, ninebark, cinquefoils, and hawthorns.

Amelanchier Medik. — Serviceberry, Shadbush, Juneberry

Small trees or shrubs, sometimes growing in clumps (surculose) as a result of suckering or in continuous patches (colonial) by the spreading of underground stolons. Stems unarmed and usually slender. Leaves alternate and simple, the margins serrate with fine to coarse teeth. Flowers white, 5-parted, often in showy racemes. Fruit a berrylike pome, sometimes juicy and edible; seeds 5–10. (*Amelanchier*—a name said to be derived from a common name of a European species, possibly the French word *amelancier* used in southern France)

Note Intergrading morphological series connect even the most distinctive of the serviceberries, supporting the concept that natural hybridization is not only frequent but that the offspring are relatively fertile. Thus, authors may differ considerably in their treatment, and decisions concerning many specimens may diverge considerably, as can be observed in any herbarium collection. (For another treatment of the serviceberries of Ontario, see S.M.McKay, 1973.)

Key to *Amelanchier*

a. Flowers solitary or 2–4 from the leaf axils; summit of ovary conical; sepals erect in fruit **_A. bartramiana_**

a. Flowers mostly 5–20 in short or elongate racemes; summit of ovary flat or slightly rounded; sepals spreading or reflexed in fruit

- b. Stems solitary or in clumps; flowers in long, lax, drooping racemes
  - c. Summit of ovary glabrous; tall erect shrubs or small trees; branchlets grayish
    - d. Leaves about half grown at flowering time, mostly glabrous, often reddish when young; lowest fruiting pedicel 2.5–5 cm long; fruit sweet and juicy **_A. laevis_**
    - d. Leaves scarcely unfolded at flowering time, pubescent to white-tomentose beneath; lowest fruiting pedicel 1–2.5 cm long; fruit dry and mealy, flavourless **_A. arborea_**
  - c. Summit of ovary pubescent; arching or straggling shrub with a few stems or a solitary stem or somewhat stoloniferous; young branchlets reddish **_A. sanguinea_**
- b. Stems usually numerous, growing in spreading colonies or patches from rhizomelike bases; flowers in short, dense, erect racemes
  - e. Apex of leaf rounded to subacute, mucronate; margins finely and sharply serrate (5–8 teeth per cm); veins irregular, unequally distant, usually with frequent, intermediate, shorter ones **_A. stolonifera_**
  - e. Apex of leaf broadly rounded to truncate or retuse; margins coarsely dentate (2–6 teeth per cm); veins conspicuous, usually straight, parallel, and close together, intermediate ones few or none
    - f. Margins of the leaf forming an angle at the apex; leaves unfolding with the flowers; sepals short-triangular or ovate, 2–4 mm long **_A. humilis_**
    - f. Margins of the leaf forming a rounded or subtruncate, rarely retuse apex; leaves flat and more than half grown at flowering time; sepals triangular-lanceolate, 1.5–2.5 mm long **_A. alnifolia_**

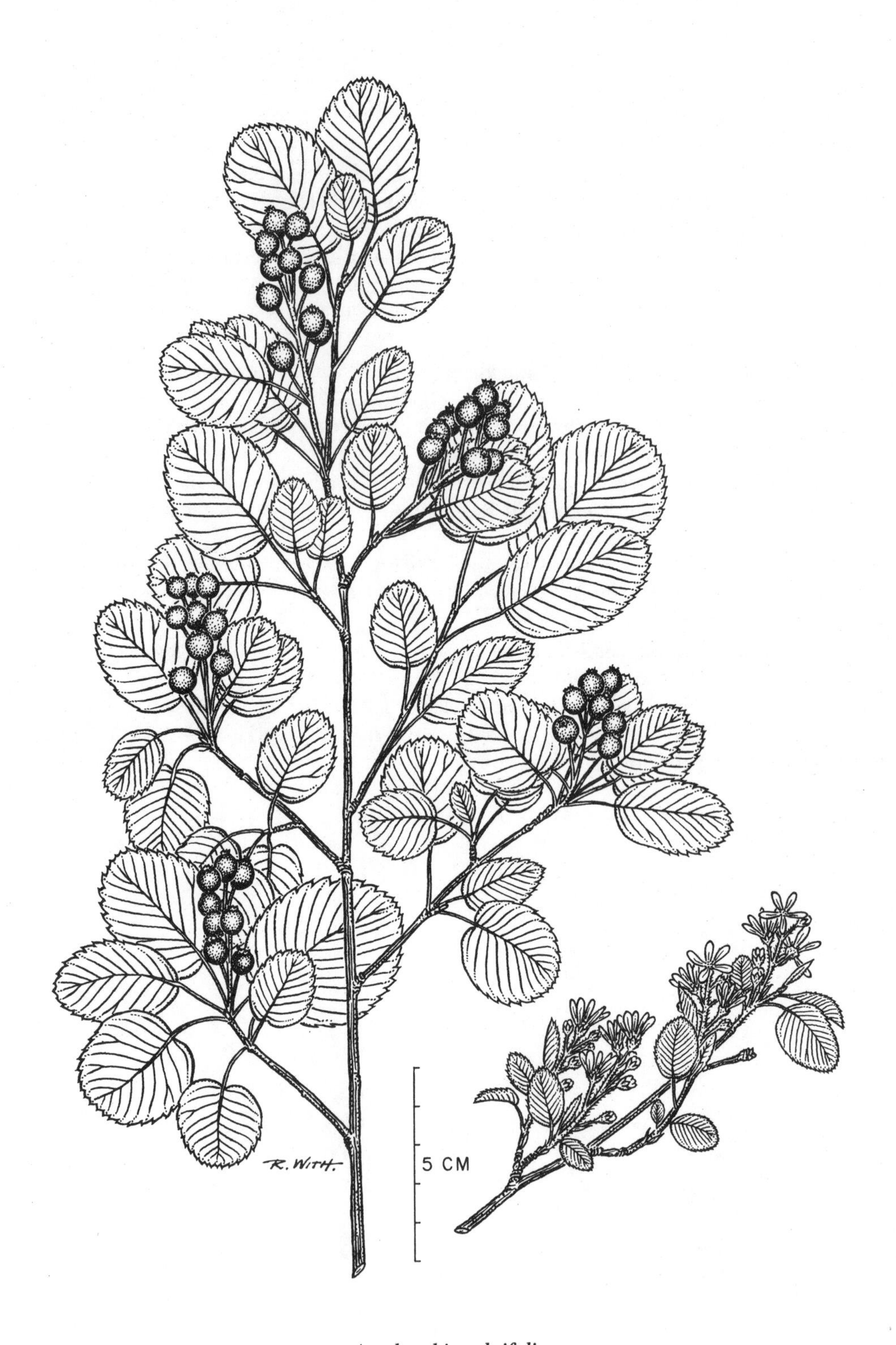

Amelanchier alnifolia

Amelanchier alnifolia Nutt.

Saskatoon berry
Serviceberry

A small to medium-sized shrub, spreading by stolons to form colonies, and varying in height from 1 to 4 m. Branchlets reddish-brown and more or less silky-pubescent at first, later becoming smooth; bark of older stems gray.

Leaves alternate, simple, and deciduous, tomentose beneath when young, usually unfolded, and more than half grown at flowering time; mature blades oval to roundish or somewhat quadrangular, blunt, rounded or truncate at the apex, and rounded, truncate, or subcordate at the base, 2-5 cm long, 1.5-4 cm wide, dark green and smooth above, paler and smooth beneath or very finely hairy along the lower part of the midvein; margins coarsely dentate above the middle, the lower one-third to one-half usually entire, the teeth ovate, acuminate, somewhat incurved, 2-5 per cm; veins conspicuous, 8-13 pairs; petioles slender, 8-18 mm long, hairy when young but becoming glabrous.

Flowers white, fragrant, 9-13 mm across, conspicuous, in dense erect racemes 3-6 cm long; rachis and pedicels covered with woolly pubescence; lower pedicels 6-10 mm long; petals 6-10 mm long; sepals triangular-lanceolate, becoming reflexed in age; June. Fruit edible, 6-10 mm in diameter, blue-purple to nearly black when ripe, usually sweet and juicy; July and August. (*alnifolia*—with leaves like *Alnus*, the alder)

In sandy thickets and borders of woods; also along rocky ridges and banks of streams.

Rare; apparently spreading or introduced from the prairies eastward into the northern and western parts of Ontario and along the north shore of Lake Superior. (n. Ont. to Yuk., south to Oreg., Colo., and Neb.)

Note It has been difficult to determine the status of this typically western shrub in Ontario and many specimens so named may as easily fit into the eastern *A. humilis*, which is very similar in habit and other characteristics. *A. alnifolia* may be distinguished from *A. humilis* by its usually broadly truncate or subtruncate leaves, the blade broadly oval or oblong-oval, the leaves fully unfolded but not full grown at flowering time, soon becoming glabrous, and also by the shorter lower pedicels of the racemes.

Field check Stoloniferous shrub forming colonies; leaves oval to roundish, blunt to truncate, and coarsely toothed above the middle, teeth 2-5 per cm; leaves well developed at flowering time; flowers conspicuous in short, dense, erect racemes; fruit edible, blackish, sweet, and juicy.

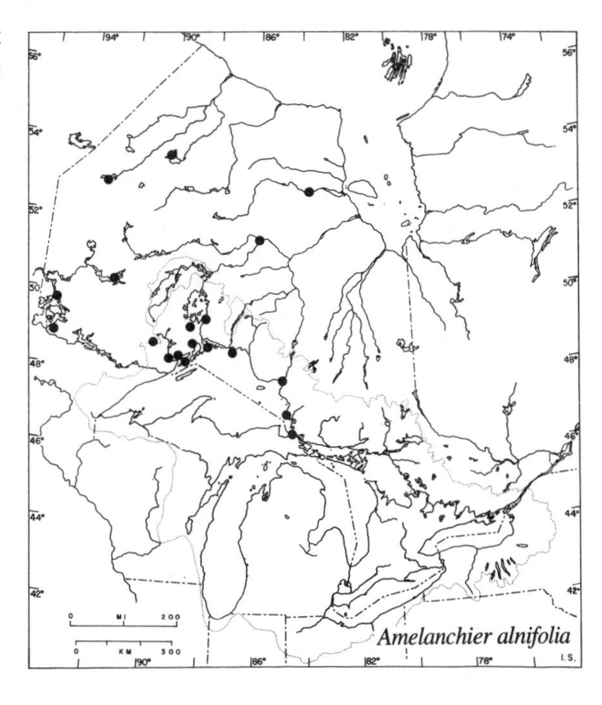

Amelanchier arborea

Amelanchier arborea (Michx. fil.) Fern. — Serviceberry, Shadbush Juneberry

A tall shrub or small tree, often with several stems in a clump, reaching a height of 10 m or more. Branchlets purplish when young, the bark of older stems and trunks light to dark gray and rather smooth.

Leaves alternate, simple, and deciduous, small, folded, and whitish-pubescent beneath at flowering time; fully developed blades oblong-ovate to slightly obovate, 3 - 8 cm long and 2 - 4 cm wide, acuminate or acute at the apex, rounded to cordate at the base, essentially glabrous on both surfaces or sparingly pubescent beneath; margins finely serrate with numerous sharp teeth (more than twice as many as lateral veins); petioles 1 - 2.5 cm long, finely hairy.

Flowers appearing with the leaves in April and May in elongate drooping racemes; rachis and pedicels silky-hairy; pedicels 5 - 20 mm long, subtended by slender, reddish, silky-hairy bracts; petals white, 10 - 15 mm long, narrowly oblong to cuneate. Fruit dark reddish-purple, dry, and insipid; June and July. (*arborea*—treelike)

Open fields, fencerows, edges of woods, thickets, wooded slopes, sandy bluffs, and rocky ridges.

Widely distributed in southern Ontario from the Ottawa and St. Lawrence valleys to Lake Erie, Lake Huron, and the eastern end of Lake Superior; not reported from the Algonquin Park highlands. (s. N.B. and Maine to Minn., south to Okla. and n. Fla.)

Field check Large shrub in clumps, or tree; leaves up to twice as long as broad, broadest below the middle, margins with many fine teeth; flowers white, in drooping racemes (April, May); fruit dry and insipid (June, July); leaves folded and pubescent beneath at flowering time, glabrous at maturity.

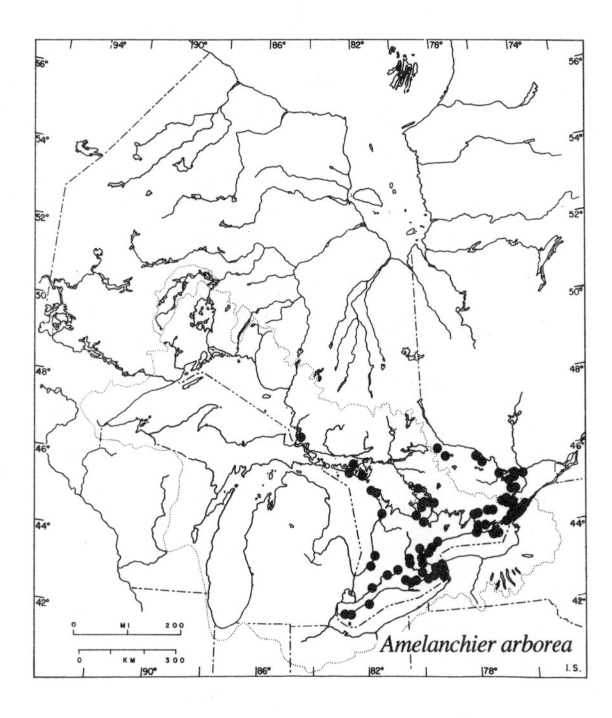

Amelanchier bartramiana

Amelanchier bartramiana (Tausch) Roem. **Oblong-fruited juneberry Mountain juneberry**

A slender shrub about 2 m tall, often forming clumps. Branchlets purplish, glabrous or essentially so.

Leaves alternate, simple, deciduous; blades ovate, elliptic, or somewhat obovate, 1.5 - 5 cm long, 1 - 2.5 cm wide, thickish, green above, paler beneath, often tinged with purplish-red when unfolding, glabrous on both surfaces, the tip blunt to acute, often mucronate, the base tapered; margins finely and closely sharp-toothed nearly to the base; petioles 2 - 10 mm long, glabrous or slightly silky.

Flowers few (1 - 4) at the ends of the short branches or a solitary terminal flower and several others from the upper leaf axils on glabrous pedicels which become 1 - 2 cm long in fruit; petals white, elliptic to oblong-oval, 6 - 10 mm long; May and June. Fruit pyriform, 1 - 1.5 cm long, becoming purple-black, edible; July and August. (*bartramiana*—named for William Bartram, 1739 - 1823, a pioneer collector of seeds and plants in the southeastern United States)

Chiefly on acid soils; sandy lake shores, stream banks, peat bogs, swamps, boggy thickets, and rocky ridges.

Common from Lake Superior to the Hudson Bay lowlands; less common in southern Ontario and rare off the Canadian Shield. (Nfld. to w. Ont., south to Minn. and Mass.)

Note Collections made in 1891 and 1893 by J.A.Morton, labelled Wingham (Huron County), have not been plotted, as confirmation is lacking for the occurrence of this species in that part of Ontario.

Field check Leaves more or less elliptic and twice as long as wide, margins finely and closely serrate; flowers and fruits solitary or few, terminal or in upper leaf axils; in acid habitats.

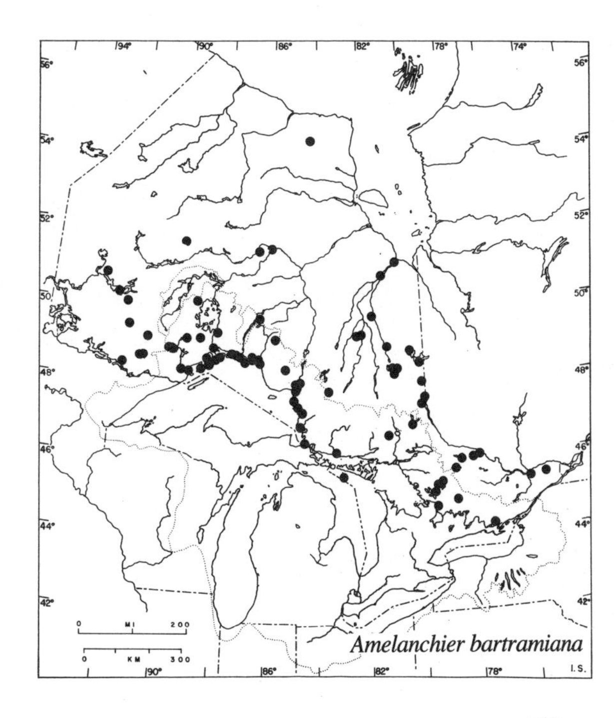

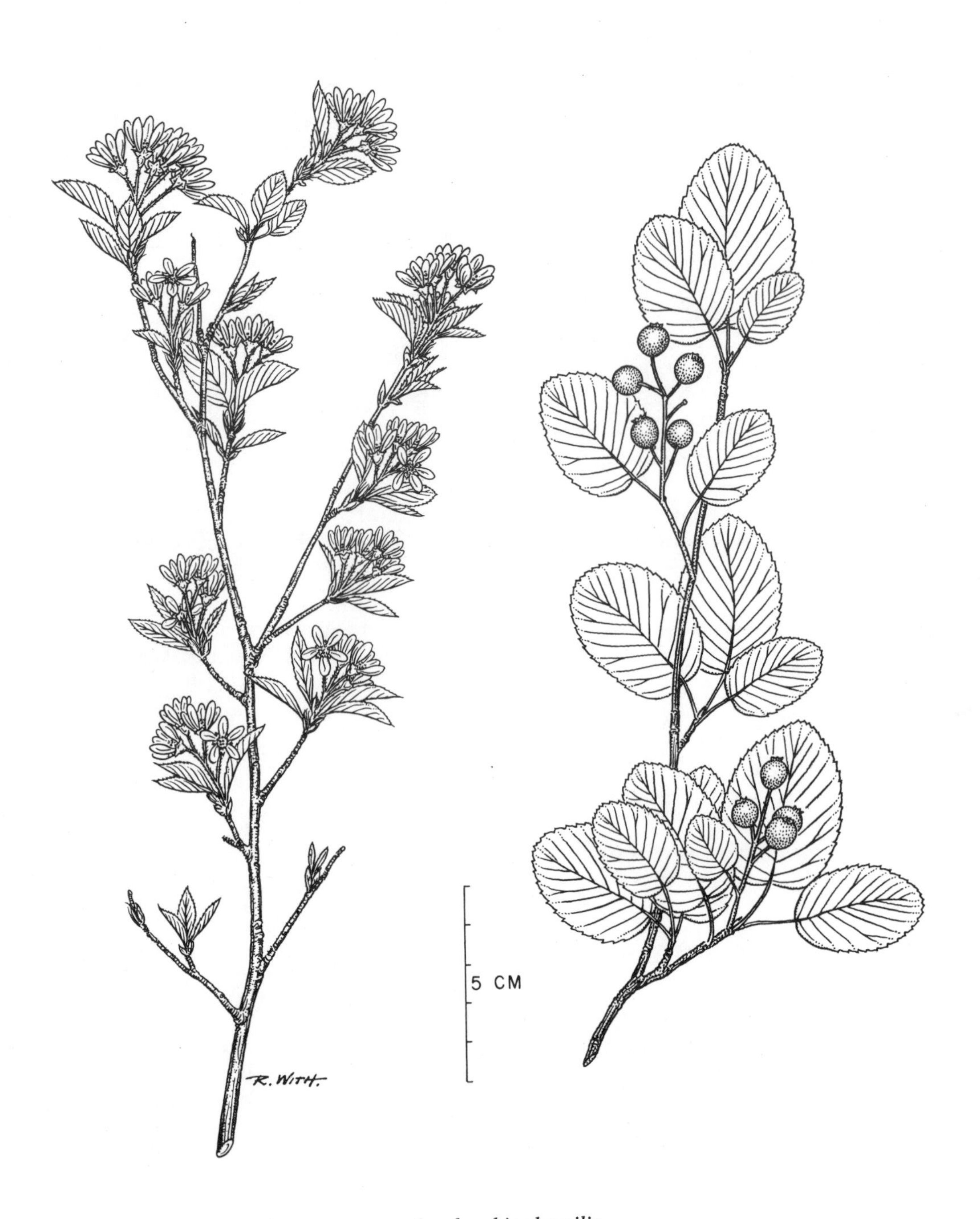

Amelanchier humilis

Amelanchier humilis Wieg.

Serviceberry, Juneberry Shadbush

A straggling or much-branched and stiffly upright shrub spreading by stolons to form patches and loose colonies, the individual stems scattered; often low (0.3 - 1.2 m), especially on exposed sites, but sometimes reaching a height of 5 - 6 m. Young branchlets densely silky-hairy, smooth, and eventually gray.

Leaves alternate, simple, and deciduous, green above when young, at first pale yellowish- to grayish- or whitish-tomentose beneath; mature blades oblong-oval to elliptic, 2.5 - 5 cm long, 2 - 4 cm wide, green and glabrous above, paler and glabrous or sometimes permanently tomentose beneath, the apex broadly rounded to subacute, the base rounded to cordate; margins somewhat coarsely serrate-dentate to below the middle with 4 or 5 (rarely 6) teeth per cm; veins conspicuous, 7 - 13 pairs, most of the veins or their branches extending into the teeth; petioles 8 - 20 mm long.

Flowers appearing with the unfolding leaves in May and June, white, small, and numerous, in dense terminal and lateral upright racemes 4 - 5 cm long; axis of the inflorescence and the pedicels silky-hairy, lower pedicels 8 - 17 mm long; petals 7 - 10 mm long; sepals triangular-lanceolate or ovate, 2 - 4 mm long, revolute from the middle when the petals fall, usually hairy on both sides; ovary densely woolly at the summit. Fruit globose to ellipsoid, almost black with a bloom, juicy, and sweet; late July and August. (*humilis* — low)

On limestone flats, lakeshores, river banks, gravel fans, rocky hillsides, ridges, and cliffs and in sandy areas, clearings, semi-open woods, and ravines.

Widely distributed throughout Ontario, the common dwarf species in calcareous habitats. (Que. to Man., south to S. Dak. and Pa.)

Field check Stoloniferous shrub forming patches; leaves unfolding with the flowers; blades more or less elliptic and little longer than broad; margins with 4 - 6 teeth per cm; fruit edible, blackish, sweet, and juicy.

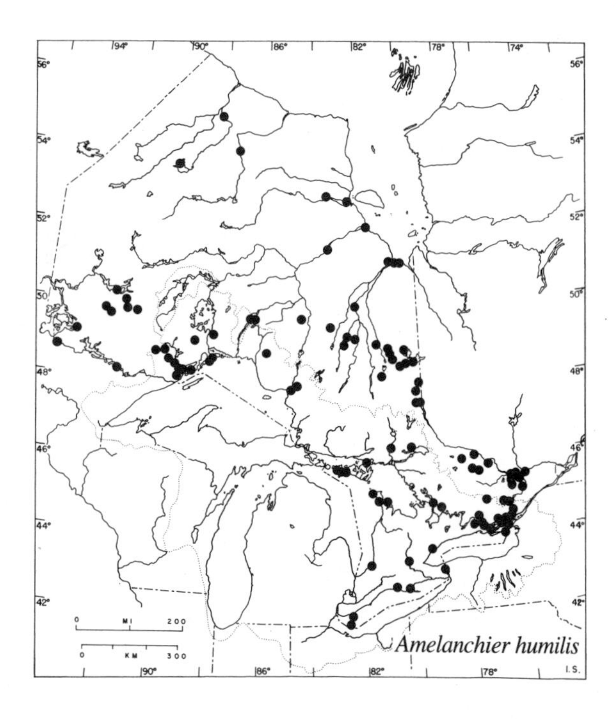

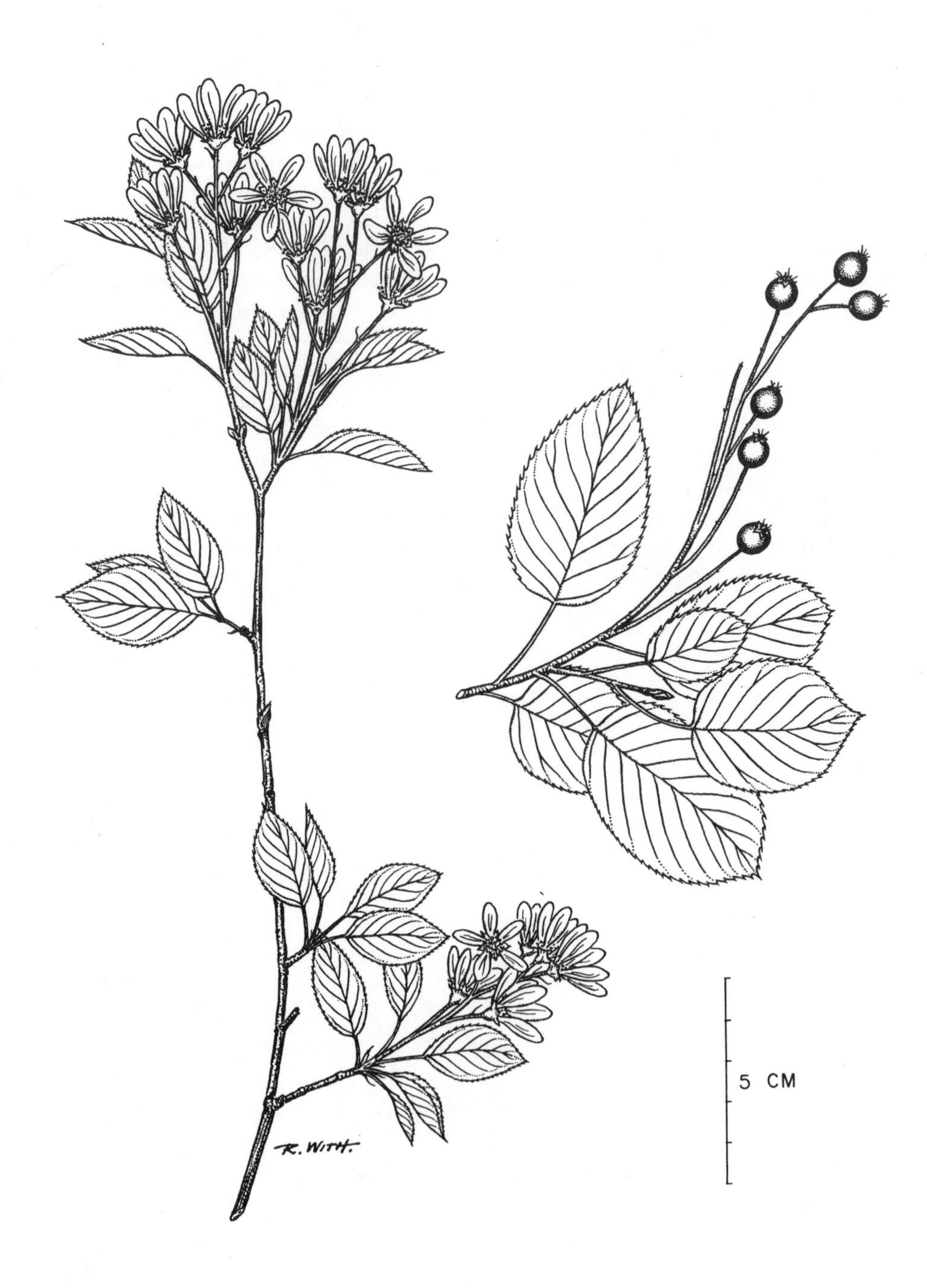

Amelanchier laevis

Amelanchier laevis Wieg. — Smooth serviceberry

An erect shrub with several stems in a clump or a small tree up to 10 m in height. Branchlets glabrous and purplish, the bark of older stems smooth and gray.

Leaves alternate, simple, and deciduous, usually glabrous, reddish, and folded or partially expanded (half grown) at flowering time; mature blades ovate to elliptic or ovate-oblong, 3–8 cm long and 2–4 cm wide, dark green above and paler or glaucescent, glabrous or with a few hairs beneath, especially along the veins; the tip acute to abruptly acuminate, the base rounded to cordate; margins finely and sharply serrate with numerous, almost callus-tipped teeth, these more than twice as many as the lateral veins; petioles 1–3 cm long, glabrous.

Flowers appearing with the unfolding and developing leaves in April and May in elongate drooping racemes; rachis and pedicels glabrous; bracts of the inforescence soon deciduous; petals linear-oblong, 1–2 cm long; ovary glabrous at the summit. Fruit juicy and edible, dark reddish-purple to black, on long pedicels, the lowest up to 4 or 5 cm long; July and August. (*laevis*—smooth)

In clearings and thickets, along the edges of woods, fencerows, and roadsides, and in coniferous and mixed woods.

Throughout southern Ontario and northward to about 51° N. (Nfld. to Minn., south to Iowa and Ga.)

Field check Large shrub or small tree with drooping racemes of white flowers; leaves reddish and smooth at flowering time, the blades mostly elliptic and usually a little less than twice as long as wide; edible fruits on long pedicels.

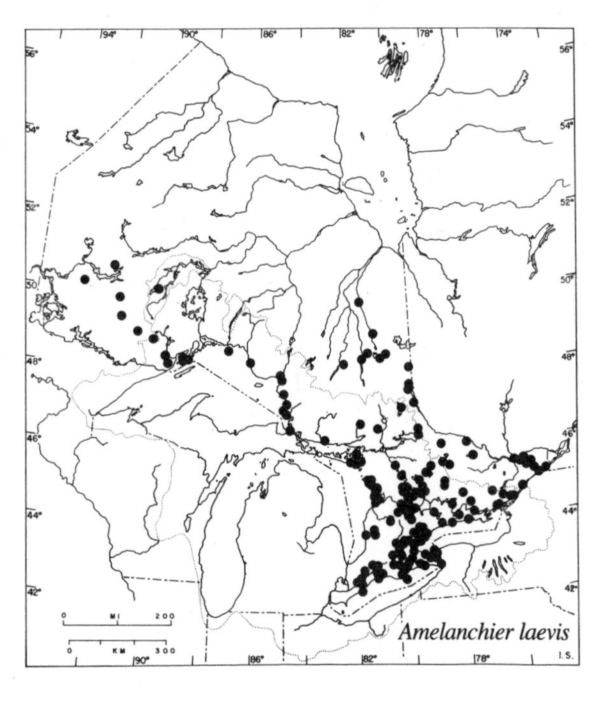

Amelanchier sanguinea

Amelanchier sanguinea (Pursh) DC. — Shadbush, Serviceberry

A straggling or erect shrub with the stems usually growing in a clump and reaching a height of from 1 to 3 m. Branchlets glabrous and reddish when young, the older stems and slender trunks grayish and smooth.

Leaves alternate, simple, and deciduous, nearly unfolded and about half grown at flowering time when the lower surface is covered with a grayish-green to whitish tomentum; mature blades oblong-elliptic to subrotund, 3-7 cm long and 2-4.5 cm wide, green and glabrous above, paler beneath and glabrous or with a few hairs along the midrib, the apex blunt, rounded, or subacute, the base rounded to subcordate; margins rather coarsely serrate at least above the middle and often nearly to the base, the teeth less than twice as many as the veins, usually with a vein or a fork going to each tooth; petioles 1-2 cm long, pubescent.

Flowers white, numerous, in loose drooping racemes 3-8 cm long, appearing while the leaves are unfolding and expanding in late May, June, and (northward) early July; the rachis and pedicels glabrous or sparsely pubescent, the deciduous slender floral bracts silky-hairy; petals 10-15 mm long. Fruit globose, dark purple, juicy, 5-10 mm in diameter, on pedicels 1-3 cm long; July and August. (*sanguinea*—blood-red, in reference to the red branchlets)

On gravelly and rocky soil of ridges, slopes, and lakeshores, in crevices of open rock faces and cliffs, in and along the edges of mixed forests.

Throughout southern Ontario and northward beyond Lake Superior to about 51° N. (s.w. Que. to w. Ont., south to Iowa and N.C.)

Field check Stems in clumps; leaves pubescent beneath at flowering time, glabrous when mature, little longer than broad; teeth fewer than twice as many as the veins; late-blooming; flowers and fruits in drooping racemes.

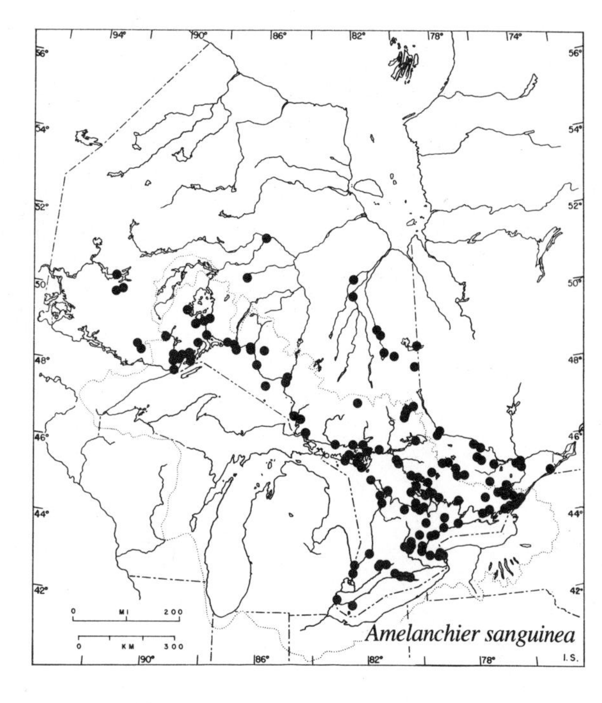

Amelanchier stolonifera

A stiff upright shrub, 0.3 - 2 m high, stoloniferous, and forming open colonies. Branchlets at first covered with silky hairs, soon becoming smooth; purplish-red and finally brownish to gray on the older stems.

Leaves alternate, simple, and deciduous, green and glabrous above and densely white-tomentose beneath when unfolding, about half grown at flowering time; mature blades broadly oval to elliptic, rarely oblong-oval or orbicular, 2 - 5 cm long, 2 - 3.5 cm wide, glabrous throughout or slightly hairy on the midrib, the apex rounded to subacute, often obtuse and mucronate, the base rounded or subcordate, rarely cuneate; margins finely and sharply serrate at least on the upper two-thirds; teeth 5 - 8 per cm; primary veins 7 - 11 (mostly 8 - 9) pairs, somewhat irregularly spaced, and curved upwards beyond the middle, becoming indistinct before reaching the margin; petioles 10 - 18 mm long, glabrous or slightly pubescent.

Flowers small, white, in short, dense, erect racemes 1.5 - 4 cm long; axis and pedicels glabrous to silky-hairy; lower pedicels 7 - 15 mm long; petals 7 - 9 mm long; sepals triangular-lanceolate, 2.5 - 3 mm long, revolute from the middle when the petals fall, tomentose on the inner surface; summit of ovary woolly; May and June. Fruit purplish-black, sweet, and juicy; July and Aug. (*stolonifera* — stolon-bearing)

On rocky ground and in crevices of rocks (chiefly acidic), along shores, on ridges, and in river gorges, and also on sandy soil of clearings and semi-open woods.

From Lake Erie to Lake Superior and in the Moose River drainage basin; apparently absent in the extreme northern and western parts of Ontario. (Nfld. to Ont., south to Minn., Mich., and Va.)

Field check Stoloniferous shrub less than 2 m tall and forming open colonies; leaves oval to elliptic, a little longer than broad, often mucronate, the margins finely and sharply serrate with 5 - 8 teeth per cm; flowers in short, dense, erect racemes.

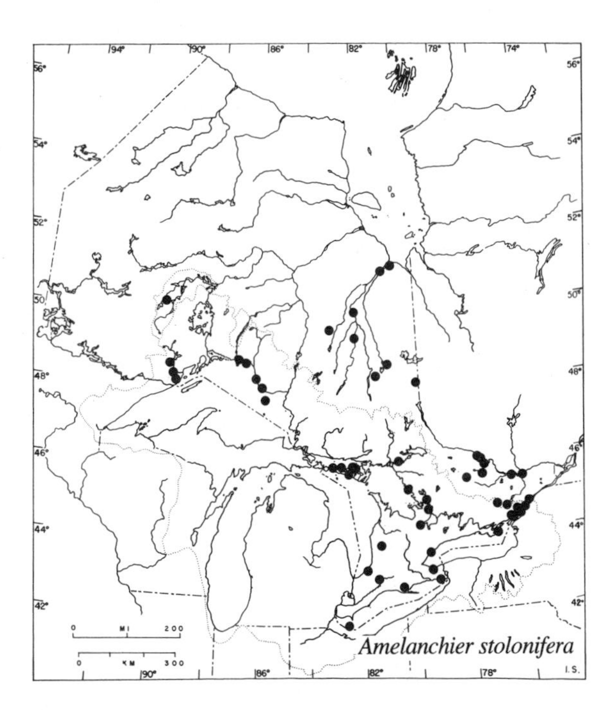

Amelanchier stolonifera

Aronia melanocarpa

Aronia Medik. — Chokeberry

Shrubs with slender stems and alternate, simple, deciduous leaves bearing glands along the midrib on the upper surface as well as gland-tipped teeth along the margins. Flowers small, white, 5-parted, in rounded or flat-topped clusters. Fruit a purple to blackish berrylike pome. The species may be treated under *Pyrus* in various manuals. (*Aronia*—said to be derived from *Aria*, part of the name of the European beamtree *Sorbus Aria* (L.) Crantz)

Aronia melanocarpa (Michx.) Ell. (including *A. prunifolia* (Marsh). Rehd., *Pyrus floribunda* Lindl.) — Chokeberry

A low to medium-sized shrub up to 2.5 m tall. Branchlets gray-brown to purplish, glabrous to pubescent.

Leaves alternate, simple, and deciduous; blades 2-8 cm long and 1-4 cm wide, obovate, oblanceolate, or oval to elliptic, dark green and glabrous above, the midrib with a row of dark hairlike glands, paler and glabrous to pubescent beneath; margins finely serrate with gland-tipped teeth; apex acute or abruptly acuminate and the base tapered to a petiole 2-10 mm long; stipules deciduous.

Flowers white, 5-parted, 5-10 mm in diameter, in stalked clusters of 5-15 at the ends of leafy branches; pedicels and calyx tubes glabrous, sparsely hairy, or densely pubescent; May and June. Fruit a purplish to blackish berrylike pome 6-10 mm in diameter; July to Sept.; the fruits sometimes persist on the bushes into the winter. (*melanocarpa*—from the Greek *melas*, black, and *karpos*, a fruit; *prunifolia*—with leaves like *Prunus*; *floribunda*—blooming profusely)

In and around peat bogs, swamps, and wet woods; also on dry sandy or rocky ridges and in pine woods.

Widespread throughout southern Ontario and northward to about 50° N. (Nfld. to Minn., south to n. Ga. and N. Eng.)

Note We have followed the treatment of *Aronia* by Hardin (1973), who accepts *A. melanocarpa* and *A. arbutifolia* (L.) Ell. as variable species, the variability arising from their past and present hybridizations. In Ontario, where *A. arbutifolia* (a coastal plain species from Nfld. to Fla. and Tex.) does not occur, the hybrid derivatives can be accommodated in *A. melanocarpa*. (*arbutifolia*—with leaves of *Arbutus*)

Field check Low or medium-sized shrubs with alternate, simple, mostly oblanceolate leaves, about twice as long as wide; glands on marginal teeth and along midrib of upper leaf surface; 5-parted white flowers and small clusters of purple to blackish berries.

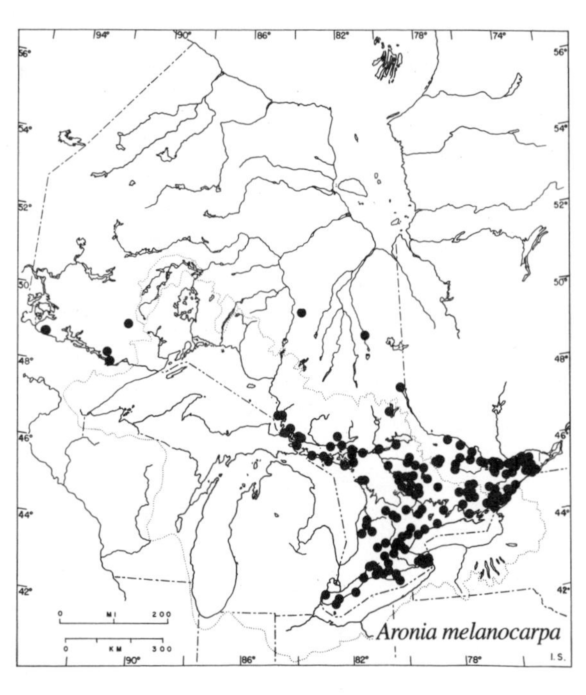

Crataegus L. — Hawthorn

A large genus of several hundred species, chiefly in the northern hemisphere and especially well developed in eastern North America. Some 1000 "species" have been described from North America but many of these may be hybrids or represent minor variations. Shrubs and small trees with numerous, slender, crooked branches often bearing sharp axillary thorns; bark on the trunks scaly and deeply fissured. Leaves alternate, mostly deciduous, simple or lobed, the margins entire to coarsely toothed; stipules usually present. The leaves of the vegetative shoots are often larger and more conspicuously lobed than those of the flowering or fruiting branches. Flowers white to pink, perfect, 5-parted, in few- to many-flowered corymbs, often showy, with a heavy disagreeable odour but attractive to many insect pollinators; petals deciduous; stamens 5-25 in 1-3 whorls, the anthers white, yellow, pink, red, or purple. The calyx-tube or floral cup surrounds the pistil of 1-5 carpels and the ovaries are partly to wholly inferior. The fruit is a red, orange, yellow, blue, or black pome containing 1-5 stones, each with a single seed; the 1-5 styles protrude from the open end of the fruit.

Abundant in disturbed and cleared areas, along streams and fencerows, and in clearings and pastures, particularly on calcareous soils. These shrubs often provide important habitats for many forms of wildlife through the protection afforded by the thorny branches and the food produced in the form of edible fruits, which in some species hang on the branches throughout the autumn and winter. Fruits on the ground and under the snow also provide food for birds, small rodents, and other wildlife.

Plants of this genus freely invaded newly cleared land as settlement spread in eastern North America and may even have begun their evolutionary diversification in areas cleared for agriculture by Indians before the arrival of European settlers. Hybridization, polyploidy, and apomixis, the latter confirmed by cytological studies at the University of Western Ontario (see Muniyamma & Phipps, 1979), have created perplexing taxonomic confusion in *Crataegus*. According to Gleason and Cronquist (1963), species that may be hybridizing include *C. punctata* with *C. chrysocarpa*, *succulenta*, *flabellata* and *crus-galli*; *C. chrysocarpa* with *C. flabellata* and *succulenta*; *C. succulenta* with *crus-galli*. Hybrid derivatives of these combinations may be expected in Ontario. (*Crataegus*—from the Greek *kratos*, strength)

Note The treatment presented here is a conservative one, covering a selection of eight species representative of eight of the Series recognized in this genus. Other species, including representatives of two or three additional Series, may be found in manuals with more extensive keys, e.g., Gleason & Cronquist (1963), and in the recent treatment by Phipps & Muniyamma (1980).

Key to *Crataegus*

a. Leaves deeply cleft (more than half way to the midrib) into 1-3 pairs of lateral lobes; margins entire in the sinuses, irregularly toothed near the tips of the lobes; veins running to the sinuses as well as to the lobes; style and nutlet solitary ***C. monogyna***

a. Leaves not deeply cleft, either without lobes or with shallow lobes; margins serrate; veins running only to the lobes and the larger teeth; styles and nutlets 2-5

- b. Thorns few, short (1-2.5 cm); fruit purplish-black ***C. douglasii***
- b. Thorns few to many, long (2.5-10 cm); fruit red to yellow
  - c. Leaves widest at or above the middle; narrowly cuneate at base
    - d. Leaves glabrous, leathery, glossy above, obscurely lobed or unlobed ***C. crus-galli***
    - d. Leaves pubescent beneath, firm, dull green, sometimes shallowly lobed towards the tip ***C. punctata***
  - c. Leaves widest at or below the middle, broadly cuneate to truncate, rounded or subcordate at the base
    - e. Calyx lobes glandular-serrate or denticulate; points of the main lobes of the leaf straight
      - f. Leaves mostly broadest below the middle; fruit pubescent, at least when young ***C. mollis***
      - f. Leaves mostly broadest about the middle; fruit glabrous
        - g. Leaf margins shallowly lobed, mostly above the middle; teeth not gland-tipped; petioles stout, winged and grooved near the leaf blade, glandless; stamens 10-20 ***C. succulenta***
        - g. Leaf margins with conspicuous lobes to below the middle; teeth gland-tipped; petioles slender, with scattered glands near the leaf base; stamens 10 ***C. chrysocarpa***
    - e. Calyx lobes narrow, entire or nearly so; main lobes of the leaf with acuminate tips, spreading or reflexed ***C. flabellata***

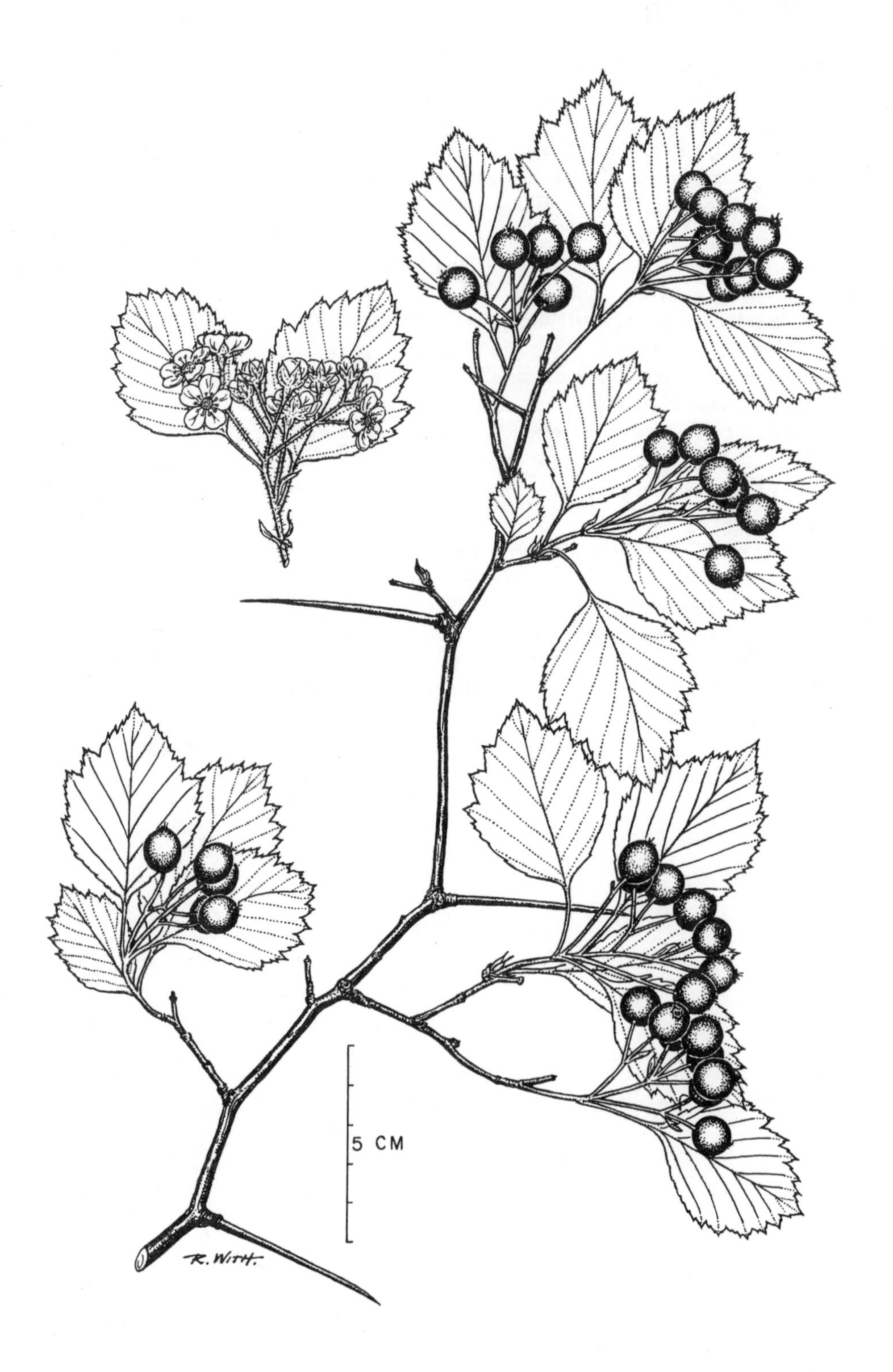

Crataegus chrysocarpa

Crataegus chrysocarpa Ashe **Hawthorn**

A much-branched shrub or occasionally a tree up to 6 m high. Branchlets stout and usually very thorny, the thorns 2-8 cm long, straight or slightly curved.

Leaves alternate, simple, and deciduous; blades 2-9 cm long, 2-7 cm wide, broadly elliptic, oval or nearly orbicular, with 4-6 triangular lobes on each side, yellow-green, firm, roughened above with short appressed hairs while young, glabrous at maturity; veins slightly impressed above; margins serrate except near the base, the teeth gland-tipped; petioles slender, glabrous or sparsely pubescent, and with small scattered glands near the blade.

Flowers 10-15 mm wide, in loose, sparingly villous corymbs with conspicuous (but soon deciduous) narrow floral bracts bearing numerous marginal glands; stamens about 10; anthers white or pale yellow; calyx lobes serrate or with stalked glands along the margins; May and June. Fruit short-oblong to nearly globose, 8-10 mm thick, dark red and rarely golden yellow but remaining green until late in the season; nutlets 3-4; late Aug. to Oct. (*chrysocarpa*—from the Greek *chrysos*, golden, and *karpos*, a fruit, not very appropriate as the fruit is more often red than yellow)

In clearings, open, rocky, or gravelly ground, and thickets, on river banks, lakeshores, and along edges of swamps.

Throughout southern Ontario and northward to the southern half of the Moose River drainage system; also between Thunder Bay and Lake-of-the-Woods. (Nfld. to Man., south to N. Mex. and N. Eng.)

Field check Leaves about as broad as long, conspicuously lobed to below the middle; teeth of leaf margins gland-tipped; the slender petioles, floral bracts, and sepal lobes with marginal glands.

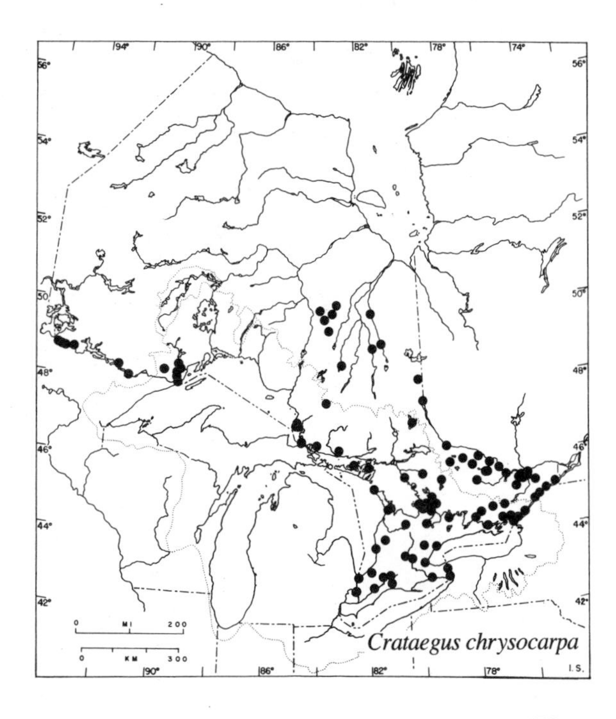

Crataegus crus-galli

Crataegus crus-galli L. — Cockspur thorn

An arborescent shrub or tree 6 - 10 m high with wide-spreading stiff branches and a somewhat depressed crown. Branchlets slender and flexuous but soon developing many, stout, sharp thorns 2 - 6 cm long, straight or slightly curved; bark gray-brown and slightly scaly.

Leaves alternate, simple, and deciduous; blades firm or leathery, dark green and glossy above, dull and paler beneath, those of the flowering branches generally obovate with a tapering cuneate base, 3 - 5 cm long and 1 - 3 cm wide, the shoot leaves often more elliptic to oblong and up to twice as large, mostly unlobed, the apex rounded or pointed; margins sharply serrate at least above the middle; petioles 3 - 12 mm long, slightly winged near the junction with the blade.

Flowers 1 - 1.5 cm across, numerous, in loose glabrous corymbs; stamens about 10; anthers pink, white, or pale yellow; calyx lobes linear-lanceolate, entire or sometimes glandular-serrate; May and June. Fruit short-ovoid to nearly globose, often 5-angled, 8 - 10 mm thick, green to dull red, often dark-dotted, the flesh thin and dry; nutlets 1, 2, or occasionally 3; Sept. and Oct. (*crus-galli* — spur of a cock)

In thickets, open woods, fields, and abandoned pastures, especially on dry rocky ground.

Throughout the main portion of the Deciduous Forest Region and in Prince Edward County. (s. Que. and s. Ont. to Minn., south to e. Tex. and S.C.)

Field check Leaves obovate with a cuneate base, thick, shiny above; fruits green to dull red, often 5-angled.

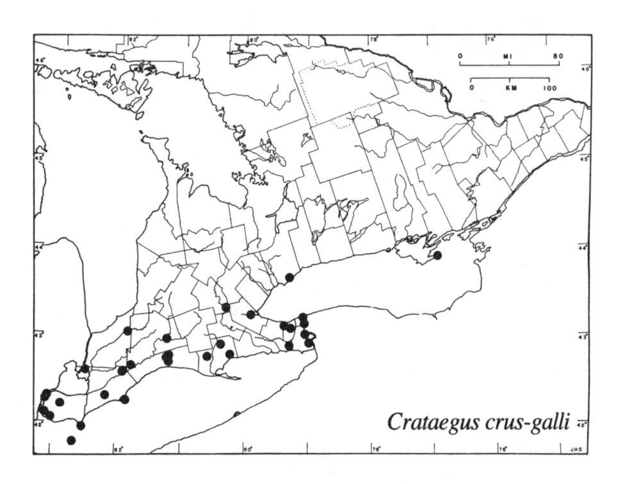

Crataegus douglasii

Crataegus douglasii Lindl. — Western hawthorn

An arborescent shrub or small tree growing to a height of 10 m or more. Branchlets slender and glabrous, either thornless or bearing scattered, short, stout, straight or slightly curved thorns 1-2.5 cm long; bark of older stems gray-brown and scaly.

Leaves alternate, simple, and deciduous; blades 2-8 cm long, 1.5-4 cm wide, oval, broadly elliptic or oblong-ovate, usually indented or shallowly lobed only above the middle or near the obtuse to acutish apex, the 2-4 pairs of lobes rounded or acute and somewhat irregular or asymmetrical; dark green above, paler beneath, smooth on both surfaces or with some short appressed hairs above when young, lustrous above at maturity; base of leaf cuneate to rounded; margins serrate with gland-tipped teeth; petioles 0.5-3 cm long, slightly winged near the blade.

Flowers 1-1.3 cm wide, in 5-12-flowered corymbs; petals about 5 mm long; calyx-lobes broadly triangular, more or less villous towards the apex, reflexed, mostly entire; stamens 10-20, anthers white or pink; styles usually 5; June. Fruit short-oblong, 8-10 mm thick, dark wine-coloured to purple-black when ripe, the 3-5 nutlets surrounded by succulent flesh; Sept. (In herbarium specimens the fruit usually dries black even when immature, a feature that is especially noticeable at the top of the collar, i.e., at the summit of the hypanthium just below the reflexed calyx lobes.) (*douglasii*—named for David Douglas, 1798-1834, collector for the Royal Horticultural Society who made some collections in southern Ontario but is better known for his work in the Pacific Northwest)

In thickets, along margins of woods, lakeshores, and river banks; on summits of cliffs and rocky ridges.

Along the north shore of Lake Superior, at Lake Nipigon and Lake Abitibi, and on Manitoulin Island and the Bruce Peninsula. (s. Alaska to Calif., east to s.w. Sask. and S. Dak.; disjunct in the upper Great Lakes region)

Field check Thorns few, short (1-2.5 cm long) or lacking; leaves somewhat longer than wide, shallowly lobed chiefly near the apex; corymbs few-flowered (5-12); fruit purple-black.

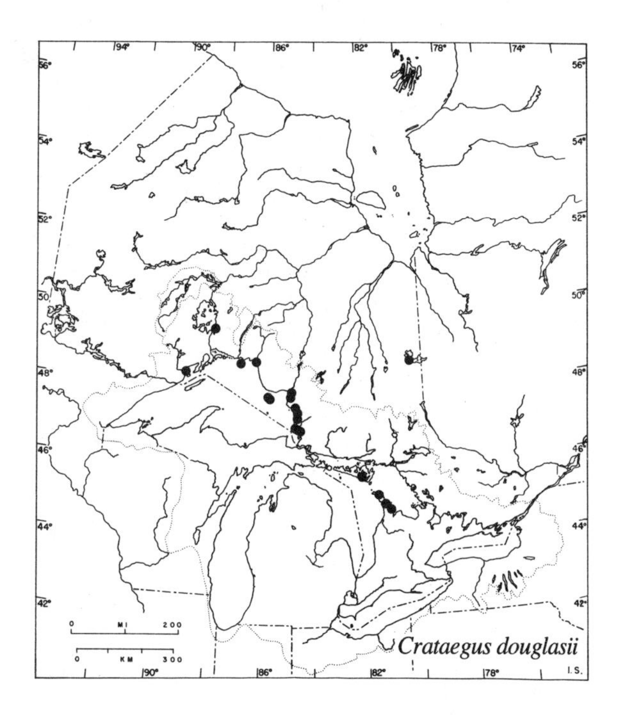

Crataegus flabellata

Crataegus flabellata (Spach) Kirchn. — Hawthorn

A much-branched shrub or rarely a tree 5 - 6 m high. Branchlets slender and glabrous, older stems stout and armed with straight or slightly curved thorns 3 - 10 cm long.

Leaves alternate, simple, and deciduous; blades 3 - 8 cm long, 2 - 7 cm wide, ovate to rhombic, acuminate at the tip and cuneate to truncate at the base, short-pilose above and sometimes villous on the veins beneath when young but usually glabrous at maturity; margins indented with 4 - 6 small acuminate lobes on each side and serrate nearly to the base; petioles slender, 1 - 3 cm long, somewhat winged and grooved near the blade; leaves of the vegetative shoots broadly ovate to nearly orbicular, more deeply lobed or laciniate, the acuminate tips of the lobes often spreading or reflexed.

Flowers 15 - 18 mm across, numerous, in loose, more or less pubescent corymbs; stamens 10 - 20; anthers pink; calyx lobes narrow, entire or nearly so; May and June. Fruit oblong to nearly globose, sometimes slightly angular, 8 - 12 mm thick, bright red, with thick mellow flesh surrounding the 3 - 5 nutlets; Sept. and Oct. (*flabellata*—fan-shaped, in reference to the shape of the leaves)

In open woods, thickets, fields, and pastures and along river banks and rocky ridges.

In southern Ontario, where essentially restricted to calcareous areas south of the Canadian Shield, and in the Ottawa Valley. (N.S., N.B., s. Que., and s. Ont., south to N.Y. and N. Eng.)

Note Related species recognized in some manuals include: *C. beata* Sarg., *C. grayana* Eggl., and *C. macrosperma* Ashe.

Field check Leaves broad, sharply lobed with the tips of the lobes acuminate, often reflexed; calyx lobes narrow, entire or nearly so; fruit bright red.

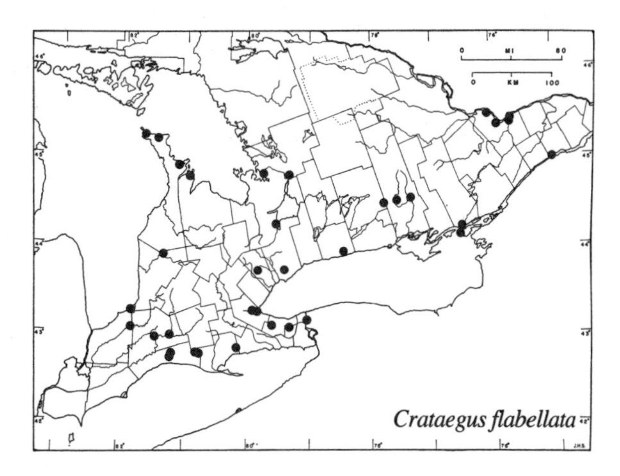

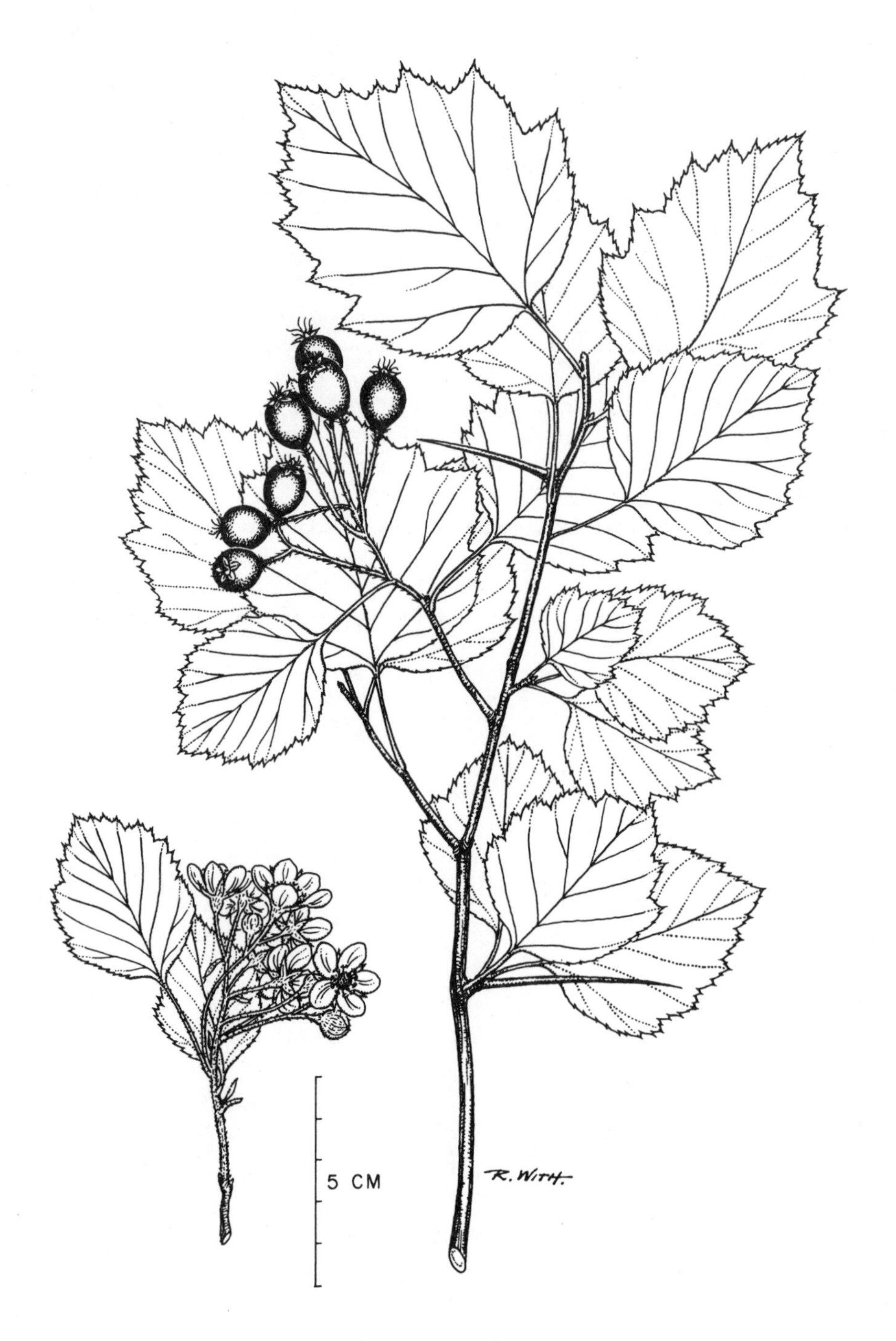

Crataegus mollis

Crataegus mollis (T. & G.) Scheele (including *C. submollis* Sarg.) — Downy hawthorn

A tree or arborescent shrub up to 8 or 10 m high. Branchlets flexuous, thorny, villous at first, becoming smooth and reddish-brown, finally grayish; thorns reddish-brown, slender, 2-6 cm long; bark brownish-gray, slightly scaly.

Leaves alternate, simple, and deciduous, thin but firm, at first covered with appressed hairs above and densely tomentose beneath, at maturity dark green and glabrous above and slightly pubescent beneath; blades ovate to oval, 4-8 cm long, 3-6 cm wide, round to abruptly cuneate at base, widest at or below the middle, the apex sharp-pointed; margins coarsely and sharply doubly serrate or shallowly incised to form 4 or 5 sharp-pointed lobes on each side; petioles 2-6 cm long, villous at first, and usually slightly pubescent at maturity, narrowly winged and grooved on the upper side near the junction with the blade.

Flowers 2-2.5 cm wide, showy, in loose, pubescent, many-flowered corymbs; stamens 10 or fewer; anthers white or pale yellow; calyx tomentose, the lobes glandular-serrate with stalked red glands, conspicuous, and usually persistent on the summit of the fruit; May and June. Fruit obovoid to pyriform, 1-1.5 cm thick, bright red, and slightly pubescent, the thin, dry or mellow flesh surrounding the (usually) 5 nutlets; Sept. (*mollis*—soft; *submollis*—somewhat soft)

On wooded hillsides, in thickets, and along roadsides and fencerows, often in moist soil.

In southern Ontario from Lake Erie north to Lake Simcoe and the Bruce Peninsula and in the east to the Ottawa and St. Lawrence valleys. (s. Ont. to Mich., south to Okla. and Ala.)

Field check Leaves little longer than broad, broadest below the middle, doubly serrate or sharply lobed above the middle; flowers showy; calyx lobes glandular-serrate; leaves and fruit pubescent at least when young.

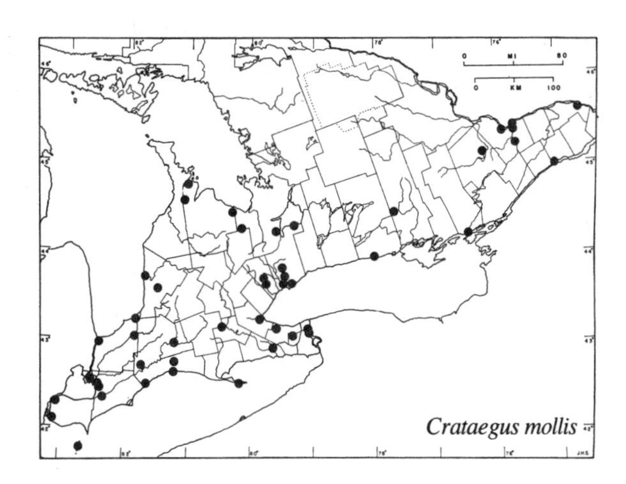

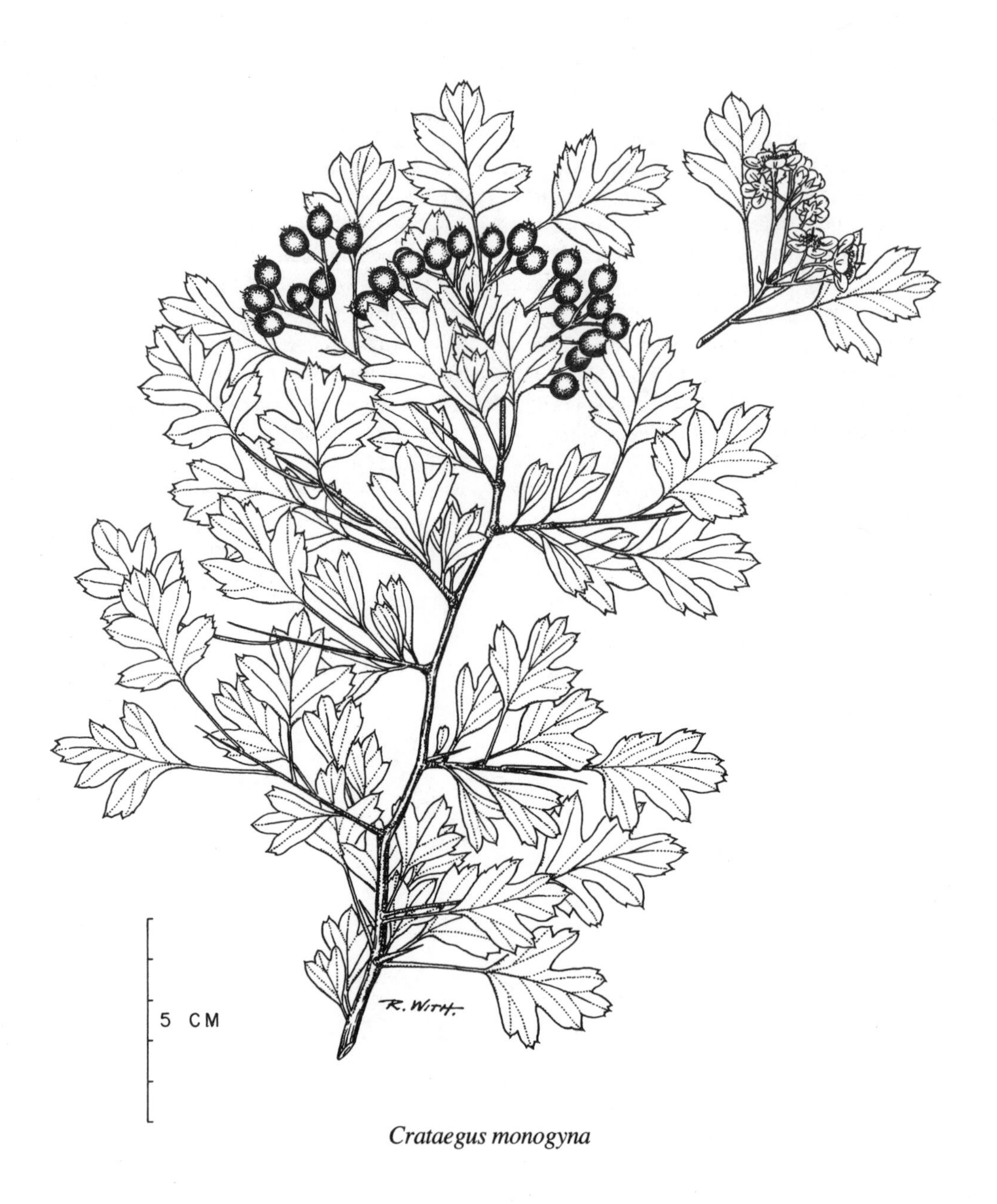

Crataegus monogyna

Crataegus monogyna Jacq. **English hawthorn**

An arborescent shrub or small tree reaching a height of 8 – 10 m. Branchlets slender and smooth, bearing short, sharp-pointed, persistent, axillary thorns usually less than 2 cm long but also developing stout, spurlike, spine-tipped branchlets 3 – 8 cm long on which leaves and secondary flowering branches arise; bark of older stems dark gray to brownish, slightly scaly.

Leaves alternate, simple, and deciduous, firm in texture and often persisting until late in the season; blades 1.5 – 5 cm long, deltoid to broadly ovate in general outline but deeply cleft with two or three lateral lobes on each side and one terminal lobe, dark green above, paler beneath, usually smooth throughout or slightly pubescent beneath when young; margins entire in the acute sinuses, the lobes irregularly toothed near the tip; lateral veins running to the sinuses as well as to the lobes; petiole slender and glandless, slightly winged near the base of the leaf, shorter than the blade or equalling it in length.

Flowers numerous, 8 – 15 mm wide, in glabrous or slightly villous corymbs; stamens about 20; anthers pink to red; style solitary; calyx lobes deltoid and entire; May and June. Fruit bright red, subglobose to ellipsoid, 5 – 8 mm thick, the flesh rather thin around the single nutlet; Sept. and Oct. (*monogyna* — from the Greek meaning with a single pistil or gynoecium)

Along roadsides, borders of woods and in open woods and fencerows.

Southern Ontario, chiefly in the calcareous areas south and east of the Canadian Shield from Lake Erie to Georgian Bay and east to the St. Lawrence River and the Ottawa district. (Native of Europe, the Mediterranean region, and w. Asia)

Note A related European species, *C. laevigata* (Poir.) DC. (*C. oxyacantha* L. in some manuals), with less deeply cut leaves and 2 or 3 styles and nutlets is planted in gardens. Paul's scarlet thorn is a double-flowered cultivar of this species with ample and handsome rose-red flower clusters, commonly planted as an ornamental, and hardy throughout most of southern Ontario. (*laevigata* — smooth and polished; *oxyacantha* — sharp-spined)

Field check Leaves 3 - 5 - (- 7 -) lobed; branchlets with short thorns (1 – 2 cm long) and longer, spurlike, spine-tipped branches (3 – 8 cm); flower usually with a single style and fruit with one nutlet.

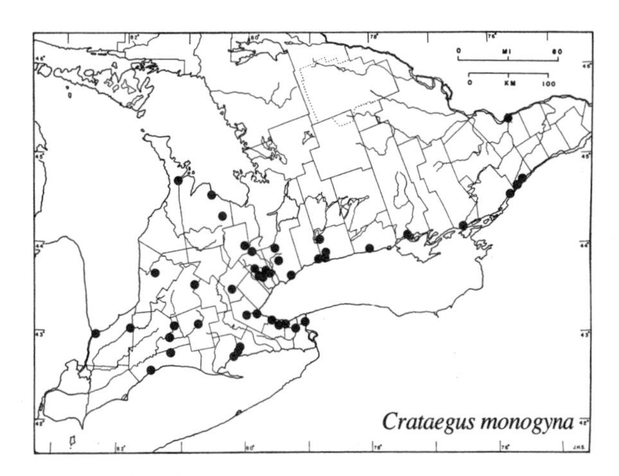

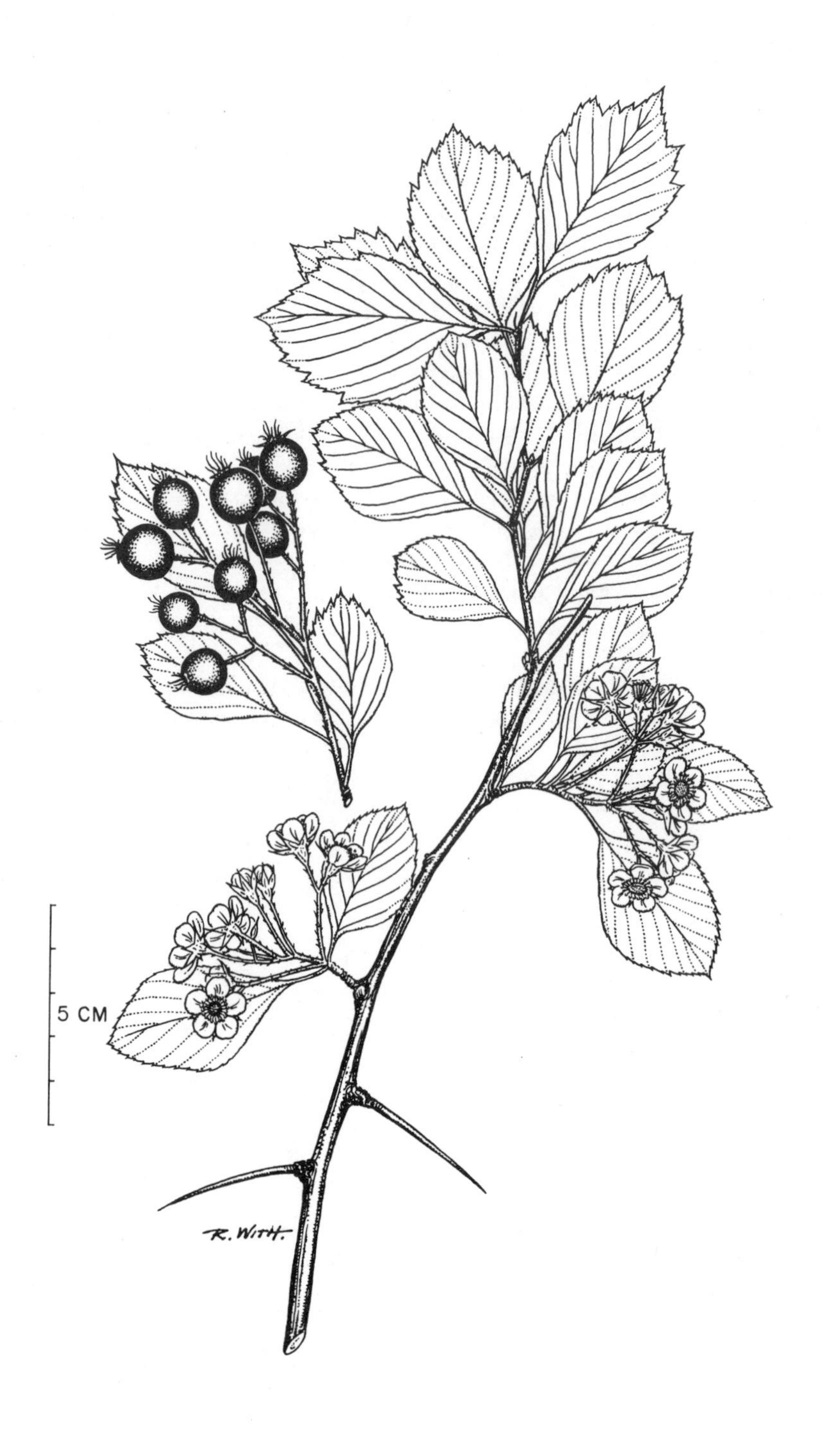

Crataegus punctata

Crataegus punctata Jacq. **Hawthorn**

A tree up to 8 - 10 m tall with an open crown of stiff spreading branches, the trunk bearing compound thorns. Branchlets at first slender, smooth, and brownish, the older stems stout, grayish, and bearing sharp thorns 3 - 6 cm long and short lateral shoots with 3 - 9 leaves; bark of trunk brownish to gray and fissured.

Leaves alternate, simple, and deciduous; blades 2 - 8 cm long, 1 - 5 cm wide, firm, dull green above and appressed hairy when young, paler beneath and slightly hairy at least along the veins, elliptic-oblong to obovate (those of the vegetative shoots much broader above the middle, sometimes nearly fan-shaped), the apex pointed, rounded, or (on vegetative shoots) shallowly incised, the base cuneate and tapered to a winged petiole; margins sharply serrate beyond the middle and usually appearing irregularly doubly serrate around the broad summit of leaves on vegetative shoots; veins distinctly impressed on the upper surface at maturity; petioles 1 - 2 cm long.

Flowers 1 - 2 cm across, numerous, in loose pubescent corymbs; stamens about 20; anthers pink to red or yellow; calyx lobes densely gray-pubescent; May and June. Fruit at first pear-shaped, becoming obovoid to nearly globose, and 1 - 1.5 cm thick when mature, dull red to orange-red, spotted with pale dots, and somewhat ridged on one side, the mellow to scarcely succulent flesh covering the 3 - 5 nutlets; Sept. and Oct. (*punctata*—dotted)

In thickets, pastures, edges of woods, and open rocky ground.

Throughout the calcareous areas of southern Ontario south and east of the Canadian Shield. (N. Eng. to s. Que. and s. Ont., south to Iowa and Ky.)

Field check Leaves longer than wide, often lobed or incised at the summit, firm in texture, dull green with clearly impressed veins above and pubescent beneath; fruit orange-red and pale-dotted.

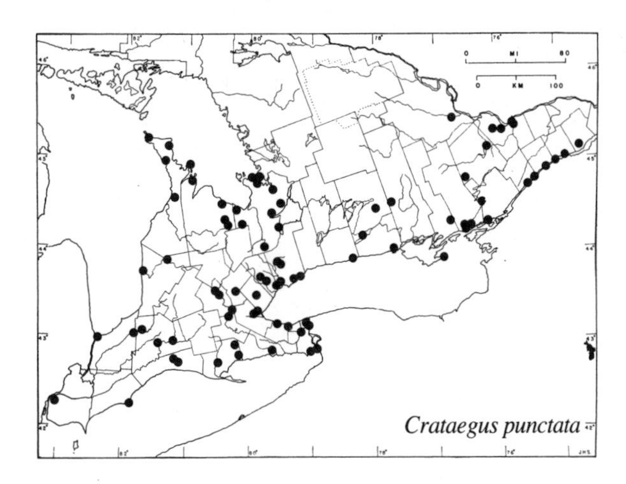

Crataegus punctata

Crataegus succulenta

A tree up to 6-8 m tall or a stout arborescent shrub. Branchlets slender, often flexuous, glabrous or rarely slightly hairy when young, the older stems stouter, usually bearing strong slender thorns up to 7 cm long.

Leaves alternate, simple, and deciduous; blades 3-6 cm long, 2-5 cm wide, broadly elliptic to ovate or rhombic, usually incised above the middle with 4 or 5 short acute lobes on each side and sharply serrate except near the base, the teeth not gland-tipped; firm to leathery, dark green above and with short appressed hairs when young, paler beneath and slightly pubescent, at least along the veins, at maturity the upper surface somewhat lustrous and with impressed veins; petioles stout, glandless, winged and grooved near the blade, especially on the leaves of vegetative shoots.

Flowers 10-18 mm across, numerous, in glabrous or slightly villous corymbs, stamens 10-20; anthers white, pale yellow, or pink; calyx lobes glandular-serrate; May and June. Fruit nearly globose, 6-12 mm thick, bright red, succulent when ripe or sometimes remaining hard and dry until late in the season; nutlets 2 or 3; Sept. and Oct. (*succulenta*—with juicy flesh)

In dry gravelly or rocky fields, thickets, along roadsides and fencerows, on beaches, in ditches and ravines.

Common in southern Ontario, less frequent north and west of Lake Superior. (N. Eng. and N.S. to s. Man., south to Iowa and N.Y.)

Note Some Ontario specimens that may key out with *C. succulenta* have been recognized by Gleason (1963) and others as belonging to a related species, *C. calpodendron* (Ehrh.) Medik., described as being similar but with the branchlets tomentose or villous when young and the leaves larger and more prominently hairy. (*calpodendron*—urntree, in reference to the shape of the fruit)

Field check Leaves ovate to rounded, shallowly lobed above the middle; margins sharply serrate, teeth not gland-tipped; petioles winged and glandless; calyx lobes glandular-serrate; fruit bright red.

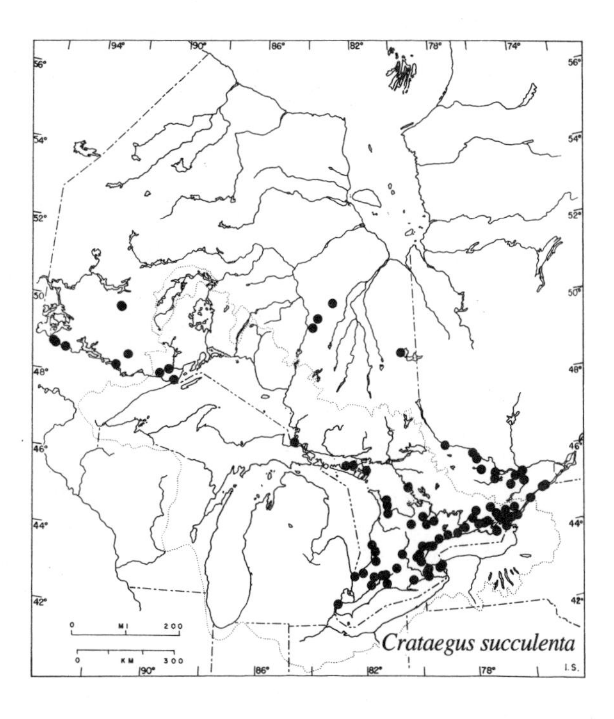

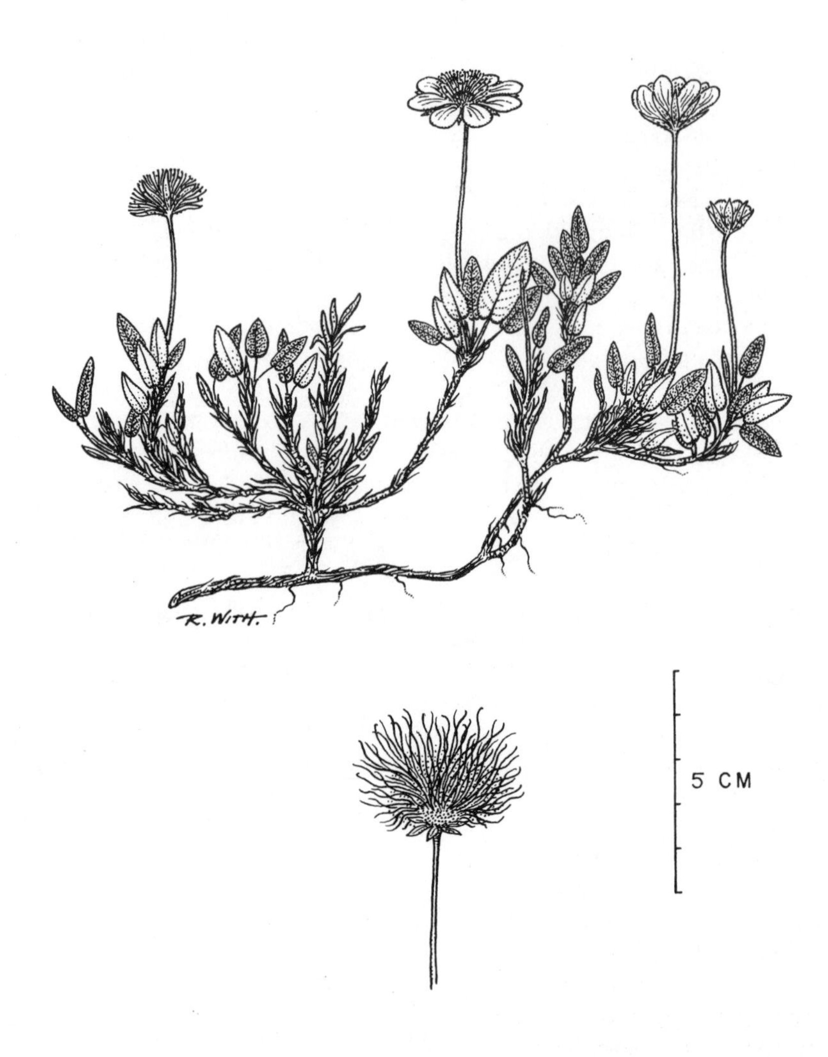

Dryas integrifolia

Dryas L. — Mountain avens

A small genus of 3 or 4 species found in arctic, boreal, and alpine regions. Low shrubs, trailing or prostrate, often forming mats. Stems and branches flexible, covered with persistent scales, and bearing numerous, small, simple, leathery leaves, dark green above and conspicuously whitened beneath. Flowers solitary on an erect naked scape. Fruit a head of achenes with long feathery styles which aid in wind dispersal. (*Dryas*—wood-nymph, dryad)

Key to *Dryas*

a. Leaves ovate to lanceolate (broadest below the middle), entire or toothed near the cordate to truncate base; sepal lobes linear-lanceolate; petals white, wide-spreading — ***D. integrifolia***

a. Leaves elliptic-obovate (broadest at or above the middle), coarsely crenate, tapered to base; sepal lobes ovate; petals yellow, erect — ***D. drummondii***

Dryas integrifolia Vahl. — Mountain avens

A prostrate shrub, mat-forming or cushion-forming, with a tough woody caudex and numerous short branches. Branchlets slender, flexible, clothed with scaly petiole bases; older stems dark brown to blackish.

Leaves alternate, simple, and evergreen, usually persisting for several years in tufts at the ends of the branches; blades firm and leathery, narrowly ovate to lanceolate-oblong (broadest below the middle), 5 - 15 mm long, dark green, rugose and glabrous or hairy above, grayish to white beneath with a layer of woolly hairs almost obscuring the midvein, the apex blunt to pointed; margins revolute, mostly entire or with several broad teeth or lobes near the truncate to cordate base; petioles slender, up to 10 mm long; stipules linear-lanceolate, at least the lower half fused to the petiole.

Flowers solitary on erect, bractless peduncles which elongate up to 10 cm in fruit; sepal lobes 8 - 10, linear-lanceolate, tomentose with scattered black hairs; petals 8 - 10, elliptic, white to cream-coloured, about 1 cm long, widely spreading; June to Aug. Fruit a cluster of achenes with feathery styles 15 - 25 mm long, forming a fluffy head; July to Sept. (*integrifolia*—with entire leaves)

In tundra habitats: sandy shores, beach ridges, dune hollows; also rocky shores and crevices of rock exposures.

Shores of James Bay and Hudson Bay; Slate Islands off the north shore of Lake Superior. (Greenl. to Alaska, south to B.C., n. Ont., Nfld., and the Gulf of St. Lawrence)

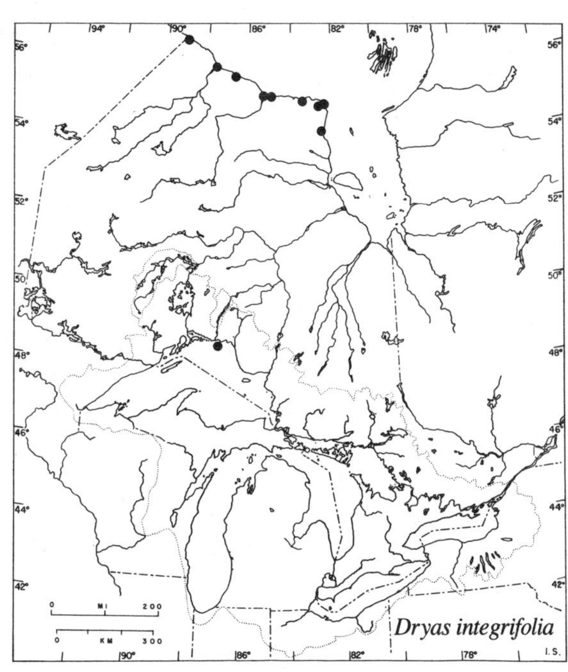

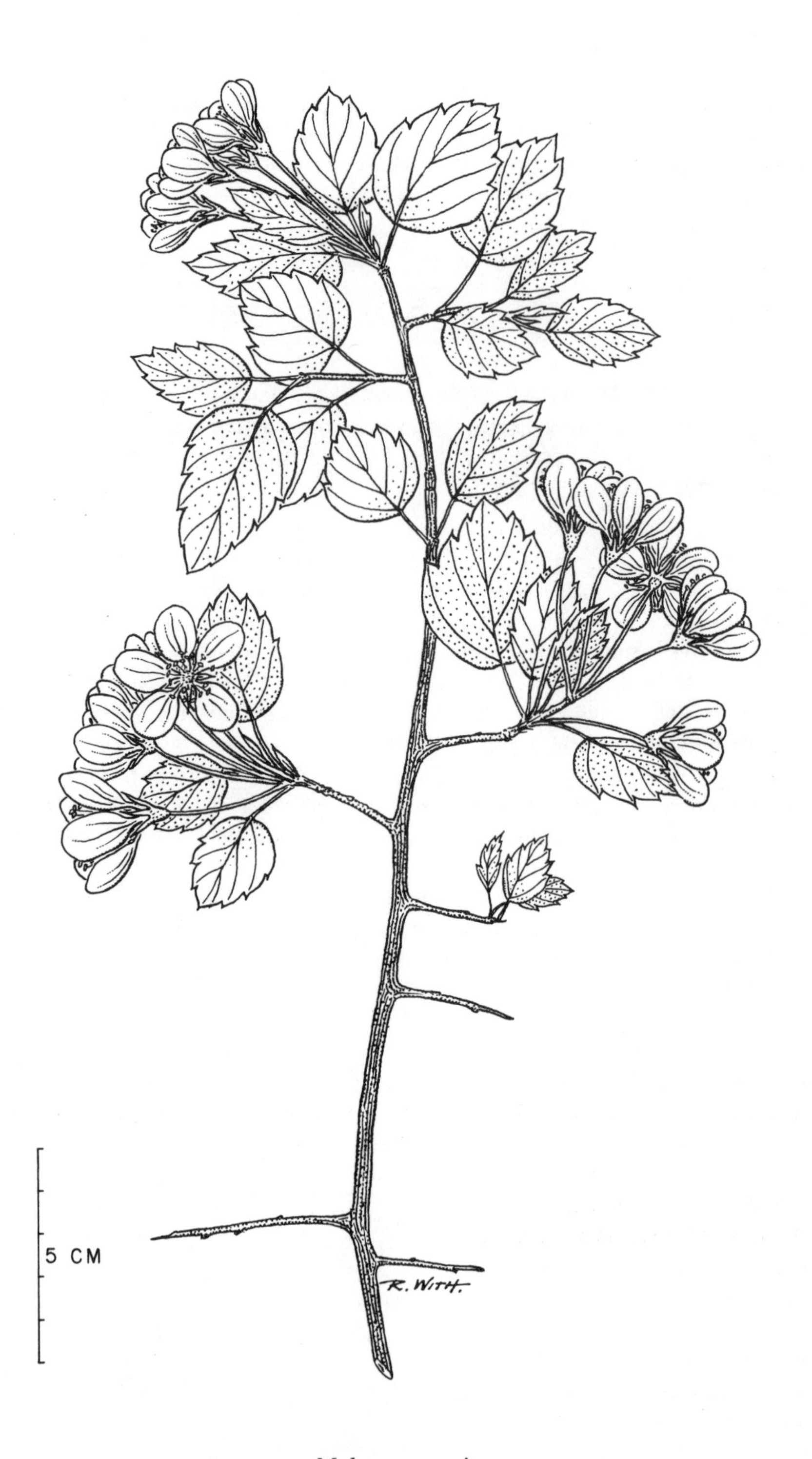

Malus coronaria

Note A second species, *D. drummondii* Richards., has been collected on Mortimer and Patterson islands (Slate Islands) along the north shore of Lake Superior. It differs in having larger, crenately-toothed, elliptic-obovate leaves (widest at or above the middle) tapered at base and yellow to orange-coloured flowers with ovate sepals and erect (rather than widely spreading) petals. (w. Nfld. to Alaska, south to Oreg. and Mont.; also at isolated stations in the east)

D. drummondii is one more species in the growing list of non-leguminous plants found to be capable of nitrogen fixation. (*drummondii*—in honour of Thomas Drummond, 1780-1865, a Scottish botanical collector in North America)

Field check Prostrate matted shrub with numerous lance-ovate leaves, dark green above and white-pubescent beneath; 8-10-parted white flowers and fluffy heads of long-plumed achenes.

Malus Mill. — Apple

A small genus of 20-30 species (sometimes treated as a subgenus of *Pyrus*) found in north temperate regions. Shrubs and small much-branched trees often with sharp thornlike lateral branches or spur shoots; leaves alternate, simple, and deciduous, the margins serrate or lobed. Flowers in simple clusters, perfect, 5-parted. Fruit a pome, the enlarged and fleshy receptacle forming the bulk of the tissue at maturity.

Several species and numerous cultivars are grown for the edible fruit, e.g., common apple (*Malus pumila* Mill.) and Siberian crabapple (*M. baccata* (L.) Borkh.), and a wide range of handsome flowering crabs have been developed which are hardy in many parts of southern Canada.

M. coronaria is the only native species in Ontario but derivatives of cultivated apples appear to be naturalized or persist without cultivation in many places where seeds have been discarded. (*Malus*—the Latin name for the apple tree; *pumila*—dwarf; *baccata*—bearing berries)

Malus coronaria (L.) Mill. (*Pyrus coronaria* L.) — Wild crabapple

A stiffly branched tall shrub or low tree up to 10 m high. Young branchlets hairy at first but later becoming glabrous; older twigs stout, purplish-gray, with sharp thornlike spur-shoots.

Leaves alternate, simple, and deciduous; young leaves sparsely hairy, at least on the veins beneath, glabrous at maturity; blades of those on the flowering and fruiting branches usually ovate to oval, 5-10 cm long, acute or acuminate at the tip, the base mostly rounded, those of the sprout growth and on non-flowering branches often more sharply triangular-ovate with 1-4 prominent forward-arching lobes on each side and a somewhat cordate base, bright green above and paler beneath; margins of mature blades serrate and more or less notched; petioles 1-3 cm long; stipules linear, reddish-tinged, soon deciduous.

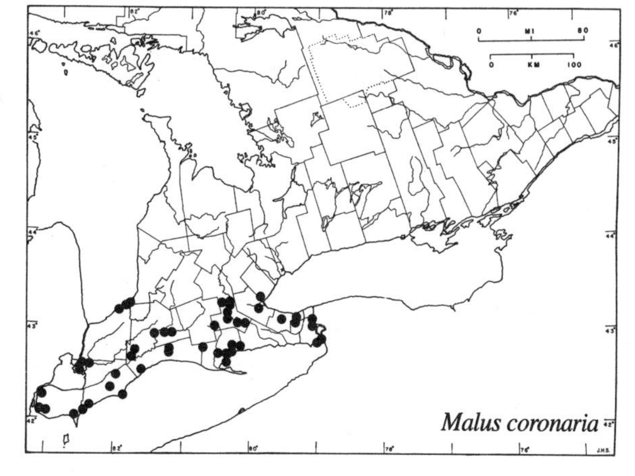

Flowers showy and fragrant, up to 3 cm across, the petals deep rose-pink in the bud, fading to nearly white on opening;

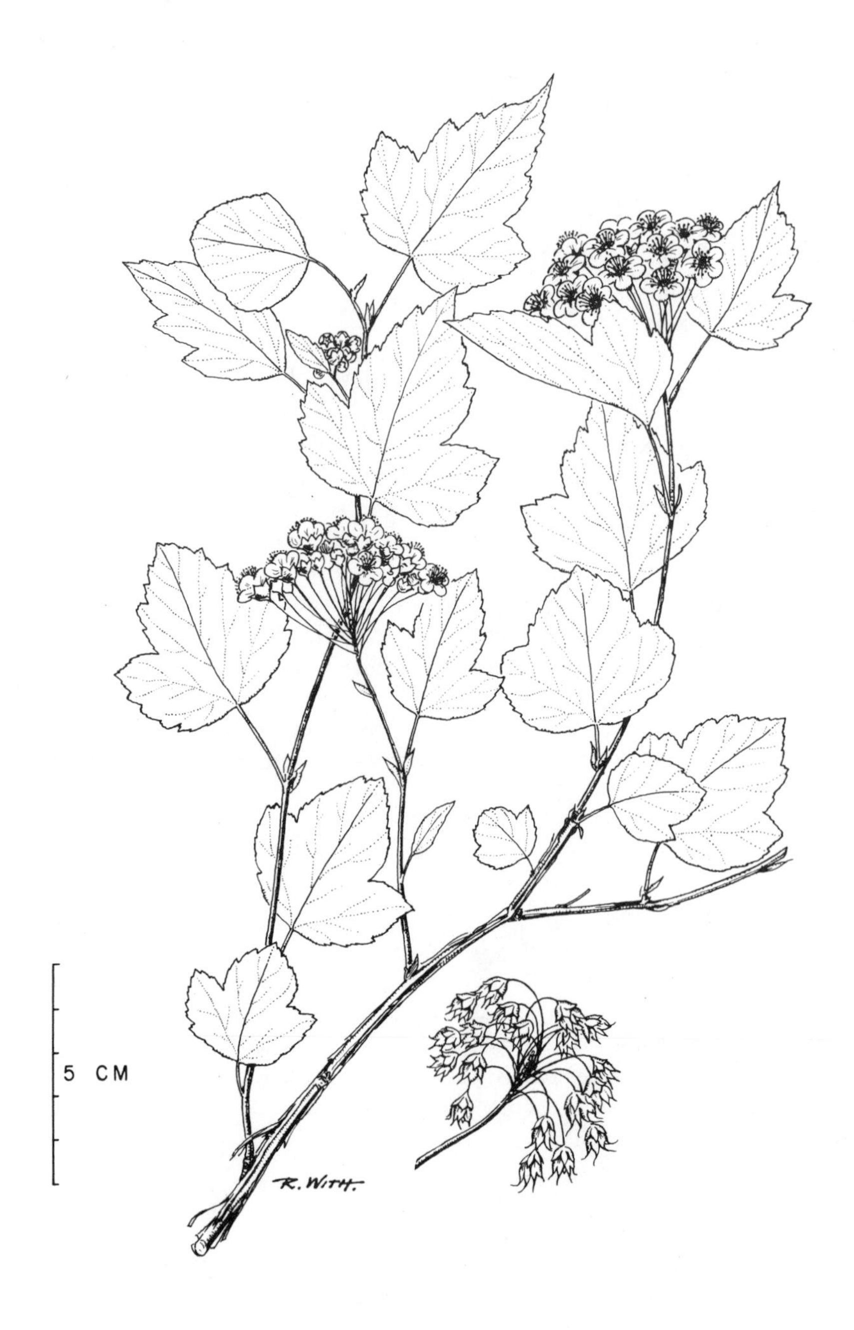

Physocarpus opulifolius

individual flowers borne on pedicels up to 3.5 cm long and forming open ample clusters mostly at the ends of short leafy side branches; calyx glabrous to sparsely pubescent on the outside and densely villous within; anthers pink or salmon-coloured; late May and early June. Fruit a yellow-green pome up to 2.5 cm across, nearly globose but a little flattened at each end, hard, and sour-tasting; Sept. and Oct. (*coronaria*—forming a crown or suitable for a wreath, in reference to the handsome flowers)

Open woods, thickets, fencerows, roadsides, open slopes, and edges of woods.

Restricted to the Deciduous Forest Region. (N.Y. and n. Pa. to Minn., south to Kans., Mo., and Ala.)

Note This is a handsome species worthy of cultivation in the southern part of the province. Two varieties may be distinguished: var. *coronaria*, with glabrous hypanthium and sepals, and var. *dasycalyx* (Rehd.) Fern., with the hypanthium persistently somewhat pilose and the sepals glabrescent. (*dasycalyx*—with hairy calyx)

The native crabapple may be distinguished from the naturalized apples and crabapples which are frequent in fencerows and clearings by the following characters: the leaves are folded longitudinally in the bud; the blades are coarsely serrate or dentate and often lobed or cleft on vigorous shoots; the anthers are red instead of yellow.

Field check Scraggly small tree or tall shrub with thornlike short branches, triangular-ovate toothed or lobed leaves, open clusters of long-stalked pink to white fragrant flowers with reddish anthers and small, hard, yellow-green, sour crabapples.

Physocarpus Maxim. — Ninebark

A small genus of 6–10 species, chiefly North American but one species occurring in northeastern Asia. Shrubs with alternate, simple, palmately lobed leaves. Flowers white, on long pedicels and crowded in showy clusters, perfect, 5-parted. Fruit a group of 2–5 follicles; seeds with a hard shiny coat. Only one species is native to Ontario. (*Physocarpus*—from the Greek *physa*, bellows or bladder, and *karpos*, fruit, in reference to the inflated pods)

Physocarpus opulifolius (L.) Maxim. — Ninebark

A coarse, spreading, much-branched shrub up to 2–3 m tall. Branchlets greenish and smooth or slightly pubescent, somewhat angled or longitudinally ridged; older stems gray-brown with conspicuously peeling papery bark which persists in long thin strips.

Leaves alternate, simple, and deciduous; blades orbicular to ovate, 3–7 cm long and 3–6 cm wide, usually 3-lobed, cordate to truncate at the base, and mostly acute at the tip; dark green above, paler and somewhat pubescent beneath or glabrous except for a few hairs in the axils of the main veins; margins crenate-dentate; petioles 1–2 cm long with a pair of elongate deciduous stipules at the base.

Flowers 4–9 mm in diameter, 5-parted, and numerous, on elongating slender pedicels in showy, corymbose, stalked clusters at the ends of the branches; late June and July. Fruit a hemispherical cluster of conspicuous reddish-brown pods 5–10 mm long, containing 3 or 4 shiny, light brown seeds, and sometimes remaining on the bush over winter; Aug. and Sept. (*opulifolius*—with leaves like *Opulus*, the name referring to *Viburnum Opulus*, the guelder rose of Europe)

On sandy, gravelly, and rocky soils, especially along banks and shores of rivers and lakes, or along edges of thickets.

Around the shores of the Great Lakes and along major rivers from Lake Erie north to at least 53° N around James Bay; not reported west of 90° in northern Ontario. (Que. to Minn., south to Colo. and S.C.)

Note Pubescence on the lower surface of the leaves and on the branchlets and pods varies considerably. Plants with permanently pubescent pods have been segregated as *P. opulifolius* var. *intermedius* (Rydb.) Robins. The closely related species *P. capitatus* (Pursh) Ktze. is found on the Pacific coast, chiefly west of the Cascades from southern Alaska to southern California. (*intermedius*—intermediate; *capitatus*—capitate)

Field check Coarse much-branched shrub with conspicuously peeling papery bark, 3-lobed leaves, and hemispherical clusters of 5-parted white flowers maturing into showy reddish-brown pods.

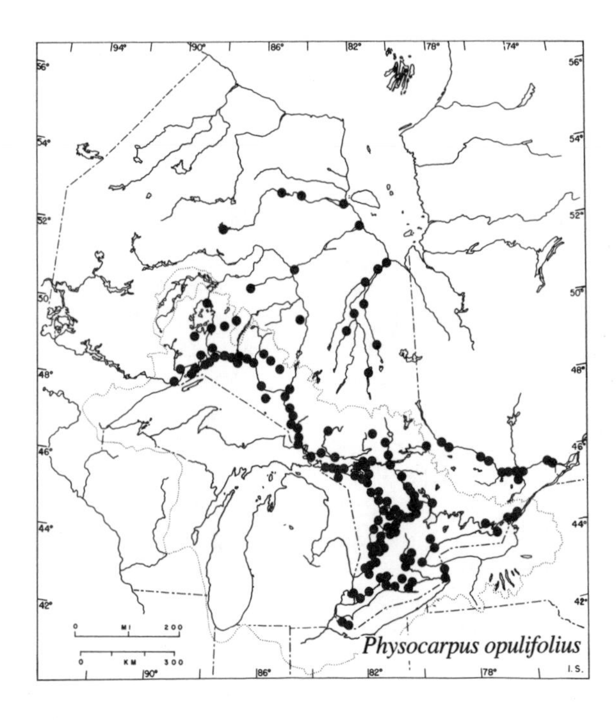

Physocarpus opulifolius

Potentilla L. — Cinquefoil, Five-finger

A large genus of about 500 species, chiefly herbs but also a few shrubs, widely distributed but most abundant in north temperate and arctic regions. Stems creeping and rooting at the nodes. Leaves alternate, and ternate, digitate, or odd-pinnate; stipules usually present. Flowers 5-parted or rarely 4-parted, the calyx flat and surrounded by an outer series of alternating bracts (the epicalyx); petals rounded, usually yellow; stamens few to many, attached outside a ring-shaped nectary; carpels numerous. Fruit a head of achenes on the dry receptacle. (*Potentilla*—from the Latin *potens*, powerful, but with a diminutive ending; certain species were reputed to have strong medicinal powers)

Key to *Potentilla*

a. Stems trailing or ascending from a woody caudex; leaves fan-shaped with 3 leaflets, each leaflet 3-toothed at the apex; flowers white **_P. tridentata_**

a. Stems erect, much-branched, up to 1 m tall; leaves pinnately compound with 3–7 (usually 5) silky-hairy leaflets; flowers bright yellow **_P. fruticosa_**

Potentilla fruticosa

Potentilla fruticosa L. — Shrubby cinquefoil

A low, much-branched, erect shrub usually less than 1 m tall. Branchlets pale brown to purplish-red and covered with long, silky, white hairs; older stems brown to gray-black with prominent exfoliating bark.

Leaves alternate, pinnately compound, and deciduous; blades 1–3.5 cm long with 3–7 (mostly 5) leaflets, the three terminal leaflets often united at base; leaflets oblong-lanceolate to elliptic, 1–2 cm long and 2–7 mm wide, pointed at both ends, dark green above and paler beneath, silky-pubescent on both surfaces or at least beneath; margins entire and usually revolute; petiole 1–2 cm long, pubescent, with a pair of conspicuous, long, pale, papery stipules at the base.

Flowers yellow, 5-parted, 1–2.5 cm in diameter when open, solitary or in close clusters at the ends of the branches; June to Sept. Fruit a compact small head of densely pubescent achenes surrounded by the persistent 10-parted calyx; achenes maturing in late summer or in autumn but persistent over winter. (*fruticosa*—shrubby)

In dry or wet habitats of river banks and lakeshores, fens and marshes; on sandy, rocky, or gravelly soils, shore dunes, rock ledges; in crevices of cliff faces and limestone pavement.

Throughout Ontario except along the western part of the north shore of Lake Erie. (Nfld. to Alaska, south to Calif., Ill., and N.J.; Eurasia)

Note This species shows a great deal of variation in growth habit, leaf width, and amount of pubescence. Its compactness, attractive foliage, showy flowers, and long period of blooming make shrubby cinquefoil a prized ornamental, widely used for low hedges and accent plantings.

Field check Low much-branched shrub with alternate, silky-hairy, pinnately compound leaves and prominent papery stipules; flowers yellow, 5-parted, blooming throughout the summer.

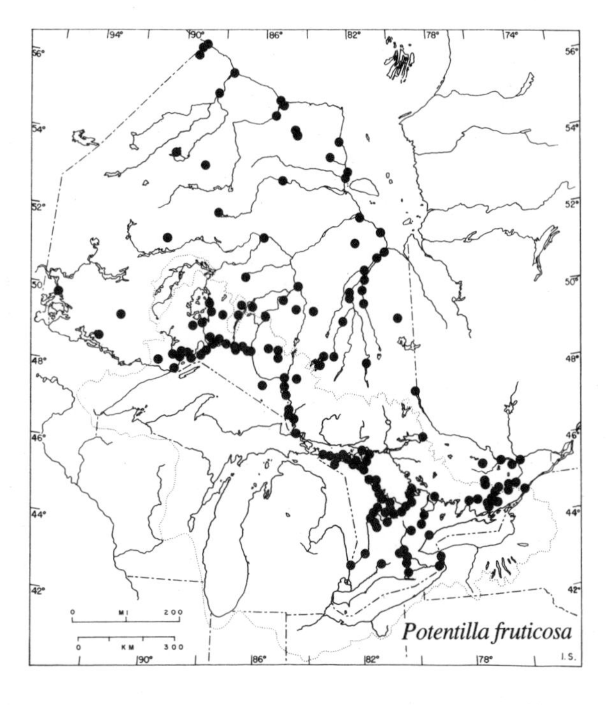

Potentilla tridentata

Potentilla tridentata Ait. — Three-toothed cinquefoil

A prostrate perennial with a woody caudex, slender stems, creeping underground or trailing, and erect or ascending, leafy, flowering branches 0.5 - 2 dm high. Stems reddish-brown to gray-black, with a scaly covering of persistent petiole bases.

Leaves evergreen, alternate, and compound, with a fanlike cluster of 3 leaflets; leaflets firm and leathery, cuneate-oblong to oblanceolate, 10 - 25 mm long, mostly 3-toothed at the blunt or truncate apex, tapering gradually to the base, dark green above, paler beneath, usually glabrous on both surfaces or the lower surface bearing short yellowish hairs, rarely hirsute on both sides; margins entire, except for the apical teeth, and barely revolute; petioles up to 3 cm long; stipules lance-linear, the lower part fused with the petiole.

Flowers 5-parted, white or rarely pinkish, 10 - 15 mm wide, in open stiffish cymes; June and July. Fruit a small cluster of densely hairy achenes surrounded by the persistent 10-parted calyx; July and Aug. (*tridentata*—three-toothed)

Chiefly in dry or exposed habitats, including cliffs, crevices of rocks, sandy, gravelly, or rocky beach ridges, rocky ledges along lakeshores, and sandy pine woods.

Common throughout the Lake Superior region and less frequent southward on the Canadian Shield to about 45° N; widely scattered in the drainage basin of Hudson Bay and James Bay. (Greenl. and Nfld. to Mack. Dist. of N.W.T., south to N. Dak., Iowa, N.Y., and the mountains of n. Ga.)

Field check Prostrate perennial with a woody caudex and erect, leafy, flowering branches 0.5 - 2 dm high; leaves evergreen, with 3 leaflets, 3-toothed at the apex; flowers white.

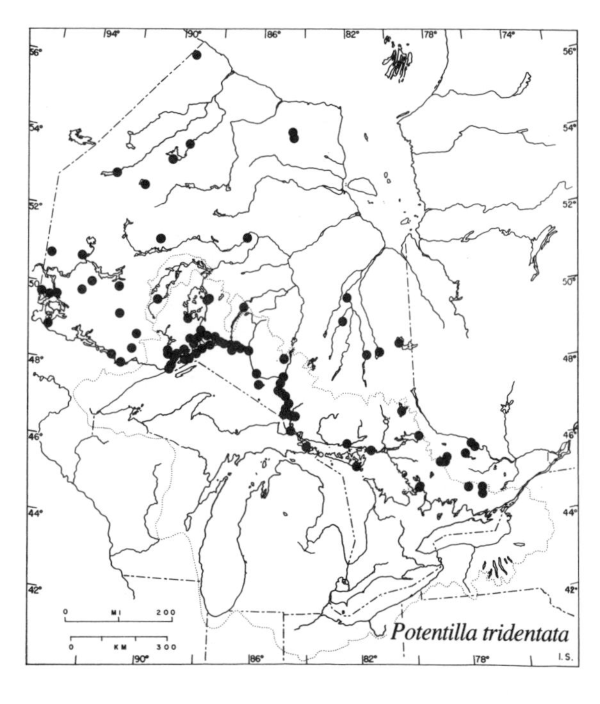

Prunus L. — Plum, Cherry

A genus of 200–400 or more species, widespread in the northern hemisphere and extending also into the Andean region of South America. Shrubs and trees, sometimes bearing thorns or spinelike short shoots. Leaves alternate, simple, and deciduous or evergreen, the margins usually serrate, and the petioles often glandular, especially near the base of the blade. Flowers perfect, 5-parted, white to pink, solitary or in lateral or terminal clusters. Fruit a fleshy drupe with a bony stone enclosing the seed. (*Prunus*—Latin name for the plum tree)

Several species are important for their edible fruits, e.g., apricot (*P. armeniaca* L.), plum (*P. domestica* L.), almond (*P. amygdalus* Batsch), peach (*P. persica* (L.) Batsch), and cherry (*P. cerasus* L.), and others are cultivated as ornamental trees and shrubs for their attractive flowers, e.g., Japanese cherry (*P. serrulata* Lindl.) and flowering almond (*P. triloba* Lindl.). (*armeniaca*—Armenian; *domestica*—at home, indigenous; *amygdalus*—a generic name for the almond; *persica*—Persian; *cerasus*—a generic name for the cherry; *serrulata*—finely serrate or saw-toothed; *triloba*—with three lobes)

Note The stems of cherries and plums are sometimes infected by a fungus, *Apiosporina morbosa* (Schw.) Arx, which results in irregularly swollen and deformed blackish portions, commonly called Black Knot.

Key to *Prunus*

a. Stems usually armed with short spine-tipped spur-shoots; leaves broadly ovate; flowers solitary or in clusters of 2–5 from the old wood; fruit a drupe with a stone somewhat compressed and longer than broad (Plums)

  b. Leaf margins sharply serrate, the teeth acuminate, not gland-tipped **_P. americana_**

  b. Leaf margins crenate-serrate, the teeth triangular-ovate, gland-tipped (only the scars may remain on old leaves) **_P. nigra_**

a. Stems unarmed; leaves elliptical, ovate-lanceolate, obovate, or cuneate; fruit a drupe with a rounded or ellipsoidal stone (Cherries)

  c. Flowers solitary or in umbels of 2–5, close to the stem on the old wood; sepals pubescent within, at least at the base

    d. Coarse shrub or tree; leaves lanceolate, with conspicuously tapered tips, the margins toothed to the base; teeth gland-tipped, at least at first **_P. pensylvanica_**

    d. Low shrub, erect or trailing; leaves broadly elliptical to cuneate with acute or blunt tips, the margins not toothed to the base; teeth not gland-tipped **_P. pumila_**

  c. Flowers in elongate racemes at the tips of the branches of the current season; sepals glabrous within

    e. Shrub, forming thickets or growing singly as a small to medium-sized tree; leaves mostly broadest at or above the middle; teeth sharp-pointed, not incurved; lower surface of leaves glabrous or with tufts of hairs in the axils of the veins **_P. virginiana_**

    e. Shrublike in some habitats but normally a large tree of deciduous or mixed woods; leaves broadest at or below the middle; teeth blunt, somewhat incurved; lower surface of leaves with white to tawny pubescence along basal portion of midvein **_P. serotina_**

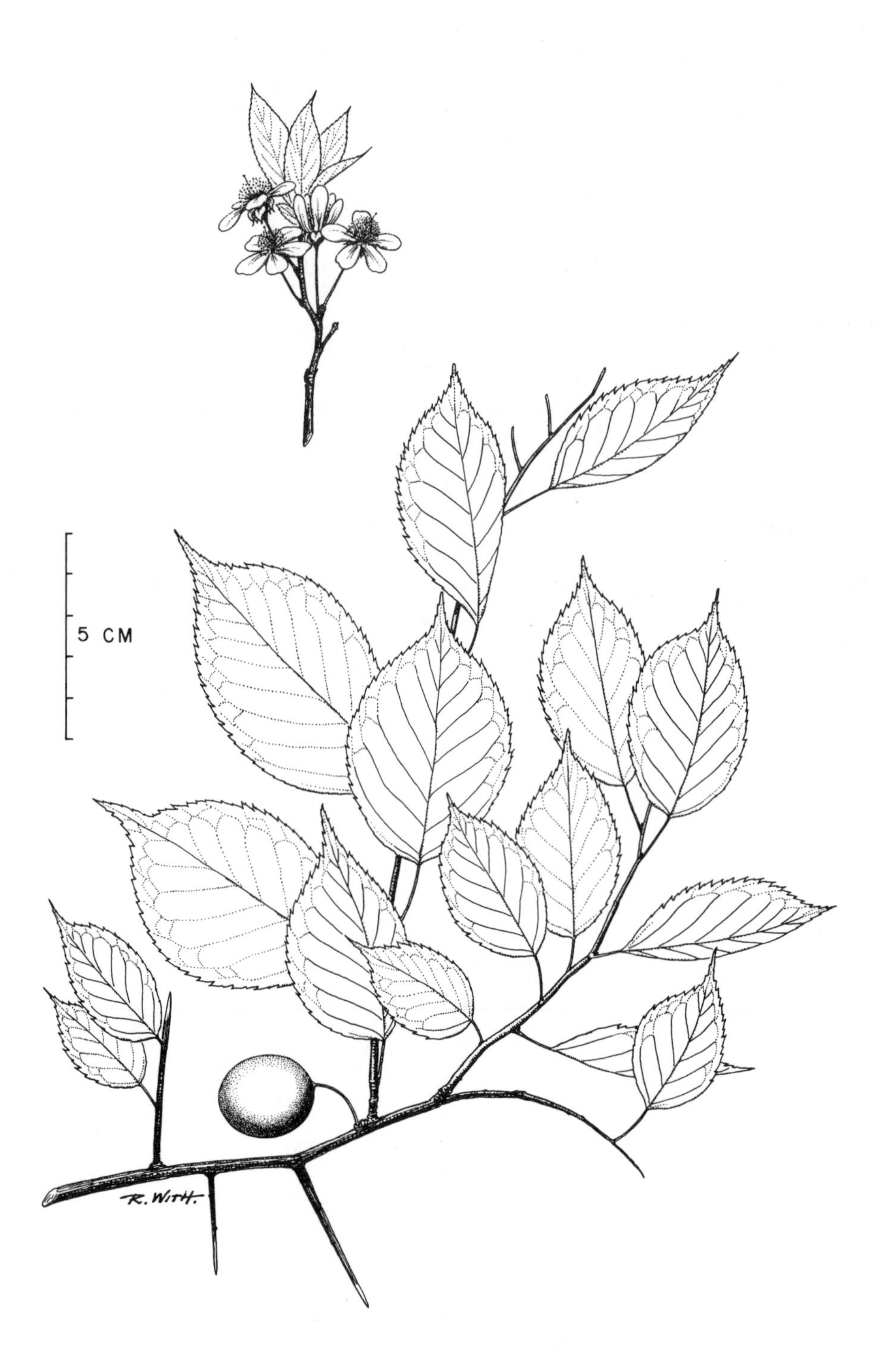

Prunus americana

Prunus americana Marsh. **Wild plum**

A coarse shrub or small tree which may reach a height of 8–10 m, often spreading by root-shoots and forming thickets. Branches more or less thorny, mostly glabrous; bark on old stems and trunks rough, dark brown or grayish, breaking up into irregular plates.

Leaves alternate, simple, and deciduous; blades 4–12 cm long and 2–5 cm wide, broadly lance-ovate, oblong, or obovate, glabrous above, slightly pubescent or glabrate beneath, the apex long-tapered; margins sharply serrate, the teeth sharp-pointed or bristle-tipped but not glandular; petiole about 1 cm long.

Flowers white and very fragrant, on pedicels 1–2.5 cm long and in clusters of 2–5, opening as the leaves expand; corolla 5-parted, 1.5–2.5 cm across; late April to early June. Fruit a red to yellow, edible, subglobose drupe 2–3 cm in diameter with a compressed central stone; Aug. to Sept. (*americana* - American)

In thickets and fencerows and along stream banks and borders of woods; often in alluvial or moist soils.

Commonest in the Deciduous Forest Region; also at the eastern end of Lake Ontario. (N. Eng. to s. Sask., south to Ariz. and n. Fla.)

Note Although the fruit has a tough astringent skin, the pleasantly flavoured and juicy flesh makes good jam or preserves.

Field check Stems somewhat thorny; leaves ovate, oblong, or obovate, sharply toothed; fragrant white flowers in spring with the expanding leaves; edible yellow to red plums, solitary or in clusters.

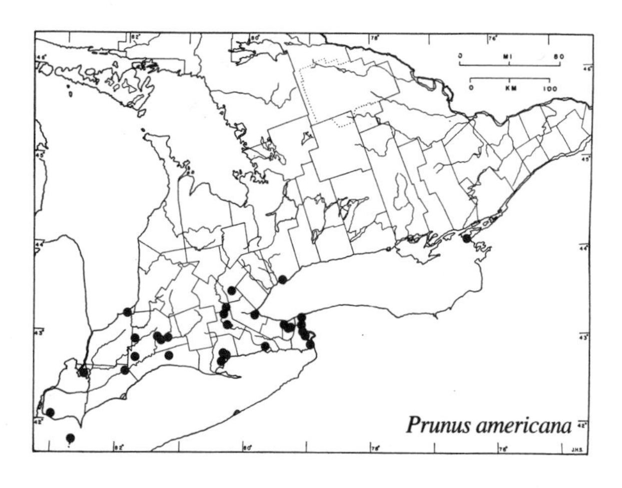

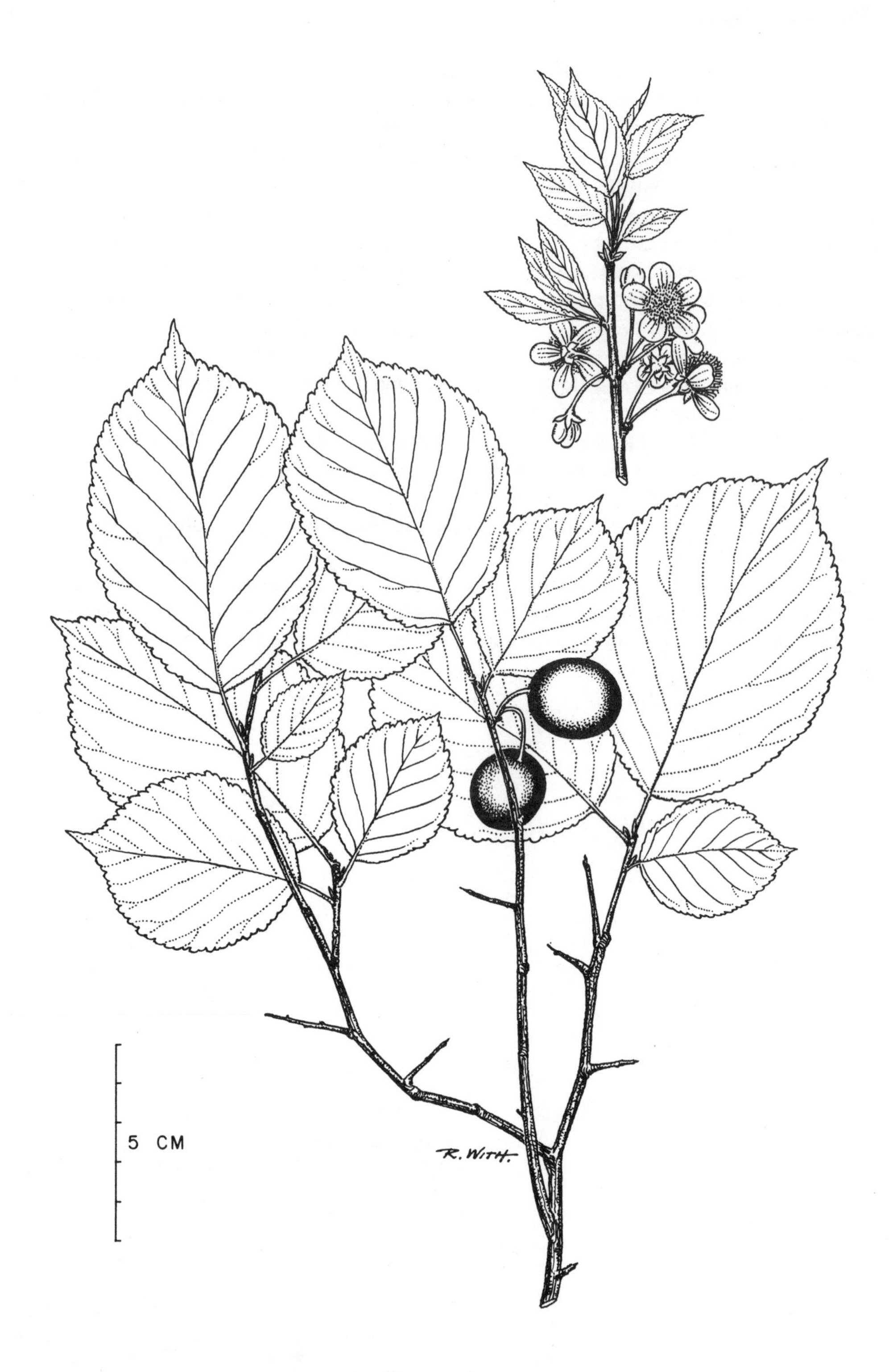

Prunus nigra

Prunus nigra Ait. **Canada plum**

A small tree or coarse thicket-forming shrub reaching a height of 7 – 10 m. Branches armed with short spine-tipped spur-shoots; bark of older stems and trunks brownish-gray, exfoliating in plates.

Leaves alternate, simple, and deciduous; blades 7 – 12 cm long and 3 – 7 cm wide, broadly oval to oblong or obovate, nearly glabrous on both surfaces but usually with a few hairs along the veins, the tip abruptly pointed or acuminate, the base rounded or slightly heart-shaped; margins with rounded or triangular-ovate, blunt, gland-tipped teeth; petioles 1 – 2 cm long and usually with one or two glands near the junction with the blade.

Flowers white and fragrant, turning pinkish with age, appearing before or with the leaves in clusters of 3 – 4 on reddish pedicels 1 – 2 cm long; corolla 5-parted, 2 – 3 cm across; late April to early June. Fruit an ellipsoidal drupe, red, orange-red, or yellowish, 2 – 3 cm long, with a compressed central stone; Aug. and Sept. (*nigra* — black, in reference to the dark branches)

In thickets, on bottomlands and hillsides; also along the edges of woods, in fencerows, and on abandoned farmland.

Common throughout southern Ontario, less so northward to Manitoulin Island and Lake Timiskaming at about 47°30′ N; also in the Rainy River and Thunder Bay districts west of 88° W. (N. Eng. and N.S. to s.e. Man., south to Iowa and Ga.)

Note Until fully ripe the fruit is tough and bitter-tasting but with the first frosts it becomes edible and is used for making jams and preserves.

Field check Stems thorny; leaves oval to oblong or obovate, with gland-tipped blunt teeth; fragrant white to roseate flowers in spring before or with the leaves; edible yellow to red plums, solitary or several in a cluster.

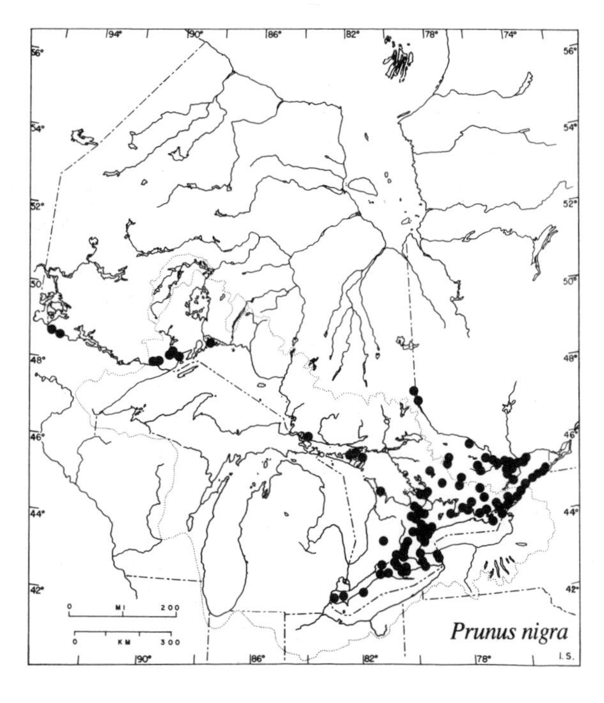

Prunus pensylvanica

Prunus pensylvanica L. fil

Pincherry, Birdcherry Firecherry

A coarse shrub or small tree up to 12 m tall, often with numerous, ascending, slender, and leafy vegetative shoots. Branchlets smooth, gray to reddish or purplish, with scattered, small but conspicuous, pale brown lenticels; bark on older branches and trunks separating horizontally into broad papery plates.

Leaves alternate, simple, and deciduous; blades oblong-lanceolate to narrowly ovate (rarely oblong, obovate, or broadly ovate, especially on vigorous vegetative shoots), 4–11 cm long, 1–2 cm wide, usually less than half as wide as long; the base obtuse to rounded and the tip acute to long-acuminate; margins finely and irregularly serrate, with rounded or blunt, somewhat incurved, gland-tipped teeth; upper surface bright green and shiny, the lower slightly paler and smooth; petioles 1–3 cm long, usually glandular near the junction with the blade.

Flowers white, in clusters of 2–6 on pedicels 1–2 cm long, scattered along the branches of the previous season or on short shoots, often crowded at the tips of the branches, flowering with the expanding leaves; late March in the south to June and early July in the north. Fruit a globose, light red, juicy drupe, 4–7 mm in diameter, with acidic flesh and a large central stone; July to Sept. (The original spelling of the specific name, although not current usage for the name of the state, should be retained, according to the International Rules of Botanical Nomenclature. *pensylvanica*—Pennsylvanian)

In dry woods, clearings, recent burns, and thickets; also on sandy and gravelly banks and shores of rivers and lakes, along trails, roadsides, fencerows, rocky ridges, cliffs, and on limestone pavement.

Common throughout Ontario from Lake Erie to Lake Superior and in the Hudson Bay lowlands to 54°30′ N. (Nfld. to B.C., south to Colo. and N.C.)

Field check Bark reddish with conspicuous lenticels; leaves lanceolate, with long-tapered tips and gland-tipped incurved teeth; 5-parted white flowers in clusters along the previous year's growth.

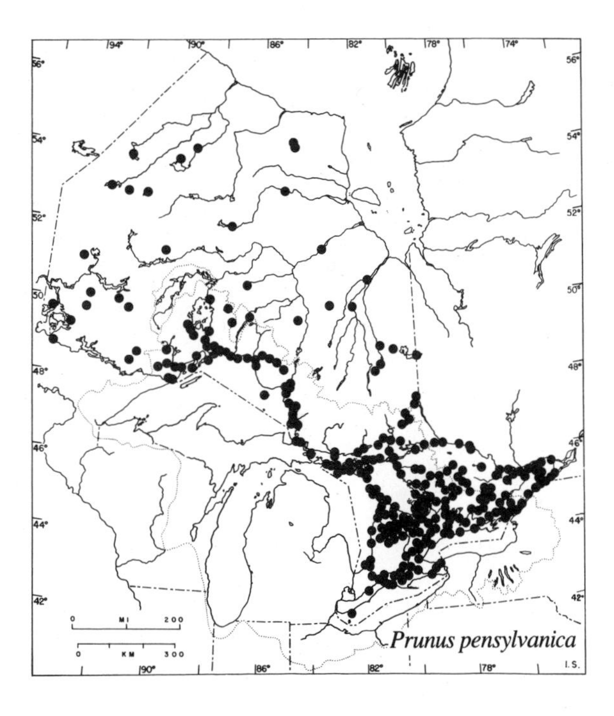

Prunus pumila

Prunus pumila L. (including *P. susquehanae* Willd., *P. cuneata* Raf., *P. depressa* Pursh) — Sandcherry, Dwarf cherry

A prostrate and trailing or decumbent shrub or a low, erect, diffusely branched shrub up to 2 m high.

Leaves alternate, simple, and deciduous; blades 4–10 cm long and 1–3 cm wide, from narrowly elliptical to oblong, firm to leathery, usually glossy above and paler or whitened beneath, acute or merely pointed at the tip and tapered at base; margins firm or cartilaginous, finely to remotely toothed, at least above the middle, the teeth usually gland-tipped and often appressed; petioles 3–10 mm long; stipules slender, up to 8 mm long, deciduous.

Flowers white, 5-parted, in clusters of 2–4 on pedicels 4–15 mm long; petals 4–8 mm long; May to early July. Fruit an edible purple to blackish cherry 10–15 mm in diameter, with an acid or astringent taste; July to Sept. (*pumila*—dwarf; *susquehanae*—of the Susquehanna River region; *cuneata*—wedge-shaped; *depressa*—low, depressed)

In both dry and wet habitats; open sandy, gravelly, or bouldery beaches of lakes and shores of rivers; in crevices of rock faces and ridges; on sand dunes, flats, and barrens; also in sedge mats of bogs and fens and in wet depressions between dunes.

From Lake Erie and Lake Ontario to Lake Superior, westward to Lake-of-the-Woods and northward to about 52° N. (N.B. and s. Lab. to s. Man., southwest to Wyo. and east to N.C.)

Note The sandcherries need further study in our area but those wishing to distinguish varieties can follow the treatment by Gleason (1952). Three of the four varieties he describes occur in Ontario, namely:

P. pumila var. *pumila*, a diffusely branched shrub with branches erect or sometimes decumbent on active sand dunes; leaves oblanceolate; especially along the shores of the Great Lakes.

P. pumila var. *depressa* (Pursh) Gleason, with stems depressed, prostrate, or trailing, often rooting at the nodes and forming mats, the new shoots often reddish and highly lustrous; leaves spatulate, oblanceolate, or narrowly obovate, whitened beneath.

P. pumila var. *cuneata* (Raf.) Bailey, with erect or diffusely branched stems; leaves oblong to oblong-obovate, prominently reticulate.

Field check Low or trailing shrub with alternate, narrow, dark green leaves, paler beneath, finely toothed, mainly above the middle; flowers 5-parted, white, in clusters of 2–4 along the stem; purple-black cherries in late summer or early fall.

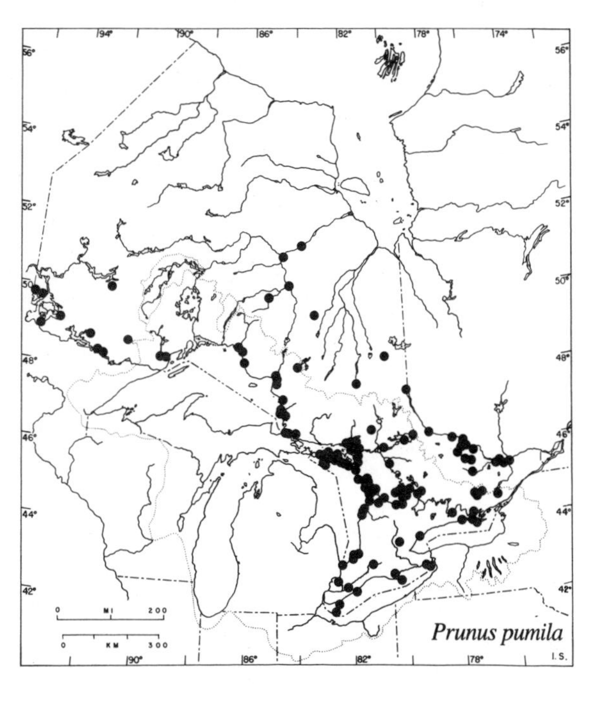

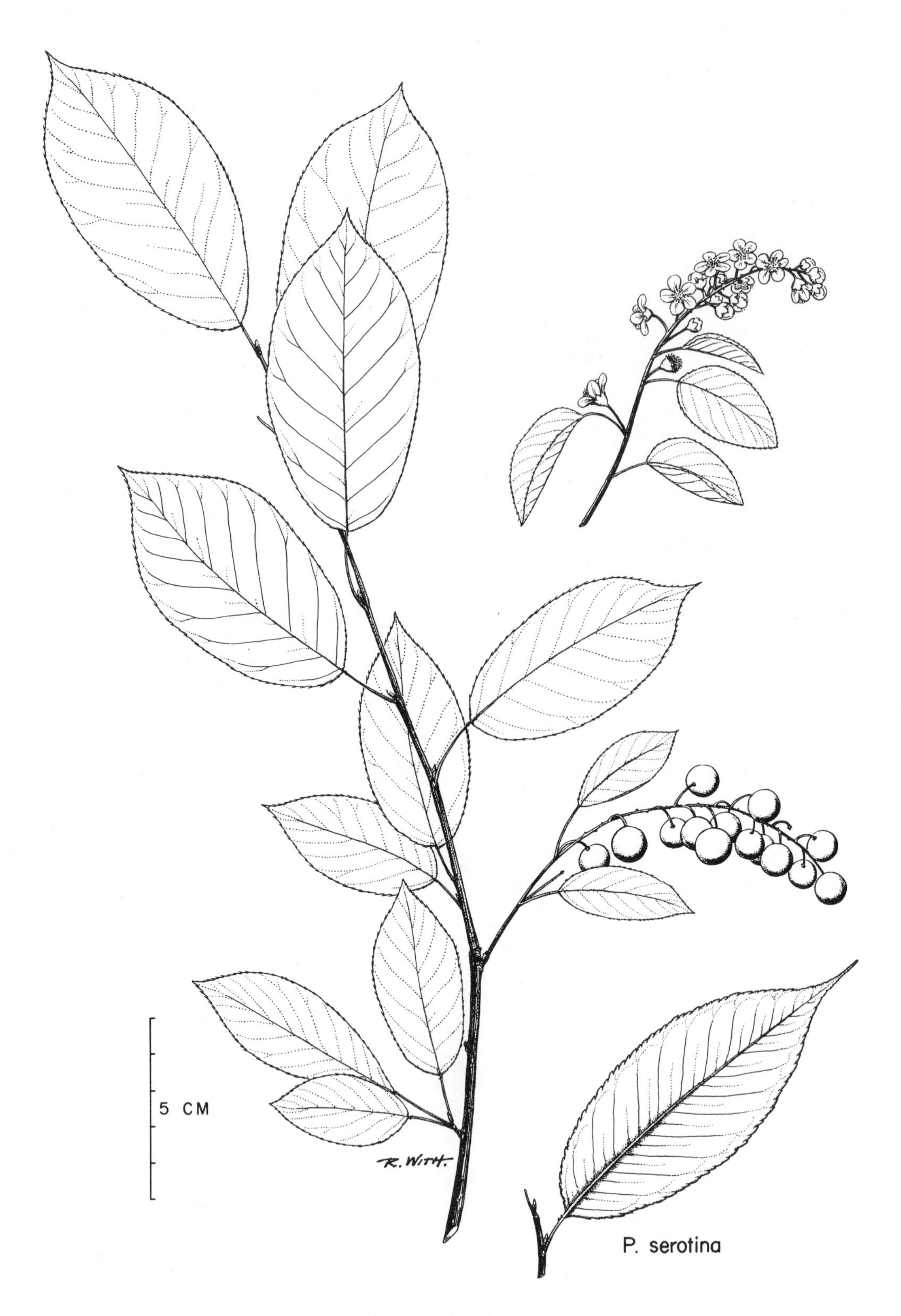

Prunus virginiana

Prunus virginiana L.fil. — Chokecherry

A large shrub spreading underground and sending up shoots which form thickets 2 - 3 m high, or occasionally a small tree up to 10 m tall with a trunk 5 - 10 cm in diameter. Branchlets glabrous or puberulent, reddish-brown to purplish-gray, with strong disagreeable odour when bruised.

Leaves alternate, simple, and deciduous; blades 4 - 12 cm long and 2 - 6 cm wide, broadly oval to ovate or obovate, usually broadest at or above the middle, thin, glabrous above and glabrous to pubescent beneath or with tufts of hairs in the vein axils, acute or abruptly acuminate at the tip, and rounded to tapered at the base; margins finely serrate, the teeth sharp-pointed, not incurved; petioles 0.5 - 2 cm long with one to several glands at or near the base of the blade.

Flowers white, small, and numerous, in elongate arching racemes terminating the leafy branches; racemes 5 - 15 cm long, the 10 - 25 flowers well spaced and borne on pedicels 4 - 8 mm long; May and June. Fruit a deep red, red-purple, or nearly black drupe 8 - 10 mm in diameter, with acid astringent flesh and a large central stone; Aug. and Sept. (*virginiana*—Virginian)

Along river banks, roadsides, fencerows, lakeshores, and the edges of woods and swamps; on hillsides, talus slopes, rocky ridges, open ledges, and gravelly and sandy soils.

Common throughout southern Ontario and northward to James Bay, reaching its limit at about 53° N. (Nfld. to B.C., south to Calif. and N.C.)

Note Although they cause a puckering of the mouth when eaten raw, the cherries can be cooked and combined with apple juice to make a delicious jelly. Chokecherry may be confused with black cherry (*Prunus serotina* Ehrh.), which is normally a large forest tree in southern Ontario but is also found with a shrubby habit in thickets and fencerows and along the margins of the woods. Black cherry may be distinguished by its firm, lanceolate, acuminate leaves, dark green above and much paler beneath, with slender or blunt incurved teeth. The midrib on the lower surface is usually lined on both sides with whitish to rusty-coloured hairs from the base of the leaf for about one-third to one-half its length (shown at lower right of illustration of *P. virginiana*).

It has been reported that leaves of chokecherry that have been injured by frost or extreme drought are poisonous to cattle. The same has been said of wilted leaves and young new growth of black cherry. The toxic principle is hydrocyanic acid. (*serotina*—late)

Field check Thicket-forming shrub or small tree; leaves oval to obovate, with fine sharp teeth and acute tips; flowers in terminal racemes; fruit a dark red-purple cherry with acid astringent flesh.

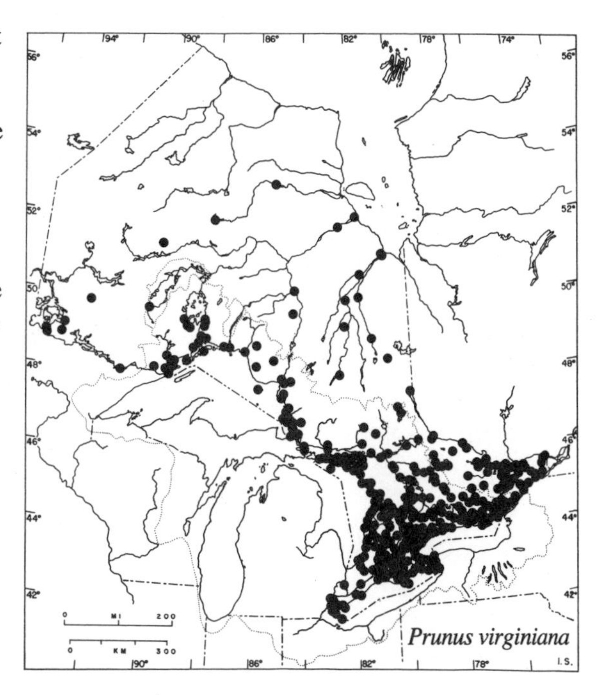

Rosa L. — Rose

A genus of at least 200 species, although more than twice that number have been described. Found chiefly in the northern hemisphere but also on mountains in the tropics. Shrubs and woody vines usually with prickles and sometimes also bristles. Leaves alternate, pinnately compound (3–11 leaflets), deciduous or persistent, the margins serrate; stipules prominent and fused to the petiole for most of their length. Flowers perfect, 5-parted, pink to red, yellow, or white; petals broad and rounded; stamens numerous in several whorls, inserted on a floral disc. Carpels numerous, borne on the inside of a deeply urn-shaped floral cup. Fruit a yellow to red "hip", the fleshy or pulpy receptacle tissue surrounding the hairy achenes. (*Rosa*—the Latin name for the rose)

Several Ontario species are of horticultural interest. *Rosa carolina*, *R. palustris*, and *R. setigera* have been recommended by Wyman (1969) for planting in Hardiness Zone 4, which includes the Deciduous Forest Region of southern Ontario. *R. blanda* and *R. acicularis* are used in breeding programs where hardiness is desired. The introduced and frequently planted *R. multiflora* is used as stock for grafting. At least two Ontario roses were in the Empress Josephine's famous Gardens of Malmaison—*R. carolina* and *R. setigera*. The latter, although rare as a native shrub in Ontario, has been extensively used in this province in rose breeding. (*multiflora*—many-flowered)

The mature hips of some species are edible, contain Vitamin C, and are often collected for jelly.

Note In the following key the hybrids and intermediates that occur in nature will not key down conclusively to one of the species treated. Hybrids that may occur in Ontario between native species are *Rosa blanda* × *carolina*, *R. blanda* × *palustris*, *R. carolina* × *acicularis*, *R. carolina* × *palustris*.

Key to *Rosa*

a. Lower surface and margins of leaflets heavily glandular **R. *eglanteria***

a. Lower surface of leaflets not heavily glandular (glands may be present on stipules, rachis, and tips of teeth)

- b. Stems trailing, leaning, or climbing, vinelike; leaflets 3 (rarely 5); styles about as long as the stamens, united into a column, exserted **R. *setigera***
- b. Stems erect, branching; leaflets 5 - 9; styles shorter than the stamens, free, forming a bunch in the throat of the floral tube
  - c. Pedicel and hypanthium glabrous; nodal prickles usually absent
    - d. Stems without prickles, at least on the upper branches; rachis of leaf pubescent; stipules entire or remotely gland-toothed **R. *blanda***
    - d. Stems with prickles throughout; rachis of leaf glandular; stipules with small dark glands on the margin **R. *acicularis***
  - c. Pedicel and hypanthium stipitate-glandular; stems with conspicuous nodal prickles
    - e. Leaves finely toothed; nodal prickles stoutish, decurved; plants of wet habitats **R. *palustris***
    - e. Leaves coarsely toothed; nodal prickles slender, straight; plants of dry habitats **R. *carolina***

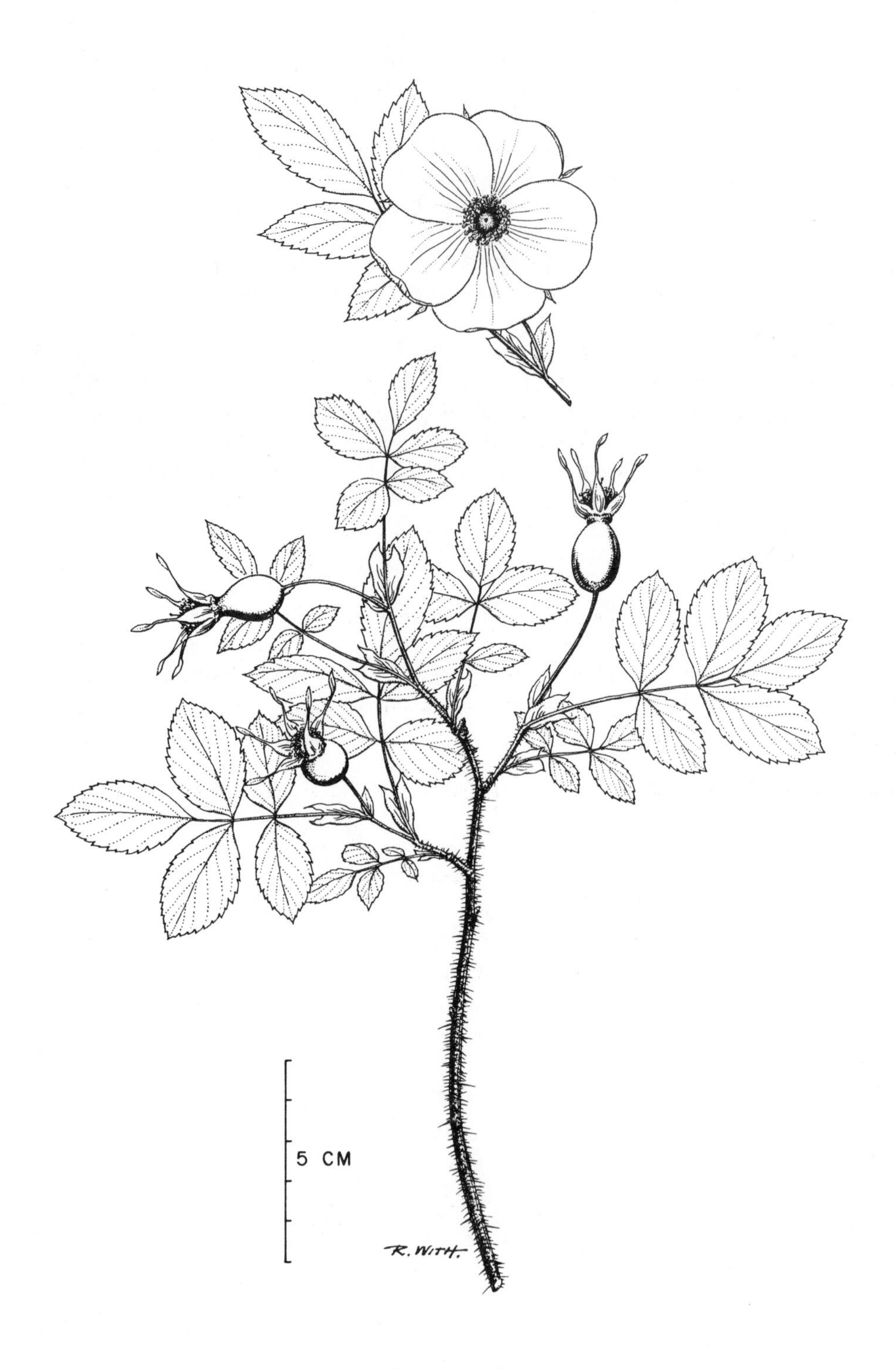

Rosa acicularis

Rosa acicularis Lindl. Prickly wild rose

A low bushy shrub usually less than 1 m tall. Branchlets reddish, covered with numerous, slender, straight prickles 3–4 mm long which persist on the older branches often down to ground level.

Leaves alternate, deciduous, and pinnately compound with 5 or 7 leaflets and a pair of conspicuous stipules; leaflets thin, oval to elliptic, 2–5 cm long, acute to obtuse at the apex, rounded to cordate at base, sessile or short-stalked, dull green and glabrous above, paler and minutely downy beneath; margins sharply serrate with round-based gland-tipped teeth, these less prominent near the leaf base; stipules narrow at the base, broadened at the free end, with somewhat spreading lobes, the margins bearing small dark glands; rachis of the leaf minutely pubescent and usually glandular.

Flowers solitary or few at the ends of the leafy branches; petals pink, 2–3 cm long and nearly as wide; sepals glandular on the outside and often white-pubescent on the inside; late May to July. Fruit ovoid, ellipsoid, or nearly globular, about 2 cm long, bright red, and glabrous, many-seeded, ripening in late summer or early fall. (*acicularis*—needlelike)

In meadows, clearings, and open woods; on rocky shelves and ridges, limestone flats, talus slopes, clay and sand banks; and along roadsides, lake shores, and river banks.

Throughout Ontario but less common south and east of Lake Huron. (Que. to Alaska and B.C., south to n.N. Mex., Pa., and N. Eng.; Eurasia)

Note A similar wild rose of the prairies, *Rosa arkansana* Porter (including var. *suffulta* (Greene) Cock.), probably extends into western Ontario near the Manitoba and Minnesota borders. That species and its variety are distinguished from *R. acicularis* by the absence of glands on the petiole, leaf rachis, and margins of the stipules. (*arkansana*—of Arkansas; *suffulta*—supported, propped up)

Field check Stems prickly; leaflets 5 or 7; leaf rachis and margins of stipules glandular.

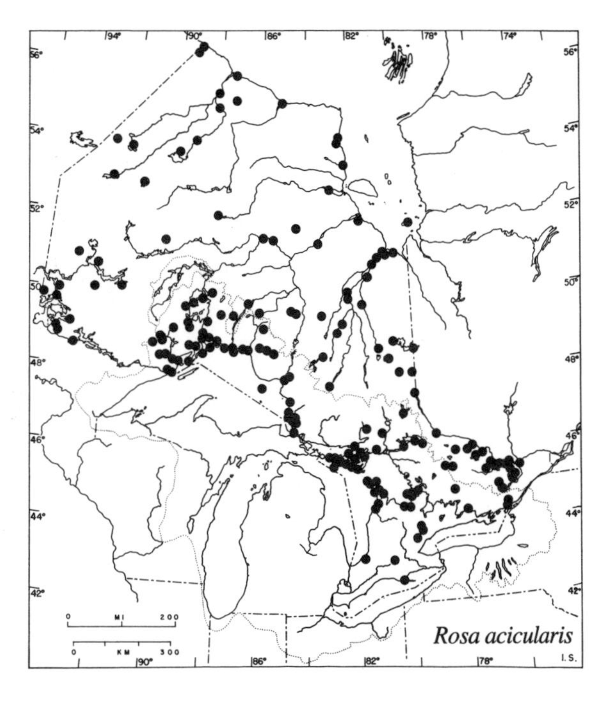

Rosa blanda

Rosa blanda Ait. — Smooth wild rose

A low shrub up to 1.5 m tall. Branchlets reddish-purple, unarmed or with a few slender straight prickles mostly near the base of the plant; vigorous shoots often densely prickly below but the upper leafy flowering branches smooth and nearly without prickles.

Leaves alternate, deciduous, and pinnately compound, with 5, 7, or rarely 9 leaflets and a pair of conspicuous stipules; leaflets thin, oval, elliptic or obovate, 1-4.5 cm long, the terminal one short-stalked and the lateral ones sessile or short-stalked, dull green and nearly glabrous above, paler and finely pubescent beneath, acute, obtuse, or rounded at the apex and rounded to cuneate at the base; margins sharply serrate to just below the middle; stipules more or less parallel-sided but gradually broadened toward the free end, the margins entire or remotely glandular-toothed; rachis of the leaf minutely pubescent.

Flowers solitary or few in clusters at the ends of the leafy branches; petals pink, 2-3 cm long, broadly wedge-shaped; sepals glandular or smooth on the outside, white-pubescent within; May to early July. Fruit subglobose, ovoid, or somewhat pear-shaped, 1-1.5 cm in diameter, red, and glabrous; Aug. to early Oct. (*blanda*—mild or bland, in reference to the scarcity of prickles on the upper leafy flowering branches)

In pastures, meadows, thickets, clearings, and open woods; on sandy and clayey banks, rocky and gravelly shores; and along roadsides.

Common in southern Ontario and northward, becoming rare along the west coast of James Bay at about 53° N. (N.B. to e. Sask., south to Neb. and Pa.)

Field check Prickles rare or absent on the upper leafy branches; leaflets 5 or 7; stipules entire or remotely glandular-toothed.

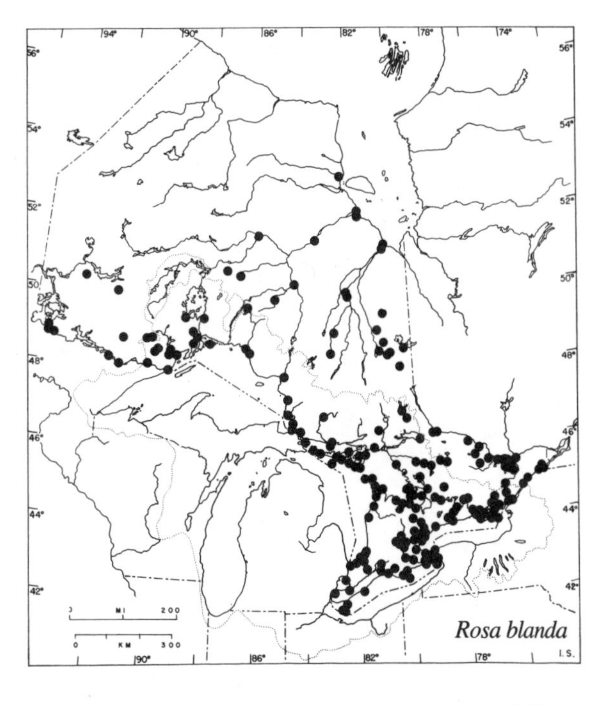

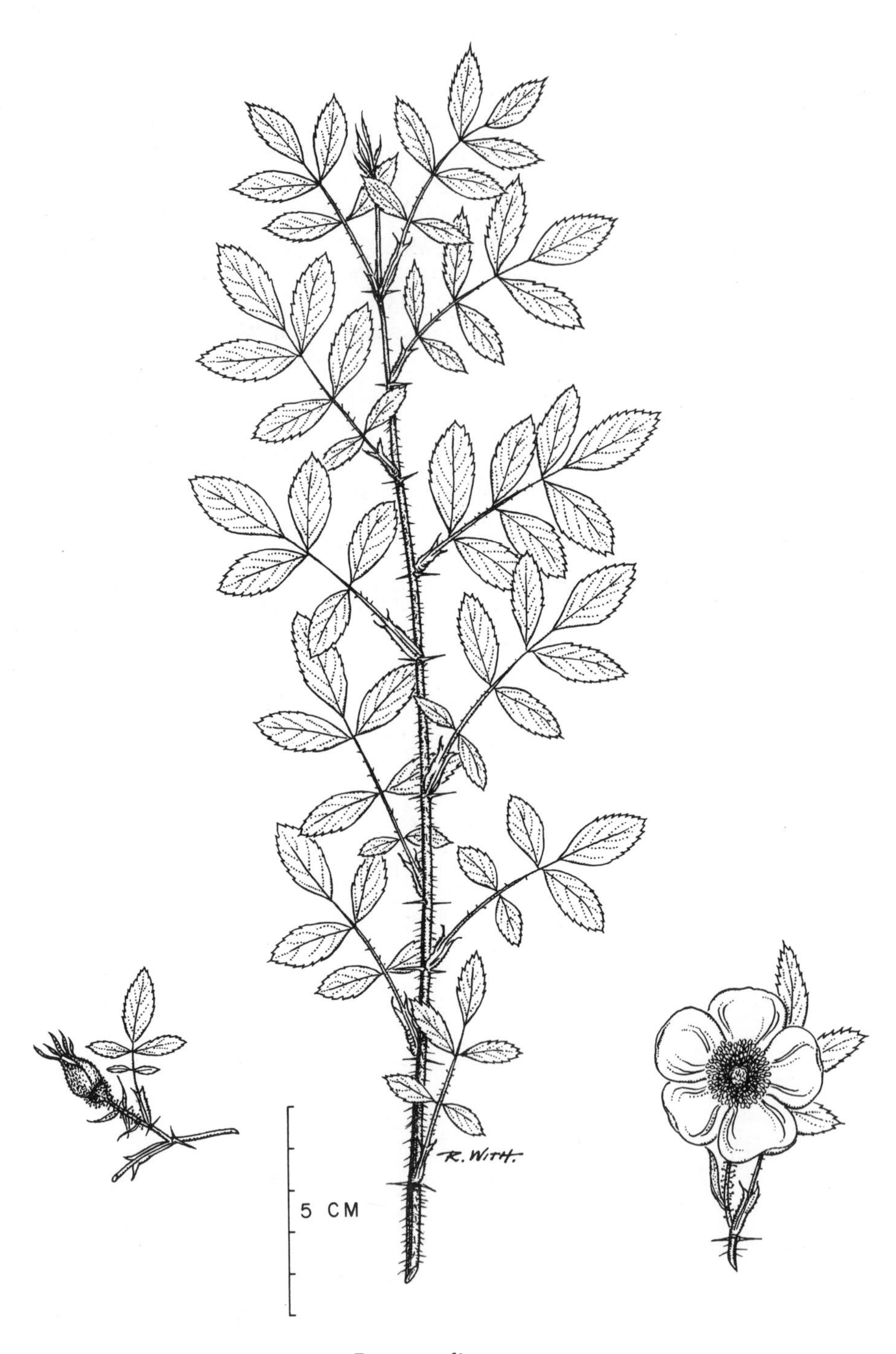

Rosa carolina

Rosa carolina L. — Pasture rose

A low slender or little-branched shrub less than 1 m tall, the stems usually arising singly from the underground rhizome. Branchlets reddish-brown, smooth or with numerous scattered slender straight prickles which, in the leafy flowering branches, may occur only as infrastipular pairs, i.e., at the base of the stipules. The prickles are round in section, not conspicuously broad-based.

Leaves alternate, deciduous, and pinnately compound with 3, 5, or 7 leaflets and narrow stipules; leaflets thin, ovate, elliptic-lanceolate to broadly oval, 1.5 - 4.5 cm long, sessile or short-stalked, dull green and glabrous above, paler and glabrous or minutely downy beneath, acute or obtuse at the apex, rounded to wedge-shaped at base; margins coarsely serrate except near the base, often with scattered glands; stipules narrow (less than 2.5 mm wide), entire to gland-toothed; rachis of the leaf minutely pubescent, sometimes glandular.

Flowers mostly solitary, up to 6 cm broad when fully open; petals pink; pedicel, hypanthium, and sepals stipitate-glandular; mid-June to July. Fruit red, subglobose, 1 - 1.5 cm in diameter; late summer to fall. (*carolina* — Carolinian)

Chiefly in dry places; sandy roadsides, thickets, borders of woods and pastures.

In the Deciduous Forest Region, including the eastern end of Lake Ontario and the upper part of the St. Lawrence Valley. (N.S. to Minn. and Neb., south to Tex. and Fla.)

Field check Low shrub of dry habitats; stems with straight slender prickles at the nodes; pedicels and hypanthium with stalked glands.

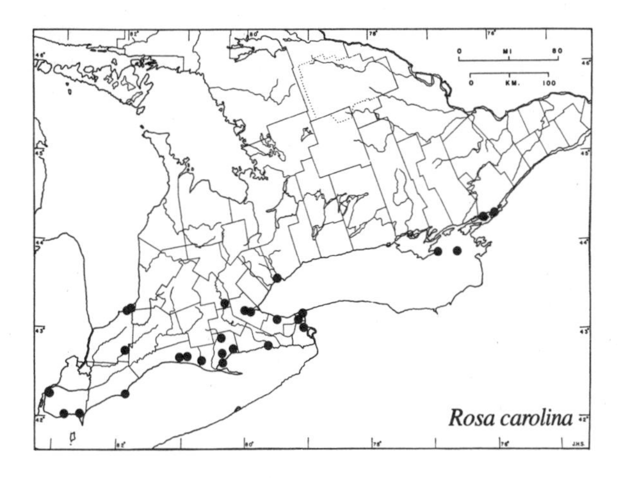

Rosa eglanteria

Rosa eglanteria L. (*R. rubiginosa* L.) — Sweetbrier, Eglantine

A strong shrub reaching a height of 2 m or more with stiffly erect or spreading stems. Branchlets brownish to gray, the coarse new shoots bearing many flattened and decurved, broad-based, hooklike prickles up to 1 cm long, these extending into the leafy flowering branches, which also bear scattered straight spinelike prickles.

Leaves alternate, deciduous, and pinnately compound with 5, 7, or 9 (mostly 7) leaflets and elongate stipules; leaflets thin to firm, aromatic, especially when crushed, with a fragrance reminiscent of fresh apples, obovate to broadly oval, 1–3 cm long, blunt or obtuse at the apex and rounded at the base, the terminal one the largest and the lateral pairs reduced in size towards the leaf base, dull green and glabrous above, scurfy and glandular beneath, especially along the veins; margins serrate with sharp divergent teeth and lined with small dark-tipped glands; stipules with marginal stipitate glands and sharp-pointed divergent lobes; leaf rachis stipitate-glandular and with scattered straight or hooked prickles.

Flowers pink, varying to white, 2–5 cm in diameter, solitary or in small clusters; sepals lacerate or with a comblike fringe and covered with stalked glands; pedicels with conspicuous stalked glands; hypanthium glabrous or with a few stalked glands; late May to early July. Fruit orange to red, subglobose, 1–1.5 cm in diameter; Aug. and Sept. (*eglanteria*—the Latin form of the old English and French names for this rose; *rubiginosa*—brownish-red)

In pastures, old fields, fencerows, thickets, and sandy woods and along roadsides, stream banks, and river flats.

Widely distributed throughout southern Ontario from Lake Erie to the northern end of Lake Huron and eastward to the St. Lawrence River and the Ottawa district; rare or absent on the Canadian Shield. (Introduced and naturalized from Europe)

Field check Aromatic foliage; coarse stems with large, hooked, flat-based prickles; pedicels, sepals, and leaves copiously glandular.

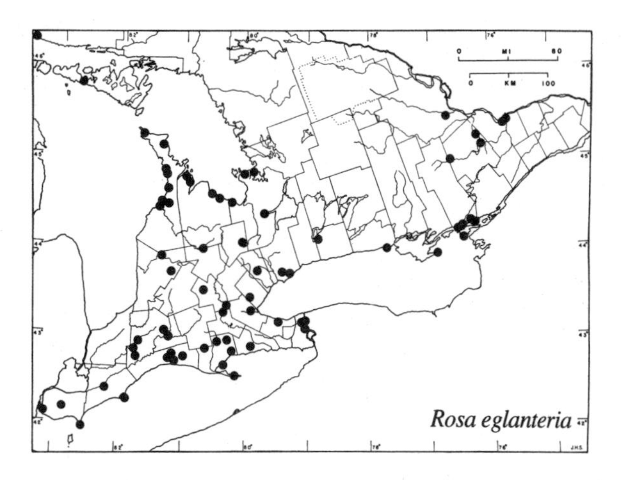

Rosa palustris

Rosa palustris Marsh. — Swamp rose

A medium-sized and much-branched shrub reaching a height of 2 m or more. Branchlets reddish-brown, smooth, with paired, broad-based, usually decurved prickles at the nodes, and lacking internodal bristles.

Leaves alternate, deciduous, and pinnately compound with 7, rarely 5 or 9, leaflets and long narrow stipules; leaflets thin, narrowly lance-oval to elliptic or obovate, 2–6 cm long, more or less acute at both ends, sessile or short-stalked, dull green and glabrous above, only slightly paler beneath and minutely pubescent, at least along the main veins which are sometimes reddish; margins with numerous, fine, and somewhat incurved teeth almost to the leaf base; stipules long and slender with toothlike lobes diverging from the petiole; leaf rachis finely hairy and often bearing small, broad-based, straight or curved prickles.

Flowers in small clusters of 2–5 or solitary at the ends of the leafy branches; petals pink, 2–2.5 cm long; pedicels, hypanthium, and sepals stipitate-glandular; late June to early Aug. Fruit subglobose, orange-red, 6–10 mm in diameter; Aug. and Sept. or later. (*palustris*—of swamps)

In wet places: swamps, wet thickets, damp shores of creeks, beaver dams, edges of marshes, floating sedge mats, and shrubby borders of quaking bogs.

Throughout southern Ontario and northward to about 47° N at Lake Timagami and along the southeastern shore of Lake Superior. (N.S. to Minn., south to Ark. and Fla.)

Field check Medium-sized shrub of wet habitats; stems with broad-based curved prickles at the nodes; leaves with 7 narrow leaflets and long narrow stipules; pedicels and hypanthium with stalked glands.

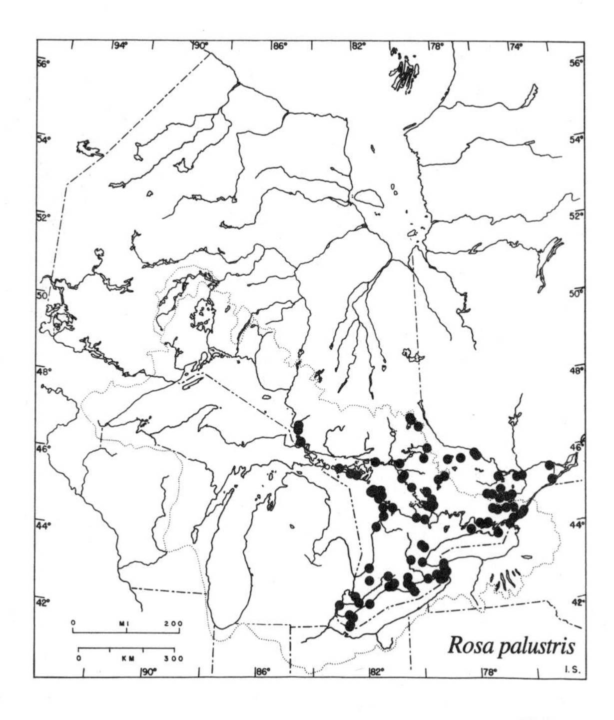

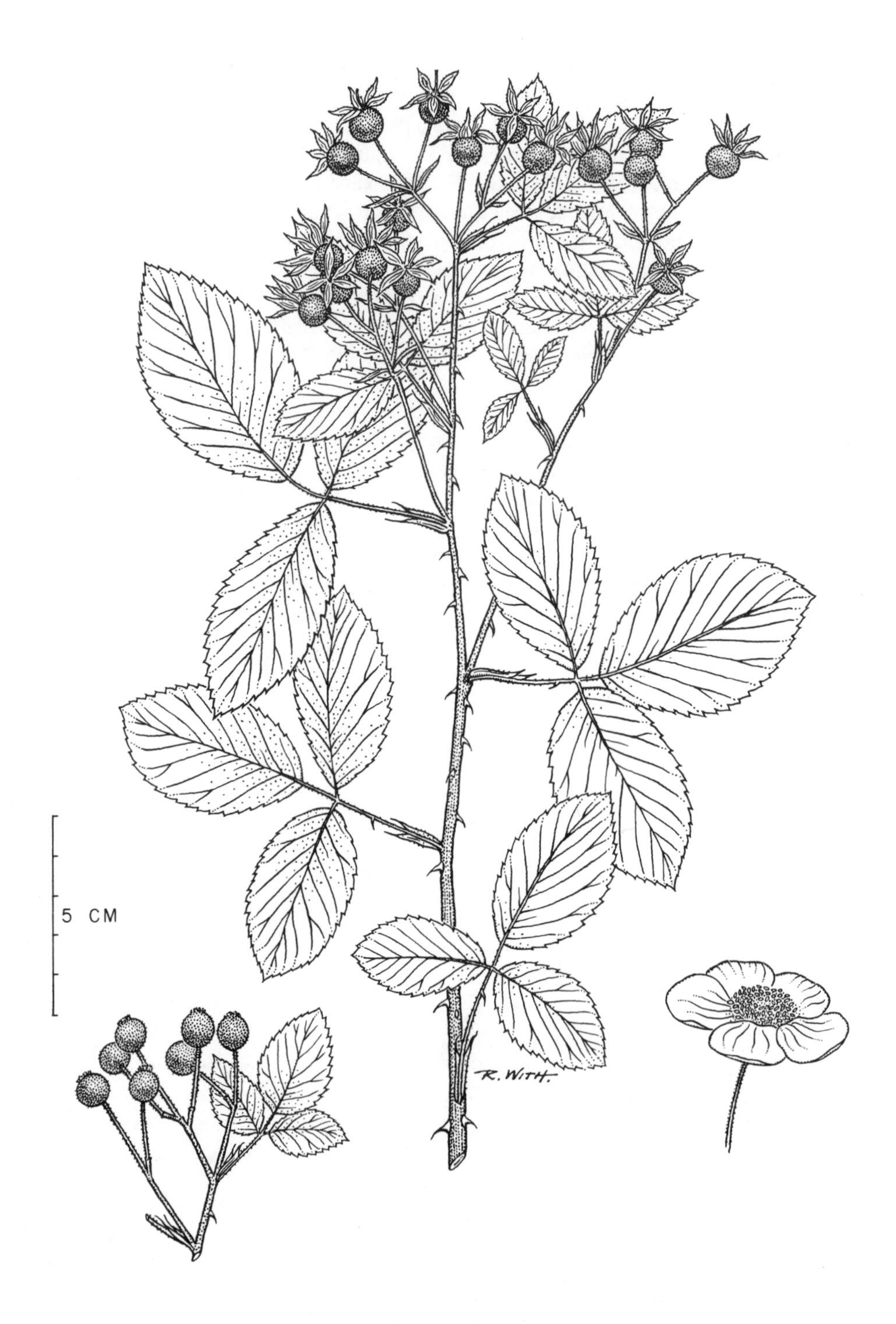

Rosa setigera

Rosa setigera Michx. **Prairie rose**

A stout coarse shrub, the stems becoming canelike and eventually up to several meters long, climbing, leaning on various supports, or forming large circular mounds in the open. Branchlets with few, remote, stout, broad-based and decurved prickles or rarely unarmed.

Leaves alternate, deciduous, and pinnately compound, up to 1 dm long; leaflets usually 3 (or 5 by the development of an extra, much smaller pair at the base), dark green, glabrous and somewhat shiny above, paler to slightly glaucous beneath, usually glabrous or pubescent along the veins or occasionally tomentose; margins of leaflets sharply to doubly serrate with gland-tipped teeth; lateral leaflets sessile, the terminal one long-stalked; petiole and rachis glandular-hispid and often sparsely prickly; stipules slender and elongate with glandular-ciliate margins and spreading lobes.

Flowers several to numerous, in one or more terminal corymbs; petals 2 - 3 cm long, rosy-pink, fading to white; sepal lobes 1 - 1.5 cm long, lance-attenuate, soon reflexed, and finally deciduous; pedicel, hypanthium, and sepals glandular-hispid; styles united into a column; June and July. Fruit red, nearly globose, 8 - 12 mm long; late July and Aug. (*setigera*— bristly, a somewhat inappropriate choice since the stems are prickly rather than bristly)

Open woods, thickets, and clearings and at the edges of swamps.

In the vicinity of Lake St. Clair and the western end of Lake Erie (Essex and Lambton counties); also in a swamp in Prince Edward County. (s.w. Ont. to Kans., south to Tex. and Fla.; naturalized in N. Eng. and e. N.Y.)

Note Some of the material from Ontario exhibits the dull upper leaf surfaces and tomentose lower surfaces of var. *tomentosa* Torr. & Gray (*R. rubrifolia* Ait.). Braun (1961) concludes that, in Ohio, segregation of such a variety is unnecessary since the plants vary from one extreme to the other. Ontario specimens are too few to assess. (*tomentosa*—densely woolly or pubescent; *rubrifolia*—red-leaved)

Field check Coarse vinelike shrub; stems with broad-based decurved prickles; leaflets 3, rarely 5; petioles and pedicels glandular-hispid; sepals soon reflexed and finally deciduous.

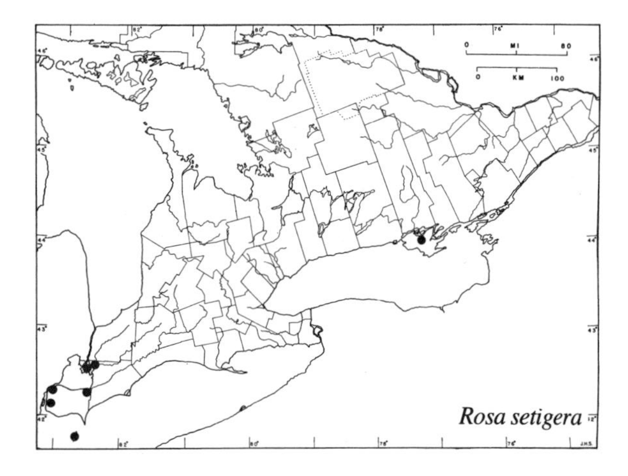

Rubus L. — Bramble

A large genus of several hundred species, widely distributed, but with its greatest diversity in north temperate regions.

Woody perennials, some herbaceous with a woody base, often shrubby with bristly or prickly stems. Vegetative reproduction by suckers, trailing stems which root at the nodes, or by arching canes which take root at the tips when these touch the ground. In some species the stems are biennial, the first year's growth, the *primocane*, developing from buds at or below ground level and bearing alternate leaves along the stem. During the second season axillary buds on the primocane give rise to leafy flowering branches with leaves usually different from those on the primocane. The stem is then called a *floricane* but this includes the woody tissues of the primocane from which the flowering branches grew. Both floricanes and primocanes are required for identification.

The leaves are alternate, mostly deciduous or rarely persistent, simple, palmately lobed, or compound, usually with stipules. Flowers 5-parted, perfect or unisexual; calyx without an extra set of bractlets; petals white to pink or rose-purple; stamens numerous, surrounding a nectariferous disc; carpels numerous, maturing into small 1-seeded drupes. The fruit, an aggregation of drupelets which are either free from the receptacle or fused with it, is generally called a berry.

The genus *Rubus* has been divided into eleven subgenera, five of which are represented in Ontario. Of these, the highly variable subgenus *Eubatus* (blackberries, dewberries) in eastern North America has been further classified into twelve sections of which five are represented in Ontario. From these sections, each of which contains a large number of equivocal species, the most clearly definable representative has been selected for inclusion here. Others that may be found in Ontario are listed under Notes and, for the Ottawa District, in a key published for that area (see Breitung, 1952). (*Rubus* —from the Latin *ruber*, red)

Key to *Rubus*

a. Stems armed with bristles or prickles or both
  b. Leaves whitish or gray-tomentose beneath; fruit, when ripe, separating readily from the receptacle (Raspberries)
    c. Stems erect or spreading, armed with stiff straight bristles or prickles; plants suckering from the roots; fruit red **R. *idaeus* var. *strigosus***
    c. Stems arching, often rooting at the tip, armed with short, recurved, broad-based prickles; plants not suckering; fruit usually black **R. *occidentalis***
  b. Leaves green on both sides, more or less pubescent, at least on the veins beneath; fruit falling with the receptacle at maturity
    d. Plants low and trailing (less than 0.5 m high); stems often rooting at the nodes; flowers solitary or few (less than 10), in a cluster; fruit red or reddish-purple (Dewberries)
      e. Stems hispid, with numerous slender bristles; leaves firm and leathery, often evergreen; petals less than 10 mm long **R. *hispidus***
      e. Stems with scattered, hooked, broad-based prickles; leaves thin, deciduous; petals more than 10 mm long **R. *flagellaris***
    d. Plants tall (up to 2 m high); stems erect and not rooting at the nodes or high-arching and rooting at the tips in contact with the ground; flowers numerous (10–20 or more) in elongate bracted clusters; fruit black (Blackberries)

f. Stems mostly smooth or with some scattered prickles; leaves glabrous on both surfaces; petioles and inflorescences usually glabrous (or minutely pilose), with a few small prickles **R. canadensis**

f. Stems glandular-pubescent with broad-based prickles or densely covered with spreading to reflexed bristles; leaves usually pubescent beneath, at least along the veins; petioles and inflorescences bristly, glandular-pubescent or with scattered prickles

g. Stems with bristles, glandular hairs, and scattered broad-based prickles; leaflets of primocanes ovate to broadly lanceolate, acute to long-acuminate at the tip, margins sharply serrate **R. allegheniensis**

g. Stems densely clothed with fine straight bristles (lacking broad-based prickles); leaflets of primocanes oblong-ovate, acute to obtuse at the tip, margins coarsely toothed **R. setosus**

a. Stems unarmed

h. Leaves all simple

i. Plants low (1–3 dm); leaves reniform, with rounded lobes and blunt teeth; flowers solitary (Cloudberry) **R. chamaemorus**

i. Plants taller (0.3–2 m); leaves not reniform, the lobes and teeth sharp-pointed; flowers in open clusters of 3–10 (Thimbleberries)

j. Flowers white; fruits orange to red, edible; stems and inflorescence more or less hairy and glandular **R. parviflorus**

j. Flowers rose-purple; fruits pink to red, dry, and unpalatable; stems and inflorescence more or less clammy, with dark glandular hairs **R. odoratus**

h. Leaves compound, with 3–5 leaflets

k. Plants low (1–3 dm); stems tufted, erect or trailing; flowers solitary or few (2–10), in a cluster; fruit red (Dwarf raspberries)

l. Flowering stems borne singly from a long creeping stem; leaflets sharp-pointed to acuminate; flowers white to greenish, rarely pink; petals 6–10 mm long **R. pubescens**

l. Flowering stems solitary or tufted from a short, branched, perennial base; leaflets abruptly pointed to rounded; flowers pink to deep rose-coloured; petals 10–20 mm long **R. acaulis**

k. Plants taller (0.3–2 m); stems erect or high-arching; flowers numerous (10–20 or more), in elongate and conspicuously-bracted clusters; fruit black **R. canadensis**

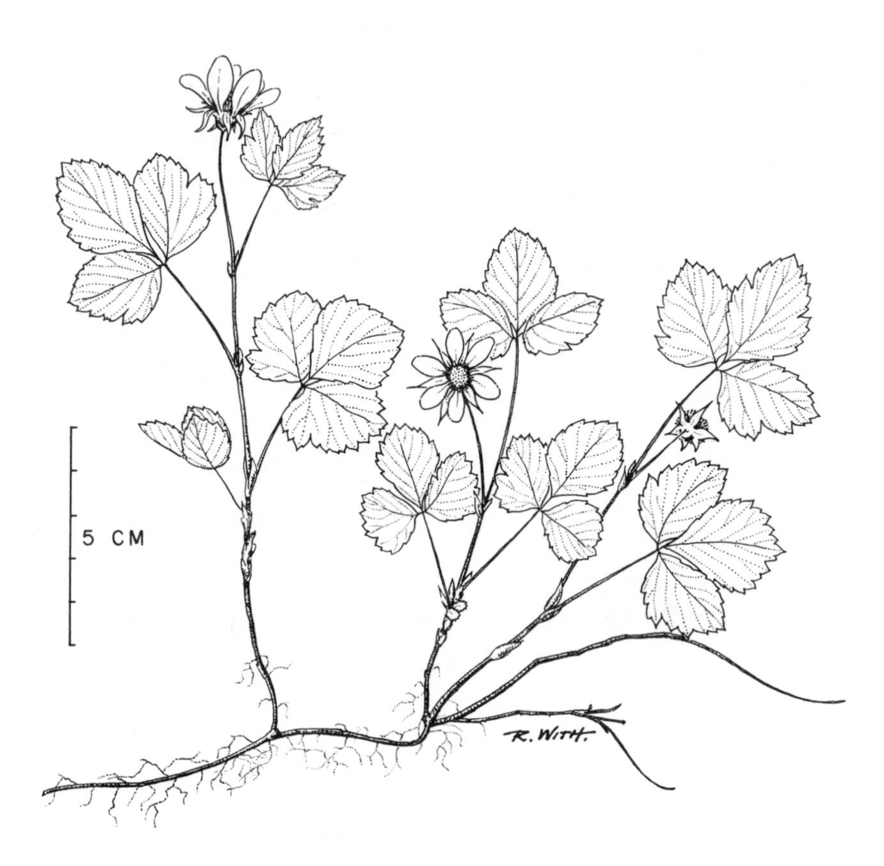

Rubus acaulis

Rubus acaulis Michx. — Northern dwarf raspberry, Nagoonberry Arctic raspberry

A low herbaceous perennial 5 – 10 cm high; stems slender, without prickles or bristles, sparingly branched, becoming woody at the somewhat tufted base; flowering branches erect, with 2 or 3 leaves and a solitary terminal flower.

Leaves alternate, deciduous, and trifoliolate; leaflets 1 – 4.5 cm long, 0.5 – 3.5 cm wide, the terminal one broadly obovate to rhombic and short-stalked, the lateral pair nearly sessile and asymmetrical, often with a partially developed lateral lobe, glabrous above, minutely pubescent beneath; margins coarsely serrate with blunt teeth, often ciliate; petioles finely hairy, usually longer than the terminal leaflets; stipules ovate, pointed, the lower pairs sheathlike.

Flowers solitary, pink to deep rose-coloured, on finely pubescent slender peduncles; petals up to 2 cm long, obviously narrowed towards the base; calyx lobes long-tapered and reflexed; June to Aug. Fruit nearly globose, about 1 cm in diameter, red, edible, composed of medium-sized drupelets; July to Sept. (*acaulis* — stemless, in reference to the lack of an obvious tall woody stem)

In sphagnum mats and lichen heath of arctic meadows, in alder and willow thickets, in black spruce forest and muskeg, and on moist banks of streams and rivers.

Throughout northern Ontario from Hudson Bay and James Bay south to the north shore of Lake Superior and the Lake Timiskaming region in eastern Ontario. (Nfld. to Alaska, south to Colo., Minn., and e. Que.)

Note This species is sometimes included in the circumpolar *Rubus arcticus* L., from which it differs in relatively minor attributes, such as, narrower stipules, single flowers below the leaves, absence of glands on peduncles, and longer petals. Ontario specimens may have a pubescent calyx and a few glands, mostly on the margins of the calyx lobes, and rarely 2-flowered peduncles. There is considerable variation in the shape of the leaf and of the calyx lobes. (*arcticus* — arctic)

Field check Unarmed, dwarf, herbaceous perennial with short, upright, leafy branches; leaves with 3 leaflets; flowers solitary, pink to red; fruit an edible red raspberry.

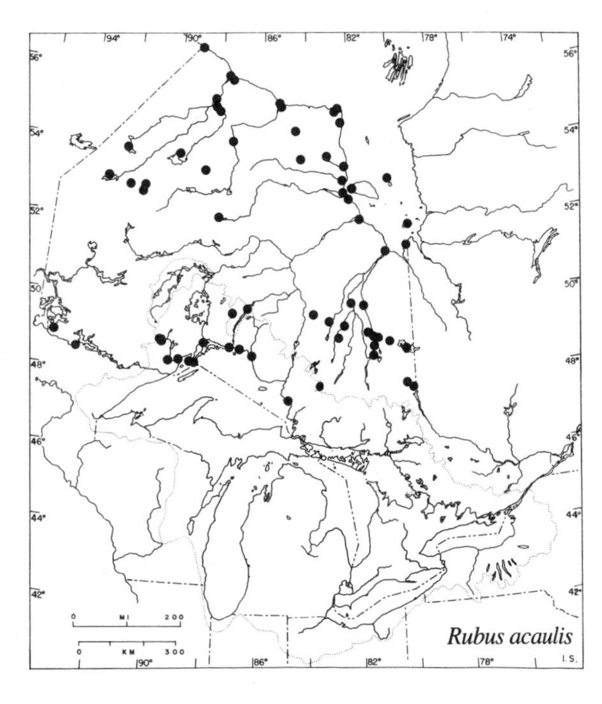

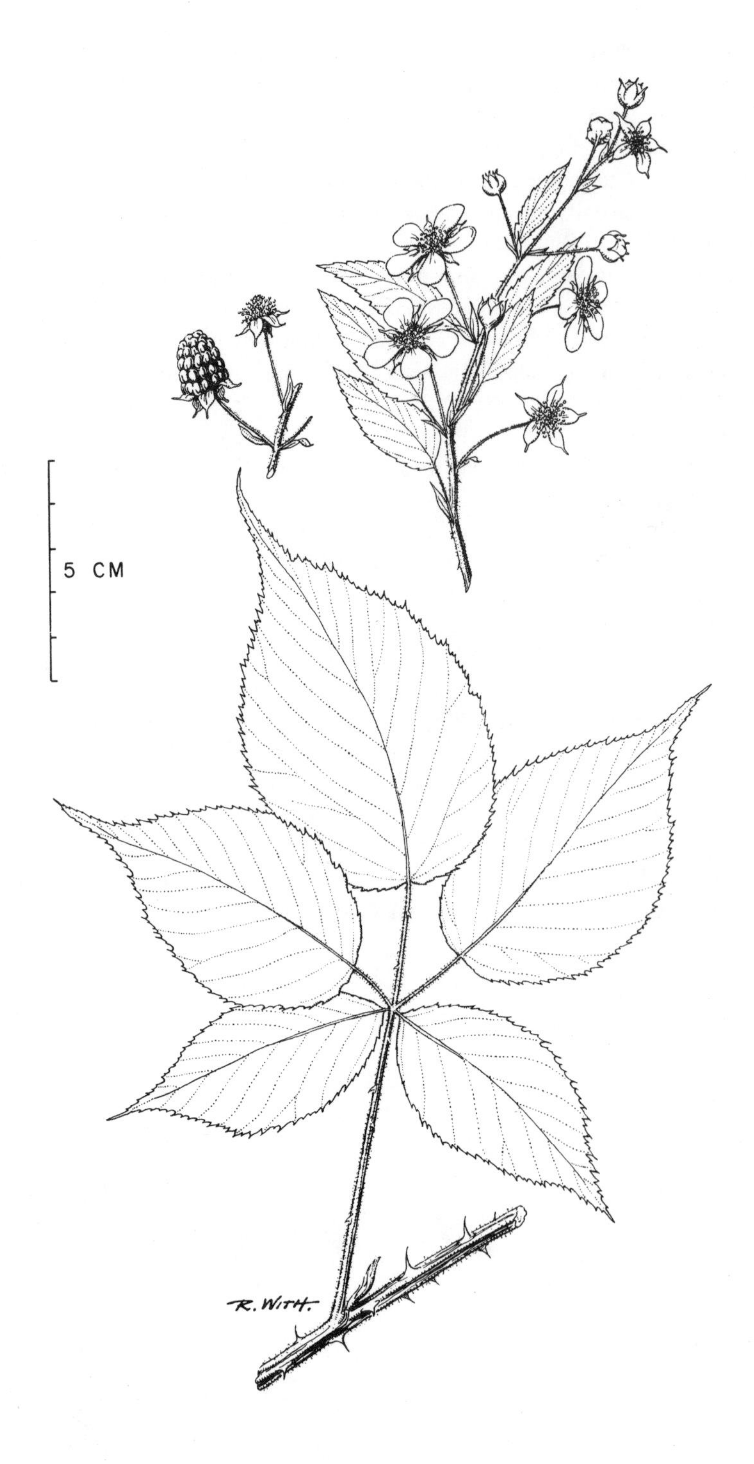

Rubus allegheniensis

Rubus allegheniensis Porter

Common blackberry
High-bush blackberry

A medium-sized shrub reaching a height of 2 m or more, with erect and high-arching canes. Stems covered with gland-tipped hairs and straight stiff bristles, floricanes brownish to purplish-red, more or less smooth and somewhat ridged, with scattered broad-based prickles and fewer small glandular hairs.

Leaves alternate, deciduous, and compound, those of the primocanes palmately 5-foliolate, those of the floricanes with 3 leaflets or reduced to a simple blade in the upper part of the inflorescence; terminal leaflet of primocane leaves 0.5–2 dm long, 3–10 cm wide, ovate to broadly lanceolate, cordate or rounded at base and acute or acuminate at the tip, the basal pair of leaflets sessile, the other leaflets stalked; all leaflets finely pubescent at least beneath and usually on both surfaces when young; margins sharply to doubly serrate; petioles and petiolules pubescent, with stalked glands and scattered coarse prickles, the latter also present along the midveins of the leaflets; stipules lance-linear to subulate, pubescent, and glandular.

Flowers numerous (up to 20 or more), on long pedicels in an elongate inflorescence with conspicuous subtending leaves and bracts, the axis and pedicels copiously glandular-pubescent and with scattered prickles; calyx glandular-pubescent on the outside, including the elongate attenuate tips of the lobes, whitish-pubescent within; petals white, 1–2 cm long, 5–8 mm wide; June and July. Fruit cylindric to thimble-shaped, up to 2.5 cm long, black, edible, varying from tart to sweet, often of good flavour, not separating from the fleshy receptacle; July to Sept. (*allegheniensis* — of the Alleghenies)

In old fields, pastures, and conifer plantations; along roadsides and borders of woods; and in thickets and fencerows.

From Lake Erie north to Georgian Bay and east to the Ottawa Valley; northward locally to 48° N on the east side of Lake Superior. (N.S. to Minn., south to Mo. and N.C.)

Note Though variable, *R. allegheniensis* is a clearly defined species, diploid, and probably reproducing sexually (see Hodgdon & Steele, 1962). It is considered to hybridize freely with other species, chiefly *R. setosus*.

R. alumnus Bailey (s.w. Que. to Minn.) and *R. attractus* Bailey (s.w. Que. to s. Mich.), which may be found in Ontario, are regarded by Gleason and Cronquist (1963) as synonymous with *R. orarius* Blanchard (Que. to Wis., south to Maine, Md., Va., and Mo.), described as being much like *R. allegheniensis*. Hodgdon and Steele (1966) interpreted *R. alumnus* as a hybrid of *R. allegheniensis* with *R. pensilvanicus* Poir., a species not found in Ontario. In view of the assessment of these authors, we have included these taxa in *R. allegheniensis*. (In *R. pensilvanicus*, the original spelling of the specific name, although not current usage for the name of the state, should be retained, according to the International Rules of Botanical Nomenclature. *alumnus* — student, nursling; *attractus* — dragged; *orarius* — of the coast; *pensilvanicus* — Pennsylvanian)

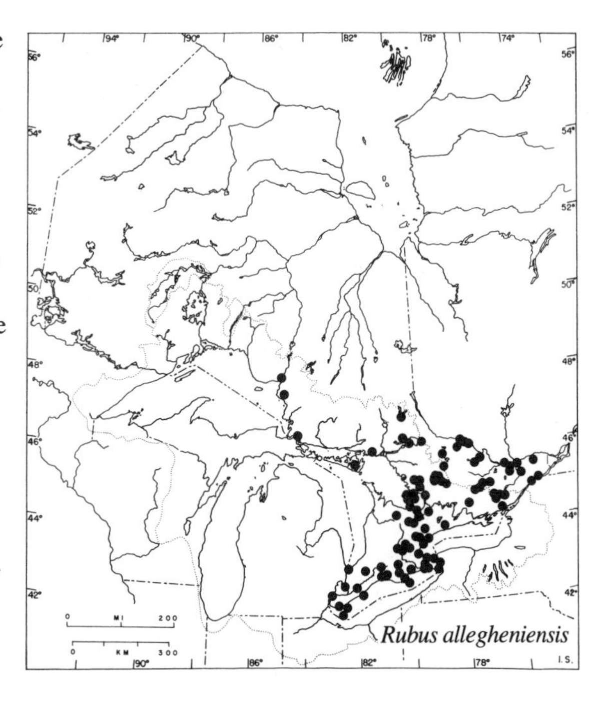

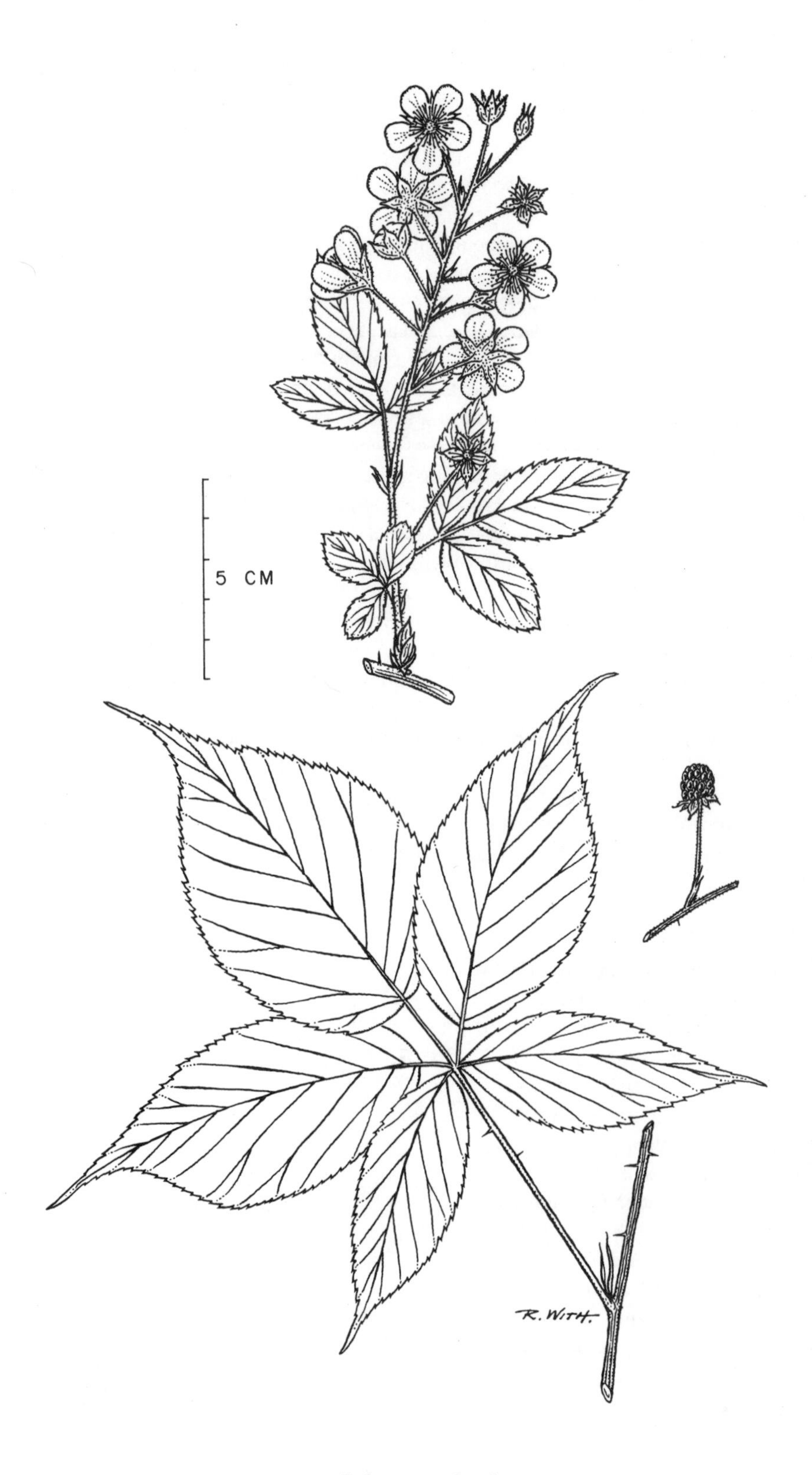

Rubus canadensis

Field check Canes glandular-pubescent, bristly, and with scattered hooked prickles; primocane leaves palmately 5-foliolate; petioles and inflorescences glandular-hairy and with scattered prickles; fruit black, elongate, cylindric to thimble-shaped, not separating from the fleshy receptacle.

Rubus canadensis L. — Smooth blackberry

A medium-sized shrub about 2 m tall, with erect or high-arching canes. Stems smooth, the older canes reddish-purple to brownish, usually angled or ridged, with or without scattered weak prickles.

Leaves alternate, deciduous, and compound, those of the primocanes palmately 5-foliolate, those of the floricanes trifoliolate or reduced to simple blades in the upper part of the inflorescence; terminal leaflet of primocane leaves 1–2 dm long with very long petiolule, the central pair of leaflets with distinct petiolules, and the basal pair nearly sessile; leaflets varying from broadly ovate to lanceolate, those of the primocane leaves long-acuminate at the tip; all leaves thin, glabrous on both surfaces; margins serrate with sharp-pointed teeth; petioles glabrous or with scattered prickles; stipules linear-subulate and soon withering on the primocanes but lance-linear and prominent on the flowering branches.

Flowers numerous (up to 25), in elongate leafy clusters, the usually glabrous (or minutely pilose and remotely prickly) pedicels exceeding the prominent stipulelike basal bracts; calyx lobes white-pubescent on the inside, the tips smooth, green, and abruptly narrowed; petals white, 1–2 cm long and 5–8 mm broad; June and July. Fruit globose to thimble-shaped, black, edible but rather pulpy and often dryish, the drupelets not separating easily from the receptacle; July to Sept. (*canadensis*—Canadian)

In open woods, thickets, and clearings, along roadsides, creek banks, and lakeshores, on talus slopes and railway ballast, and in fencerows.

Southern Ontario and northward to the Ottawa Valley, along the east shore of Lake Superior; also at the south end of Lake Nipigon. (Nfld. to Ont., south to Minn., Tenn., and Ga.)

Note Einset (1951) found *R. canadensis* to be a triploid apomict and probably pseudogamous. Few hybrids should be expected but any that do occur may establish closely similar apomictic taxa ("microspecies").

Field check Canes smooth or remotely prickly; primocane leaves palmately 5-foliolate and glabrous; flowers numerous, prominently stalked and bracted, in long clusters; fruit black, thimble-shaped, not separating easily from the receptacle.

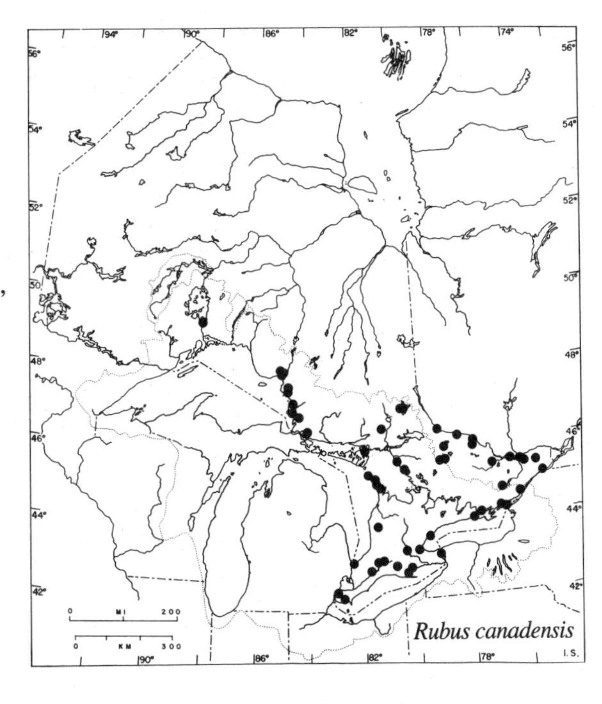

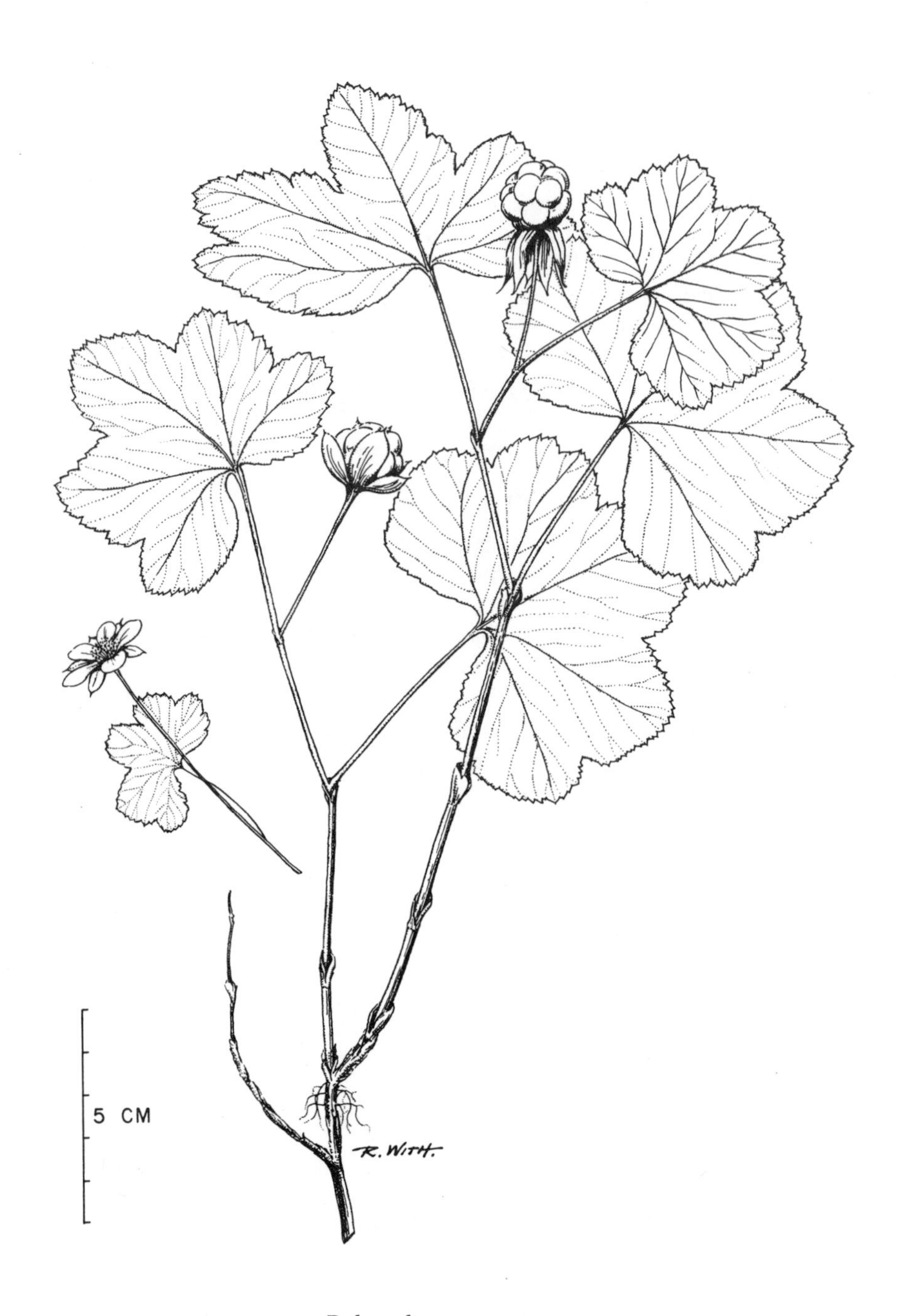

Rubus chamaemorus

Rubus chamaemorus L.

Cloudberry, Bake-apple Baked-apple berry

A low herbaceous perennial with slender, creeping, woody rhizomes covered with brownish papery bark; dioecious. Stems upright, unbranched, and unarmed, 1–3 dm high, bearing leaves at the upper 1–3 nodes and stipules only at the lower nodes.

Leaves alternate simple, and deciduous, the somewhat leathery blades rounded to reniform, 4–11 cm wide, with 5–7 rounded lobes and rather shallow sinuses; margins serrate with blunt teeth; stipules sheathlike, especially the lower pairs; petioles 2–8 cm long.

Flowers unisexual, 2–3 cm in diameter, solitary, and terminal on long peduncles; calyx lobes pubescent and glandular, sometimes lacerate or fringed; petals white and wide open in the mature flower; June and July. Fruit 1–2 cm in diameter, composed of a few large-seeded drupelets, at first reddish, then turning amber to yellowish, soft and translucent when ripe, falling quickly from the dry receptacle, edible, with a pleasant distinctive flavour. (*chamaemorus*—from the Greek *chamae*, on the ground, and the Latin *morus*, the mulberry tree, in reference to the dwarf habit and mulberrylike fruit)

Chiefly in sphagnum mats and hummocks, in black spruce bogs and muskegs, and on mossy tundra.

Restricted to northern Ontario from the shores of Hudson Bay and the islands in James Bay (N.W.T.) south to the north shore of Lake Superior; not collected in Ontario south of 48° N. (Greenl. to Alaska, south to B.C., and east to Ont. and N. Eng.)

Field check Low perennial with unarmed stems and 2 or 3 broadly rounded 5–7-lobed leaves; flower solitary; fruit edible, large, pale, raspberrylike, and composed of a few yellowish drupelets.

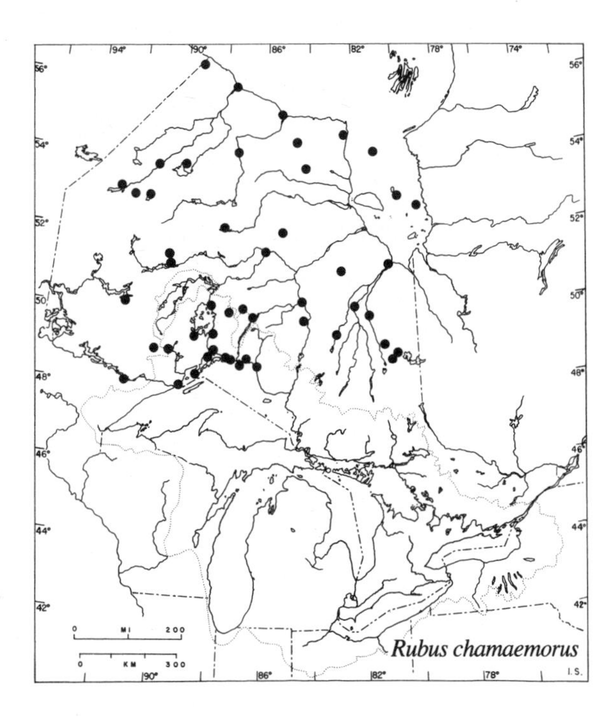

Rubus flagellaris

Rubus flagellaris Willd. — Northern dewberry

A prostrate shrub with long-trailing whiplike branches 2–4 m long which often root at the tip. Stems brownish to purplish-red with scattered, broad-based curved prickles.

Leaves alternate, deciduous, and compound; leaves of the primocanes with 3–5 leaflets, the terminal leaflet 2–7 cm long, 1–5.5 cm wide, ovate to subelliptic, abruptly narrowed to an acuminate tip, often with small lobes above the middle, rounded at the base. The lateral leaflets, when one pair, asymmetrical or deeply lobed, and when two pairs, the intermediate pair close to the basal pair; leaves of the floricanes usually smaller, with 3 leaflets or the uppermost leaf simple and often lobed, the blade of such simple leaves, or the terminal leaflet of the compound leaves, obovate to oblanceolate, blunt or abruptly acuminate, the lateral leaflets when present broadly elliptic, asymmetrical, with a rounded or tapered base; blades of all types of leaves thin, green on both surfaces, essentially glabrous or with appressed hairs along the veins beneath; petioles finely pubescent, with scattered hooked prickles; stipules linear, persistent.

Flowers 2–9, on nearly erect prolonged pedicels which are minutely pubescent and have scattered small prickles or the flowers sometimes solitary; calyx minutely pubescent, the lobes contracted to a narrow darkish tip, reflexed after the flower has opened; petals white, 10–15 mm long; June. Fruit red, globose or slightly elongate, composed of rather large juicy drupelets and usually of a rich flavour, not separating readily from the receptacle; July and Aug. (*flagellaris*—whiplike, referring to the long, slender, whiplike trailing stems)

In dry woods and thickets, often on sandy soil.

Most abundant along the shores of Lakes Erie, Ontario, and Huron; also in Simcoe County and Muskoka District, and along the Ontario-Minnesota border in western Ontario. (N.B. to Minn., south to Ark. and Ga.)

Note Some other related northern dewberries that may be identified from Ontario are *R. baileyanus* Britt., *R. jaysmithii* Bailey, *R. maltei* Bailey, and *R. recurvicaulis* Blanch. (See Fernald, M.L. 1950, 825–833.) (*baileyanus*—in honour of Liberty Hyde Bailey, 1858–1954, American student of the genus; *jaysmithii*—in honour of Stanley Jay Smith, 1915–1979; *maltei*—in honour of Malte Oscar Malte, 1880–1934, Canadian botanist; *recurvicaulis*—with recurving stem)

Einset (1951) determined from samples of *R. flagellaris* studied at Cornell University that this species is apomictic, probably pseudogamous, and 9-ploid, i.e., with nine sets of chromosomes (genomes). Steele and Hodgdon (1963, 1970), working with specimens from the New England states, considered *R. flagellaris* and related taxa to hybridize freely among themselves. In pseudogamous apomicts occasional sexually produced seeds could lead to the production of new stable taxa differing little from the mother plant from which they were derived. *R. recurvicaulis*, usually thought to belong here, has been assigned by Gleason and Cronquist (1963) to a section not accepted by the present authors as occurring in Ontario; they have included *R. baileyanus* in *R. flagellaris* but have not recognized either *R. jaysmithii* or *R. maltei*.

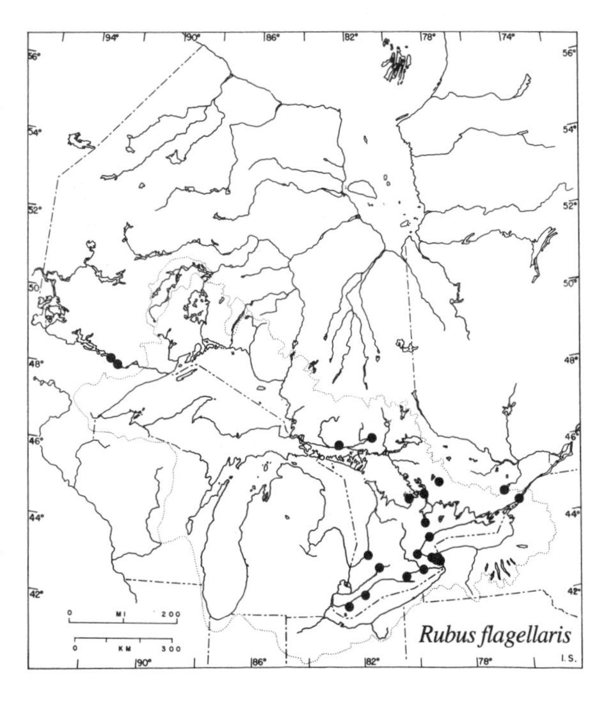

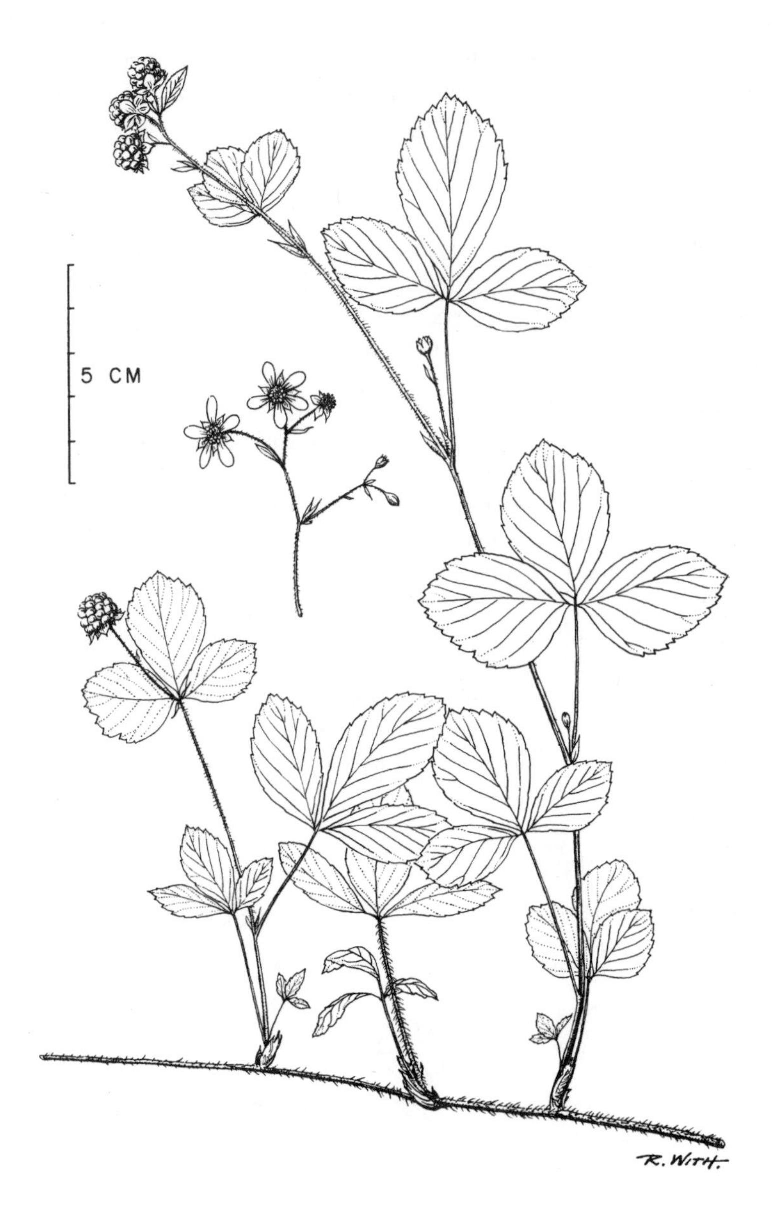

Rubus hispidus

Field check Trailing shrub with whiplike primocanes and short, upright, few-flowered branches, both bearing broad-based hooked prickles; leaves 3-5-foliolate; fruit red, composed of large juicy drupelets, not separating readily from the receptacle.

Rubus hispidus L. — Swamp dewberry

A low trailing shrub with prostrate or low-arching stems often rooting at the tip. Stems armed with slender, straight to curved, sometimes gland-tipped bristles or spines with scarcely enlarged bases.

Leaves alternate and compound, some usually persisting through the winter; leaflets 3 or rarely 5, rhombic-ovate to obovate or nearly orbicular, 2-5 cm long, 1-3.5 cm wide, somewhat leathery, dark green and a little glossy above, only slightly paler beneath, mostly glabrous or with a few hairs on the veins beneath, blunt to pointed at the tip and rounded to cuneate at the base; margins coarsely blunt-toothed; petioles finely hairy and prickly; stipules linear, persistent.

Flowers solitary in the upper leaf axils or 2-8 in an open lax inflorescence at the end of a short upright branch, the axis and pedicels minutely pubescent and with scattered small prickles; calyx minutely pubescent, the broadly ovate lobes terminating in a small dark gland; petals white, 5-10 mm long; June and July. Fruit less than 10 mm in diameter, reddish-purple, with relatively few dryish sour drupelets, not separating from the receptacle; July to Sept. (*hispidus* —bristly)

In swampy ground and peat bogs; on damp sandy and gravelly shores; also in sandy woods.

Northward from Lake Erie and westward from the Ottawa and St. Lawrence valleys to Georgian Bay; also northwestward at several sites along the shore of Lake Superior. (N.S. to Wis., south to Mo. and N.C.)

Note This is our only blackberry that invades acid peat bogs. Its leaves often turn attractive shades of bronze, red, russet, and purple in the fall.

Other dewberries that may be identified in Ontario are *R. trifrons* Blanch. and *R. tardatus* Blanch., but both of these have been included in *R. hispidus* by Gleason and Cronquist (1963). Hodgdon and Steele (1966) consider *R. hispidus* and *R. setosus* to hybridize freely, giving rise to variants that have been described as species, e.g., *R. trifrons* and *R. tardatus*. (*tardatus*—delayed; *trifrons*—with three fronds or leaves)

Field check Trailing bristly stems with over-wintering, glossy, trifoliolate leaves; flowers few; fruits red-purple, not separating from the receptacle.

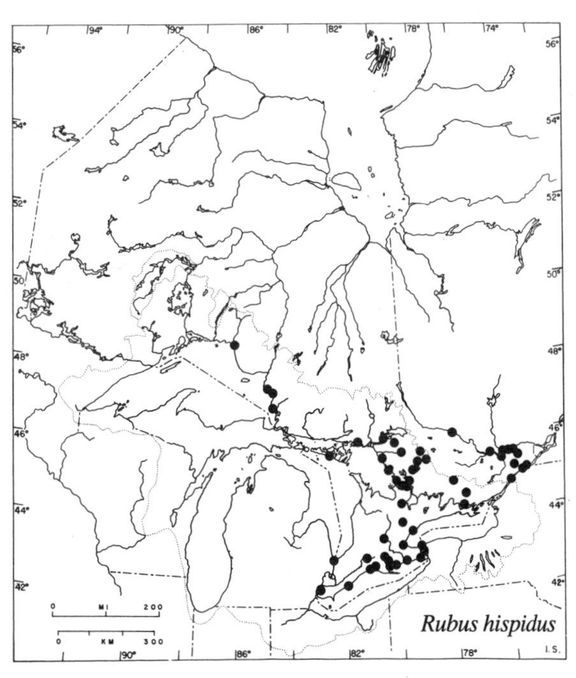

Rubus hispidus

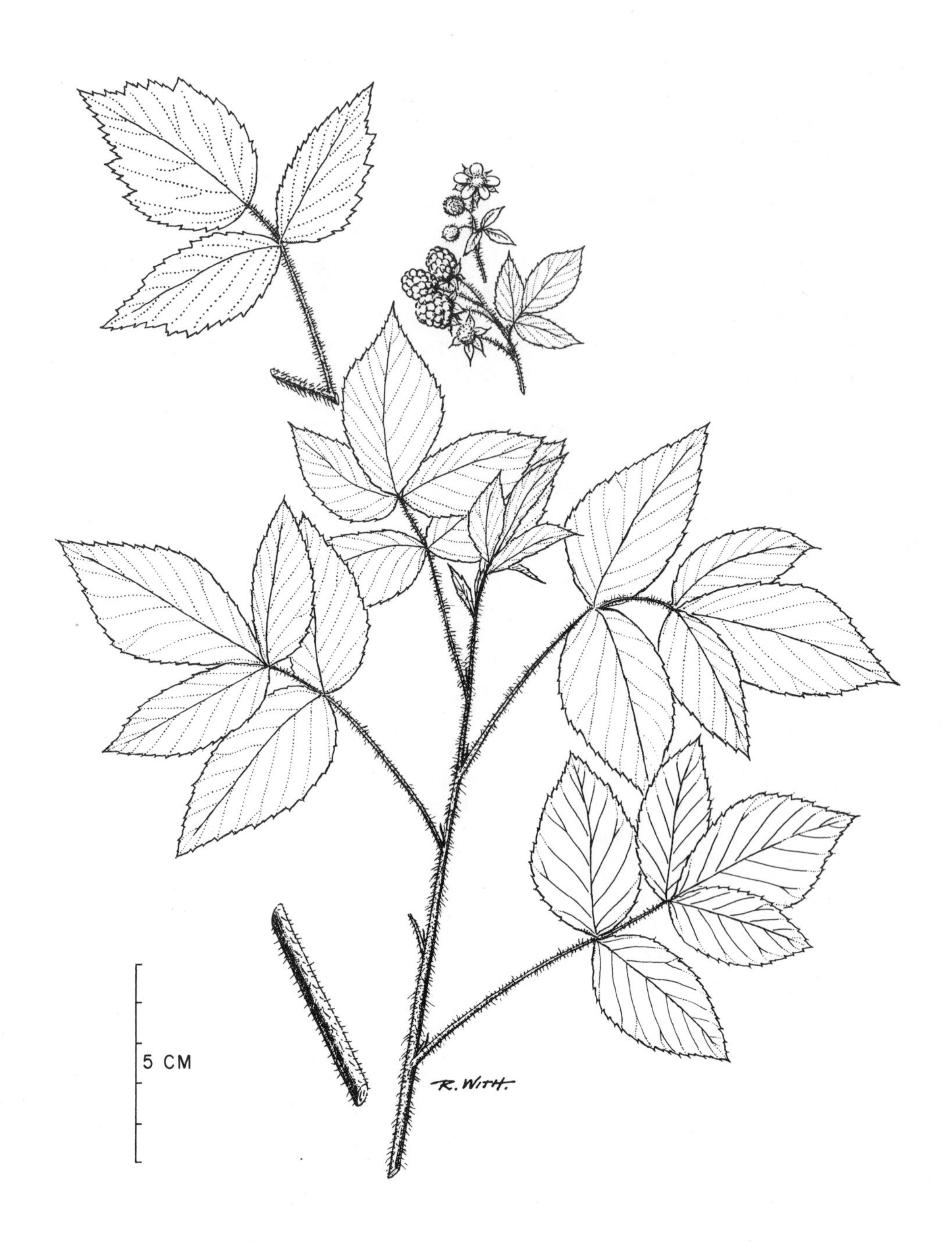

Rubus idaeus var. *strigosus*

Rubus idaeus L. var. *strigosus* (Michx.) Maxim. (*R. strigosus* Michx.; *R. idaeus* ssp. *sachalinensis* (Levl.) Focke) — Wild red raspberry

A stoloniferous shrub with erect, arching or spreading stems up to 1.5 m high. Stems biennial; young branches sparsely to densely bristly with stiff, slender, usually gland-tipped hairs; older stems brownish, smooth, and striate after the outer papery bristle-bearing layer has loosened.

Leaves alternate, deciduous, and pinnately compound: leaves of the primocanes with 3, 5, or rarely 7 leaflets; leaves of the floricanes mostly trifoliolate; leaflets varying from broadly ovate to narrowly lanceolate, the terminal leaflet of primocanes often with a pair of lateral lobes. When the leaves have 5 leaflets, the middle pair is closer to the terminal one than to the basal pair. Leaflets dark green and glabrous or sparingly hirsute above, gray to white-pubescent beneath, sharp-pointed and acuminate at the apex, rounded to tapered at the base; margins irregularly or doubly serrate; petioles bristly-hispid; stipules slender, soon withering.

Flowers in terminal clusters of 2–5 and solitary in the upper leaf axils; calyx glandular-hairy, the grayish-pubescent lobes with slender dark tips; petals white and narrow; June and July. Fruit usually red, rarely yellow to amber-coloured, about 1 cm in diameter, edible, falling intact from the dry receptacle; July and Aug. (*strigosus*—with stiff bristles; *idaeus*—named for Mount Ida; *sachalinensis*—from Sakhalin)

Chiefly in open areas; talus slopes, edges of woods, thickets, roadsides, clearings, burns, and waste places.

Widespread throughout Ontario from Lake Erie northward to Lake Superior and Lake-of-the-Woods; less common in the Hudson Bay drainage basin. (Nfld. to Alaska, south to Calif., east to Ariz. and N.C.; Asia)

Note Our wild red raspberry is closely related to the cultivated red raspberry of the garden, a species of European origin (*R. idaeus* L.). The eastern North American plants may be considered as a variety of the latter, under the name *R. idaeus* var. *strigosus* (Michx.) Maxim. Another interpretation (Hitchcock et. al., 1955–69) has been to recognize a circumboreal complex with several subspecies, the eastern North American material then belongiııg to *R. idaeus* ssp. *sachalinensis* (Levl.) Focke var. *sachalinensis*.

Field check Canes with prickles and bristlelike hairs; leaves pinnately compound with 3–5 leaflets, gray or white-pubescent beneath; fruit an edible red raspberry falling intact from the receptacle.

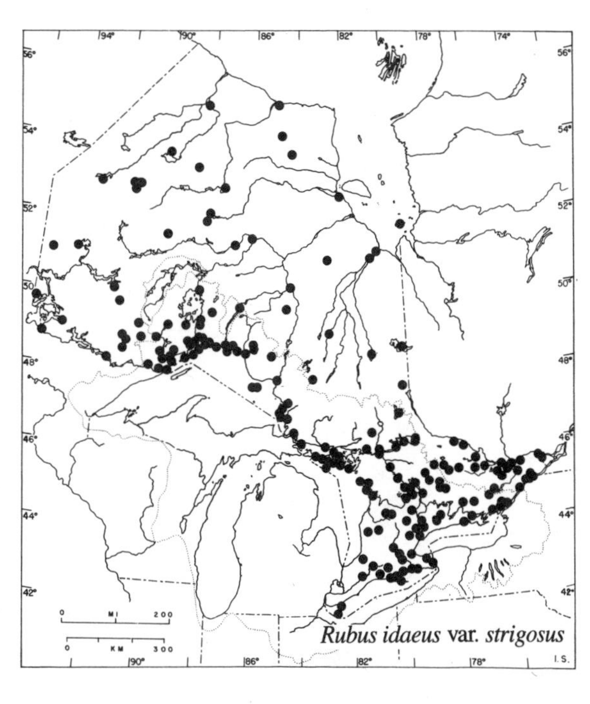

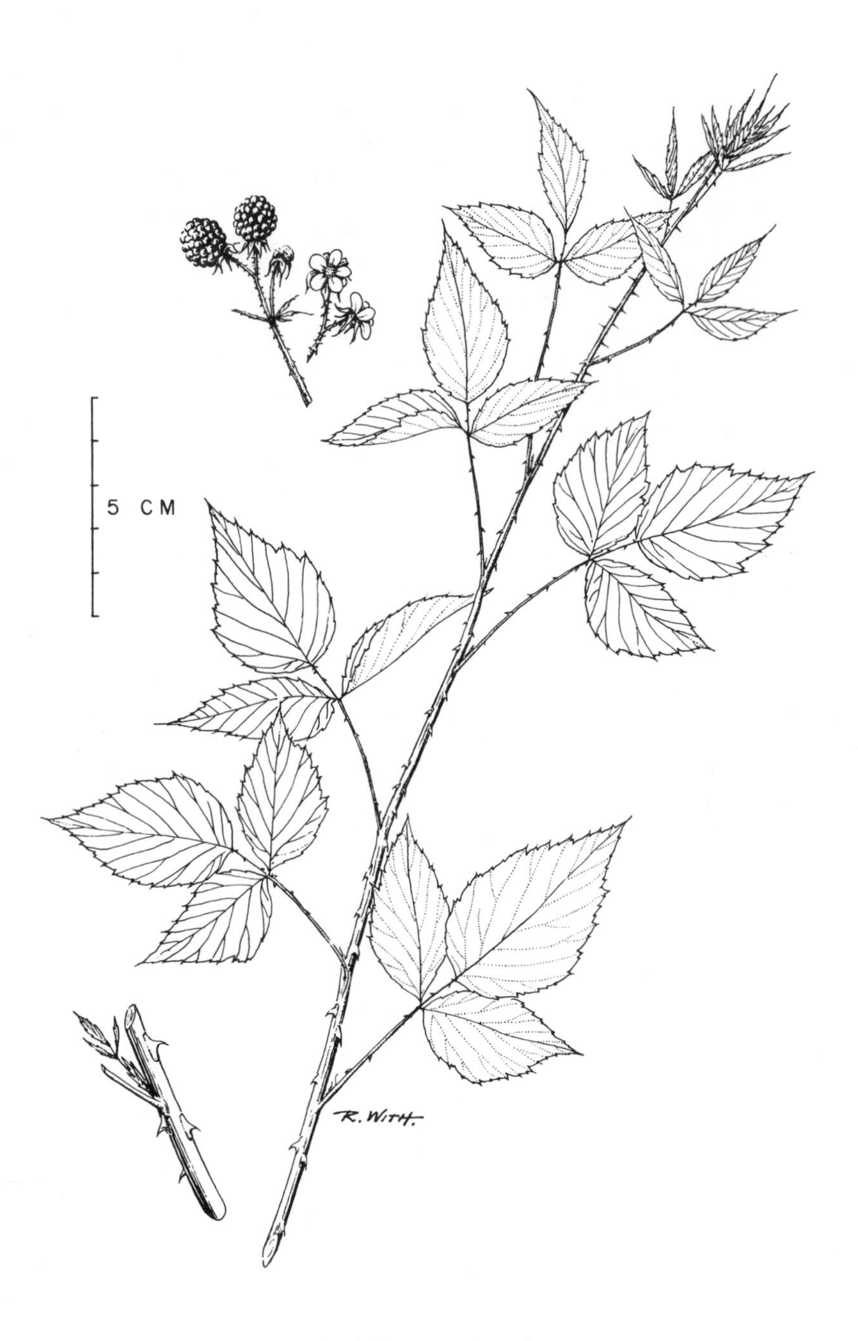

Rubus occidentalis

Rubus occidentalis L. — Black raspberry

A medium-sized shrub 1 - 2 m tall, with prominently arching canelike stems arising in clumps, the tips sometimes rooting when they touch the ground. Stems biennial, purplish-brown, the young ones usually glaucous, armed with hooked prickles which on older stems are stout and broad-based.

Leaves alternate, deciduous, and pinnately compound, those of the floricanes trifoliolate, those of the primocanes with 3 - 5 leaflets. When there are 5 leaflets, the intermediate pair is usually closer to the basal pair than to the terminal leaflet; when trifoliolate, the lateral leaflets may be deeply lobed; leaflets broadly ovate to ovate-lanceolate, 3 - 12 cm long, 2 - 7.5 cm wide, the terminal one often with lateral lobes, sharp-pointed to abruptly acuminate at the apex, cordate, rounded, or tapered at base, dark green and nearly glabrous above, whitened and downy beneath; margins coarsely doubly serrate; petioles mostly glabrous, with small scattered hooked prickles; stipules linear-subulate, deciduous.

Flowers in terminal and axillary clusters of 3 - 10 or at the tips of a series of short branches standing erect from an arching cane; peduncles and pedicels downy and prickly; calyx tomentose, with reflexed acuminate lobes; petals elliptical, shorter than the sepals; May and June. Fruit at first bright red but becoming black, hemispherical, about 1.5 cm wide, juicy and edible, separating easily from the receptacle when ripe; July. (*occidentalis*—western, i.e., with its home in the western hemisphere)

In open woods and thickets, along the edges of woods, roadsides, and in fencerows.

Restricted to southern Ontario, chiefly south and east of the Canadian Shield. (Que. to N. Dak. and e. Colo., south to Ark. and Ga.)

Note Hybrids of *R. occidentalis* and *R. idaeus* var. *strigosus* are cultivated as purple raspberries. The hybrid that occurs between these two species in the wild has been named *R.* × *neglectus* Peck. (*neglectus*—neglected)

Field check Arching canes with hooked and broad-based stout prickles; primocanes often glaucous; leaves 3 - 5-foliolate, downy and whitened beneath; edible black raspberries on prickly pedicels.

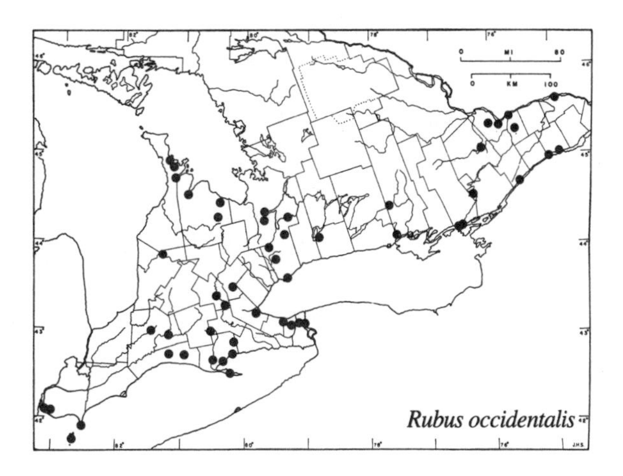

Rubus odoratus

Rubus odoratus L. — Purple-flowering raspberry

Erect or arching shrub 1–1.5 m tall, spreading by underground shoots and forming thickets. Stems unarmed, green to purplish, and glandular-sticky when young; older stems brown to gray, with exfoliating papery bark.

Leaves alternate, simple, and deciduous; blades 3–5-lobed, 10–20 cm broad and about as long, nearly smooth or thinly hairy on both surfaces but usually pubescent, at least on the veins beneath, the lobes sharp-pointed, the base cordate to deeply cleft; margins irregularly jagged-serrate; stipules slenderly lanceolate, pale, and soon withering; petioles 5–15 cm long, clammy-pubescent.

Flowers rose-purple (turning a deeper purple on drying), 3–5 cm across, in terminal clusters of 3–10, the pedicels and calyx covered with reddish-purple stalked glands; calyx lobes with long tail-like tips; June and July. Fruit a shallowly cup-shaped raspberry, 1–2 cm in diameter, pink to red when ripe but dryish, acid, and unpalatable; July to Sept. (*odoratus*—with an odour, in reference to the fragrance of the flowers, which, however, is not always noticeable)

In clearings, thickets, and ravines, along roadsides, lakeshores, riverbanks, and borders of woodlands.

Common in southern Ontario from Lake Erie to the Ottawa Valley; local in the vicinity of Georgian Bay; reported in the vicinity of Sault Ste. Marie but no recent collections seen from that area. (N.S. to Mich., south to Tenn. and Ga.)

Note Both *R. odoratus* and *R. parviflorus* are used as ornamentals in European gardens. These two species are extremely similar in growth habit and foliage but are easily distinguished by the colour of their flowers and by their fruit, dry and unpalatable in the former and edible but somewhat insipid in the latter species.

Field check Unarmed shrub with shredding bark, 5-lobed leaves, and rose-purple flowers in clusters; fruit a dryish pink to red cap-shaped raspberry.

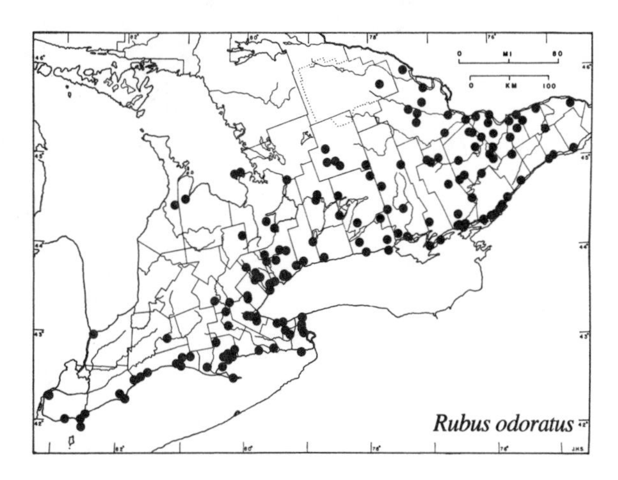

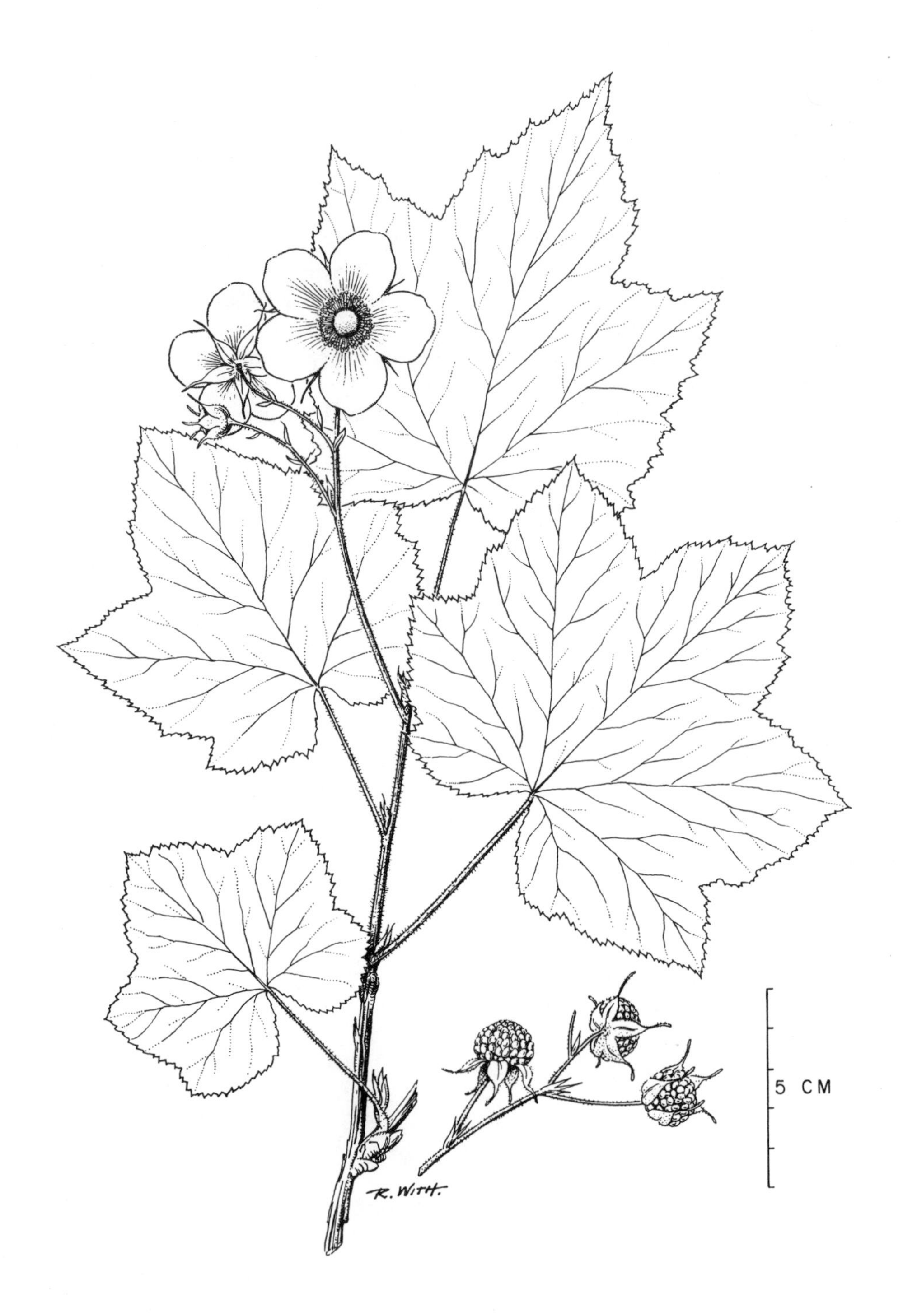

Rubus parviflorus

Rubus parviflorus Nutt. — Thimbleberry

Erect shrub 1–2 m high, often much-branched, and forming clumps or thickets. Stems unarmed, green to brownish, pubescent and glandular, sticky when young; older stems purplish-red to gray-brown, with exfoliating papery bark.

Leaves alternate, simple and deciduous; blades usually 5-lobed (3-lobed small leaves occur in the inflorescence and the larger leaves may have an extra lobe on each side near the base), 10–20 cm across and about as long, soft-hairy or with at least scattered hairs on the lower surface, the lobes sharp-pointed, the base cordate or deeply cleft; margins conspicuously and irregularly sharp-toothed; stipules lanceolate and glandular; petioles glandular-pubescent, equalling or shorter than the blades.

Flowers white, 5-parted, 3–5 cm in diameter, in clusters of 3–10 at the ends of the branches; sepal lobes with a long tail-like tip; stamens and carpels numerous; June and July. Fruit a thin, cap-shaped, pink to red raspberry, very soft when ripe, edible but somewhat insipid. (*parviflorus*—small-flowered, an inappropriate name since the flowers are larger than those of most species of *Rubus* in our area)

Mostly in open habitats, such as, roadsides, clearings, edges of woods, lakeshores, and talus slopes, and open rocky woodland; also in thickets.

Restricted to the shores of Lake Superior and locally southward on the islands of Lake Huron to the Bruce Peninsula. (Ont., Minn., S. Dak., and Alta. to Alaska, south to Calif., Ariz., and Mex.)

Note Many forms of this species have been described that are based on different character combinations within the large series of variations, particularly in pubescence and glandularity.

The possible occurrence of both this species and *R. odoratus* in the Sault Ste. Marie region needs to be investigated more thoroughly (see maps) since both species occur in the Upper Peninsula of Michigan.

Field check Unarmed shrub with shredding papery bark, sticky stems, 5-lobed leaves, white flowers in clusters, and cap-shaped pink to red raspberries with insipid taste.

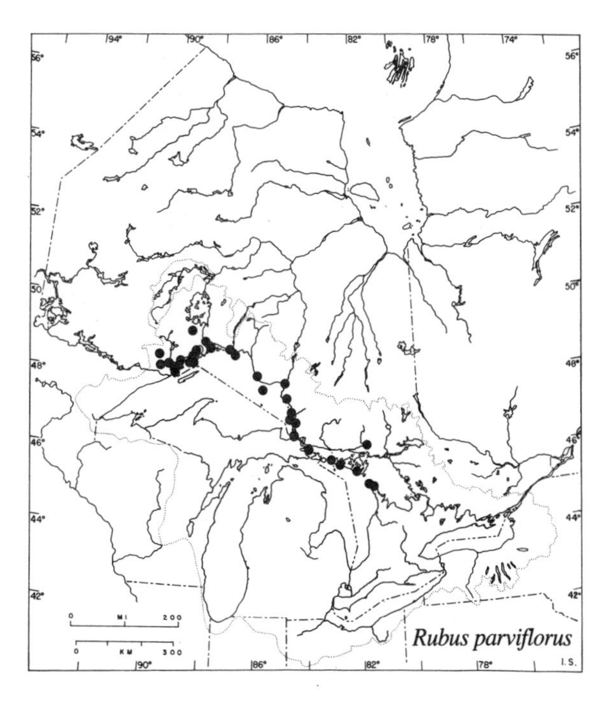

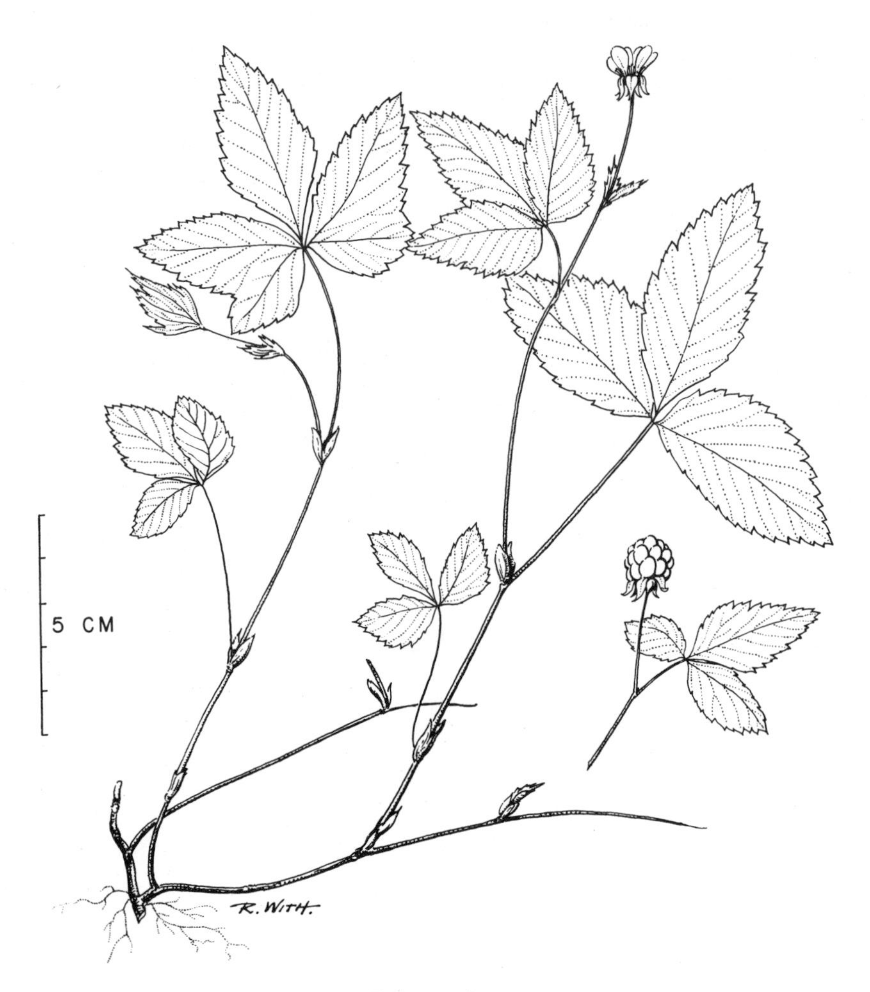

Rubus pubescens

Rubus pubescens Raf. (*R. triflorus* Richards.; *R. americanus* Britt.) — Dwarf raspberry

A low plant 1–3 dm high, with perennial runnerlike stems and herbaceous, short or long, leafy branches. Stems slender, pubescent but without prickles, the basal portion becoming woody, with smooth reddish-brown bark; flowering branches erect, with few leaves; sterile branches arching to long-trailing, with numerous leaves and whiplike ends, often rooting at the nodes, persisting as bare wiry stems.

Leaves alternate, deciduous, and compound, trifoliolate; leaflets 2–7 cm long, 1–4.5 cm wide, short-stalked or nearly sessile, rhombic-oval, and sharp-pointed, the lateral pair somewhat rounded at the base, the terminal leaflet narrowly cuneate, glabrous or sparingly pubescent on both surfaces; margins coarsely toothed; petioles pubescent, about as long as the terminal leaflet; stipules prominent, ovate, and sharp-pointed.

Flowers on slender glandular-pubescent peduncles, solitary or several in a loose cluster; petals white to pale pink, 6–10 mm long; calyx lobes sharp-pointed, reflexed; May and June. Fruit globose, 10–15 mm long, bright red, edible, composed of large juicy drupelets, not separating easily from the spongy receptacle; July to Sept. (*pubescens*—downy or covered with soft hairs; *triflorus*—three-flowered; *americanus*—American)

In deciduous, coniferous or mixed woods, in clearings, on hummocks in sphagnum bogs, and along creek banks.

Widely distributed throughout Ontario but less common in the Hudson Bay drainage basin. (Nfld. to Mack. Dist., N.W.T., and B.C., south to n. Colo. and N.J.)

Field check Trailing, unarmed, herbaceous perennial with whiplike leafy stems; leaves with 3 leaflets; flowers white to pink; fruit an edible red raspberry.

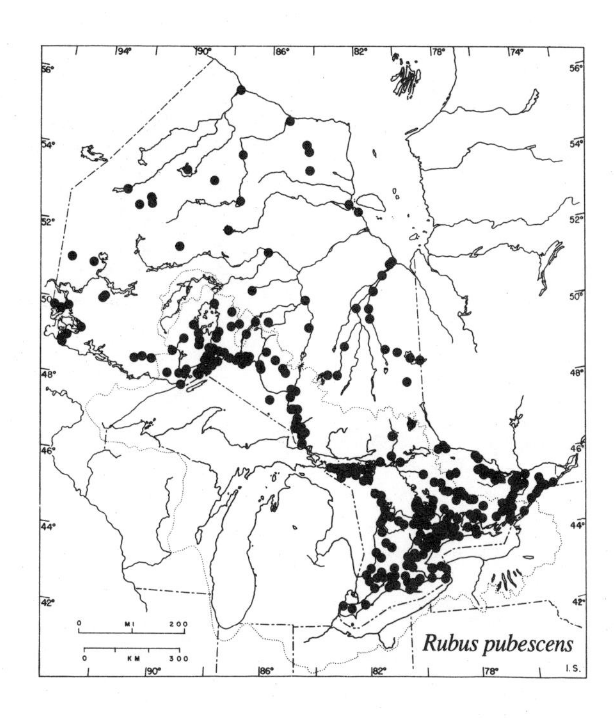

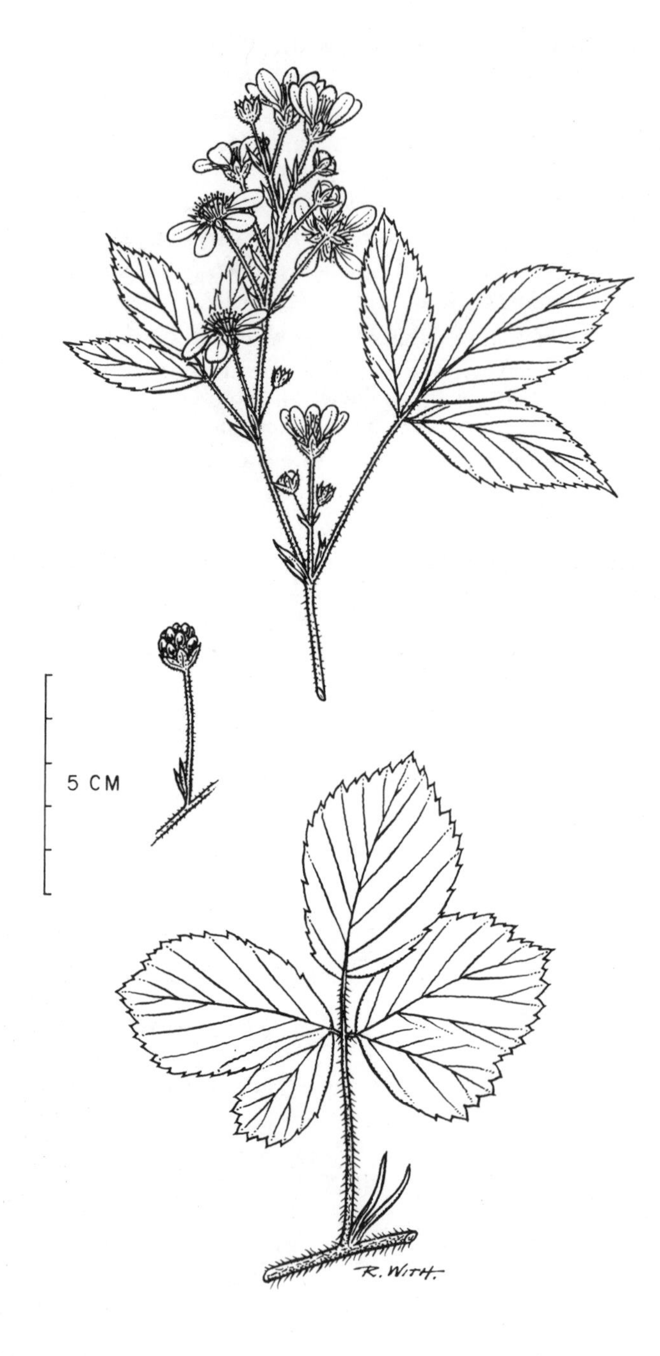

Rubus setosus

Rubus setosus Bigel. **Bristly blackberry**

A low shrub with erect ascending or arching stems up to 1.5 m high. Branchlets densely covered with innumerable spreading or reflexed bristles 1–4 mm long; older canes reddish-brown, with longitudinal ridges, the tips of the branches not rooting when they touch the ground.

Leaves alternate, deciduous and compound. Leaves of the primocanes mostly 5-foliolate but often 3-foliolate, with or without an imperfectly developed lobe on the basal side of each lateral leaflet; leaflets lance-ovate or oblong-ovate to obovate or nearly rhombic, acute to obtuse at the tip; margins doubly serrate or dentate; terminal leaflet 3–8 cm long, 1–4 cm wide, long-stalked. Leaves of the floricanes 3-foliolate, narrowly obovate to oblong-rhombic, acute to acuminate at the apex; blades pale green, dull, glabrous on both surfaces but usually pubescent on the veins beneath; petioles more or less bristly; stipules linear, glandular-bristly, 1–2 cm long.

Flowers few to many, in terminal, more or less elongate clusters with reduced leaves or bracts throughout the inflorescence; pedicels and calyx lobes glandular-bristly; petals white, 7–10 mm long, 2–4 mm wide; late June to Aug. Fruit globose, up to 1 cm in diameter, red when immature but becoming black; rather dry and of inferior taste; late Aug. and Sept. (*setosus*—bristly)

Chiefly in moist habitats, swamps, damp thickets; also along borders of woods and fields, and in sandy or rocky ground.

Occasional in southern Ontario from the Niagara River northeastward to the Ottawa Valley and Nipissing District. (s. Que. to Wis., south to Ill., N. Eng., and W. Va.)

Note Many closely related species have been described, at least three of which have been reported from Ontario, i.e., *R. groutianus* Blanch., *R. vermontanus* Blanch., and *R. univocus* Bailey (see Fernald, 1950, p. 846–851). These variants, which Gleason and Cronquist (1963) have included in *R. setosus*, may have originated from hybrids between *R. setosus* and *R. vermontanus* or between *R. setosus* and *R. hispidus*, *R. allegheniensis*, or other species of blackberry. (*groutianus*—in honour of the pioneer Grout family; *vermontanus*—Vermontan; *univocus*—with one voice)

Field check Low erect or arching canes, not rooting at the tips; stems densely covered with innumerable fine bristles; leaves 3–5 foliolate, mostly glabrous but pubescent along the veins beneath; fruit nearly globose, black, and rather unpalatable when mature.

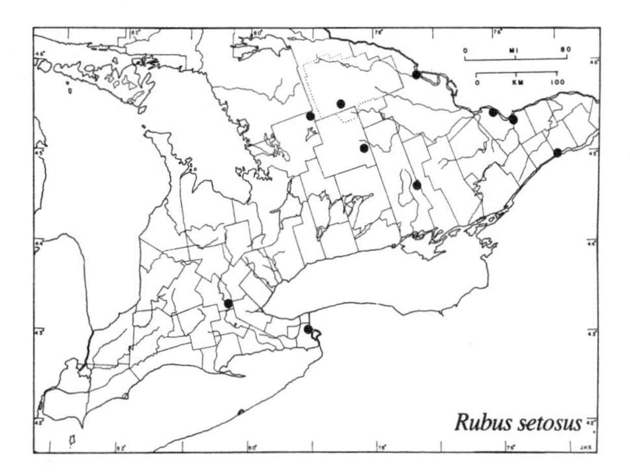

Sorbus americana

Sorbus L. — Mountain-ash

About 80 or more species, chiefly of north temperate regions. Trees or shrubs with alternate, simple or (in ours) pinnately compound, deciduous leaves with 11–17 serrate leaflets. Flowers perfect, 5-parted, small, white, and numerous, in showy corymbose clusters at the ends of the branches or on short lateral shoots. Fruit a globose, ovoid, or pyriform pome usually less than 1 cm diameter. (*Sorbus* — said to be a classical name for some European species)

The genus *Sorbus* is submerged by Fernald and others in the genus *Pyrus*, along with *Aronia* and *Malus*. (*Pyrus* — from classical Latin *pirus*, the pear tree)

Key to *Sorbus*

a. Leaflets taper-pointed to long-acuminate, 3–5 times as long as wide; petals obovate, 3–4 mm long; fruit 5–6 mm in diameter — ***S. americana***

a. Leaflets obtuse to abruptly short-acuminate, 2–3 times as long as wide; petals orbicular, 4–5 mm long; fruit 8–10 mm in diameter

- b. Leaflets glabrous and whitened beneath; branches of the inflorescence and pedicels glabrous or nearly so; winter buds sticky, their outer scales glabrous — ***S. decora***
- b. Leaflets soft-hairy beneath; branches of the inflorescence and pedicels villous; winter buds scarcely sticky, their outer scales villous — ***S. aucuparia***

Sorbus americana Marsh. (*Pyrus americana* (Marsh.) DC.) — Mountain-ash

A shrub or small tree up to 10 m high. Branchlets glabrous or becoming so, greenish-brown to reddish at first; epidermis soon loosening as a thin outer layer; lenticels elongate, pale, and prominent on the young stems; older stems reddish-brown; winter buds sticky, the scales glabrous or the inner ones ciliate.

Leaves alternate, deciduous, and pinnately compound; leaflets 11–17, usually 13–15, lanceolate to narrowly oblong, taper-pointed to long-acuminate at the apex, blades 5–10 cm long and 1–2.5 cm wide, rather thin, light green above, pale and usually glabrous beneath; margins sharply serrate.

Flowers white, small, and numerous, 5-parted, borne in showy flat-topped to roundish clusters 5–15 cm in diameter; petals obovate, 3–4 mm long, usually longer than the stamens; June and July. Fruit a bright orange-red, nearly globose pome 4–6 mm in diameter; Aug. and Sept. (*americana* — American)

Chiefly in moist or shaded habitats, mixed woods and thickets, conifer woods on sphagnum; also on rock outcrops.

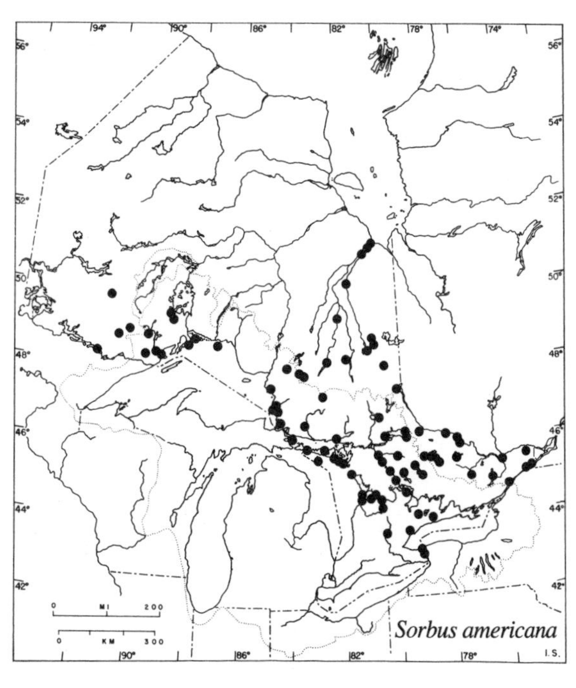

Sorbus americana

From Lake Ontario and the St. Lawrence River northward and westward to Lake

Sorbus decora

Huron, Lake Superior, and the Moose River drainage system. (Nfld. to Ont., south to Minn., Ill., Md., and the uplands of Ga.)

Field check Shrub or small tree with alternate pinnately compound leaves; leaflets taper-pointed to long-acuminate; flowers small and numerous in a showy inflorescence; fruit a cluster of bright red, small, berrylike pomes, each 5-6 mm in diameter.

Sorbus decora (Sarg.) Schneid. (*Pyrus decora* (Sarg.) Hyland) — Mountain-ash

A shrub or small tree up to 10 m tall. Branchlets as in *S. americana*; winter buds sticky, the main scales glabrous on the back but the inner scales usually conspicuously ciliate or villous.

Leaves alternate, deciduous, and pinnately compound; leaflets usually 13, 15, or 17, oblong or narrowly oval to oblong-elliptic, blunt or rounded and abruptly short-acuminate at the tip, blades 3-8 cm long and one-third to one-half as wide, firm, blue-green above, paler and glabrous to sparingly hairy beneath; margins sharply serrate.

Flowers white, 5-parted, a little larger than in *S. americana* and in more open clusters; petals broadly rounded, 4-5 mm long; flowering about a week later than *S. americana*; June and July. Fruit a bright red drupe, 8-12 mm in diameter; Aug. and Sept., often persisting through the winter. (*decora*—handsome or comely)

In moist or dry situations; in open woods and thickets; and characteristically on rocky shores of rivers and lakes.

Common from the Bruce Peninsula and Manitoulin Island, Lake Huron, to the north shore of Lake Superior and northward to about 54° N; less abundant southward on the Canadian Shield; rare south of 45° N. (Greenl. to Sask., south to Iowa, N.Y., and N. Eng.)

Note The European mountain-ash or rowan tree, *S. aucuparia* L., has been planted widely in Ontario and can be found growing in natural plant associations, probably as the result of dispersal of seed by birds. It is very similar to our two indigenous species but may be distinguished by its more pubescent young branchlets and lower surface of its leaflets and by the white-villous but not glutinous winter buds. (*aucuparia*—bird-catching, in reference to the attraction of the fruit)

Field check Shrub or small tree with alternate pinnately compound leaves; leaflets blunt or abruptly short-acuminate; fruit a cluster of bright red berrylike pomes 8-12 mm in diameter.

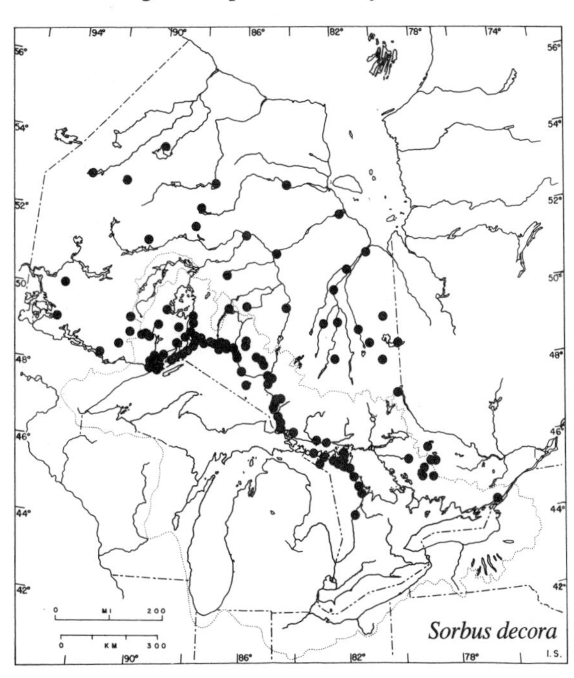

Sorbus decora

Spiraea alba

Spiraea L. — Spirea, Spiraea

A genus of about 100 species characteristic of north temperate regions mainly but also found in Mexico and the Himalayas. Shrubs with alternate, deciduous, simple leaves lacking stipules. Flowers small, white to pink or rose-purple, usually numerous in showy terminal or lateral clusters (panicles or corymbs). Fruit a follicle with small seeds.

Several European and Asian species and cultivars are commonly planted in our gardens and as part of shrubbery plantations along our highways. (*Spiraea*—from the Greek *speira*, wreath)

Note A plant known as false spiraea, *Sorbaria sorbifolia* (L.) A.Br., may be mentioned here because it is somewhat similar to *Spiraea*. It is a slightly woody herbaceous perennial native of eastern Asia and commonly planted in gardens and in public parks. It has been found growing outside cultivation in about a dozen localities from the St. Lawrence and Ottawa valleys northwestward to the north shore of Lake Superior (Cody, 1962). It tends to form thickets and the stems have abundant branches, reaching a height of about 1 metre or more. The flowers are small, white, 5-parted, and similar to those of *Spiraea*, but the leaves are compound, up to 4 dm long, and have 13–21 coarsely and doubly serrate tapering leaflets. (*Sorbaria*—from *Sorbus*, the mountain-ash; *sorbifolia*—from *Sorbus* and the Latin *folia*, leaves, indicating the resemblance of the leaves to those of the mountain-ash)

Key to *Spiraea*

a. Leaves covered beneath with a close, dense, white to tawny coating of woolly hairs; flowers rose-pink or rarely white — ***S. tomentosa***

a. Leaves glabrous or nearly so on both surfaces; flowers white or pale pink

- b. Leaves narrowly lance-oval, one-quarter to one-third as wide as long, finely toothed, pointed at the apex (the sides forming an angle of less than 60°); inflorescence narrowly pyramidal, finely pubescent — ***S. alba***
- b. Leaves broadly oblong, one-third to one-half as wide as long, blunt at the apex (the sides forming an angle of more than 60°), coarsely toothed; inflorescence broadly pyramidal with elongate lower branches, mostly glabrous — ***S. latifolia***

Spiraea alba Du Roi — Narrow-leaved meadowsweet

An erect shrub up to 1.5 m tall, often becoming much-branched and twiggy. Branchlets more or less angled or longitudinally ridged, yellowish to reddish-brown, and glabrous or minutely pubescent; older stems purplish-gray, the outer bark peeling off in narrow papery strips.

Leaves alternate, simple, and deciduous, numerous and often crowded along the stems; blades 3–6 cm long, 1–2 cm wide, about one-quarter to one-third as wide as long, narrowly lance-oval to elliptic or oblanceolate, mostly acute at the tip (with the sides forming an angle of less than 60°), glabrous on both surfaces; margins finely and sharply toothed; petioles 2–6 mm long.

Flowers small (5–8 mm in diameter), white, 5-parted, and numerous, in narrowly pyramidal compound inflorescences at the ends of the branches, the secondary branches arising toward the base and leafy near their attachment to the main stem; main axis, branches of the inflorescence, and short pedicels finely pubescent; June to Sept. Fruit a cluster of 5–8 small follicles with a few seeds in each, the fruiting branches often persisting over winter. (*alba*—white)

Spiraea latifolia

Primarily in low moist fields, sedge meadows, swamps, and roadside ditches and along the shores of lakes, streams, and ponds.

Widespread from Lake Erie to Lake Superior and Lake-of-the-Woods; becoming rare north of 50° N in the Hudson Bay lowlands. (Que. to Alta., south to N. Dak., Mo., and N.C.)

Note A related Eurasian species, *S. salicifolia* L., with leaves broadest below the middle and pink flowers, sometimes escapes from cultivation and may be found in southern Ontario. (*salicifolia* — willow-leaved)

Field check Much-branched shrub with numerous, crowded, finely and sharply toothed, narrowly lance-oval to oblanceolate leaves and finely pubescent slenderly pyramidal clusters of small white flowers.

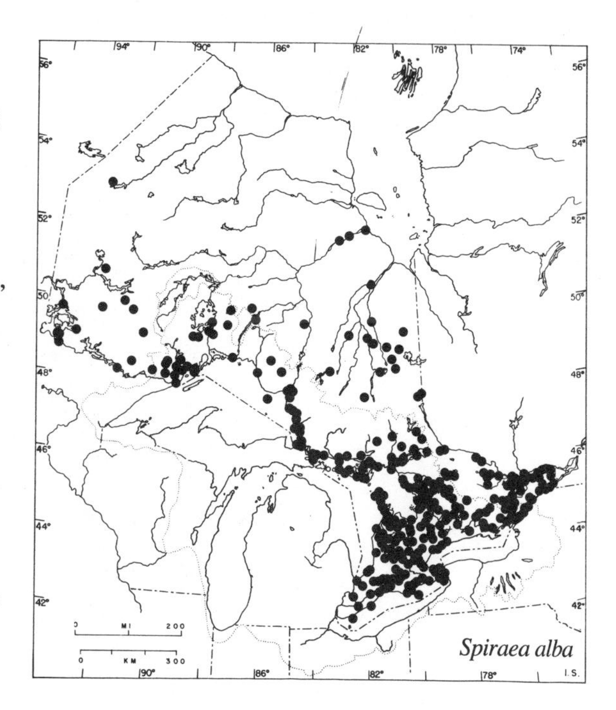

Spiraea latifolia (Ait.) Borkh. — Broad-leaved meadowsweet

A rather coarse shrub up to 1.5 m tall. Branchlets somewhat angled or longitudinally ridged, reddish- to purplish-brown and usually glabrous; older stems with grayish or silvery-gray outer bark peeling off in long papery strips.

Leaves alternate, simple, and deciduous, numerous and often overlapping on the side branches blades broadly oval or elliptic to oblanceolate, 3 – 8 cm long and 1 – 3 cm wide, about one-third to one-half as wide as long, acute or mostly blunt at the tip (the sides forming an angle of 60° or more), glabrous on both surfaces; margins coarsely and prominently toothed; petioles 2 – 8 mm long.

Flowers 4 – 5 mm in diameter, white or pale pink, 5-parted, and numerous in open, broadly pyramidal, freely branched inflorescences; branches and pedicels glabrous; July to Sept. Fruit a cluster of 5 – 8 small few-seeded follicles often persisting over winter. (*latifolia* — broad-leaved)

In sandy or rocky, usually moist, situations: river banks, edges of bogs and thickets, rock crevices and margins of rock pools, rocky pastures, roadside ditches, and open low ground.

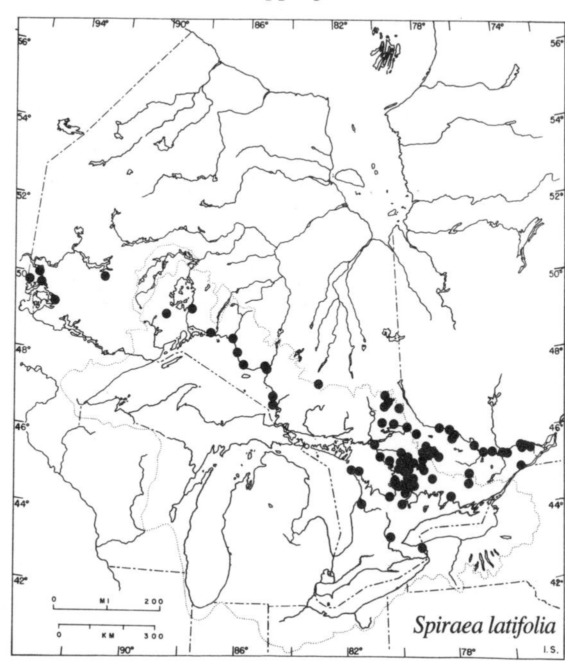

Spiraea tomentosa

From the St. Lawrence and Ottawa valleys to Georgian Bay, Lake Timagami, and the north shore of Lake Superior; also in the Lake-of-the-Woods region; rare and local south of latitude 44° N. (Nfld. to Man., south to Mich. and N.C.)

Note The frequency with which specimens intergrading between *S. alba* and *S. latifolia* occur suggests that the latter may not be specifically distinct and could be treated, as some botanists have done, as a variety of the former under the name *S. alba* var. *latifolia* (Ait.) Ahles.

Field check Much-branched shrub with numerous, often overlapping, coarsely and sharply toothed, broadly oblong leaves; inflorescence glabrous, open, and broadly pyramidal, with secondary and tertiary leafy branches bearing the numerous small white or pale pink flowers.

Spiraea tomentosa L.

Steeple-bush
Hardhack

A sparsely branched erect shrub growing to 1 m or more in height. Branchlets green to brownish, slightly ridged, and covered with a close, rusty, woolly tomentum, later becoming smooth and purplish-brown or reddish-brown.

Leaves alternate, simple, and deciduous, sessile or with a petiole 1–4 mm long; blades ovate-elliptic to oblong or lanceolate, 1–5 cm long and 0.5–2 cm wide, dull green and usually smooth above, gray-green to tawny beneath from a close felt-like pubescence which, however, does not obscure the prominent veins; the tip acute to blunt, the base usually tapered and the margins coarsely serrate.

Flowers small, 5-parted, pink to deep rose, numerous, and in compound spirelike terminal inflorescences with lateral branches at the base passing into separate axillary spikes in the axils of upper reduced leaves; branches of the inflorescence, the short pedicels, and the calyx tubes covered with a close rusty tomentum; July and Aug. Fruit a group of tiny pubescent follicles ripening in late summer or fall and often persisting through the winter. (*tomentosa*—densely woolly, in reference to the pubescence of the stems and leaves)

In sandy, marshy, or rocky, usually acid soils; especially along the edges of ponds and swamps; also in fields and fencerows and along roadsides.

Common in eastern Ontario from the St. Lawrence and Ottawa valleys west and north to Lake Huron; rare in the Deciduous Forest Region; known from a few locations north of 46° N but not reported west of 83° W. (P.E.I. to Minn., south to Ark. and Ga.)

Field check Sparsely branched erect shrub with coarsely crenate-serrate leaves, tawny beneath; pink to deep rose-coloured tiny flowers in steeplelike terminal clusters.

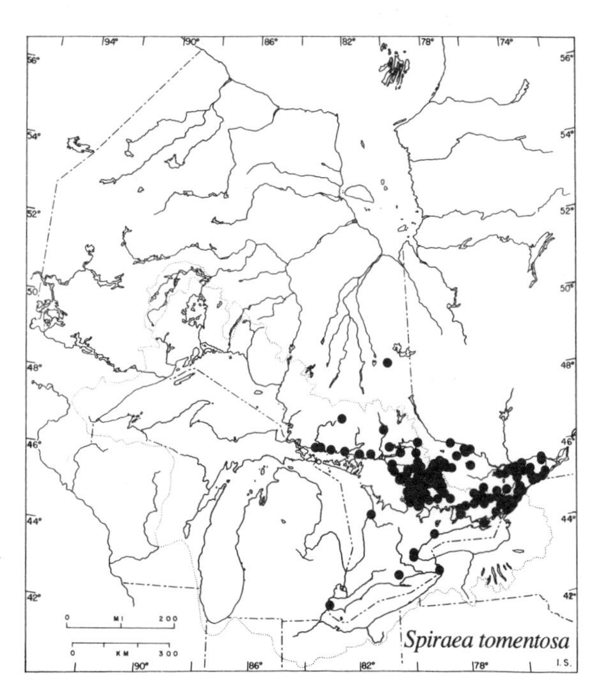

Ptelea trifoliata

RUTACEAE – RUE FAMILY

A large family of about 150 genera and 1600 species, widely distributed in tropical and temperate regions, especially in Australia and South Africa. Trees and shrubs with alternate simple or compound leaves lacking stipules. Plants often aromatic due to the presence of oil glands in various parts of the plant; leaves often gland-dotted. Inflorescence usually cymose. Flowers perfect or unisexual, 4- or 5-parted, with a disc below the pistil. Fruit a drupe, berry, samara, or schizocarp.

Important commercial products are derived from genera in this family, in particular the common citrus fruits—orange, lemon, lime, grapefruit, and citron (*Citrus* spp.). Several genera yield valuable timber, and a few trees and herbs are used as ornamentals. The name of the family is based on the genus *Ruta*, the Greek name for the common rue, a plant of Mediterranean Europe. (*Citrus*—the Latin name for an African tree with fragrant wood, later transferred to citron)

Ptelea L. – Hoptree

Three species of shrubs or small trees native to North America. Leaves alternate, petiolate, and trifoliolate. Flowers small, greenish-white to yellowish, in terminal compound cymes. Fruit a 2-locular, flat, almost circular, broadly winged, samara. (*Ptelea*—the Greek name for an elm, transferred to this genus by Linnaeus because of the resemblance of the fruit to that of the elm)

Ptelea trifoliata L. — Hoptree, Wafer-ash, Stinking-ash

A tall shrub or small tree growing to a height of 5 m or more. Branchlets stout and smooth, with a large white pith; bark reddish-brown and shiny at first, later turning gray and becoming roughened. Polygamous.

Leaves alternate, trifoliolate, and deciduous; leaflets rather thick in texture, dark green and shiny above, much paler beneath, and malodorous when crushed, ovate-lanceolate to obovate or elliptic, up to 10 cm long and 5 cm wide, both surfaces dotted with tiny black glands but these more easily seen on the upper surface; margins entire or shallowly and often remotely serrulate; petioles stout, 7 cm or more in length.

Flowers greenish-white, in large cymose terminal clusters; unisexual or perfect; mid-June. Fruit a flat, waferlike, gland-dotted, nearly circular samara 2–2.5 cm in diameter, with a broad veiny wing around the central part which contains 2 seeds, borne in dense stalked clusters which frequently remain on the plant over winter. (*trifoliata* —with three leaflets)

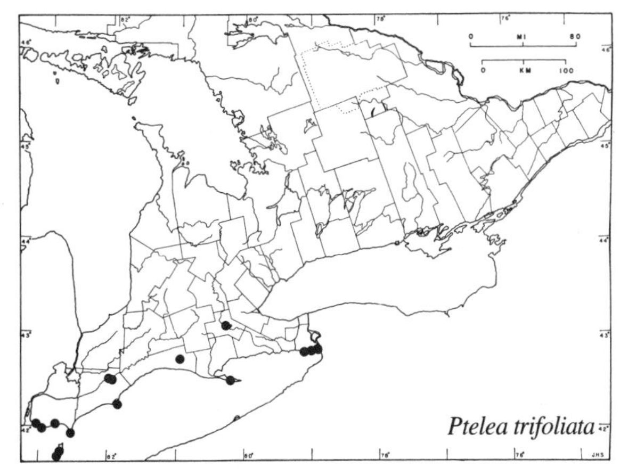

In sandy soil; on sand dunes, in thickets or open woods, especially along the shores of lakes and rivers.

Confined to the Deciduous Forest Region along the north shore of Lake Erie; probably indigenous at the western and

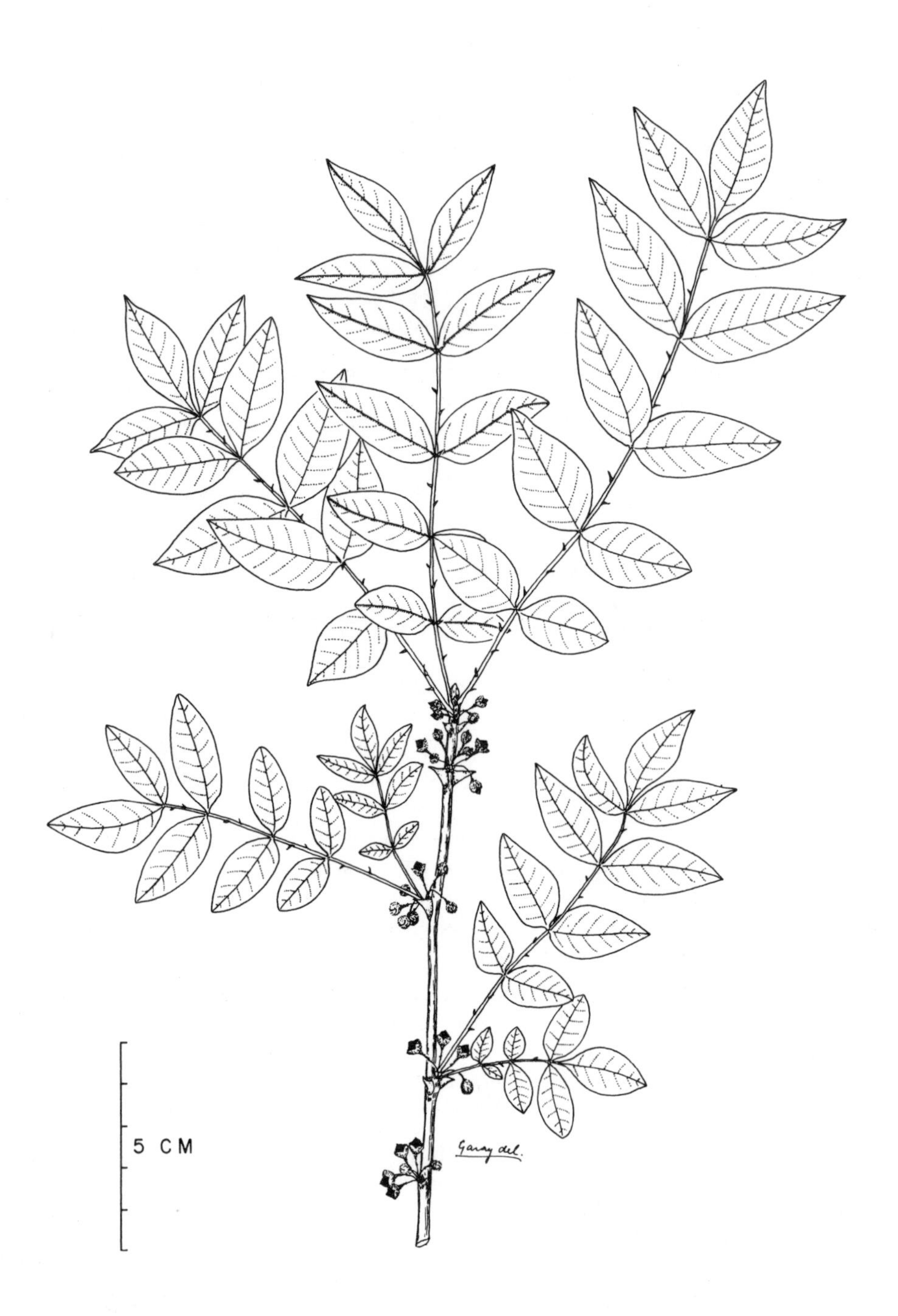

Zanthoxylum americanum

eastern ends of Lake Erie, the intermediate stations possibly introductions by Indians or early settlers. (s. Ont. and N.Y. to Lake Michigan and Neb., south to Mex. and Fla.)

Note The common name, stinking-ash, refers to the unpleasant odour of the leaves when crushed or bruised.

Field check Alternate trifoliolate leaves, malodorous when crushed; fruits in clusters of round waferlike samaras. In Ontario native only in the Deciduous Forest Region along the north shore of Lake Erie.

Zanthoxylum L.

A genus of about 200 species found in temperate and subtropical regions of North America and eastern Asia. Shrubs and small trees with prickly stems and alternate pinnately compound leaves. Flowers small, greenish or whitish, pollinated by bees or flies. Fruit a somewhat fleshy follicle with 1 or 2 seeds in each locule. (*Zanthoxylum* or *Xanthoxylum*— from the Greek *xanthos*, yellow, and *xylon*, wood; hence one of the common names, yellow-wood)

Zanthoxylum americanum Mill. — Prickly ash, Toothache tree, Yellow-wood

An upright, much-branched, prickly shrub often spreading to form dense impenetrable thickets about 3 m in height; dioecious. Branchlets stout, with a pair of strong, sharp-pointed, persistent prickles with broad flat bases flanking the petiole attachment of each leaf; bark slightly ridged, often reddish to purplish at first, turning brown to gray.

Leaves alternate, deciduous, and pinnately compound, with 7–11 leaflets; dark green above, much paler beneath, dotted with translucent glands; fragrant when crushed; complete leaf up to 25 cm long and 10 cm wide, the individual leaflets up to 6 cm long and 4 cm wide, elliptic to ovate-oblong; margins entire or shallowly crenulate; petioles with small prickles, especially in pairs where the leaflets are attached to the rachis.

Flowers small, greenish, unisexual, in close clusters on the old wood, appearing just as the leaves are unfolding in early spring; Apr. and May. Fruit a reddish-brown, round, aromatic pod less than 6 mm in diameter, splitting into two sections and exposing one or two glossy black seeds; late July to early Sept. (*americanum*— American)

In pastures, open rocky places, thickets, and fencerows, and along roadsides and edges of woods.

Throughout those parts of southern Ontario lying west, south, and east of the Canadian Shield and northward in the Ottawa Valley to about 46°N. (s.w. Que. to N.Dak., south to Okla. and Ga.)

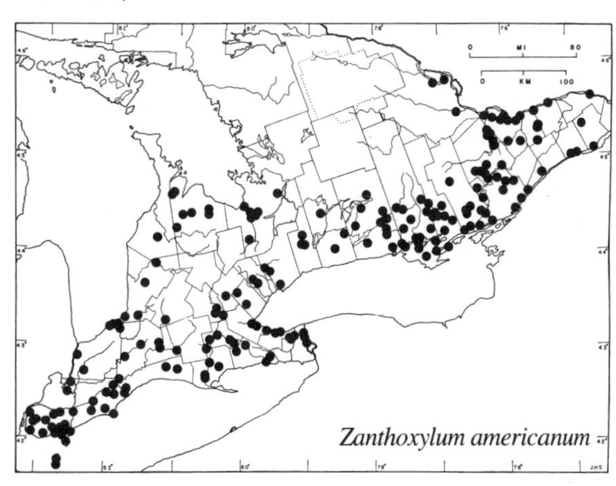
Zanthoxylum americanum

Note In colonial times the volatile aromatic oil was extracted from this shrub and used as an emergency treatment for toothache, hence one of the common names. This prickly colonial shrub is

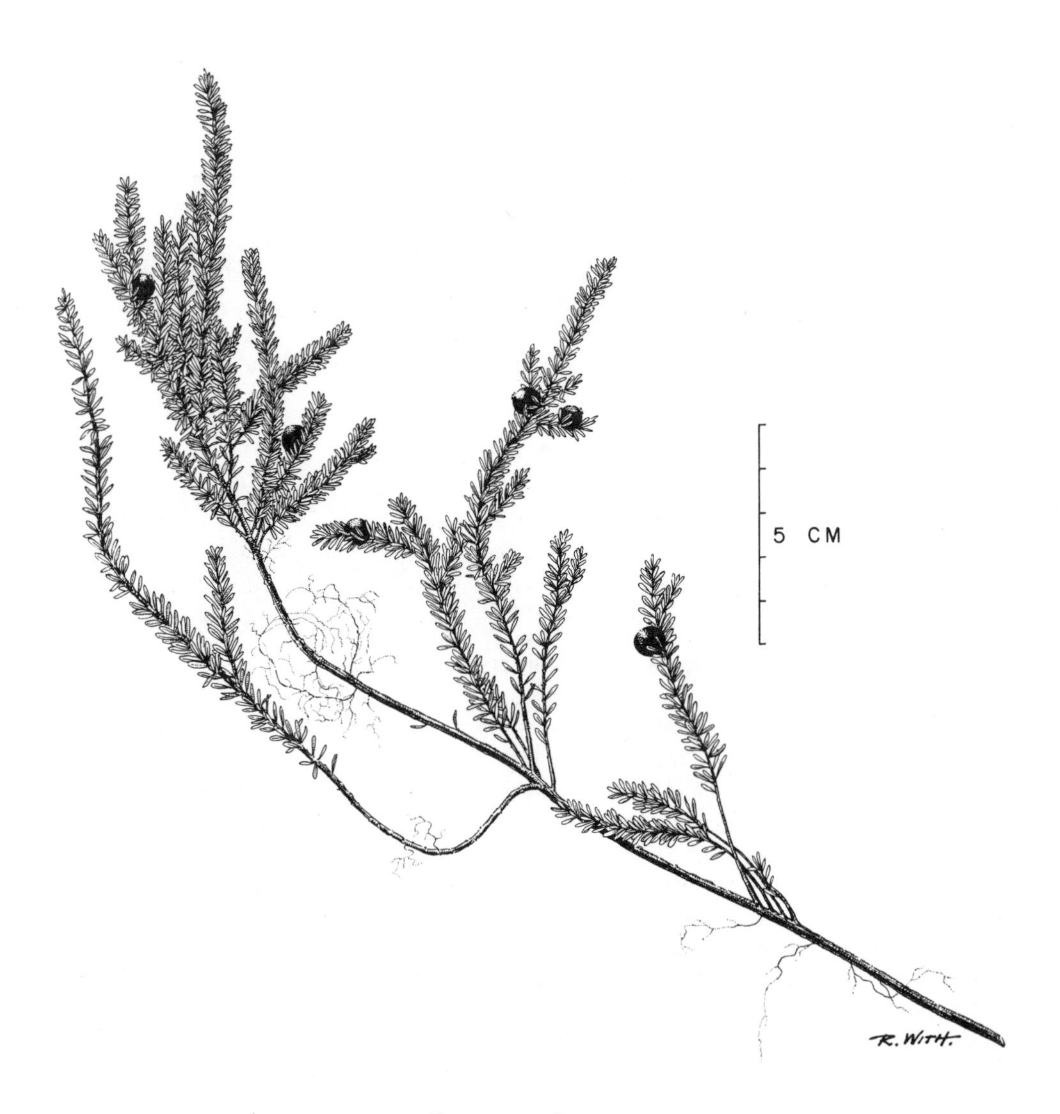

Empetrum nigrum

avoided by browsing cattle and often becomes a serious weed in fields and pastures if not rooted out immediately.

Field check Upright thicket-forming shrub; stems with stout-based prickles in pairs; pinnately compound, alternate, prickly leaves which are gland-dotted and aromatic when bruised.

EMPETRACEAE – CROWBERRY FAMILY

Evergreen shrubs of heathlike growth habit, with alternate simple revolute leaves lacking stipules; flowers small and inconspicuous. Fruit a drupe with 2–9 stones. Three genera and about 15 species in north temperate regions and in South America and islands of the south Atlantic.

Empetrum L. – Crowberry

Much-branched dwarf shrubs, often mat-forming, with small permanently revolute leaves. Flowers usually unisexual, 3-parted, axillary and nearly sessile, greenish to purplish. Two or 3 species, sometimes split into as many as 15 species. (*Empetrum*—from the Greek *en*, upon, and *petros*, a rock, in reference to a common type of habitat of these species)

Empetrum nigrum L. — Crowberry

A low heathlike evergreen shrub with spreading prostrate stems and short upright branches, often forming extensive mats 1 m or more in width but less than 2 dm high; dioecious or monoecious, sometimes polygamous or the plants with only bisexual flowers. Young branchlets slender, pubescent, and pale brown, becoming dark brown to gray or blackish with loose flaky bark.

Leaves evergreen and needlelike, scattered singly or sometimes in small clusters along the stem, even appearing whorled; blades linear-oblong, 3–7 mm long, the tip blunt or rounded and the base abruptly narrowed to an evident but short stalk; dark green and leathery; margins strongly revolute, sometimes leaving only the midrib of the lower surface exposed.

Flowers tiny and inconspicuous, pink to purplish, solitary and nearly sessile in the axils of some of the uppermost leaves, unisexual or perfect; late June and July. Fruit a purplish-black to jet-black berrylike drupe, 4–6 mm in diameter, somewhat juicy, and with 6–9 stonelike nutlets; Aug. to Oct. (*nigrum*—black, referring to the fruit colour)

In rock crevices and on sandy and rocky soil along shores, edges of woods, and in clearings; also in sphagnum bogs and lichen-moss mats.

Around Hudson Bay and James Bay and on islands in James Bay (N.W.T.); also around the shores of Lake Superior and at Lake Nipigon. (Greenl. to Alaska, south to Calif. and N.Y.; Eurasia)

Note *Empetrum nigrum*, as used here in the broad sense, includes specimens that have been classified by some botanists under *E. atropurpureum* Fern. & Wieg. or *E. hermaphroditum*

(Lge.) Hag. The distribution of monoecious, polygamous, and hermaphroditic plants in Ontario and elsewhere is unknown. The berries are edible but not pleasing to most tastes. Crowberry was one of the plants collected along the north shore of Lake Superior in 1848 by Louis Agassiz, the famous Swiss-American geologist and naturalist (see Soper & Voss, 1964). (*atropurpureum* — dark purple; *hermaphroditum* — hermaphroditic)

Field check Creeping and mat-forming evergreen shrub with numerous crowded, linear-oblong, strongly revolute leaves, tiny pink-purple flowers, and single, black, berrylike drupes in the leaf axils.

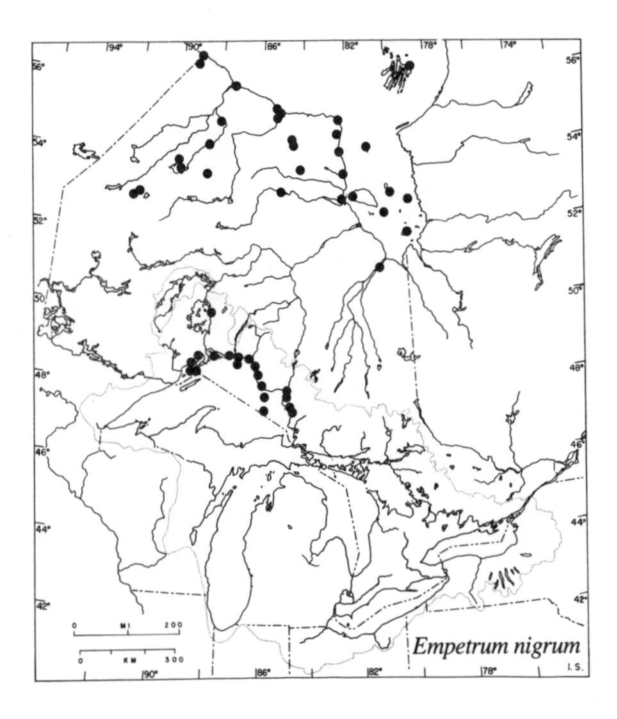

ANACARDIACEAE – CASHEW FAMILY

Trees, shrubs, and rarely vines, with alternate leaves, the stems often with resin canals, and the fruit usually a drupe. About 60 genera primarily of the tropics but also in the warm temperate zones of North America, eastern Asia, and the Mediterranean region. Included are species producing edible fruits and nuts, such as, the mango (*Mangifera*), cashew (*Anacardium*), and pistachio (*Pistacia*), and various lacquers (*Rhus, Melanorrhoea*) and tannins (*Schinopsis*). Some are cultivated as ornamentals (*Cotinus*). A few cause acute dermatitis (*Rhus*). The only genus native to Ontario is *Rhus*. (*Mangifera*—mango-bearing; *Anacardium*—from the Greek *kardia*, heart, in reference to the shape of the nut in the cashew; *Pistacia*—from an ancient Persian name, *Pista*; *Melanorrhoea*—from the Greek *melas*, black, and *rhoia*, a flow, perhaps referring to a dark resin or latex; *Schinopsis*—resembling *Schinus*, the mastic tree; *Cotinus*—Greek *Kotinos*, the name for a tree with red wood)

Rhus L. – Sumach, Poison ivy

Shrubs, small trees, and woody vines with trifoliolate or odd-pinnate compound leaves, both perfect and unisexual, small, 5-parted flowers, and drupelike fruits. The flowers are visited by several kinds of insects and the seeds are distributed by birds. About 250 species, 6 indigenous in Ontario. (*Rhus*—a name of Greek and Latin origins, possibly from *rhous* or *rhoys*, ancient name of the sumach of Sicily)

Key to *Rhus*

a. Leaflets 3
  b. Terminal leaflet stalked; flowers yellowish-green in loose clusters from the axils of the leaves, appearing after the leaves; fruits white
    c. Vine trailing or climbing by aerial roots **R. radicans** var. **radicans**
    c. Shrub with erect sparsely branched stems, spreading by underground stolons **R. radicans** var. **rydbergii**
  b. Terminal leaflet tapered at base but not stalked; flowers yellow, in tight clusters, opening before or with the leaves; fruits red, hairy **R. aromatica**
a. Leaflets 5–31
  d. Margins of leaflets toothed
    e. Twigs and petioles hairy **R. typhina**
    e. Twigs and petioles glabrous **R. glabra**
  d. Margins of leaflets entire
    f. Rachis winged between the pairs of leaflets; drupes red, in dense terminal clusters; dwarf shrub of dry habitats **R. copallina** var. **latifolia**
    f. Rachis not winged; drupes whitish, in arching, open, persistent clusters; medium to tall shrubs of wet habitats **R. vernix**

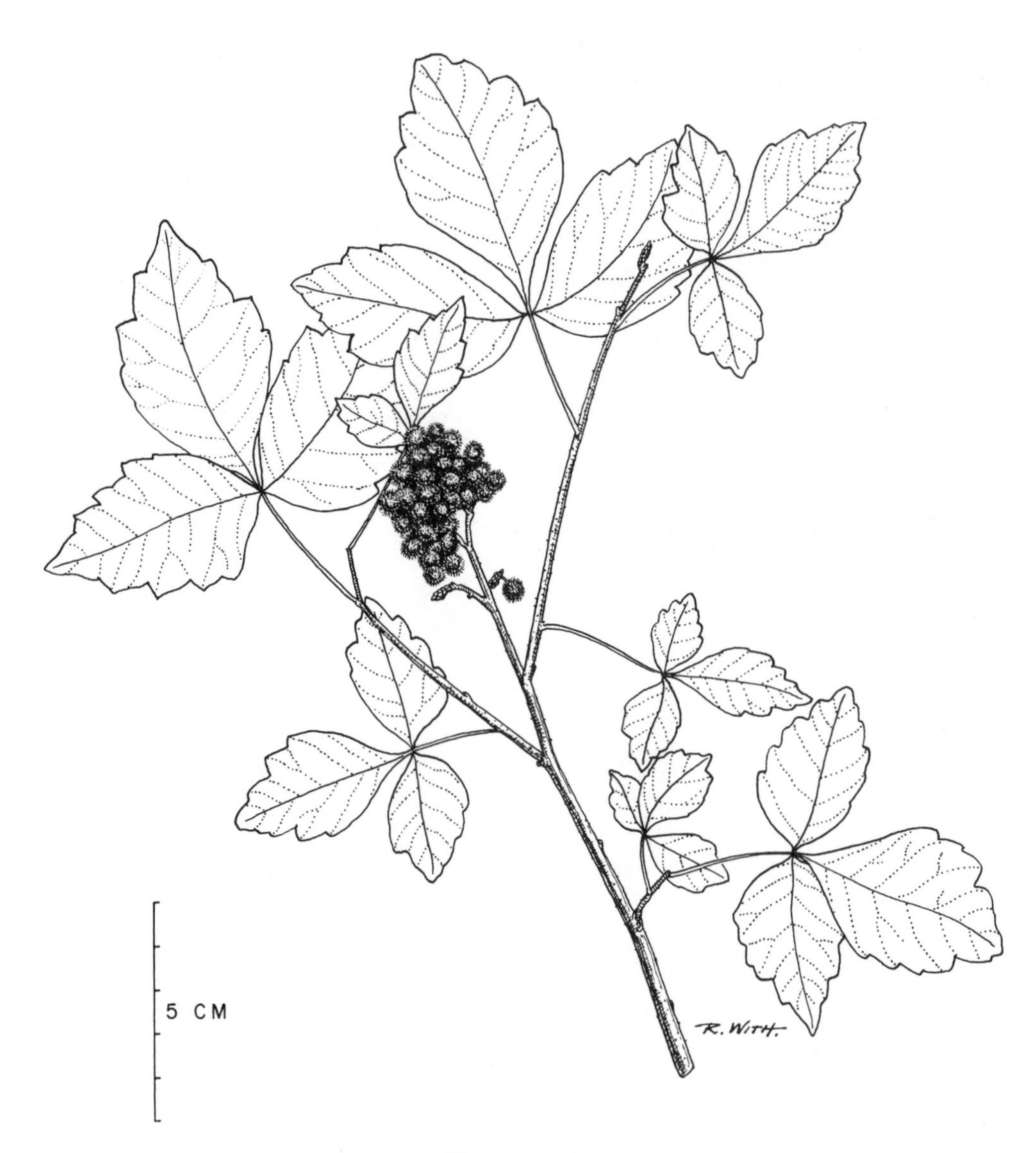

Rhus aromatica

Rhus aromatica Ait. — Fragrant sumach

A low shrub with spreading or ascending branches and aromatic foliage, often forming extensive mounds or low thickets, and usually less than 1.5 m high. Branchlets minutely hairy becoming glabrous, brownish-gray to purplish-gray, somewhat zigzag, with prominent nodes and scattered lenticels.

Leaves alternate, deciduous, and trifoliolate, 6–10 cm long, including a petiole up to 4 cm in length; leaflets sessile, terminal one 5–7 cm long and 2.5–4 cm wide, rhombic-ovate to oval with a pointed or rounded tip and tapered at the base; lateral leaflets slightly smaller and usually more rounded at the base; margins coarsely toothed above the middle with rounded and abruptly pointed teeth; leaflets dark green above, paler beneath, downy on both surfaces at first but becoming glabrous above when full grown; margins and veins of the lower surface ciliate.

Flowers small, yellowish, and in dense clustered spikes 0.5–2 cm long developing from small, slender, pointed, reddish-brown cones which are formed in late summer near the ends of the branches; flowering occurs before the leaves unfold in late Apr. and early May or with the leaves in May and early June. Fruit a densely hairy, reddish, berrylike drupe, 6–9 mm in diameter, in clusters; late July and Aug. (*aromatica*—with a fragrance)

In dry sandy or rocky places; sand dunes, limestone flats and crevices or in open pastures and clearings.

Along the shores of the lower Great Lakes, along the southern boundary of the Canadian Shield and northward to Manitoulin Island on the west and to the Ottawa Valley on the east; northern limit at about 46°N. (Vt. and s.w. Que. to Ill., south to Kans. and La.)

Field check Low thicket-forming shrub; branchlets and alternate trifoliolate leaves aromatic when bruised; margins of leaflets coarsely toothed, especially above the middle; small yellow flowers in close clusters, mostly before the leaves; fruit a compact cluster of reddish berrylike drupes covered with short stiff hairs.

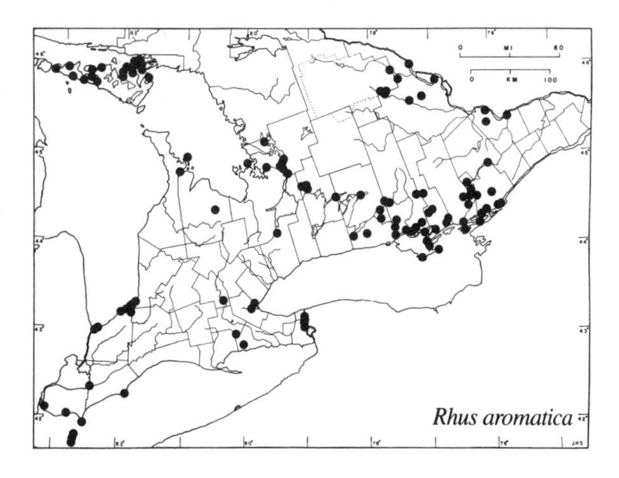

Rhus copallina var. *latifolia*

Rhus copallina L. var. *latifolia* Engler — Dwarf sumach, Shining sumach, Wing-rib sumach

A low shrub with spreading to ascending branches reaching a height of 9–12 dm. Branchlets greenish-brown to purplish-gray, downy at first but becoming glabrous; older stems somewhat ridged and warty.

Leaves alternate, deciduous, and pinnately compound, 15–23 cm long, with 5–11 oblong or oval pointed leaflets; petiole 2.5–7.5 cm long, wingless, but the rachis with lateral winglike borders of leaf tissue between the pairs of opposite leaflets; leaflets sessile, 6–7.5 cm long and 0.5–3 cm wide, the terminal one tapered at the base, the lateral ones more rounded; upper surface dark green, glabrous, and glossy, the lower surface paler and dull; margins entire.

Flowers tiny, greenish-yellow, in elongate terminal clusters; June and July. Fruit a densely hairy, red, berrylike drupe, in clusters; Aug. and early Sept. (*copallina*—like copal or resin, referring to the shiny upper surface of the leaves; *latifolia*—broad-leaved)

In dry sandy or rocky ground, along edges of woods, roadsides, and in open fields.

Rather local north of Lake Erie; north of the eastern end of Lake Ontario and along the upper St. Lawrence River. (Maine to Ill., south to Tex. and Fla.)

Note Dwarf sumach makes an attractive ornamental, prized for its fine autumn colour, but is rather slow-growing. It is hardy only in the southern part of Ontario.

Field check Dwarf shrub with alternate pinnately compound leaves; rachis of leaf winged between the pairs of glossy leaflets; fruits small, red, berrylike, in dense elongate terminal clusters.

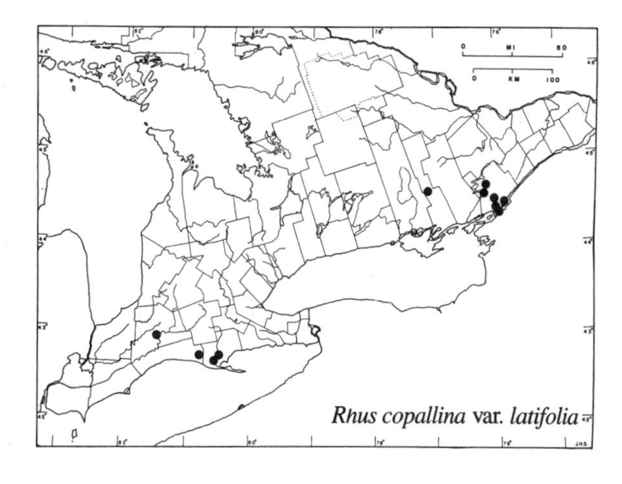

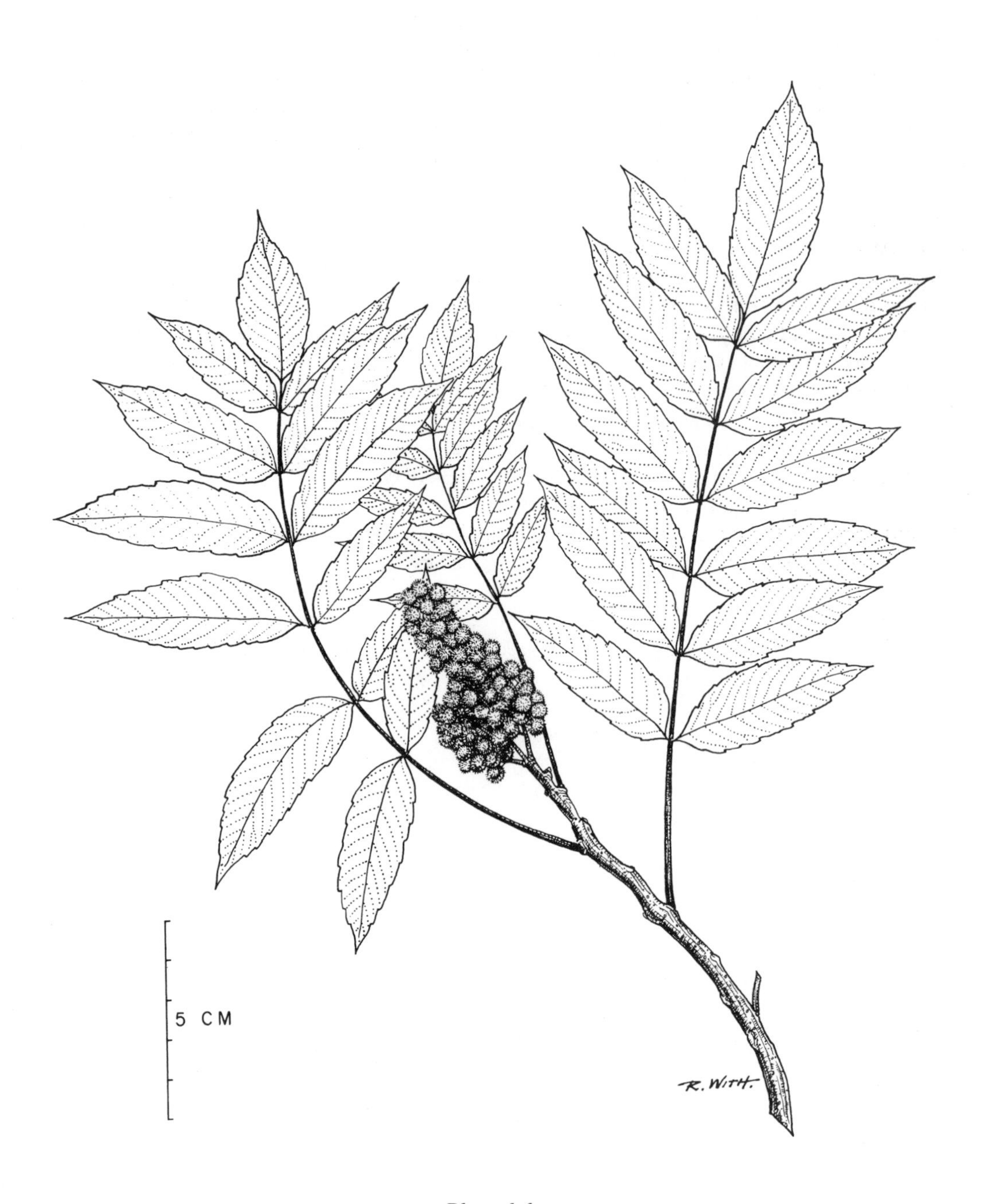

Rhus glabra

Rhus glabra L. **Smooth sumach**

A tall thicket-forming shrub growing to a height of about 3 m or occasionally taller, much like staghorn sumach except for its glabrous branches, stems, and leaves.

Leaves alternate, deciduous, and pinnately compound, 13-32 cm long, including a petiole 1.2-7.5 cm in length and a glabrous rachis with 5-14 pairs of opposite sessile leaflets and a stalked terminal leaflet; leaflets 5-9 cm long and 1.5-3.5 cm wide, narrowly oblong to elliptic, the tip long and tapering, the base rounded or oblique; upper surface dark green and glabrous, lower surface strongly whitened; margins sharply and coarsely toothed.

Flowers small, yellowish-green, in large terminal clusters up to 13 cm long; June and July. Fruit a densely short-hairy, red, berrylike drupe, in long branched clusters, persistent over winter; Aug. and early Sept. (*glabra*—smooth)

In dry sandy or rocky ground; along hillsides and escarpments.

Rather local in the areas north of Lake Erie and Lake Ontario; along the southern margin of the Canadian Shield from the upper St. Lawrence River to Georgian Bay; Manitoulin Island; east and west shores of Lake Superior to Lake-of-the-Woods. (Maine and s.w. Que. to B.C., south to Mex. and Fla.)

Note Like staghorn sumach, smooth sumach makes an attractive ornamental for natural effects in a spacious garden or on large estates.

A hybrid between *R. glabra* and *R. typhina* has been described under the name *R.* × *borealis* (Britton) Greene: it has short pilose or puberulent branches and has been reported from Algoma District. (*borealis*—northern)

Field check Tall shrub with few divergent branches; stems and leaves glabrous; leaves pinnately compound with 11-29 leaflets; fruits red, berrylike, bristly-hairy, in long branched terminal clusters.

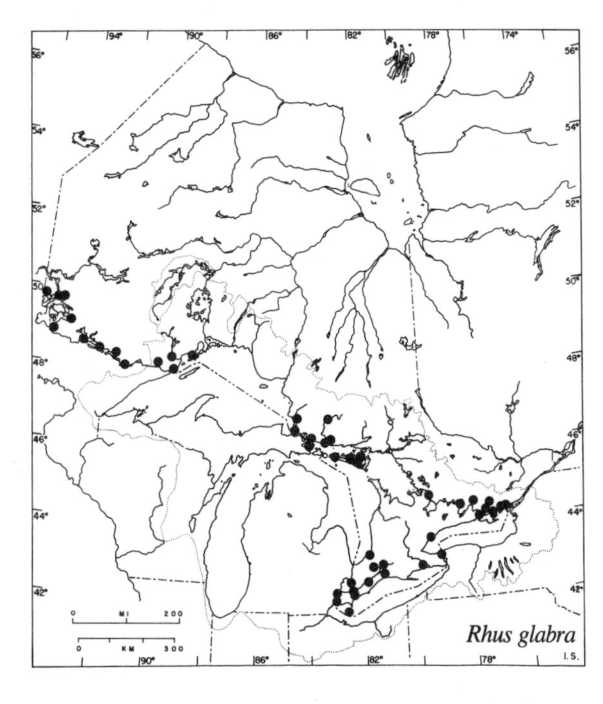

Rhus radicans

Rhus radicans L. **Poison ivy**
(including *Toxicodendron radicans* (L.) Kuntze ssp. *negundo* (Greene) Gillis; *T. rydbergii* (Small ex Rehd.) Greene)

Poison ivy occurs in Ontario as two varieties of different growth habit and geographical limits. One variety, *R. radicans* var. *radicans*, is a trailing, scrambling, or high-climbing woody vine with elongate stems freely rooting at the nodes. This variety grows along the ground, climbs over rocks and logs, and clings to vertical surfaces such as tree trunks by means of aerial roots. It ascends forest trees to a height of 15 m or more and may develop a main stem or trunk up to 10 cm in diameter. The second variety, *R. radicans* var. *rydbergii* (Small) Rehd., is a low shrub usually less than 1 m high, with erect or ascending branches, forming patches and low thickets. Branchlets in both varieties minutely hairy, becoming glabrous, brownish-gray, somewhat ridged, and warty.

Leaves alternate, deciduous, and compound; leaflets 3, varying in texture from thin to thick or leathery, the leaflets usually drooping from the end of a conspicuous petiole 5 - 25 cm long; terminal leaflet with a distinct stalk, the two lateral ones nearly sessile or with a short stalk; leaflets ovate, 5 - 15 cm long and 2.5 - 6 cm wide, the tip abruptly or gradually tapered, the base rounded; margins smooth, wavy, coarsely toothed or even lobed; upper surface dark green and glabrous, dull or shiny, the lower surface somewhat paler and minutely hairy, at least on the veins.

Flowers tiny, greenish-white to yellowish, in loose clusters from the axils of the leaves; June and early July. Fruit a small, dry, round, and beadlike drupe with a wrinkled papery coat, ivory-white to straw-coloured, conspicuous after the leaves have fallen, and persistent over winter; late July and Aug. (*radicans*—rooting, in reference to the aerial roots of the climbing variety; *rydbergii*—in honour of P.A. Rydberg, 1860 - 1931, author of manuals on Western flora; *Toxicodendron*—poison-tree; *negundo*—from *Acer negundo* L., Manitoba maple, in reference to the similarity of the leaves of the climbing poison ivy to those of Manitoba maple)

In dry or moist situations; in sandy, gravelly, or rocky soil or in clay loam; on sand dunes, talus slopes, cliffs, and rocky ridges or scrambling over banks, walls, and fences; in moist woods, ravines, or swampy thickets; on tree trunks, poles, and brick walls.

The climbing variety (var. *radicans*) is restricted to southern Ontario, chiefly south of the Canadian Shield but also along the St. Lawrence River and in the Ottawa district; the low shrubby variety (var. *rydbergii*) is common throughout most of southern Ontario and less frequent northward to about 51°N. (N.S. and Que. to s. B.C., south to Mex. and Fla.)

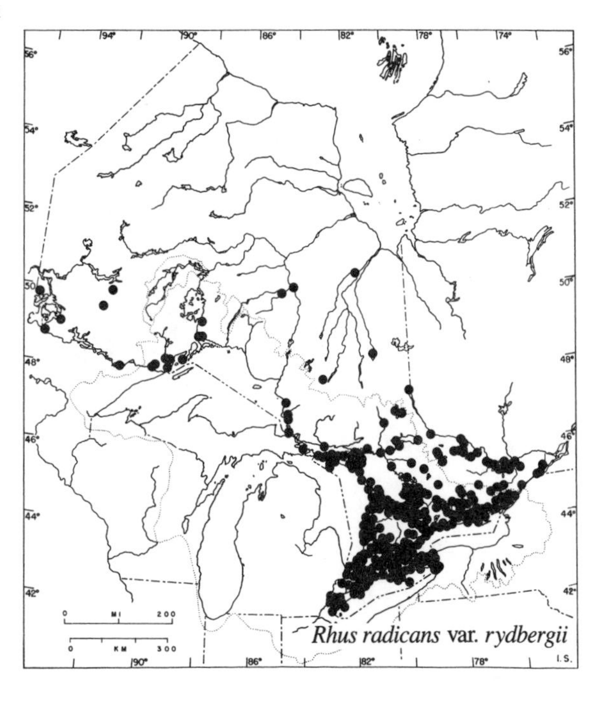

Rhus radicans var. *rydbergii*

Note In the detailed study of poison ivy and poison oaks by Gillis (1971), the names used for our shrubby and climbing varieties of poison ivy are, respectively, *Toxicodendron rydbergii* (Small ex Rehd.) Greene and *T. radicans* (L.) Kuntze ssp. *negundo* (Greene) Gillis. We are retaining the combinations under the genus *Rhus* as they are still in wide use. Both varieties of poison ivy are capable of producing in many people a serious skin rash (dermatitis) by direct contact with the plant

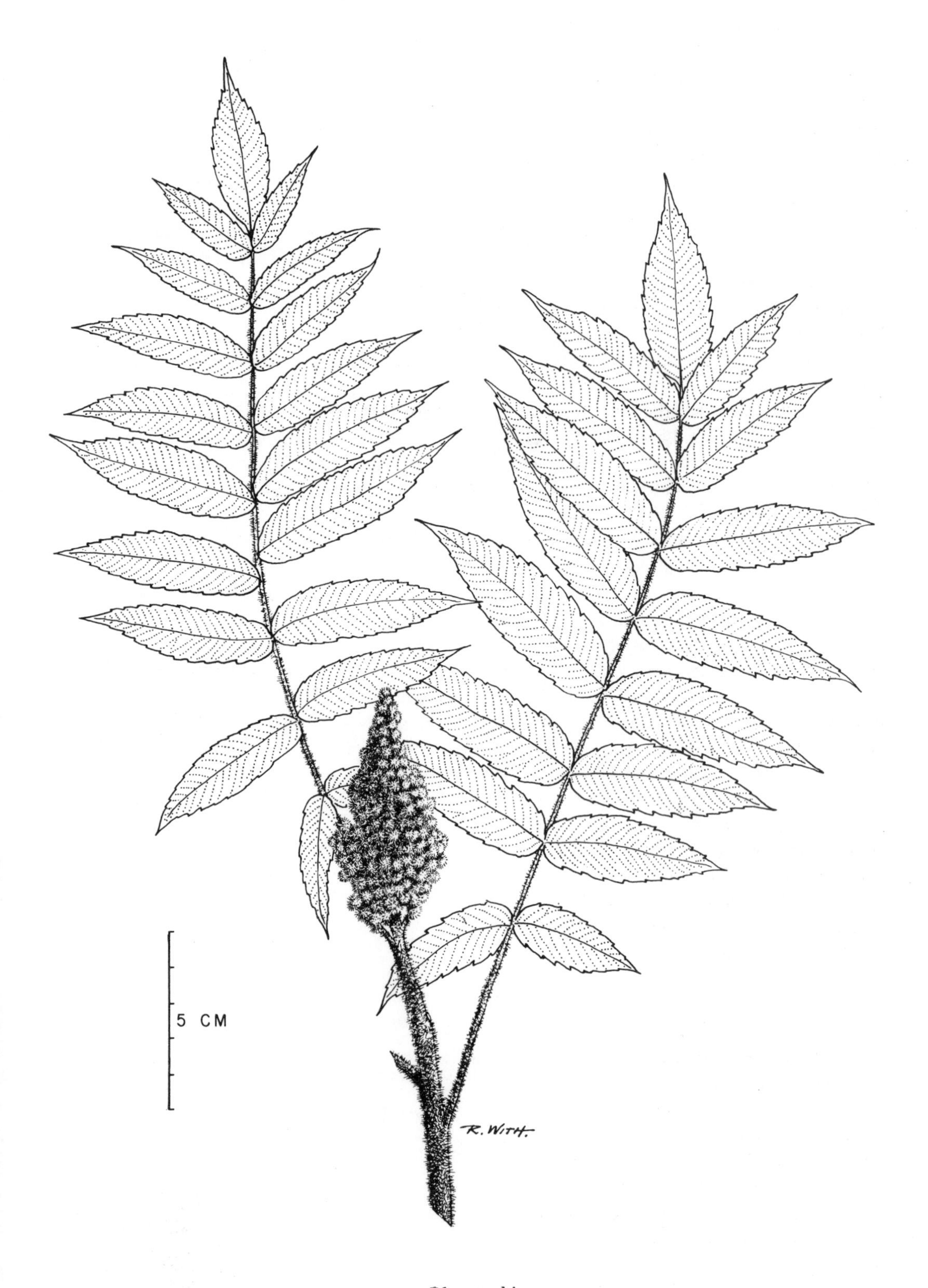

Rhus typhina

or by indirect means, such as, through the smoke from burning plants during eradication procedures, or through contact with clothing or an animal that has brushed against the plants. The active ingredient is a volatile oil known as toxicodendrol; the effect is usually the production of tiny blisters which cause itching, and if broken by scratching, spread the irritation to other parts of the body.

Poison oak is a name that should not be applied to plants in the Great Lakes region. It more properly belongs to two species that do not occur in Ontario, one in the southern states and the other on the Pacific coast.

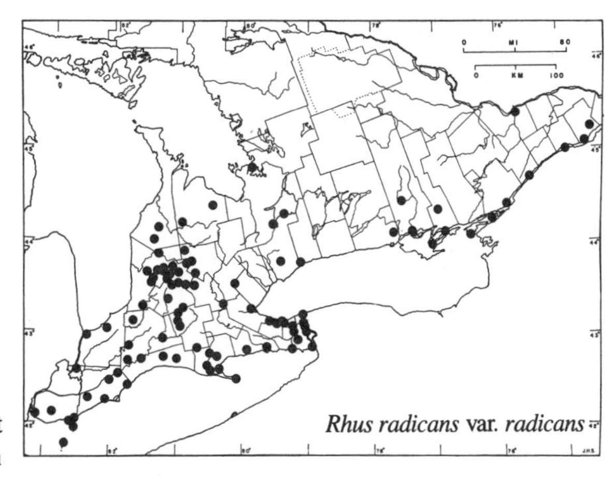

The downy woodpecker, crow, flicker, and starling are among the many familiar birds that disperse this species by feeding upon the drupes.

Field check Present in two varieties, a low shrub and a woody vine trailing or climbing by aerial roots; leaves trifoliolate on long petioles; leaflets usually drooping, the surface shiny or dull, margins entire, toothed or lobed; fruit a cluster of dry, wrinkled, whitish, berrylike drupes.

Rhus typhina L. — Staghorn sumach

A tall thicket-forming shrub or small tree up to 6 m or more in height, with few, widely diverging branches. Branchlets and older stems densely velvety-hairy; pith orange and wood greenish in colour.

Leaves alternate, deciduous, and pinnately compound with 11 - 31 leaflets, the lateral ones sessile, the terminal one stalked; rachis and petiole hairy, the latter 2.5 - 15 cm long; complete leaf 25 - 50 cm long; leaflets 7.5 - 13 cm long and 1 - 3 cm wide, rounded at the base, sharp-pointed at the usually prolonged and somewhat curved tip; the upper surface dark green and nearly glabrous, the lower surface paler to strongly whitened and hairy; margins sharply to coarsely toothed.

Flowers minute, greenish-yellow, and in terminal clusters 13 - 25 cm long; late June and July. Fruit a small, red, berrylike drupe with a coating of bristly hairs; in clusters which persist over winter; July and Aug. (*typhina* — like *Typha*, the cattail, from the resemblance of the surface of the branches to the velvety texture of the spike of the cattail)

In open fields, at the edges of woods, on river banks, slopes of ravines, and rocky ridges.

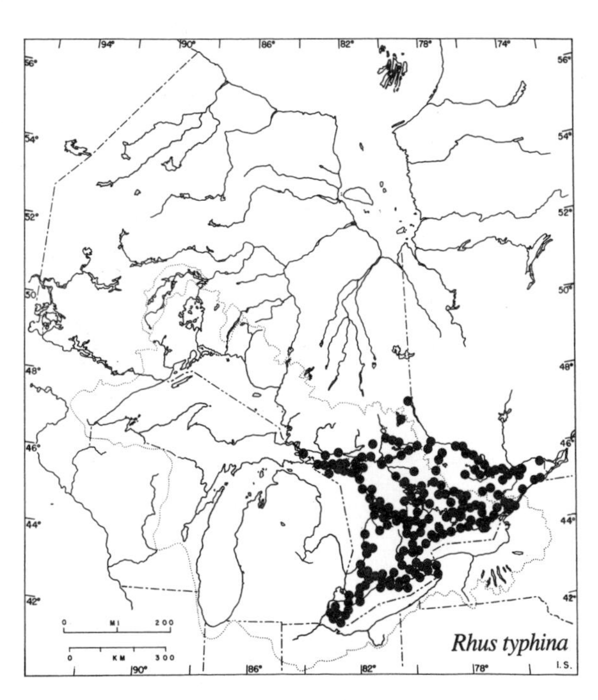

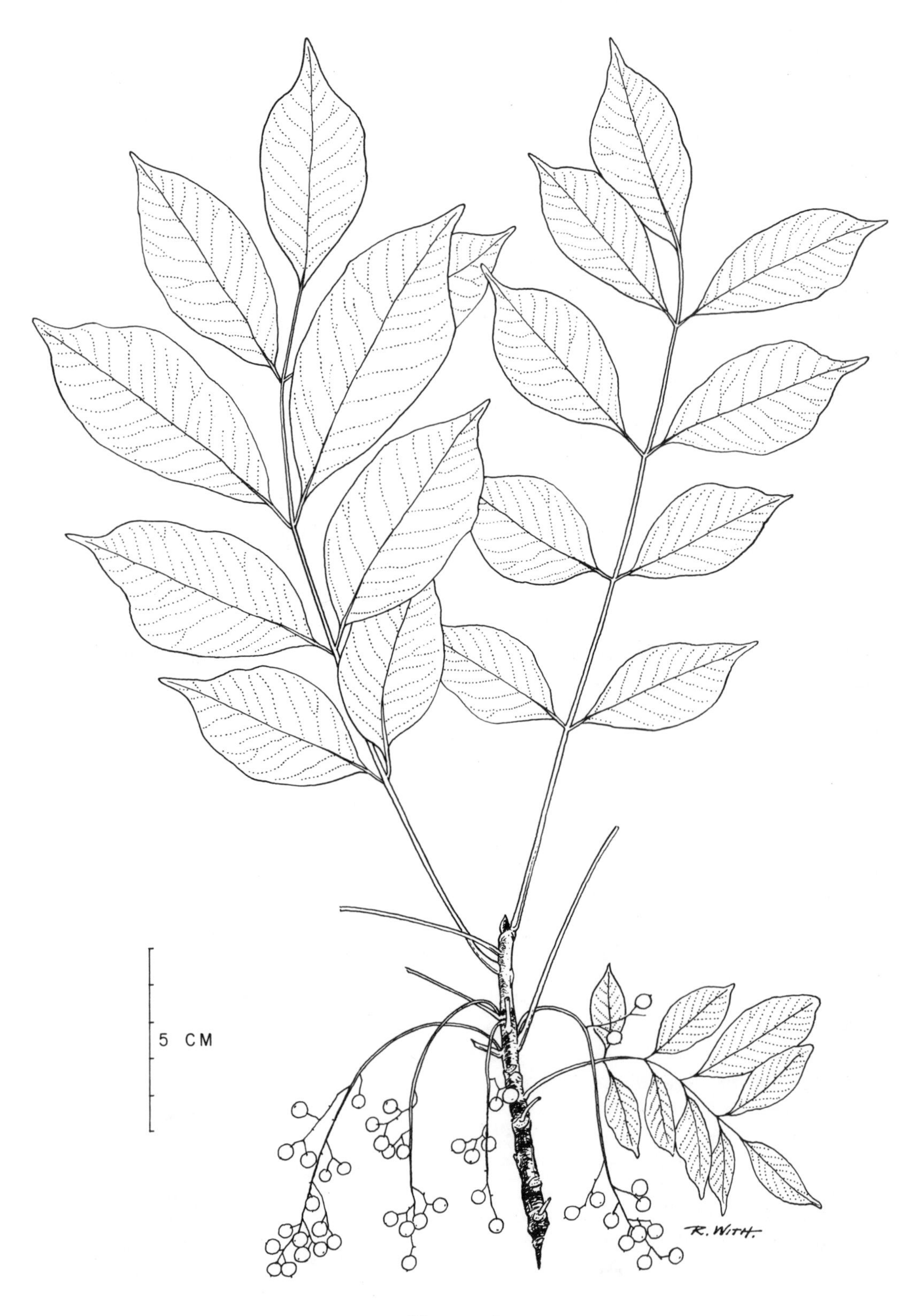

Rhus vernix

Common throughout southern Ontario; less frequent north of 46°N to the Lakes Timagami-Timiskaming region. (N.S. to Minn., south to Iowa and N.C.)

Note Staghorn sumach has attractive autumn colouring and is useful as an ornamental for planting in large gardens or on estates. It spreads underground and produces numerous suckers, a habit that makes it unwelcome in a small home garden or lawn setting.

The glandular hairs on the fruits are pleasantly acid-tasting and, if cleaned of insects, the fruits can be used to prepare a refreshing summer drink.

The wood is used in turnery.

Field check Tall shrub with few divergent branches resembling the horns of a stag; stems and petioles densely velvety-hairy; leaves alternate, large, pinnately compound with 11-31 leaflets; fruits red, berrylike, bristly-hairy, in dense steeple-shaped terminal clusters.

Rhus vernix L. — Swamp sumach / Poison sumach

A low and much-branched shrub in the open but in crowded stands reaching a height of 3-5 m with comparatively few branches near the ground. Branchlets gray-brown, glabrous, somewhat ridged, and warty, with conspicuous leaf scars and exuding a sticky milky juice when cut or broken.

Leaves alternate, deciduous, and pinnately compound, glabrous or nearly so, 15-30 cm long, including a petiole 2.5-7.5 cm long; rachis with a stalked terminal leaflet and 3-6 pairs of short-stalked opposite leaflets; leaflets ovate to oval, abruptly pointed at the tip, rounded or tapered at the base; margins entire or wavy; upper surface dark green and glabrous, lower surface paler, with a few scattered hairs along the margin; leaves often crowded near the ends of the branches, forming arching umbrellalike clusters.

Flowers tiny, greenish-yellow, in open elongate clusters from the axils of the leaves; June and early July. Fruit a dry, roundish, beadlike, papery-covered, whitish drupe; in elongate, arching or pendulous, open clusters persisting over winter; late July and Aug. (Linnaeus mistakenly attributed the Asian source of lacquer to this species when, in fact, it came from *R. verniciflua* L., a related eastern Asian species. *vernix* — varnish; *verniciflua* — flowing with varnish)

In swamps, in wet woods, and around boggy ponds.

Chiefly in the Deciduous Forest Region; also in the region around the southern end of Georgian Bay and at several stations in eastern Ontario. (Maine and Que. to s. Minn., south to e. Tex. and Fla.)

Note Poison sumach may cause a skin rash on contact similar to that caused by poison ivy. It may be distinguished from related species and from some similar shrubs or tree saplings as follows: smooth sumach (*R. glabra*) and staghorn sumach (*R. typhina*) have narrower, toothed leaflets; dwarf sumach (*R. copallina*) has a winged leaf rachis; all three related sumachs occur in drier habitats; elderberries (*Sambucus* spp.) and all ashes (*Fraxinus* spp.) have opposite compound leaves, not alternate leaves; mountain-ash (*Sorbus* spp.) has sharply toothed and

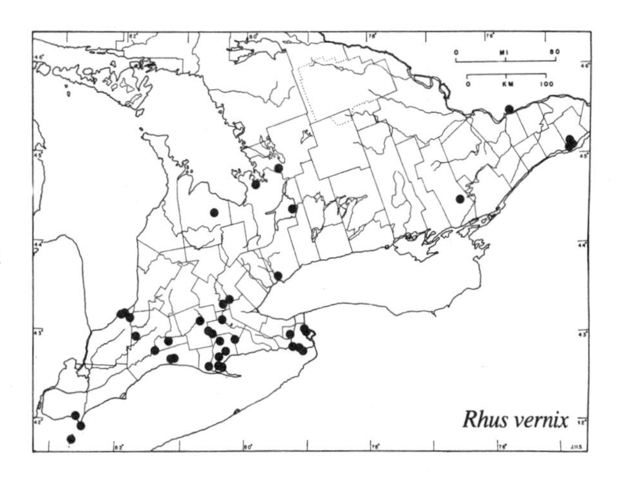

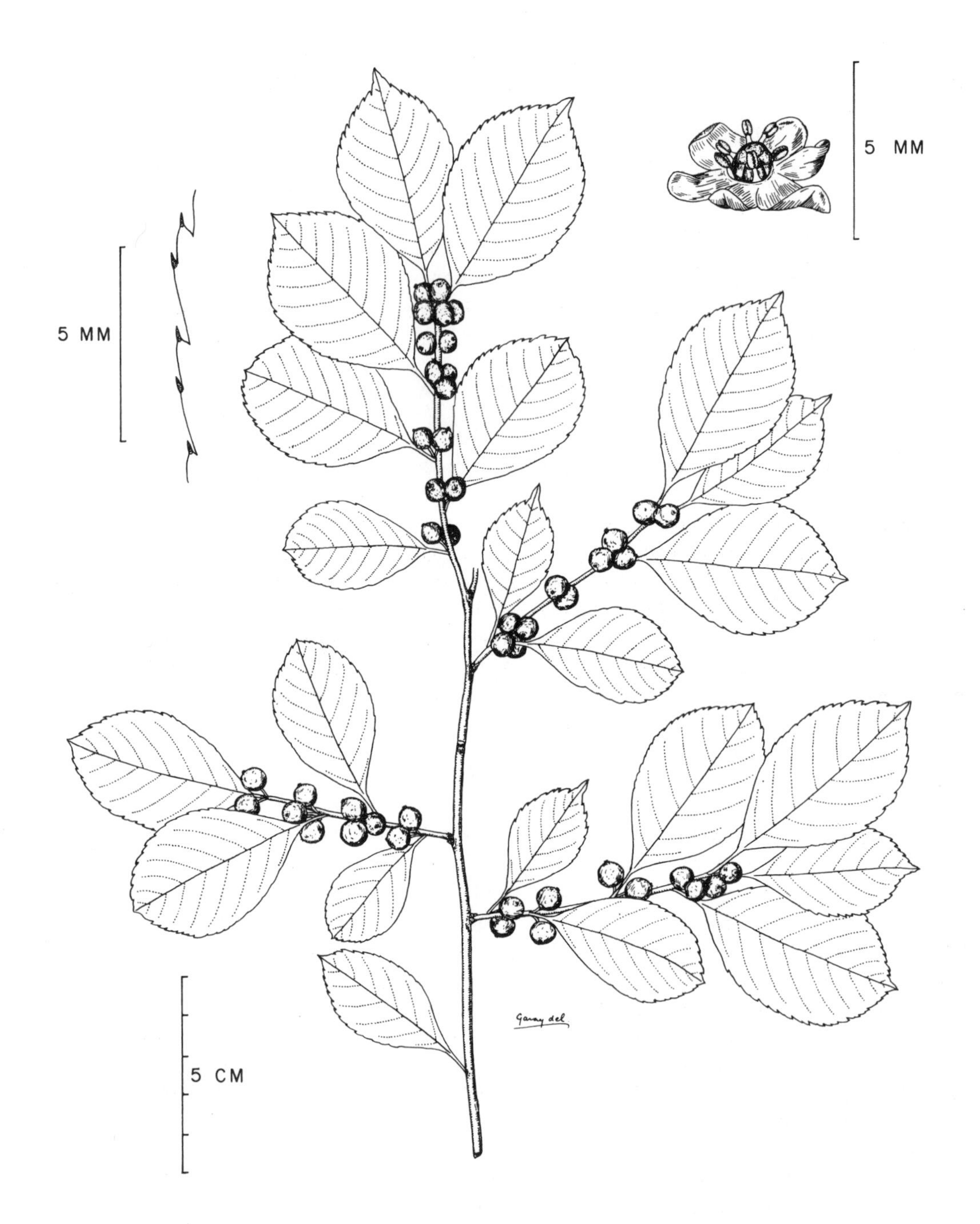

Ilex verticillata

sharply pointed leaflets; prickly-ash (*Zanthoxylum americanum*) has stout-based prickles in pairs along the stems and smaller prickles on the leaf stalks.

Field check Shrub of wet habitats, with alternate pinnately compound leaves; leaflets 7 – 13 with entire margins; fruits dry, whitish, and berrylike, in persistent, arching, or pendulous, open clusters.

AQUIFOLIACEAE – HOLLY FAMILY

Trees and shrubs, often on wet acid soils, mostly with alternate simple leaves, flowers small, axillary, 4 – 9-parted, and the fruit a berrylike drupe. Three genera, but most of the more than 400 species are in *Ilex*.

Ilex L. – Holly

Although many species of holly are evergreen, the one occurring as a native shrub in Ontario is deciduous. Pollination is reported to be by insects, especially bees, and seed dispersal by birds. The leaves of a South American species, *I. paraguariensis* St. Hil., are used in a tealike beverage called maté or Paraguay tea. The holly used at Christmas for decoration is *I. aquifolium* L. of Europe, a species with leathery persistent (evergreen) leaves which have sharp spines on their margins; it is not hardy in Ontario. (*Ilex*—the Latin name for the holly oak of the Mediterranean region, *Quercus ilex*, but used by Linnaeus for this genus; *paraguariensis* —Paraguayan; *aquifolium*—with pointed leaves, the Latin name for holly)

Ilex verticillata (L.) Gray — Winterberry, Black alder

An erect shrub 3 – 4 m in height. Branchlets stout, smooth, and finely ridged; bark brownish at first, turning gray to blackish, with warty lenticels, often mottled light and dark gray as small, thin, pale portions peel off and are shed.

Leaves alternate, simple, and deciduous, 3 – 9 cm long and 1 – 4 cm wide; blades elliptic to oval or obovate, acute to abruptly acuminate at the tip, tapered to the base, dull green above, smooth or downy and paler beneath, usually hairy along the veins; margins sharply serrate with incurved teeth; petioles about 1 cm long, grooved, and hairy, with a pair of small, narrow, and usually deciduous stipules.

Flowers small, greenish to yellowish-white, unisexual or sometimes perfect, on short stalks in the axils of the leaves; male flowers in crowded clusters, the female solitary or few in a cluster; opening before the leaves have fully expanded; late May. Fruit a bright orange to red, globular, berrylike drupe about 6 mm in diameter, containing 3 – 5 smooth bony nutlets, solitary or two or three together on short stalks in the leaf axils, remaining on the shrub well into the winter; Aug. to Oct. (*verticillata*—whorled)

In moist situations such as swampy woods and thickets, peat bogs, or lowland bordering swamps, bogs, and roadsides.

Widespread throughout southern Ontario and north to about 48°N; also along the Minnesota border of western Ontario. (Nfld. to Minn., south to Tenn. and Ga.)

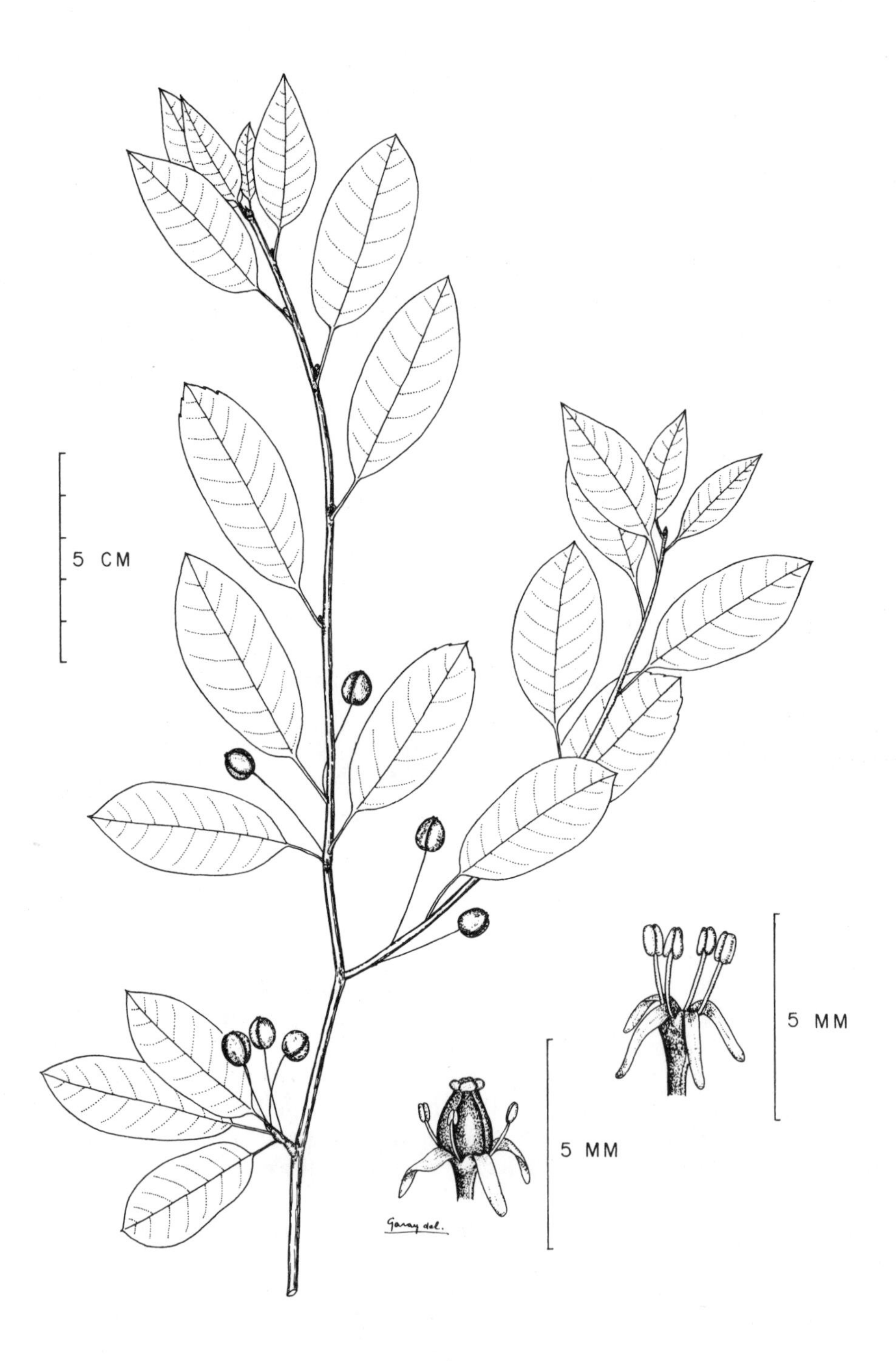

Nemopanthus mucronata

Note The common name black alder draws attention to the confusion that may arise from the use of common names. The many differences between this shrub and the green alder or the speckled alder show clearly that *Ilex* and the true alders (species of *Alnus*) are not at all related as the common name would imply.

Winterberry is one of several species of Ontario shrubs with a distribution predominantly in eastern Ontario and a disjunct area of more limited range west of Lake Superior along the Minnesota-Ontario boundary.

Field check Shrub of moist habitats; leaves alternate, elliptic to oblanceolate, with a pointed tip; sessile orange-red berrylike fruits, long-persistent.

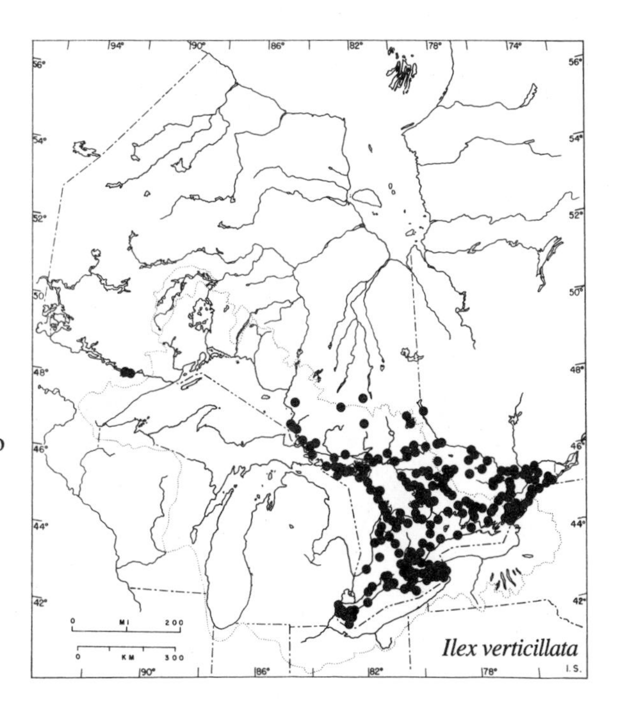

Nemopanthus Raf. — Mountain-holly

The single species in this monotypic genus is restricted to eastern North America. (The name refers to the threadlike stalk (foot) that bears the flower, and is a contraction of the etymologically more acceptable, if less easily pronounceable, *nematopodanthus*. *Nemopanthus* —from the Greek *nema*, thread, *pous*, foot, and *anthos*, flower)

Nemopanthus mucronata (L.) Trel. Mountain-holly

An erect much-branched shrub of moist places growing to a height of about 3 m. Branchlets of two types, the long shoots slender and purplish, with widely spaced leaves, the short shoots stouter, purplish or gray, with leaves more crowded and often appearing whorled; bark gray, rather smooth, the lenticels numerous and pale.

Leaves alternate, simple and deciduous; the blade up to 7 cm long and 2.5 cm wide, rather thin and smooth, bright green above, dull and paler beneath, elliptic, oblong-elliptic or ovate, the apex mucronate, the base rounded or narrowed; margins usually entire but there may be a few scattered sharp-pointed teeth; petioles slender, purplish, and about 1 cm long.

Flowers very small, mostly unisexual and on separate plants but sometimes with perfect flowers, solitary or in small clusters on threadlike pedicels 2.5 cm or more in length, arising in the axils of the leaves; opening before the leaves have fully expanded; late May. Fruit a purplish-red to crimson berrylike drupe about 6 mm in diameter on a slender stalk and containing 4 or 5 bony nutlets; Aug. to Sept. (*mucronata*—with a mucro, i.e., a short sharp point)

In moist situations, low places, swamps, and damp woods, especially the borders of sphagnum bogs and tamarack swamps.

Common throughout all but the southwesternmost part of southern Ontario, becoming

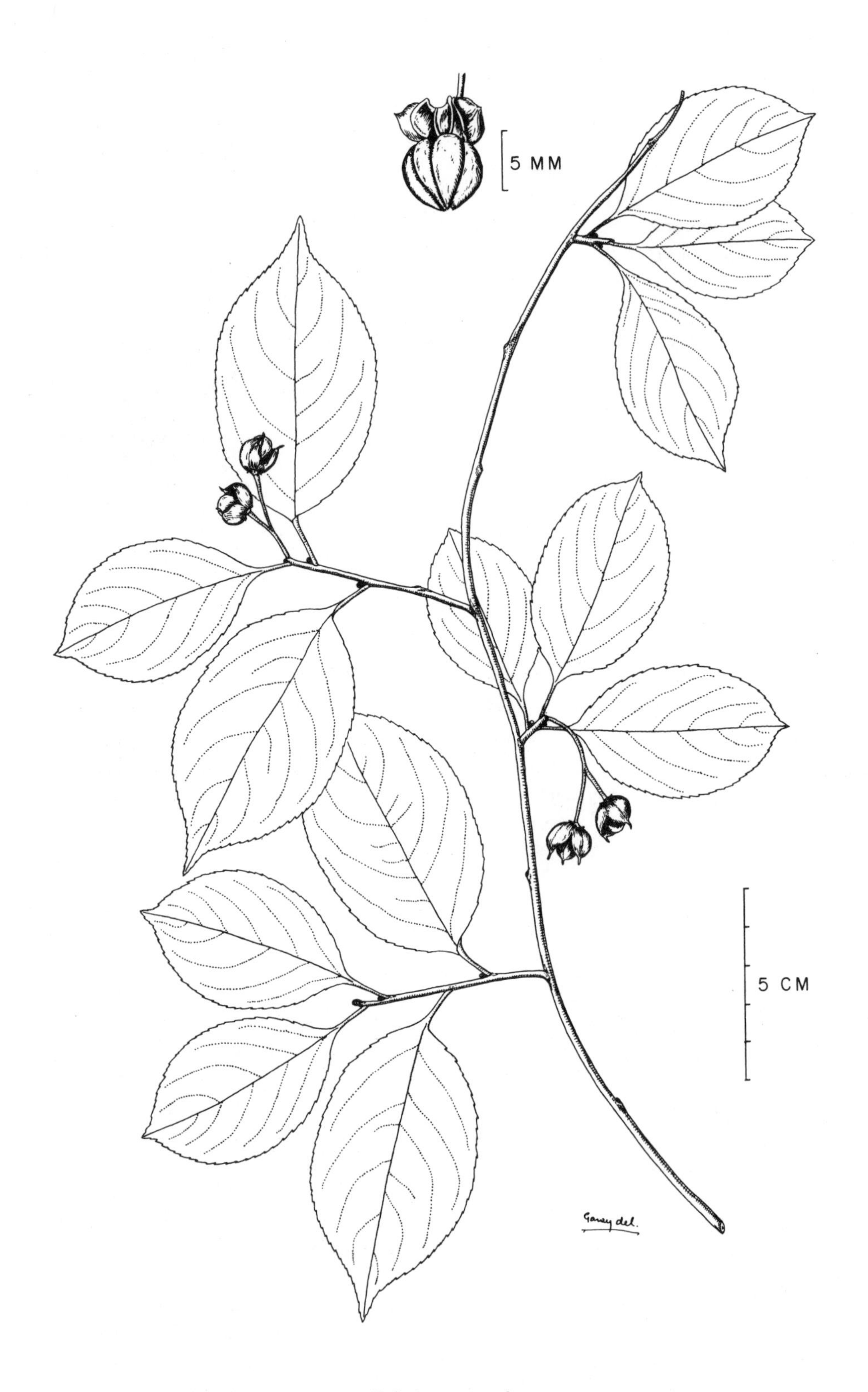

Celastrus scandens

infrequent north of 47° N and not reported beyond 48° 30′ N. (Nfld. to Minn., south to Ill. and Va.)

Field check Alternate, entire, mucronate, and mostly elliptic leaves with slender purplish petioles; fruit purplish-red, berrylike, on long threadlike stalks; moist habitats.

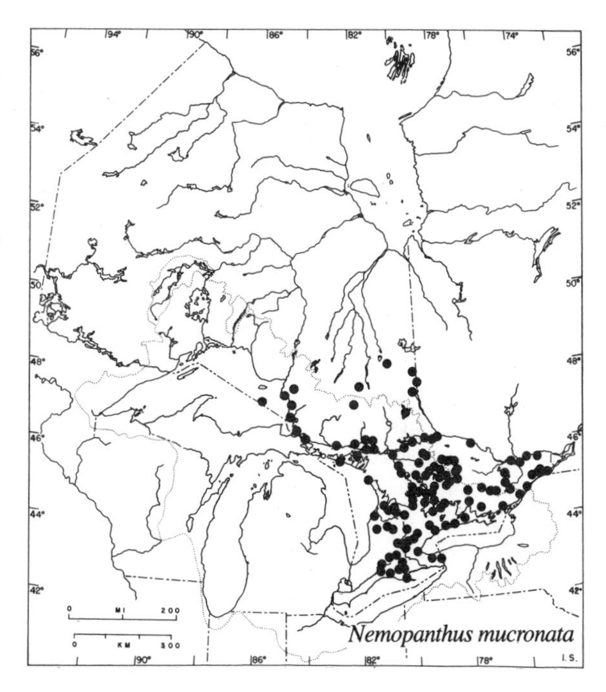

CELASTRACEAE — STAFFTREE FAMILY

Shrubs and trees or rarely woody vines with simple alternate or opposite leaves bearing stipules. Flowers small, regular; fruit a capsule, drupe, berry, or samara. Seeds usually with a brightly coloured aril. About 55 genera and 800–850 species in both tropical and temperate regions.

Celastrus L. — Bittersweet

Twining woody vines with alternate leaves; flowers in axillary or terminal clusters. Thirty species in the tropics and subtropics, one in temperate North America. (*Celastrus*—Greek *kelastros*, the ancient name for an evergreen tree but applied to this genus by Linnaeus)

Celastrus scandens L. Bittersweet

A woody vine or twining shrub forming tangled masses in open ground or climbing fences, shrubs, or trees to a height of 7 m or more. Branchlets slender, green, smooth, and flexuous; bark gray or reddish-brown.

Leaves alternate, simple, and deciduous; the blade thin, smooth, and light green, oval to oblong or ovate-lanceolate to obovate, up to 13 cm long and 5 cm wide, acute or acuminate at the apex and rounded or tapered at the base; margins crenulate-serrate; petioles 6–20 mm long.

Flowers small, greenish-yellow, in terminal elongate clusters, unisexual, the two sexes usually on separate plants but sometimes perfect flowers also present; June. Fruit a globose orange-yellow capsule about 12 mm across which splits open when ripe and exposes the orange-red arils covering the seeds; Sept. and Oct. (*scandens*—climbing)

In both dry and moist situations; sandy or rocky woods, swampy thickets, roadsides, talus slopes, and fencerows.

Common in eastern Ontario both south and east of the Canadian Shield; northward in the Ottawa Valley and through the Bruce Peninsula to Manitoulin Island; Lake Timagami; also near the eastern and western ends of Lake Superior and in the Lake-of-the-Woods region. (Que. to s.e. Sask., south to Okla., La., and Ga.)

Note The growth of saplings and young trees is sometimes restricted by the close twining of this tough and vigorous woody vine which has a strangling effect on the supporting living stems. The winter buds are firm and somewhat hooklike, which may keep the vine from slipping on its support.

An Asian species, *Celastrus orbiculatus* Thunb., differing from *C. scandens* chiefly in its more rounded leaves, lateral (not terminal) flower clusters, and smaller fruits, is sometimes found in semi-natural situations, as in woods near St. Williams, Regional Municipality of Norfolk-Haldimand. (*orbiculatus*— spherical, circular in outline)

Field check Tough twining woody vine with alternate, pointed, finely toothed, light green leaves and attractive clusters of orange and red fruits.

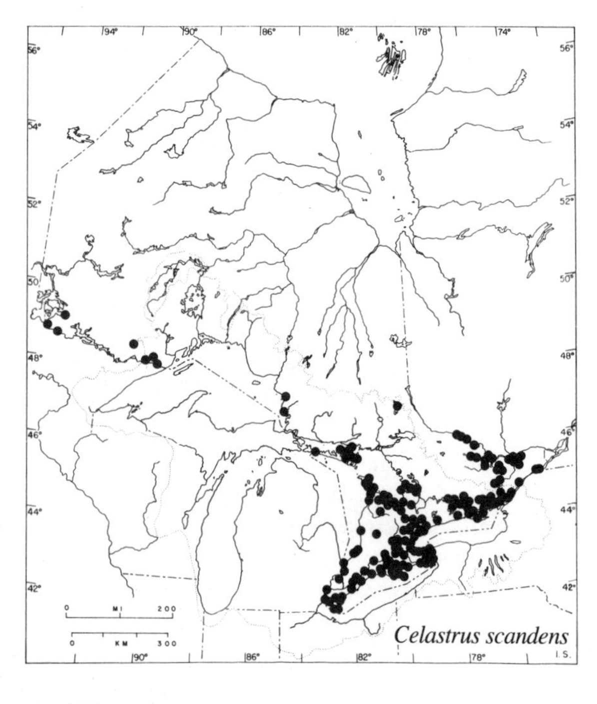

Celastrus scandens

Euonymus L. — Spindletree

Decumbent or erect shrubs and trees with opposite leaves; flowers in branched clusters or fascicles; fruit a 3–5-lobed capsule; seeds with a scarlet aril. A large genus with about 170 species centred in eastern Asia (China, Japan, and the Himalayan region). Some species are used in ornamental plantings. The common name refers to the use of the wood in turnery to make spindles. (*Euonymus*—from the Greek *eu*, well, and *onoma*, a name, hence "well-reputed" or "famous", supposed to be an ironic reference to the bad reputation possessed by some species for poisoning cattle)

Key to *Euonymus*

a. Erect shrub 3–6 m tall with spreading branches, rarely treelike; leaves mostly sharp-pointed or caudate at the tip; petiole more than 10 mm long; flowers 4-parted; capsule deeply lobed, not tuberculate-roughened **_E. atropurpureus_**

a. Prostrate or trailing shrub less than 3 dm high with freely rooting stems; leaves mostly blunt or rounded at the tip; petiole less than 5 mm long; flowers 5-parted; capsule shallowly lobed, spiny or tuberculate-roughened on the outer surface **_E. obovatus_**

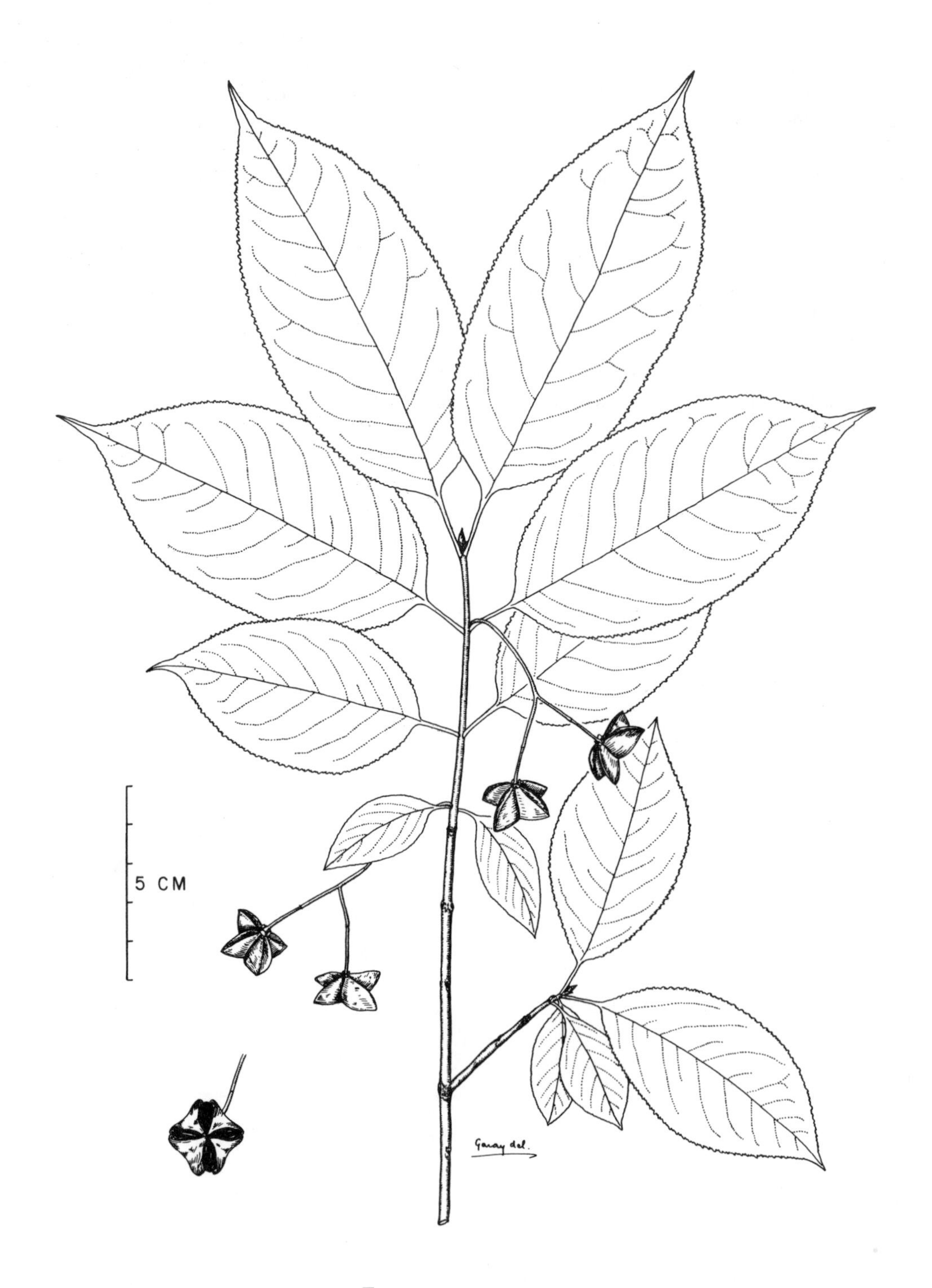

Euonymus atropurpureus

Euonymus atropurpureus Jacq.

Burning bush
Wahoo

An erect or spreading shrub, sometimes treelike, reaching a height of 4.5 – 6 m. Branchlets smooth, greenish, and usually somewhat four-sided; bark green to gray, often streaked with reddish-brown.

Leaves opposite, simple, and deciduous; blades up to 13 cm long and 5 cm wide; blade oblong-ovate to obovate, abruptly pointed at the apex and rounded at the base or narrowly to broadly elliptic and tapered at both ends; margins finely serrate; petioles about 1 cm long.

Flowers small, usually 4-parted, purplish-maroon, and rather inconspicuous, on slender stalks in few-flowered clusters from the lower leaf axils; late June and early July. Fruit a deeply 4-angled capsule which turns pink on ripening; the splitting of the capsule reveals the scarlet arils that surround the individual seeds; the colourful fruits remain on the plant after the leaves have fallen; Sept. (*atropurpureus*—dark purple)

In low places, particularly in thickets along streams, in rich alluvial soil; also in damp sandy or rocky woods.

Native in Ontario only in the Deciduous Forest Region. (s.w. Que. (naturalized) and s. Ont. to Mont., south to Okla, Ala., and e. Va.)

Field check Shrub or small tree with somewhat 4-angled stems; finely serrate, opposite, broadly elliptic, pointed leaves; persistent pink and red fruits somewhat like those of bittersweet.

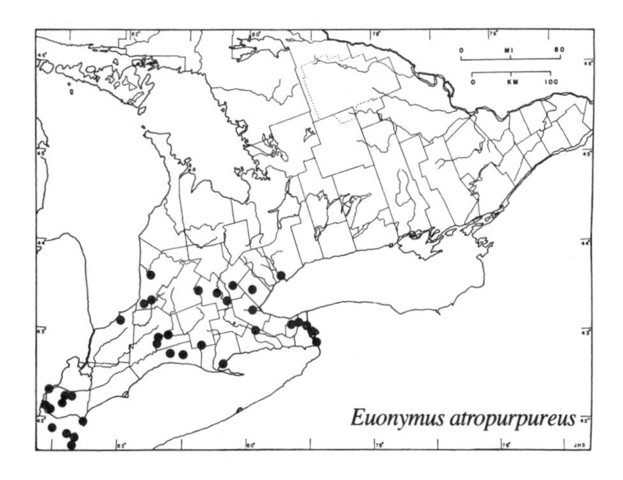

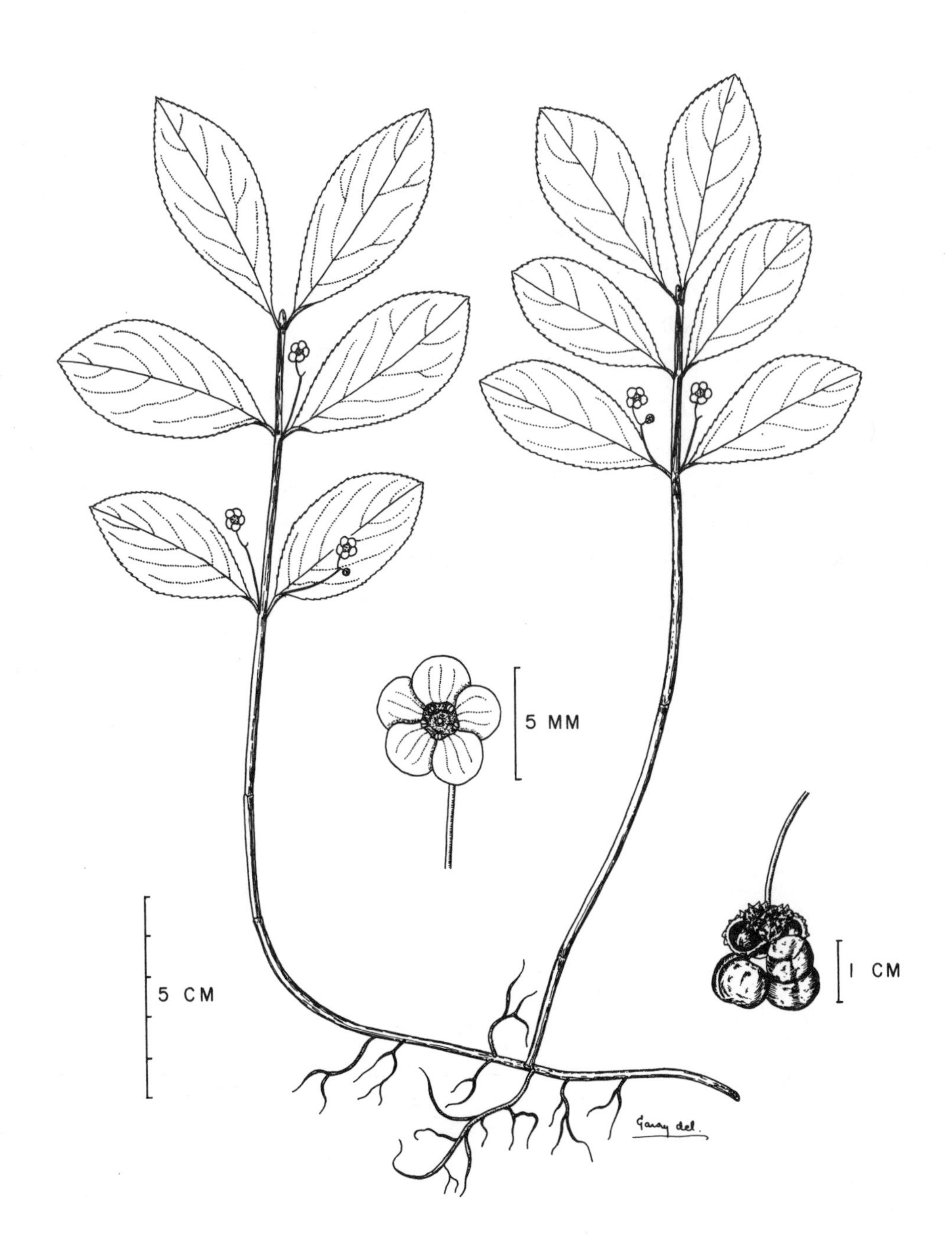

Euonymus obovatus

Euonymus obovatus Nutt. — Running strawberry bush

A low trailing shrub with prostrate freely rooting stems from which arise short erect or upturned leafy branches seldom more than 3 dm above the ground level. Branchlets smooth, greenish, and often four-sided or conspicuously angled; bark gray-green to brownish.

Leaves opposite, simple, and deciduous, 2–5 pairs on each branch, the terminal ones usually the largest; blades elliptic to oblong or obovate, thin and smooth, up to 7 cm long and 3.5 cm wide, rounded or slightly pointed at the tip and tapered at the base to a short grooved petiole less than 5 mm long; margins very finely serrate.

Flowers inconspicuous, greenish-purple, about 6 mm across, with 5 round overlapping petals and 5 short bright orange-yellow stamens protruding from a translucent green disc which encircles the pistil between the stigma and the ovary; flowers solitary or in groups of two or three, on long thin stalks from the axils of the leaves; May and June. Fruit a 3-lobed, spiny or warty, orange-pink to crimson capsule, splitting when ripe into three segments and exposing the orange to scarlet arils covering the seeds; seeds usually 3, eventually hanging suspended on slender threads from the pendent capsule; Sept. (*obovatus*—inversely egg-shaped, i.e., broadest above the middle)

In shaded habitats; ravine slopes, rocky woods, wooded talus slopes, or wooded river banks.

Chiefly restricted to the Deciduous Forest Region but recently collected on the west side of Lake Simcoe. (w. N.Y. to Mich. and Ill., south to Mo. and Tenn.)

Field check Low trailing habit; short ascending or erect branches with opposite broadly elliptic leaves, tapered at both ends; showy, bittersweetlike, spiny fruits with seeds dangling on threads.

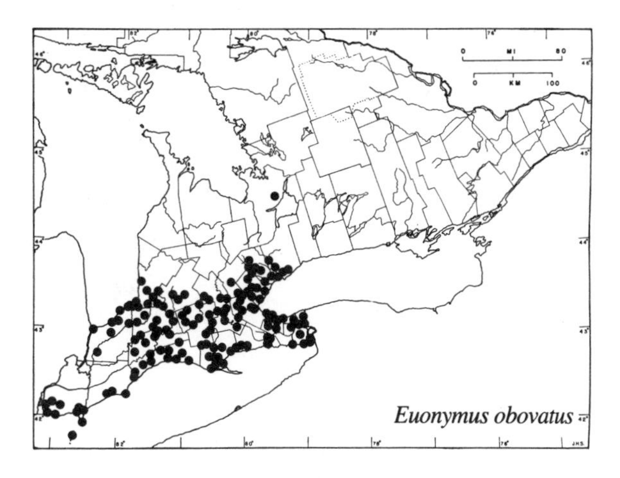

Euonymus obovatus

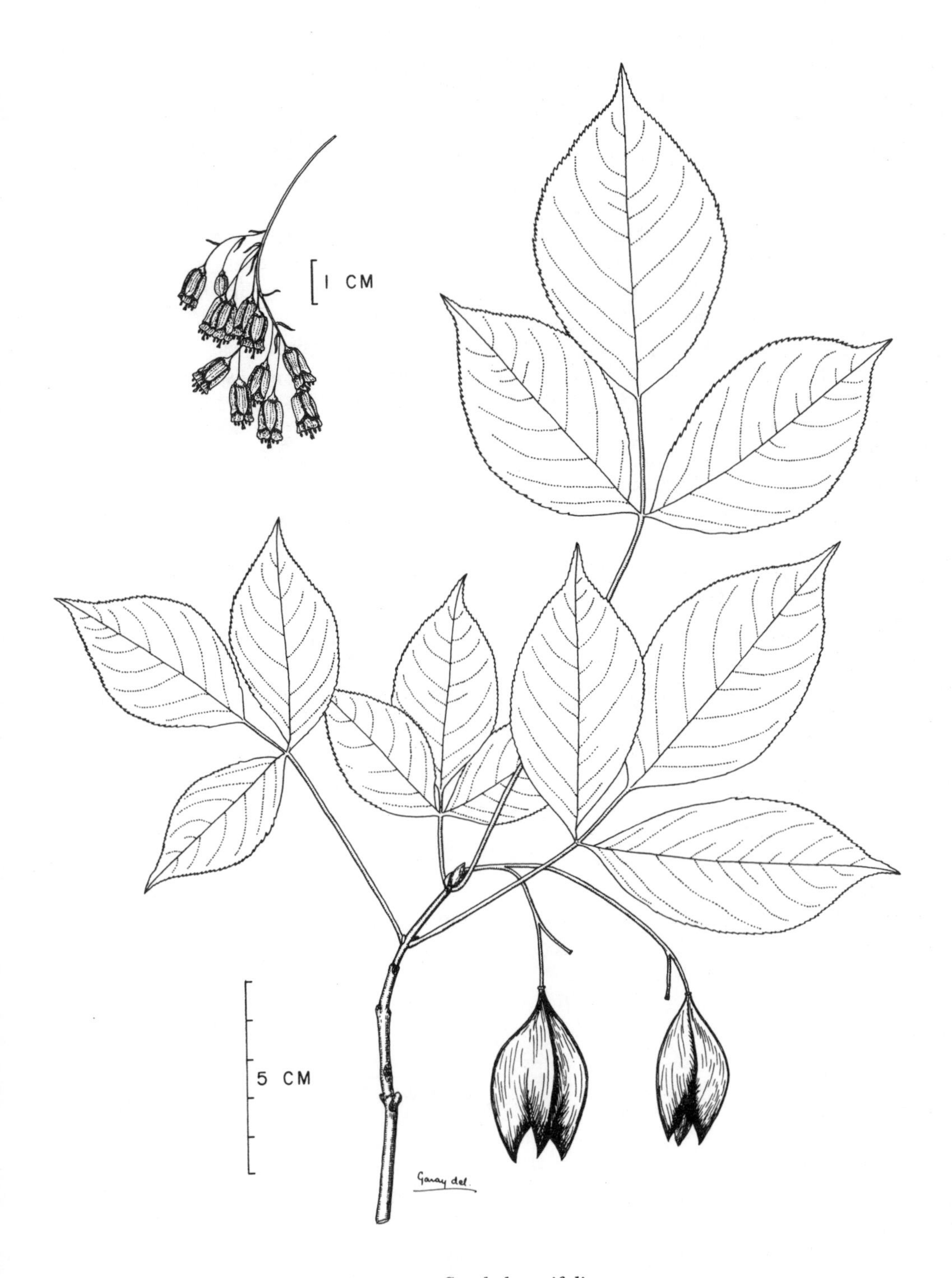

Staphylea trifolia

STAPHYLEACEAE — BLADDERNUT FAMILY

A small family of 5 genera and 50–60 species found in the north temperate regions and in tropical America and Asia. Shrubs or small trees with mostly opposite pinnate or trifoliolate leaves; stipules and stipels caducous. Flowers regular, usually perfect, 5-parted, in panicles or racemelike clusters; sepals and petals distinct; stamens borne around the edge of a large disc; ovary superior. Fruit a capsule with one to several seeds in each locule.

The name of the family is based on the genus *Staphylea*. A single species is native in Ontario.

Staphylea L. — Bladdernut

About 15 species of erect shrubs or small trees native to the temperate parts of North America and Eurasia. Leaves opposite, trifoliolate or pinnate with five serrate leaflets. Flowers greenish-white to cream-coloured in clusters at the ends of the branchlets. Fruit a large, inflated, three-angled capsule. (*Staphylea*—from the Greek *staphyle*, a bunch of grapes, in reference to the drooping clusters of flowers)

Staphylea trifolia L. — Bladdernut

An erect, rather stiffly branched shrub to about 5 m high. Branchlets stout, green, and mottled with lines or stripes, older branches turning gray or brown and slightly ridged or warty.

Leaves opposite, trifoliolate, and deciduous, up to 25 cm in length including the petiole; terminal leaflet on a long stalk, lateral leaflets up to 10 cm long and 5 cm wide, sessile or short-stalked; leaflets oval to ovate-lanceolate with an obtuse or acute and abruptly sharp-pointed apex and a rounded or tapered base, dark green and smooth above, much paler and somewhat pubescent beneath at least along the veins; margins closely and sharply serrate; petioles 10 cm or more in length with prominent, linear-filiform, hairy stipules up to 2.5 cm long, and at the base of the lateral leaflets similar but smaller stipels, both stipules and stipels soon shed.

Flowers white or cream-coloured, perfect and cylindrical, about 8–9 mm long, in drooping racemelike terminal clusters, opening shortly before the leaves have fully expanded; May. Fruit a much-inflated, prominently three-angled, veiny, brown capsule, globular to ellipsoidal, 5–8 cm long, the tip of each of the 3 lobes with a conspicuous point prolonged into a short threadlike tail (the persistent style); each chamber of the capsule contains 1–4 pale brown seeds which become loose when ripe and rattle around inside; capsules persisting on the plant over winter. (*trifolia*—three-leaved)

In rocky woods, on river banks, alluvial flats, hillsides, talus slopes, and occasionally on wooded sand dunes or ridges.

Common in the Deciduous Forest Region but also along the eastern half of the north shore of Lake Ontario to the St. Lawrence and Ottawa valleys and sporadically in the counties around the southern end of Georgian Bay. (s.w. Que. to Minn., south to Okla. and Ga.)

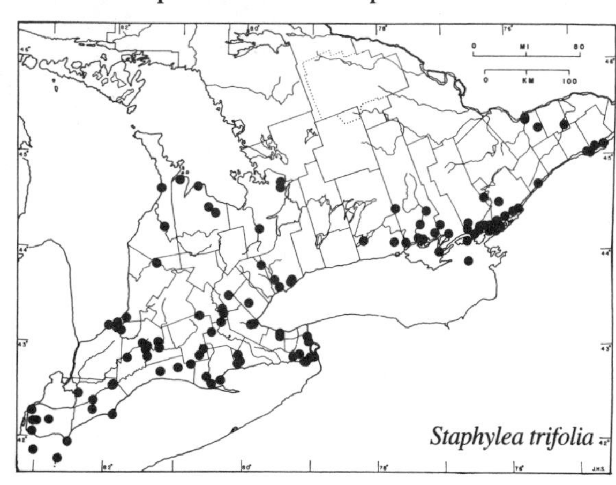

Staphylea trifolia

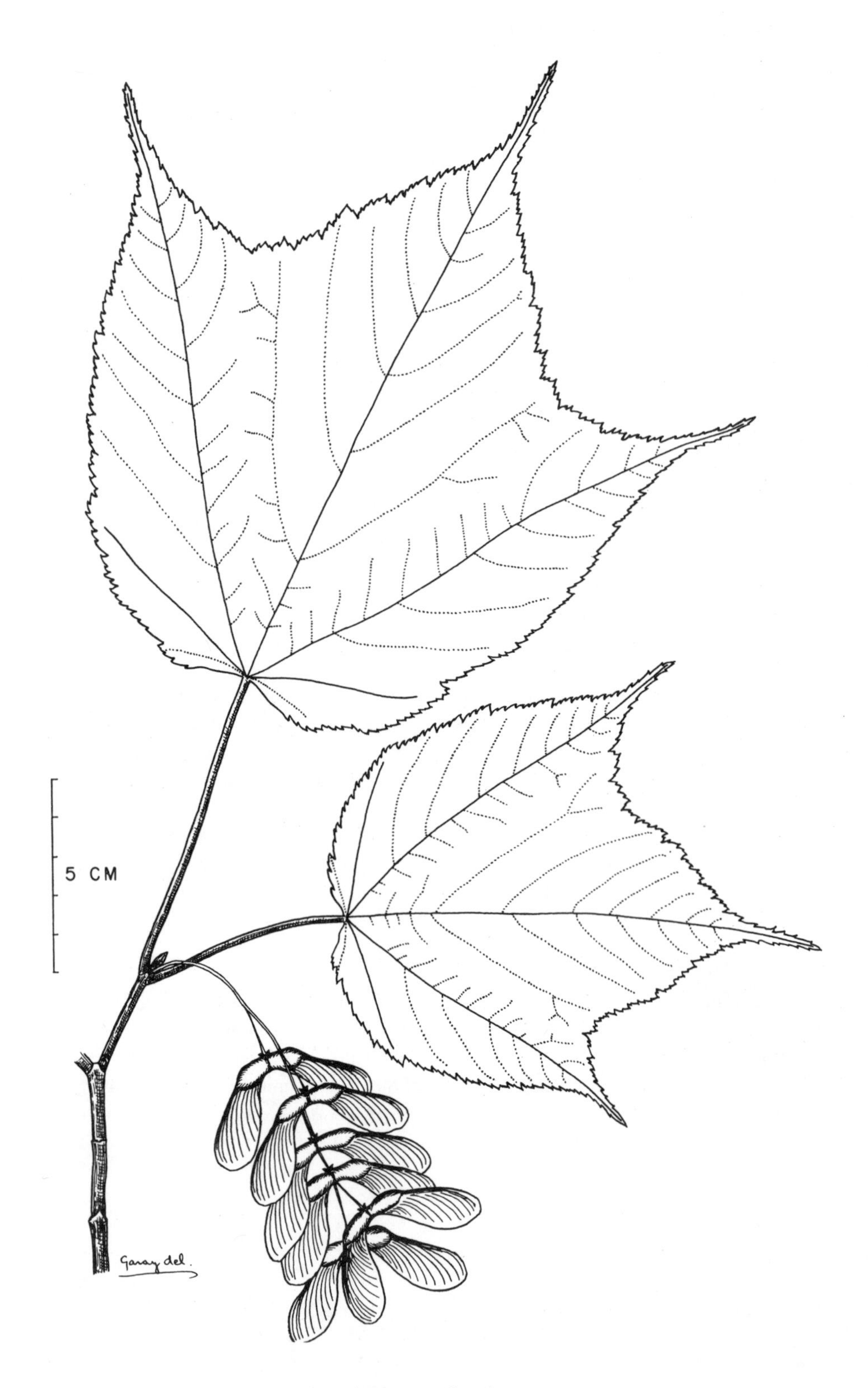

Acer pensylvanicum

Note A variant with pear-shaped fruits, *S. trifolia* forma *pyriformis* W.G. Dore, has been collected near Billings Bridge on the Rideau River at Ottawa. (*pyriformis*—pear-shaped)

The stigmas of bladdernut flowers are protogynous, thus providing an opportunity for cross-pollination. Pollination is accomplished by long-tongued insects.

Field check Opposite leaves with 3 leaflets; margins of leaflets finely serrate; inflated, three-angled, persistent, and pendent brown capsules.

ACERACEAE — MAPLE FAMILY

A small family of 2 or 3 genera, chiefly trees, most abundant in the north temperate zone but also found in mountains of the tropics; our representatives are in the genus *Acer*.

Acer L. — Maple

Trees and shrubs characterized by opposite palmately-lobed simple leaves (rarely palmately or pinnately compound), small mostly 5-parted flowers, and dry fruits, each consisting of a pair of winged single-seeded samaras joined at the base but finally separating and known as maple "keys". Among the 150–200 species are included ornamentals for street and garden planting, some forest trees with valuable timber, and one that is the source of maple syrup and maple sugar, *A. saccharum* Marsh., a species characteristic of the Deciduous Forest Region of eastern North America. The highly variable flowers are either insect-pollinated or pollinated by wind, and the samaras are distributed by wind. (*Acer*—the Latin name for the European maple, possibly derived from *acer*, sharp, in reference to the sharp-pointed lobes of the leaves)

Key to *Acer*

a. Bark of branches and young trunk with prominent pale stripes; leaves 3-lobed, the lobes long-tapering; flowers in loosely arching glabrous-stalked clusters; petals inconspicuous, samaras of each pair widely divergent **_A. pensylvanicum_**

a. Bark of branches without prominent pale stripes; leaves 3-lobed, often with 2 extra lobes near the base, the lobes abruptly pointed; flowers in erect, slender, pubescent-stalked clusters; petals conspicuous; samaras only slightly spreading, often reddish **_A. spicatum_**

Acer pensylvanicum L. — Striped maple, Moosewood

A large coarse shrub or small tree growing to a height of about 10 m. Branchlets greenish to reddish-brown and glabrous; older stems with conspicuous pale stripes on the bark, the characteristic to which one of the common names refers.

Leaves opposite, simple, and deciduous; blades roundish, 3-lobed, 10–18 cm long and nearly as wide, sometimes with smaller lobes near the base, the main lobes sharp-pointed and with long tapering tips, the base rounded or heart-shaped; upper surface bright light green and glabrous, lower surface slightly paler; margins with numerous sharp teeth; petioles 2.5–8 cm

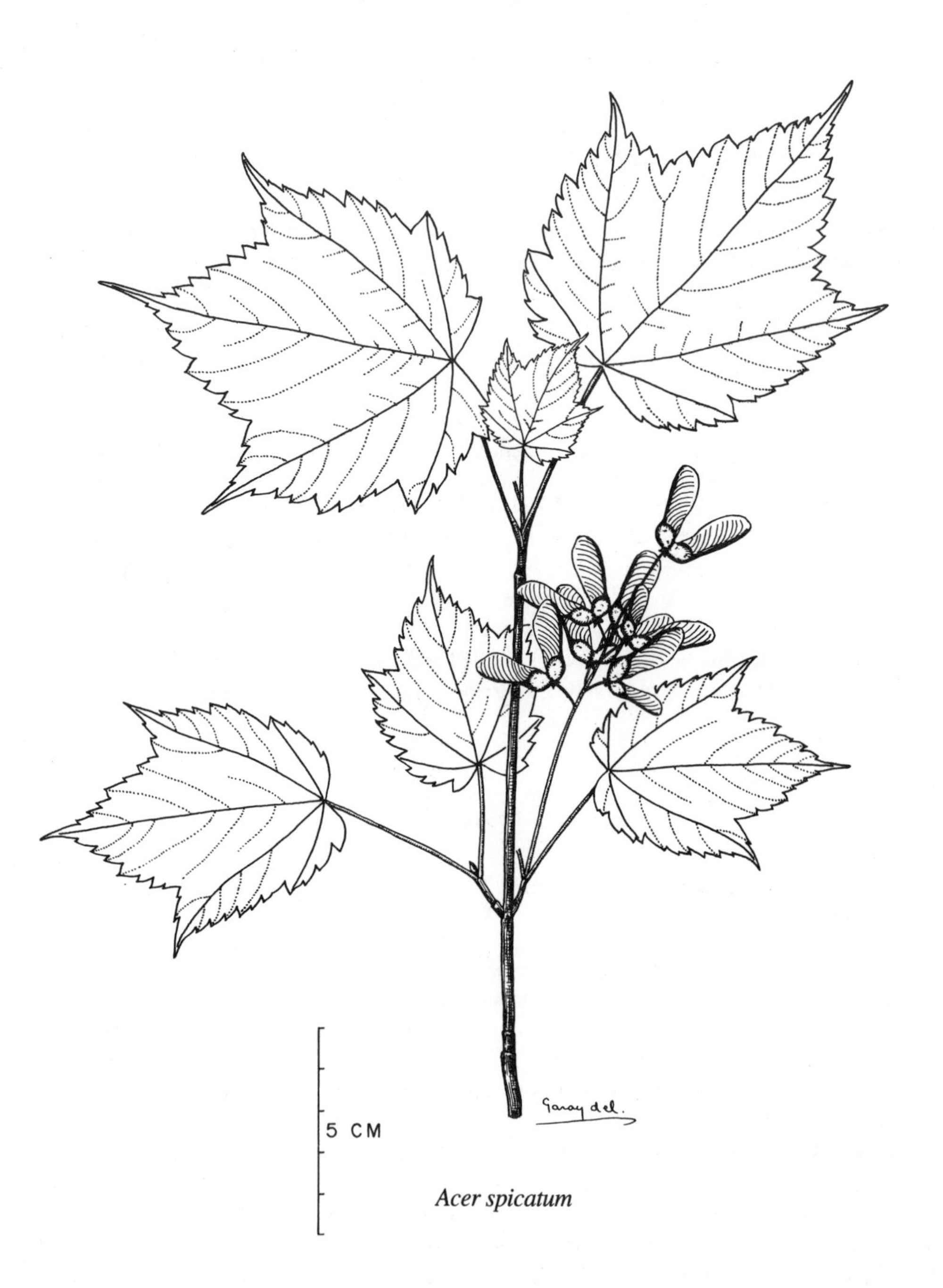

Acer spicatum

long. Occasionally some of the leaves have the lateral lobes suppressed and the blades are about twice as long as wide.

Flowers greenish-yellow and about 3 - 6 mm in diameter, on slender stalks in loosely arching clusters 7 - 14 cm long, each cluster usually containing either staminate or pistillate flowers and sometimes also perfect flowers; late May and early June. Fruit a stalked pair of widely divergent samaras or winged nutlets (keys), each 2.5 - 3 cm long, in drooping clusters; late July and Aug. (The original spelling of the specific name, although not current usage for the name of the state, should be retained, according to the International Rules of Botanical Nomenclature. *pensylvanicum* — Pennsylvanian)

A shade-tolerant species of cool moist woods.

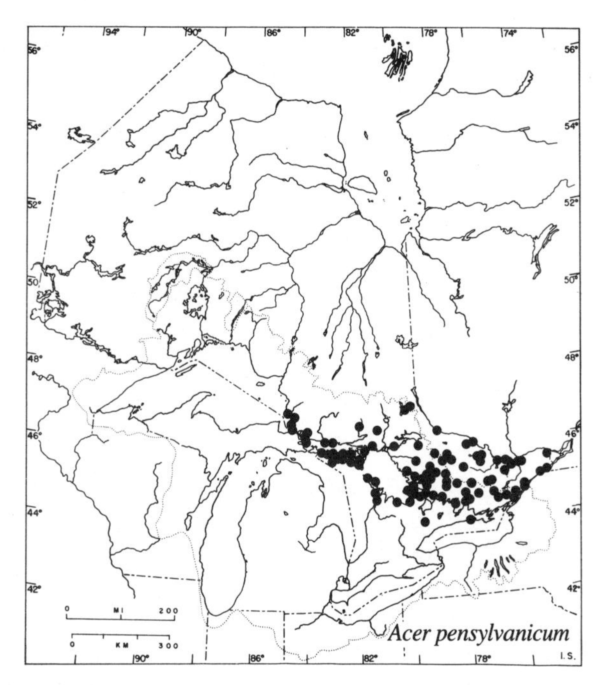

From the upper part of the St. Lawrence River to the Bruce Peninsula and northwest to the eastern end of Lake Superior; also from the Ottawa Valley northwestward to Lake Timagami; chiefly absent south of 44° N in Ontario. (N.S. to Ont. and n. Mich., south to Tenn. and n. Ga.)

Note The closest relatives of striped maple are not in our own indigenous Canadian species but among species native to eastern Asia, notably *A. nipponicum* Hara. (*nipponicum* — Japanese)

The bladdernut (*Staphylea trifolia*) also has striped bark and opposite leaves but is easily distinguished by its trifoliolate leaves and 3-lobed inflated capsules. The highbush cranberry (*Viburnum trilobum*), like the striped maple, has opposite 3-lobed leaves but its bark is not striped and its fruits are orange-red berrylike drupes.

Field check A shade-tolerant shrub or small tree with prominently striped bark; leaves opposite, roundish, simple, with 3 long-tapering lobes; fruit a drooping cluster of pairs of widely divergent samaras.

Acer spicatum Lam. — Mountain maple

A tall shrub from 3 - 5 m in height. Branchlets purplish-gray, minutely hairy, and without stripes; bark of older stems greenish-gray to blackish.

Leaves opposite, simple, and deciduous; blades 5 - 12 cm long and 5 - 10 cm wide with 3 prominent lobes and usually 2 additional smaller lobes near the base, the main lobes with abruptly pointed tips, the base heart-shaped; upper surface dark green and glabrous, the lower surface paler or whitened with fine hairs; margins sharply toothed; petioles 6 - 13 cm long, often as long as, or longer than, the blade and frequently with a reddish tinge.

Flowers small, greenish-yellow, in long-stalked erect terminal clusters 6 - 10 cm long, both unisexual and perfect flowers present on the same plant; late May to early July. Fruit a pair of stalked slightly diverging samaras or winged nutlets (keys), each about 2 cm long and usually reddish tinged, borne in clusters; July and Aug. (*spicatum* — spiked, referring to the upright clusters of flowers and fruits)

In damp soil of mixed woods, along rocky ridges and cliff bases, in swamps and thickets.

Common throughout southern Ontario and northward to Lake Abitibi and to the southern end of James Bay; westward around Lake Superior; less common northwestward to the upper part of the Severn River drainage basin at about 53°N and reaching but not entering the Boreal Forest and Barren Region except in the Moose River drainage basin. (Nfld. to Sask., south to Iowa and n. Ga.)

Note All the other maples in the section to which *A. spicatum* belongs are native to Europe and Asia, and a closely related species, considered by some botanists as merely a variety, occurs in Japan (see Li, 1952).

Mountain maple may be distinguished from the following shrubs which also have opposite 3-lobed leaves: striped maple (*Acer pensylvanicum*) has pale stripes on the stems and older branches, and the two keys of each pair are widely divergent; maple-leaved viburnum (*Viburnum acerifolium*) has downy branchlets, shorter petioles, often with slender stipules, and scattered resinous dots on the lower surface of the leaves; highbush cranberry (*Viburnum trilobum*) has shorter petioles with small club-shaped glands near the blade and leaf margins entire to sparingly toothed.

Field check Tall shrub or small tree; opposite long-stalked leaves with 3 prominent lobes and frequently 2 small additional lobes near the base; flowers and keys in erect slender clusters; keys slightly divergent and reddish.

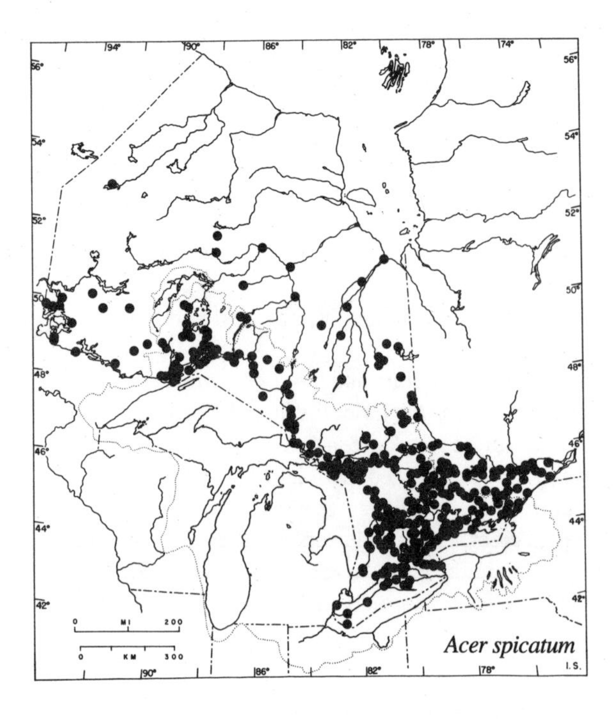

RHAMNACEAE – BUCKTHORN FAMILY

About 60 genera and 900 species widely distributed in tropical and temperate regions. Mostly trees and shrubs, some woody vines, often with thorns or stipular spines. Leaves simple, alternate, deciduous or evergreen; stipules small or absent. Flowers small and regular, 4-parted or 5-parted, usually with a well-developed floral disc. Fruit a capsule or a drupe with one seed in each locule. The family name is based on the genus *Rhamnus*.

Ceanothus L. – Redroot

This genus is native only in North America and most of the 50–55 species are found in the western part of the continent. Low shrubs or subshrubs, some of which are known to have mycorrhizal root relationships. Leaves 3-nerved, at least at the base. Flowers numerous, in axillary and terminal racemes; petals white (in our species), clawed. Fruit a leathery 3-lobed capsule. A number of species are cultivated as ornamentals for their showy flower clusters. (*Ceanothus*—Greek *keanothos*, said to be the name of some prickly plant mentioned by Theophrastus but inappropriate for an American genus)

Key to *Ceanothus*

a. Leaves ovate to ovate-oblong; flower clusters borne at the ends of branches that are mostly axillary and naked, or with a few bracts or small leaves near the ends — ***C. americanus***

a. Leaves narrowly elliptic or elliptic-lanceolate; flower clusters mostly terminating regular leafy branches — ***C. herbaceus***

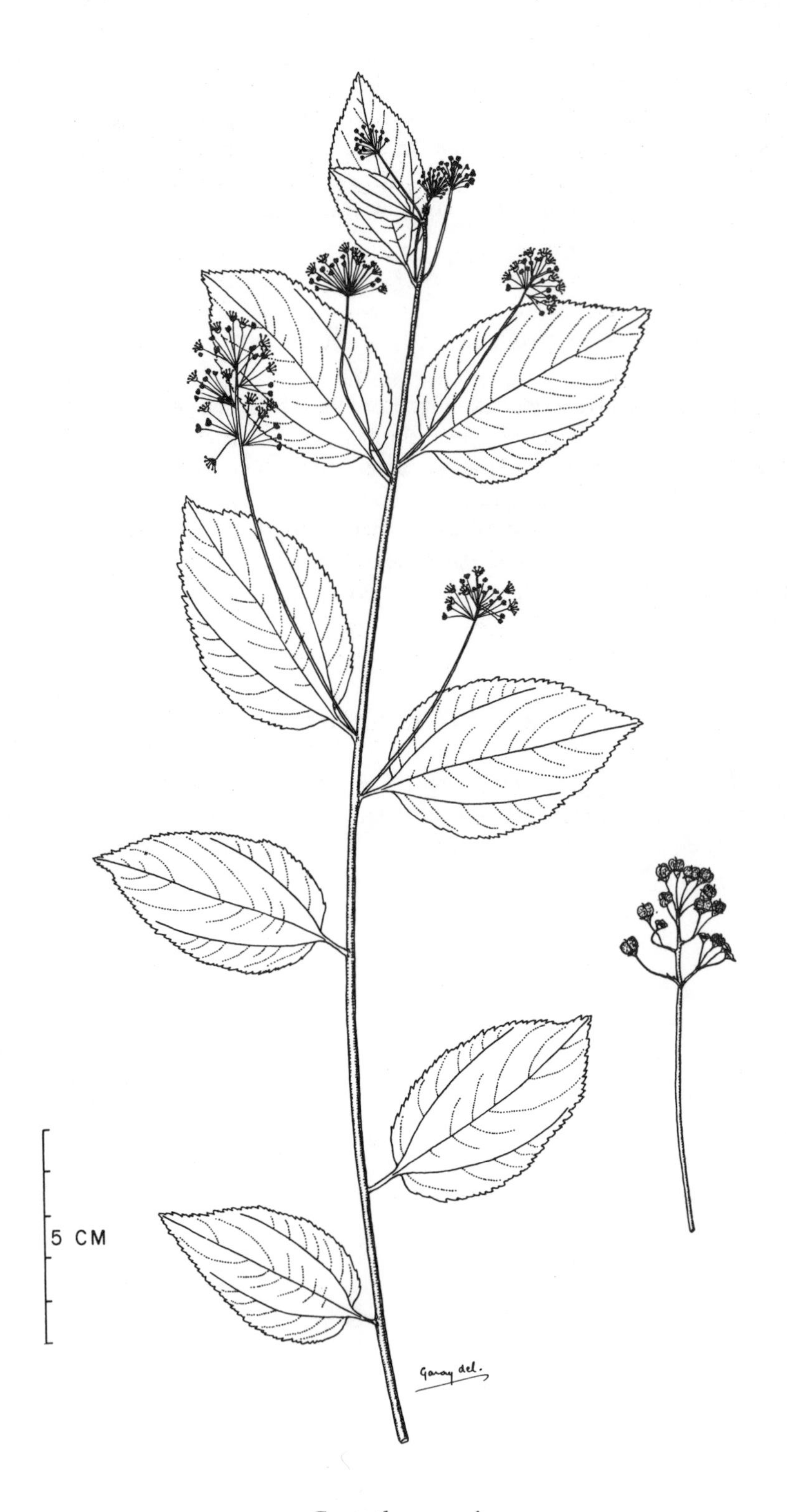

Ceanothus americanus

Ceanothus americanus L. — New Jersey tea

A low, branching shrub less than 1 m in height. Branchlets reddish-brown to gray, smooth or finely pubescent.

Leaves alternate, simple, and deciduous; blades ovate or oblong to ovate-lanceolate, up to 9 cm long and 5 cm wide, prominently 3-ribbed, smooth and dark green above, gray-green and minutely hairy beneath, at least along the veins, apex acute or acuminate and base subcordate, rounded, or rarely tapered; margins finely serrate with gland-tipped teeth; petioles grooved and about 1 cm long; stipules small and densely long-hairy but soon deciduous.

Flowers about 3 mm in diameter, white, perfect, and on slender stalks up to 1 cm long, in rather showy thimble-shaped clusters on long stout stems up to 15 – 18 cm in length arising in the leaf axils; flowering stems progressively longer from the upper part of the branch downward, and naked for most of their length or with 1 or 2 small narrow leaves just below the crowded terminal flower clusters; late June to early Aug. Fruit a small, 3-lobed, roundish, brown capsule, 3 – 6 mm in diameter on a stalked saucerlike base; in clusters at the ends of the branches; late Aug. and Sept., the fruit bases persisting through the winter. (*americanus* — American)

Usually in dry situations; sandy or rocky soil in clearings at the edges of woods, on river banks and lakeshores, in open woods, and on shaded hillsides.

Common in the Deciduous Forest Region and northeastward to the eastern half of the Lake Ontario region and to the Ottawa and St. Lawrence valleys; also on the southeastern shore of Georgian Bay. (Maine and s.w. Que. to s. Ont., south to Tex. and Fla.)

Field check Low, branching shrub with 3-ribbed glandular-serrate alternate leaves; long-stalked, mostly leafless flower clusters; small brown capsules with persistent saucerlike bases.

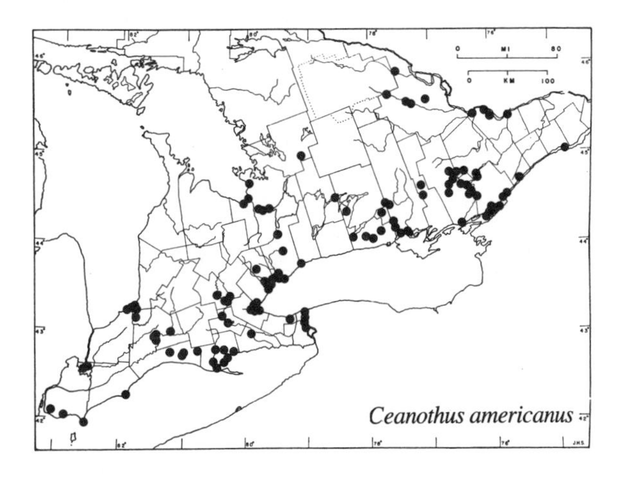

Ceanothus herbaceus

Ceanothus herbaceus Raf. (*C. ovatus* Desf.) — Narrow-leaved New Jersey tea

A low, erect, much-branched shrub usually less than 1 m in height. Branchlets smooth, purplish-brown to gray in colour.

Leaves alternate, simple, and deciduous; blades narrowly elliptic to narrowly oval or ovate-lanceolate, up to 5 cm long and 0.5 – 1.5 cm wide, prominently 3-ribbed, smooth and green above, the lower surface somewhat paler and smooth to densely hairy, apex rounded or obtuse, the base tapered; margins finely serrate with gland-tipped teeth; petioles less than 1 cm long with small hairy deciduous stipules.

Flowers white, perfect, and about 3 mm in diameter, on slender stalks not more than 1.5 cm long, in rather showy round-topped clusters at the ends of normal leafy shoots; mid-June. Fruit a small, 3-lobed, roundish, dark brown capsule about 3 mm across on a stalked saucerlike base, in clusters; late July and Aug. (*herbaceus* — herbaceous; *ovatus* — ovate, egg-shaped)

In dry habitats: sandy oak and pine woods, rocky limestone barrens.

Rather local along or near the shores of the Great Lakes and in the Ottawa Valley; on barrens at the northern end of the Bruce Peninsula and on Manitoulin Island; also in the region between Lake-of-the-Woods and Thunder Bay, Lake Superior. (Maine and w. Que. to s.e. Man., south to Tex. and Ga.)

Field check Low much-branched shrub with narrow, chiefly narrowly elliptic, 3-ribbed, glandular-serrate, alternate leaves; rounded flower clusters at the ends of leafy branches; small dark brown capsules with persistent saucerlike bases.

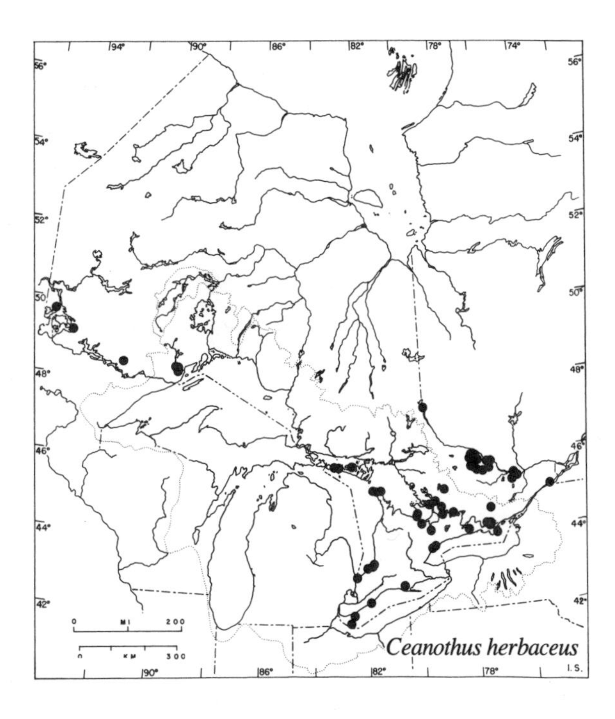

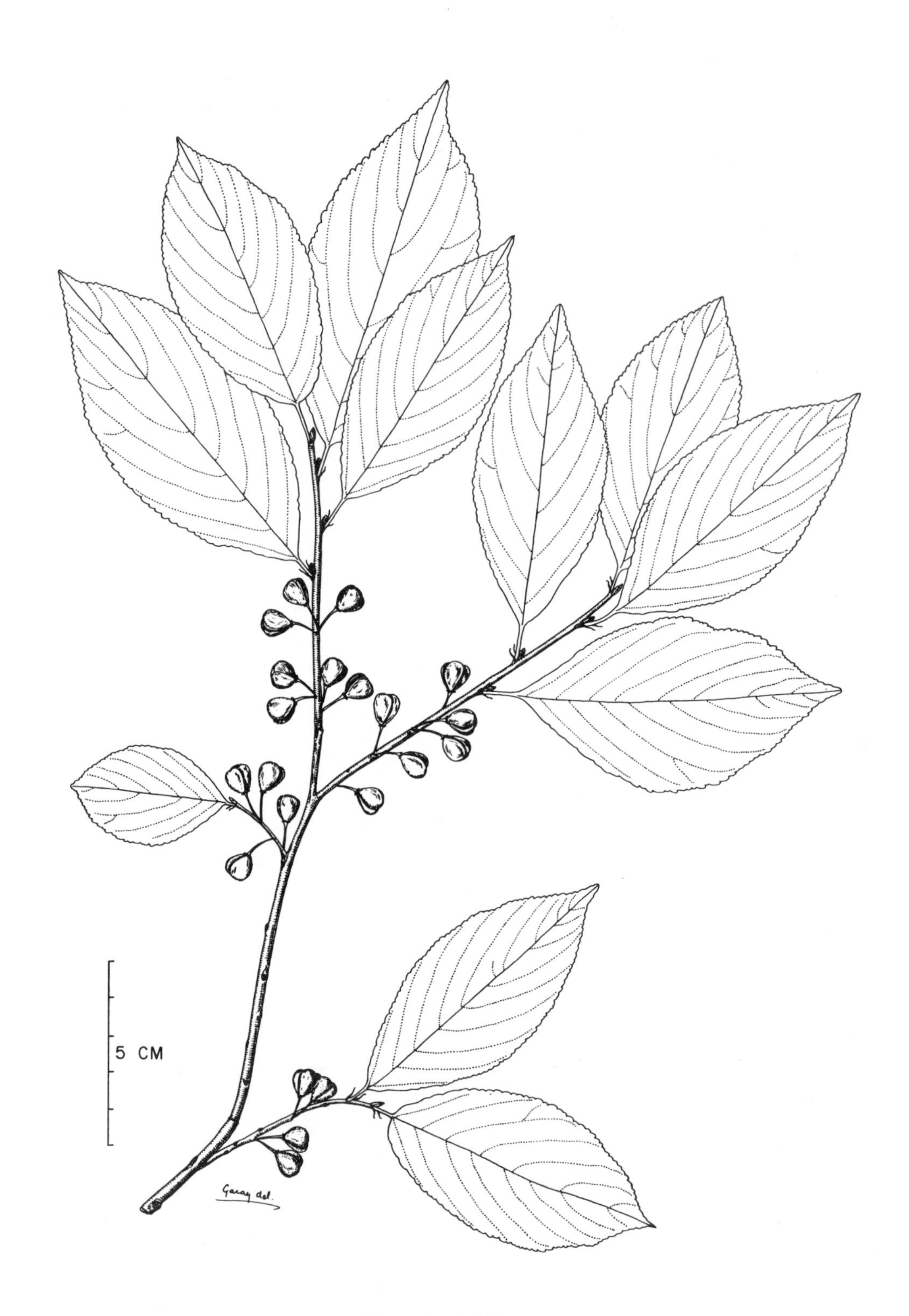

Rhamnus alnifolia

Rhamnus L. — Buckthorn

A genus of wide range with 100–160 species, mostly in the north temperate zone. Shrubs with alternate or opposite pinnate-veined leaves. Flowers in few-flowered clusters, the superior ovary surrounded by a large glandular disc. Fruit a fleshy, nearly black, subglobose drupe attractive to birds, with resulting effective dispersal of the seed. Certain species are the source of cathartics in medicines, green dyes, and charcoal. (*Rhamnus*—Greek *rhamnos*, the name for some species of this genus)

Key to *Rhamnus*

a. Stems armed with terminal and lateral thornlike short shoots; branchlets glabrous; leaves opposite or subopposite, the tip of the blade slightly folded and recurved **R. cathartica**

a. Stems unarmed, branchlets minutely pubescent; leaves alternate, the tip of the blade flat

- b. Low shrub usually less than one metre high; leaves toothed along the margins **R. alnifolia**
- b. Tall shrub or small tree up to 6 metres high; leaves with entire or slightly wavy margins **R. frangula**

Rhamnus alnifolia L'Hér. — Alder-leaved buckthorn

An upright shrub, sparsely branched and thornless, usually less than 1 m in height and sometimes spreading to form low thickets or patches in wet places. Branchlets stout, at first green and minutely hairy, later purplish-red to gray and finely ridged.

Leaves alternate, simple, and deciduous, green above and somewhat paler gray-green beneath, elliptic to ovate or obovate with crenate-serrate margins; larger leaves towards the ends of the branches up to 10 cm long and 5 cm wide with the tip of the blades acute to tapered or pointed, the smaller lower leaves with more rounded tips, the base on all blades commonly narrowed to a grooved petiole 6–12 mm long; stipules linear, 6–9 mm long, conspicuous early in the season but falling before the fruits mature.

Flowers yellowish-green, about 3 mm in diameter, on short stalks in small clusters from the axils of the lower leaves, perfect but functionally unisexual, i.e., flowers contain both stamens and pistils but one or the other is vestigial or reduced, the two kinds occurring on separate plants; late May to early June. Fruit a purplish-black, globose to ovoid, 1–3-seeded, berrylike drupe about 6 mm in diameter, slightly longer than wide, in small clusters; Aug. and Sept. (*alnifolia*—with leaves like *Alnus*, the alder)

In moist situations: low woods and thickets, swampy depressions in woods, shores of small lakes and banks of streams, sphagnum bogs and cedar swamps.

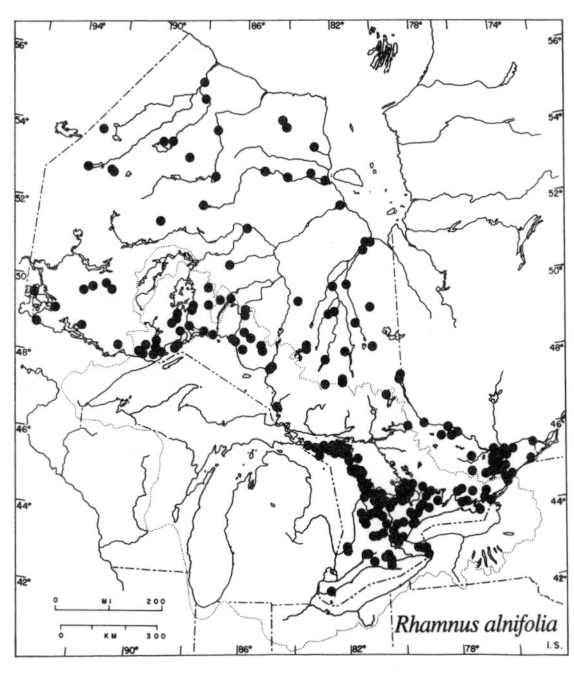

Rhamnus alnifolia

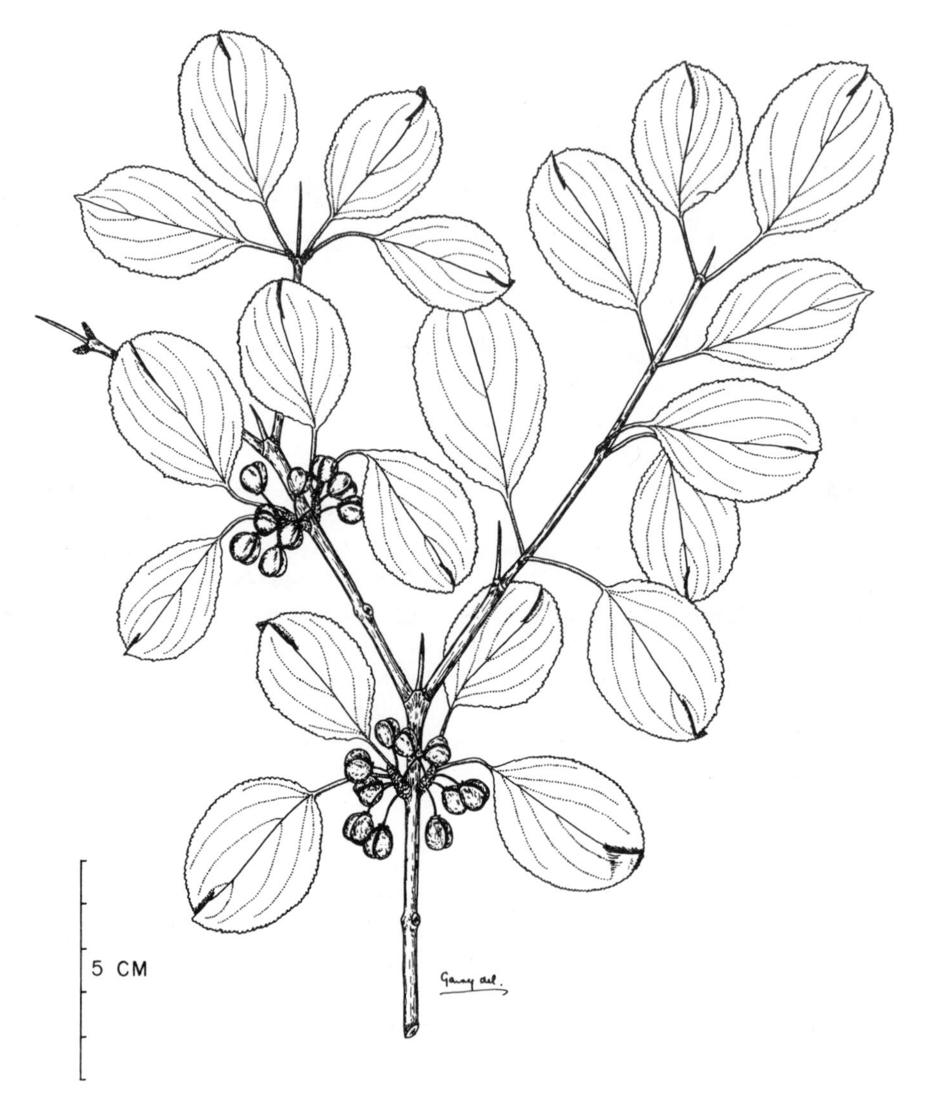

Rhamnus cathartica

Widespread in southern Ontario except on the Canadian Shield; also around Lake Superior and north to the southern boundary of the tundra along Hudson Bay. (Nfld. to s.e. B.C., south to Calif., Ind., and N.J.)

Note Alder-leaved buckthorn is an alternate host for the fungus that causes oat rust.

Field check Low erect shrub of wet places; alternate, elliptic to obovate, veiny leaves with crenate-serrate margins; black berrylike fruits.

Rhamnus cathartica L. — Common buckthorn

A large coarse shrub or small tree up to 6 m in height with some of the branches ending in a sharp thorn. Branchlets stout, purplish-red to gray-brown, and of two kinds, the long shoots smooth and somewhat angled, the short shoots rough-warty with crowded leaf scars.

Leaves chiefly opposite but some subopposite or alternate, simple, and deciduous, but often remaining green and not falling until late in the autumn; blades up to 8 cm long and 4 cm wide, elliptic to ovate, the tip sharply pointed and slightly folded to recurved, the base subcordate, rounded, or tapered; smooth and green on both surfaces, with 3 - 5 pairs of strongly curved veins conspicuous on the lower surface; margins minutely crenate-serrate; petioles grooved, pubescent, up to 2.5 cm in length with a pair of deciduous linear stipules 3 - 6 mm long at the base.

Flowers less than 6 mm in diameter, greenish-yellow, on threadlike stalks 6 mm or more in length, in rather dense clusters from the axils of the lower leaves; early June. Fruit a 3 - 4-seeded, purplish-black, globose, berrylike drupe 5 - 6 mm across, in dense clusters; Aug. and Sept. (*cathartica*—purging, in reference to the fact that certain substances in the bark, leaves, and berries have strong purgative power if eaten)

In both dry and moist habitats: open pastures, fencerows, roadsides, clearings; also in low woods, in rocky woods, and on the slopes of ravines, where it often appears as if indigenous to the area.

In southern Ontario both east and southwest of the Canadian Shield. (Native of Europe and thoroughly naturalized in many parts of North America from N.S. to Sask., south to Mo. and Va.)

Note Common buckthorn is a good example of the numerous introduced trees and shrubs that retain their leaves until late in the autumn in sharp contrast to the promptly deciduous habit of almost all native trees and shrubs. Our common beech (*Fagus grandifolia* Ehrh.) is a notable exception among native species, the young trees retaining their leaves in a dried bleached state throughout the winter.

Common buckthorn is an alternate host for the fungus causing oat rust.

Field check Coarse shrub or tree with spine-tipped short shoots; leaves opposite or subopposite with strongly curved veins; black berrylike fruits in dense clusters.

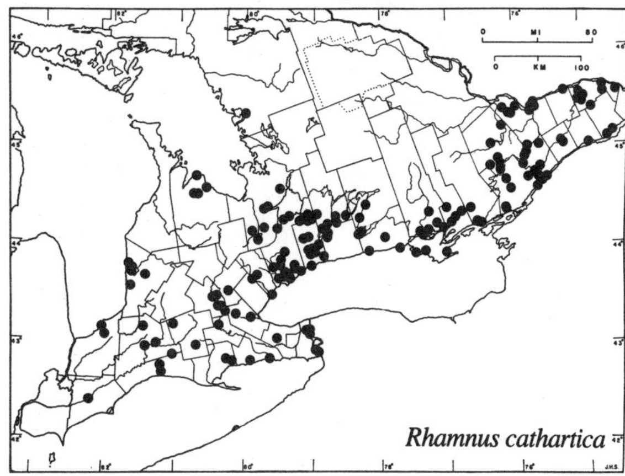

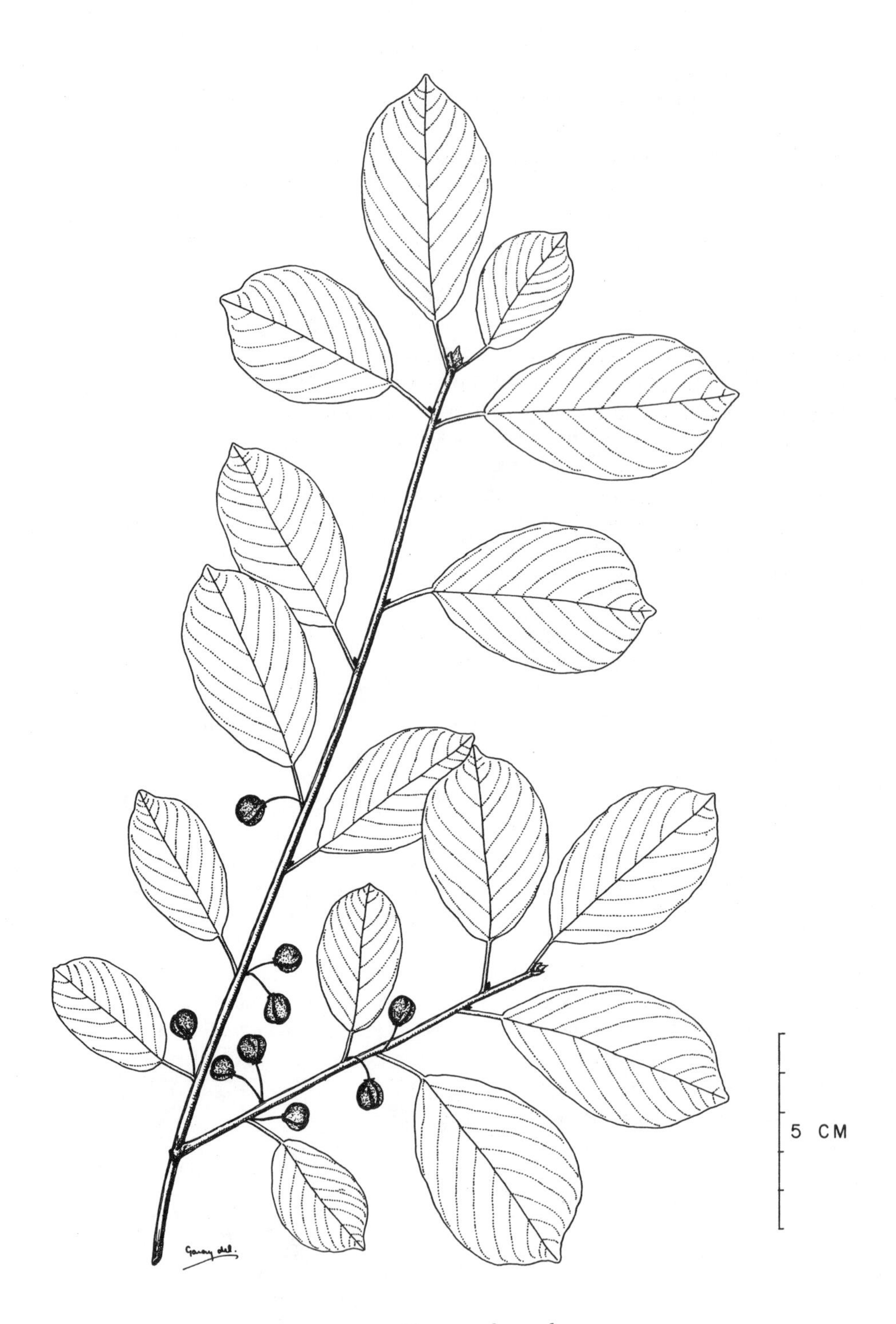

Rhamnus frangula

Rhamnus frangula L. — Glossy buckthorn

A shrub or small tree up to 6 m in height. Branchlets stout, brownish to gray, minutely pubescent, mottled with conspicuous, elongate, pale lenticels; winter buds without scales.

Leaves deciduous, simple, and mostly alternate, although some may be subopposite; blades up to 8 cm long and 5 cm wide, oval, elliptic, or obovate, with 5 - 10 pairs of rather straight veins, apex rounded, blunt, and abruptly pointed, base rounded or tapered; margins entire and faintly wavy; petioles stout, up to 2 cm long, with small, slender, caducous stipules.

Flowers greenish-yellow, less than 6 mm in diameter, solitary or in groups of 2 - 8 in sessile umbels in the axils of the lower leaves; June. Fruit a purplish-black, 2- or 3-seeded, globose, berrylike drupe about 7 mm in diameter, in small clusters; Aug. and Sept. (*frangula*—the Latin name for this plant, possibly referring to the brittle branchlets)

In mixed woods, shaded ravines, or around the edge of sphagnum bogs.

Rather local, chiefly near some of the cities in southern Ontario: in the vicinity of Ottawa, Kingston, Toronto, Guelph, Galt, and London; also in Huron County and the Midland Peninsula of Simcoe County. (Introduced from Europe and rather local from N.S. to s.Man., south to Minn., Ill., and N.J.)

Field check Shrub or small tree with alternate, entire, shiny, oval leaves with conspicuous straight veins; small clusters of black, berrylike fruits.

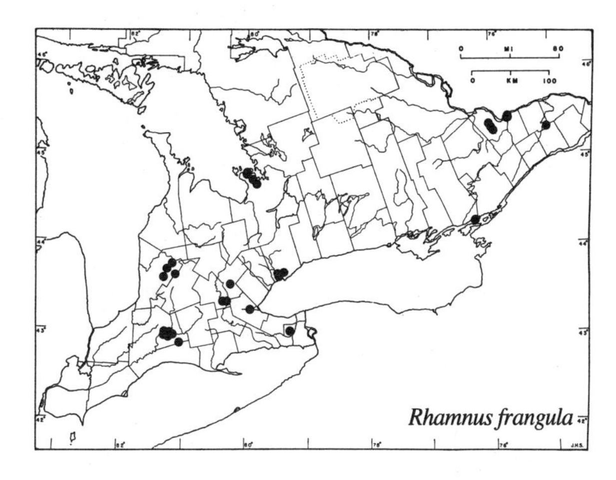

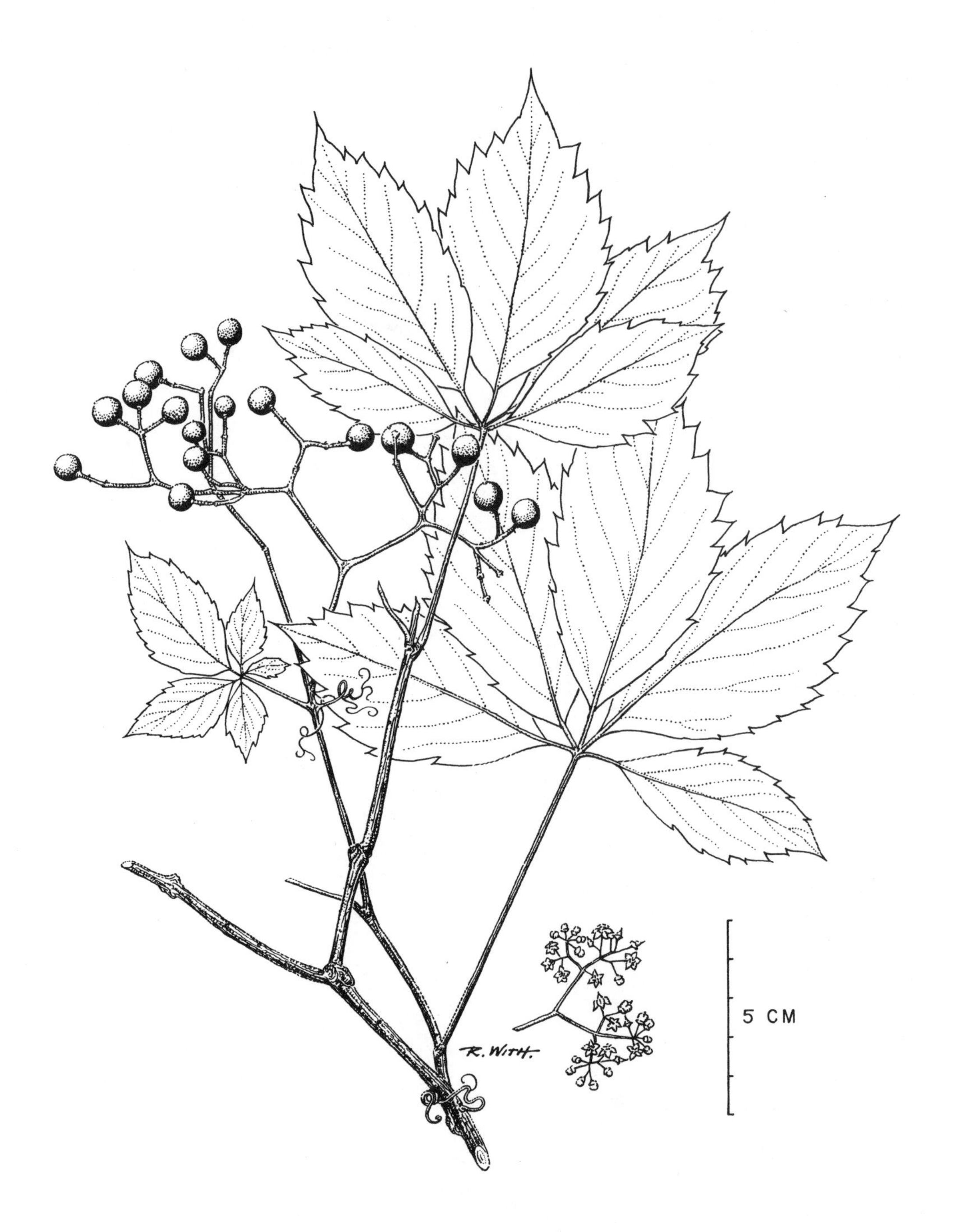

Parthenocissus vitacea

VITACEAE — GRAPE FAMILY

About 12 genera and some 700 species, mostly tropical but extending into temperate regions of the northern hemisphere. Mostly woody vines or rarely erect shrubs with climbing stems, often swollen at the nodes. Leaves alternate, simple and palmately veined or palmately compound, usually two-ranked; stipules deciduous; tendrils and inflorescences produced opposite certain leaves of each shoot; tendrils of some species develop small suckerlike discs. Flowers small, in cymose or paniculate clusters, regular, 4-parted or 5-parted, perfect or unisexual, the calyx very small and cup-shaped; stamens 4 or 5, attached at the base of a disc that surrounds the ovary; pistil of 2 fused carpels. Fruit a 1-4-seeded berry.

The genus *Vitis* is important economically as the source of edible grapes, raisins, and currants, and as the basis of the wine industry. Several species of *Parthenocissus* are cultivated in the garden, and several of *Cissus*, as indoor foliage plants. (*Cissus* — Greek *kissos*, ivy)

The name of the family is based on the genus *Vitis*.

Parthenocissus Planch. — Virginia creeper

A small genus of 10-15 species in temperate North America and eastern Asia. Deciduous vines, rarely evergreen, trailing or high-climbing. Stems with white pith; tendrils intermittent, sometimes with adhesive discs at their tips. Leaves palmately 3-7-lobed or compound, coarsely serrate. Flowers small, 5-parted, the petals expanding before they fall; stamens developing before the stigma is receptive, favouring cross-pollination. The flowers are visited by several kinds of bees and by various flies and beetles. Fruit a berry with a rather thin layer of flesh around the seeds. (*Parthenocissus* — from the Greek *parthenos*, virgin, and *kissos*, ivy, a translation of the common name Virginia creeper)

Key to *Parthenocissus*

a. Tendrils sparsely branched, their tips slender or flattened, usually without adhesive discs; flowers and berries in a broad, open and widely forking or dichotomously branched system without a main central axis ... ***P. vitacea***

a. Tendrils much-branched, their tips swollen or with adhesive discs; flowers and berries in closely grouped clusters along a prolonged central axis ... ***P. quinquefolia***

Parthenocissus vitacea (Knerr) Hitchc. (*P. inserta* (Kerner) K. Fritsch) — Virginia creeper

A woody vine scrambling over the ground or climbing on exposed rocks, or on trees and shrubs, sometimes reaching a length of 10 m or more. Young stems green and sparingly pubescent but becoming brown or gray in age; older stems ridged and warty with prominent leaf scars; tendrils with slender twining or spirally coiling branches usually without adhesive discs at their tips, developing opposite some of the leaves.

Leaves alternate, palmately compound, and deciduous, the petiole as much as 15-20 cm long and usually bearing 5 stalked leaflets 5-12 cm long; leaflets elliptic to obovate, dark green and shiny above and usually glabrous beneath, coarsely toothed, at least beyond the middle, the tips often long-tapering.

Flowers about 5 mm across, greenish, in umbel-like clusters on a widely branched inflorescence which forks dichotomously (lacks a main central axis); June and early July. Fruit a bright blue berry 8–10 mm in diameter with 3 or 4 seeds, the supporting stalk for each berry often bright red when the fruit is ripe; Aug. and Sept. (*vitacea*—like *Vitis*, the grape, in reference to the similar berries; *inserta*—inserted)

In woods and thickets, along banks, on cliffs and talus slopes, and on open ground such as beaches and fields.

Throughout southern Ontario but less common northward; not reported north of about 47° N except west of Thunder Bay. (Que. and N.Eng. to Man., southwest to Mont. and Calif. and southeast to Tex.)

Note The closely related *P. quinquefolia* (L.) Planch. has a more southern and eastern distribution and is doubtfully native in Ontario although it may escape from cultivation and persist. It differs from *P. vitacea* in its high-climbing habit, the presence of adhesive discs at the tips of its much-branched tendrils, and in the elongate shape of the inflorescence which has closely grouped clusters of flowers from a prolonged central axis. (*quinquefolia*—five-leaved)

Virginia creeper is sometimes confused with poison ivy in spite of the fact that the latter has only three leaflets and white or straw-coloured dry fruits. In the autumn the vines of Virginia creeper often make a spectacular display with the brilliant colours of their foliage, as do also, in some situations, the leaves of poison ivy.

Field check Woody vine with alternate palmately 5-foliolate leaves, slender-branched tendrils (usually without adhesive discs) and blue berries.

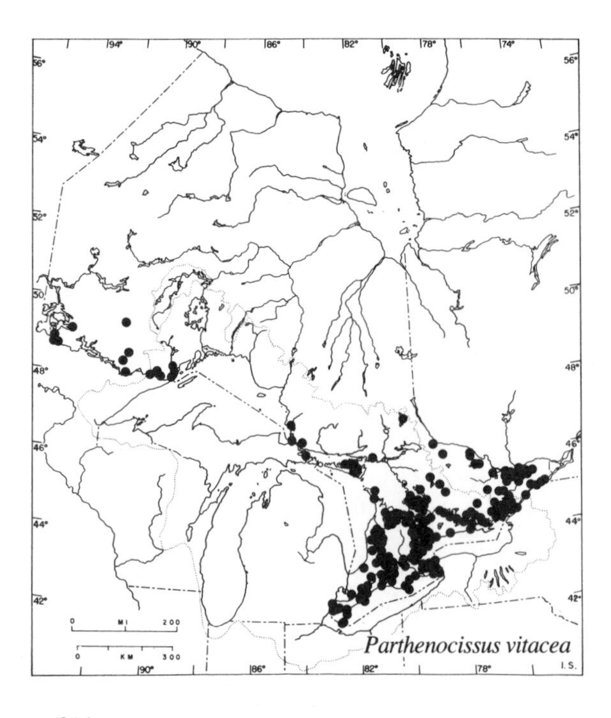

Vitis L. — Grape

A genus of the northern hemisphere with 60–70 species. Woody-stemmed vines with brown pith, the bark of old wood loose and shredding. Stems climbing by means of coiling tendrils. Leaves simple, mostly rounded or cordate, palmately veined, often deeply lobed. Inflorescence a panicle of small fragrant 5-parted flowers; calyx minute or obsolete; petals cohering at the tip and falling as a unit when torn from their base by the expanding flower bud; base of ovary surrounded by a ring of small nectary glands. The flowers are pollinated by bees but other insects are also attracted by the fragrance and nectar.

The cultivated grapes originated from *Vitis vinifera* L., a species of northwestern India and the Orient extensively cultivated in the Mediterranean regions. During the nineteenth century, a North American root aphis reached Europe, causing heavy losses in the vineyards. The wine industry was saved by the introduction of the resistant American grapes which were used as grafting stock for the susceptible cultivated varieties. Many selections of North American species and some hybrids are now used as stock for planting in the vineyards in various parts of Europe, California, and the Niagara Peninsula of Ontario. The American species differ from the European in that the fruit falls more readily and the seeds are not readily separable from the pulp. Raisins are the dried fruits of the grape, and currants are obtained from the Corinthian variety. (*Vitis*—the Latin name for the vine)

Key to *Vitis*

a. Lower leaf surface glabrous or conspicuously whitened and the hairs if present loose and soon disappearing completely, or remaining only in patches, when the leaves are fully developed
- b. Leaves smooth and greenish beneath when fully developed, often with tufts of hairs in the vein axils ... ***V. riparia***
- b. Leaves strongly whitened beneath, the prominent veins hairy ... ***V. aestivalis***

a. Lower leaf surface with complete and persistent feltlike coating of whitish to rusty-coloured hairs ... ***V. labrusca***

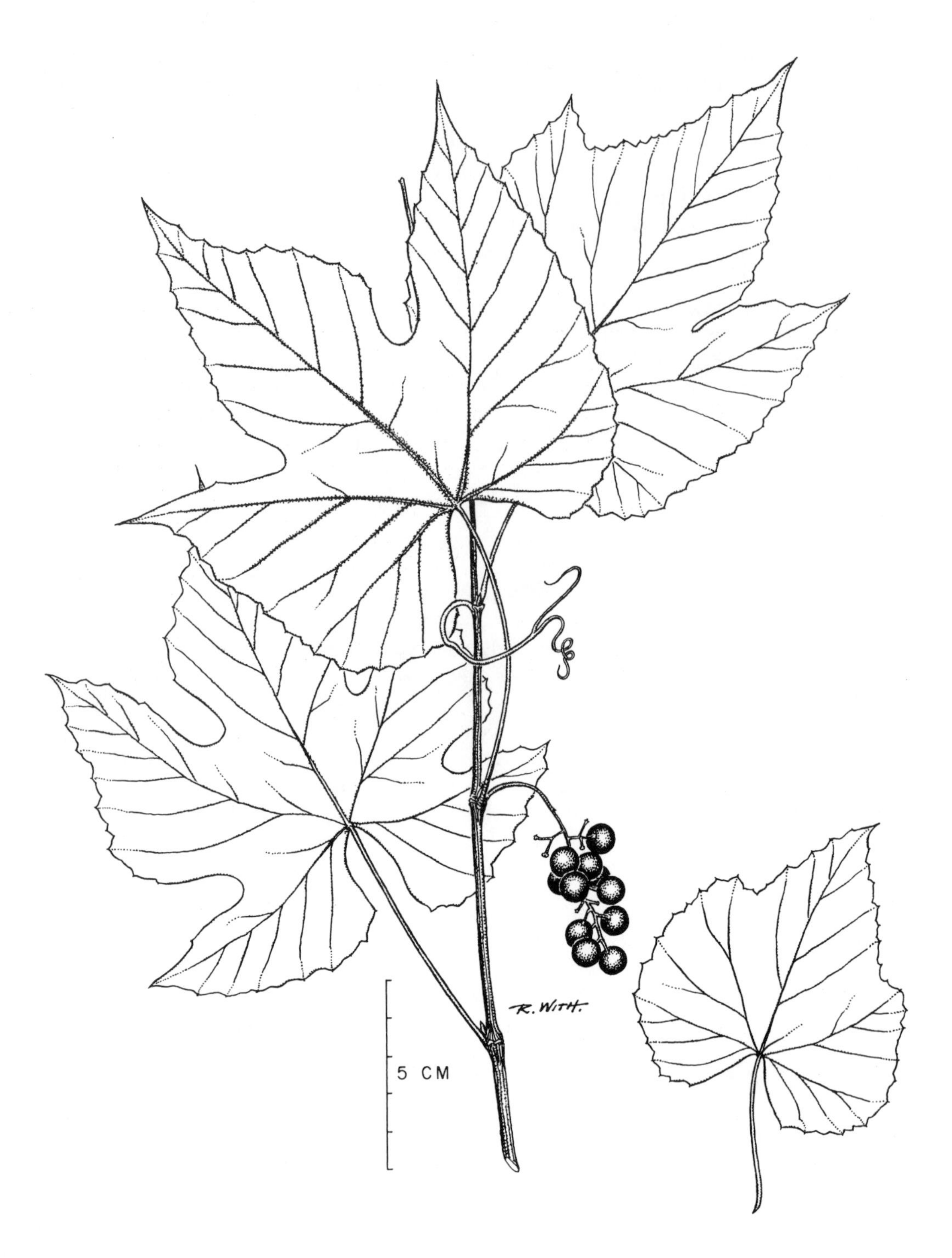

Vitis aestivalis

Vitis aestivalis Michx. **Summer grape**

A woody vine with a stout stem or trunk up to 20 cm in diameter, climbing over shrubs and into the branches of trees to a height of 5 - 10 m. Young branches often pubescent with rusty-red hairs but generally smooth at flowering time; one-year old stems brownish and finely ridged.

Leaves alternate, simple, and deciduous, covered at first with a reddish or rusty tomentum; blades cordate to round-ovate, up to 20 cm across, about as broad as long, without lobes, with low lobes, or with 3 - 5 deep lobes and rounded or pointed sinuses; green and smooth above at maturity but strongly glaucous beneath (blue-green to silvery) with persistent pubescence along the prominent veins; margins coarsely and irregularly toothed; petiole often as long as, or longer than, the blade; branched and coiled tendrils developing opposite some of the leaves and twining around the available supports.

Flowers numerous, tiny, greenish-yellow to whitish, in elongate cylindrical inflorescences 5 - 15 cm long, arising opposite one or more of the leaves; late spring. Fruit an edible but sour, blue to blackish berry, 5 - 12 mm across and covered by a thin persistent bloom; in clusters; Sept. and Oct. (*aestivalis*—of the summer)

In moist thickets and on sandy soil in open woods and along roadsides.

In the Deciduous Forest Region from Grand Bend to Toronto. (Mass. to s. Ont., west to Minn. and south to Tex. and Ga.)

Note The material examined from Ontario is quite variable and some specimens with glabrate stems and petioles and the lower leaf surface strongly whitened have been referred by Scoggan to the variety *argentifolia* (Munson) Fernald. (*argentifolia*—silver-leaved)

Field check Woody vine with alternate, palmately veined, 3 - 5-lobed leaves, often deeply lobed, green above, whitish to silvery beneath; fruit a cluster of blue-black grapes with a bloom.

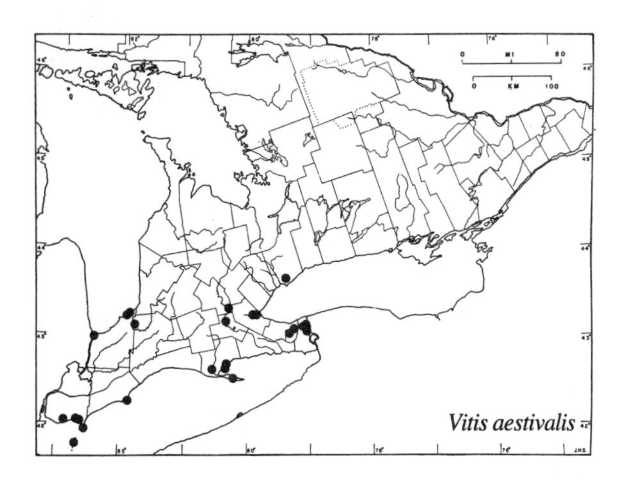

Vitis labrusca

Vitis labrusca L. **Fox grape**

A coarse, trailing or climbing, woody vine often 6 m or more in length. Young stems clothed with a dense felt of rusty-coloured hairs; one-year-old stems brownish and finely ridged.

Leaves alternate, simple, and deciduous, the blade cordate to round-ovate, up to 20 cm across, merely shouldered or with somewhat forward-pointing lobes, a triangular apex, and an irregularly toothed margin; upper surface dark green, lower surface covered with a permanent firm coating of reddish or rusty-coloured hairs; petiole downy and about as long as the blade; coiled branched tendrils or a flowering or fruiting cluster usually opposite several successive leaves.

Flowers greenish-yellow to whitish, tiny, and numerous, in small sparsely branched clusters; May and June. Fruit a cluster of dull red to purple-black berries (grapes) 1–2 cm in diameter with 2–4 large seeds; the grapes are covered with a bloom, have a strong musky scent, and a sweetish or astringent taste; Sept. and Oct. (*labrusca*—the Latin name for the wild grape-vine)

In woods or thickets and along roadsides.

Along the Niagara River, at Hamilton, and bordering the north shore of Lake Erie; also the upper part of the St. Lawrence Valley; perhaps indigenous only in the Niagara region. (s. Maine to s. Mich., south to Ky., Tenn., and Ga.)

Note This species is the source of popular cultivars, such as, Concord, Worden, Champion, and Chautauqua, and by crossing, the many others that are sometimes recognized in horticulture under the name *V. labruscana* Bailey.

Field check A woody vine with alternate, palmately veined, broadly cordate to 3-lobed leaves covered beneath with a firm coating of rusty-coloured hairs; tendrils numerous; grapes few but large, in short sparsely branched clusters.

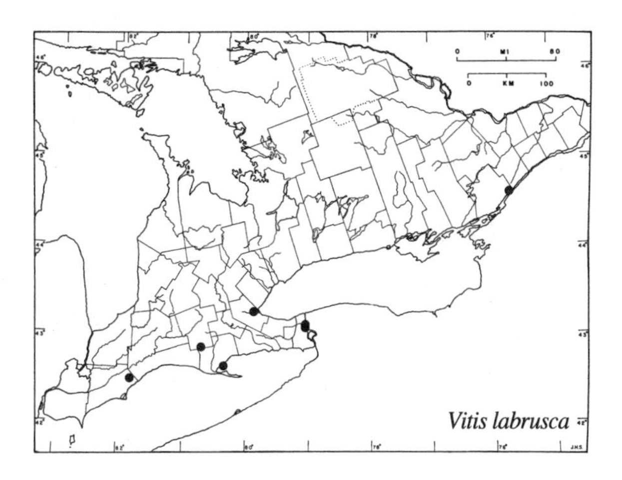

Vitis riparia

Vitis riparia Michx. (including *V. vulpina* L.) — Riverbank grape, Frost grape

A moderately high-climbing woody vine, often covering shrubs and fences or growing into the branches of trees to a height of 5 m or more. Young branches green or reddish, often pubescent, especially at the nodes, but becoming smooth and finely ridged.

Leaves alternate, simple, and deciduous, the blade cordate-ovate and 7–15 cm wide, with a deep broadly U-shaped sinus at the base and usually with a prominently triangular tip and two shorter lateral lobes; upper surface smooth and bright green, the lower surface paler and thinly pubescent, at least along the veins or in their axils; margins ciliate, with irregular forward-pointing teeth; petiole usually shorter than the blade and with small stipules at the base.

Flowers tiny, fragrant, greenish-white to cream-coloured, numerous, in much-branched and prominently stalked clusters; May and June. Fruit a dark blue to black berry, 6–12 mm in diameter, with a bloom; borne in elongate clusters; the grapes juicy but sour until fully ripe; late Sept. and Oct. (*riparia*—along the bank of a stream; *vulpina*—foxy)

On river banks, in thickets, and along roadside woods; also on open sandy or rocky (usually calcareous) ground.

Widely distributed in southern Ontario, especially east and southwest of the Canadian Shield; northward in the Ottawa Valley to Lake Timiskaming and in Lake Huron to Manitoulin Island and St. Joseph Island; not recorded north of about 48° N except in the Rainy River District. (N.S. to s. Man., south to N.Mex., Tex., and Va.)

Note The description given above includes plants that have been treated in some manuals under the name *V. vulpina* L.

Leaves on long ground shoots are often deeply lobed with prolonged teeth and conspicuously rounded sinuses. They have sometimes been mistaken for *V. palmata* Vahl, a species that is not indigenous in Ontario. (*palmata*—palmate, the type of venation and lobing of the leaves)

The fruits are eaten by flickers and by songbirds such as cardinals and thrushes. They are also gathered for the making of wine and jelly.

Field check Woody vine with alternate, palmately veined, cordate to sharply 3-lobed leaves, green on both sides; leaf margins ciliate and sharply toothed.

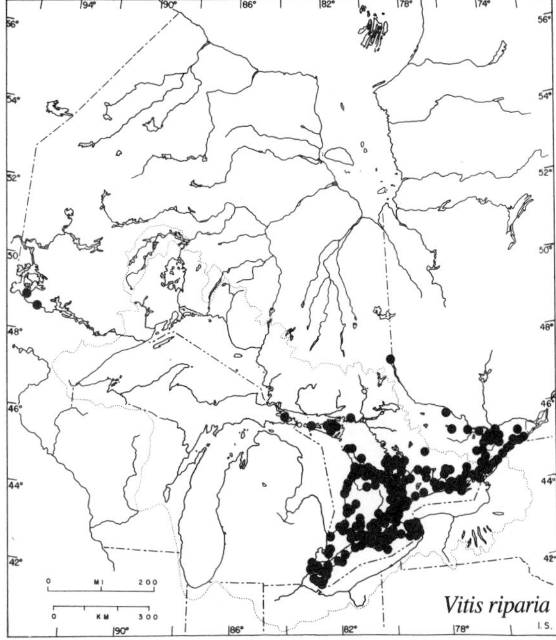

Hypericum kalmianum

GUTTIFERAE — ST. JOHN'S-WORT FAMILY

About 40 genera and some 1000 species, chiefly tropical but with representation also in temperate regions. Trees, shrubs, and herbs with simple, entire, and mostly sessile opposite leaves lacking stipules; oil glands or ducts usually present, often as glandular dots on the leaves and petals; flowers in cymes, stamens numerous, free or united in bundles; fruit a capsule, berry, or drupe. The family name is derived from the Latin *gutta*, a drop or speck, and *ferre*, to bear or carry, in reference to the glandular dots. Hypericaceae is used in some manuals, a name based on the genus *Hypericum*.

Hypericum L. — St. John's-wort

A genus of 350–400 species found in temperate regions and on mountains in the tropics. Glabrous perennial herbs or shrubs with simple, opposite, deciduous, gland-dotted leaves and cymes of mostly yellow 5-parted flowers. (*Hypericum*—Greek *Hyperikon*, the ancient name for St. John's-wort)

Key to *Hypericum*

a. Low shrub less than 1 m high; leaves usually sessile; flowers in rather open clusters at the ends of the branches; capsules ovoid, usually with 5 styles, rarely 3, 4, or 6 ***H. kalmianum***

a. Medium-sized shrub up to 2 m tall; leaves with short petioles; flowers in somewhat compact clusters both at the ends of the branches and in the axils of the upper three or four pairs of leaves; capsules subcylindric to lance-ovoid, usually with 3 styles, rarely 4 ***H. prolificum***

Hypericum kalmianum L. — Kalm's St. John's-wort

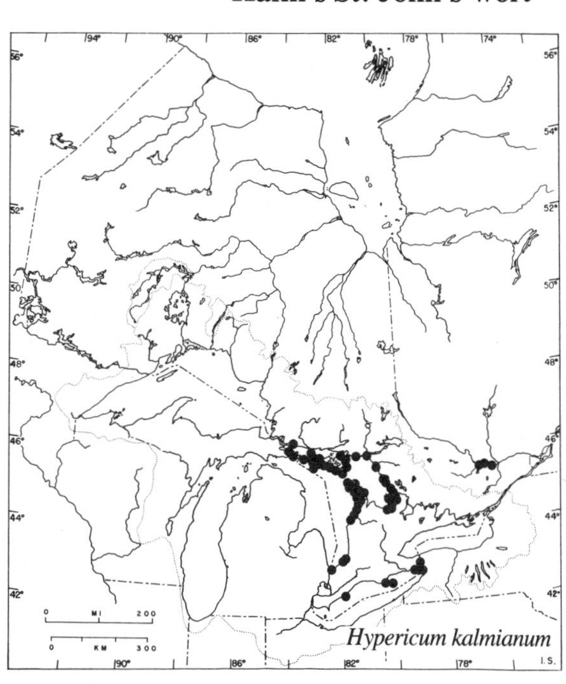

A low shrub with numerous erect branches rarely exceeding 1 m in height. Young branchlets glabrous and angled with winglike ridges from the leaf bases; older stems grayish to reddish-brown with conspicuously peeling papery bark.

Leaves simple, opposite, and deciduous, firm or leathery, 2.5–5 cm long and 3–9 mm wide, linear to narrowly oblong; rounded, blunt, or pointed at the tip, slightly narrowed at the sessile base; margins revolute; upper surface dark green with tiny glandular dots which are translucent when the leaf is held to the light, the lower surface paler with a conspicuous midrib. Smaller leaves commonly occur in tufts in the axils of the principal leaves.

Flowers small, bright golden-yellow, showy, 1.5–2.5 cm across, in open, rather

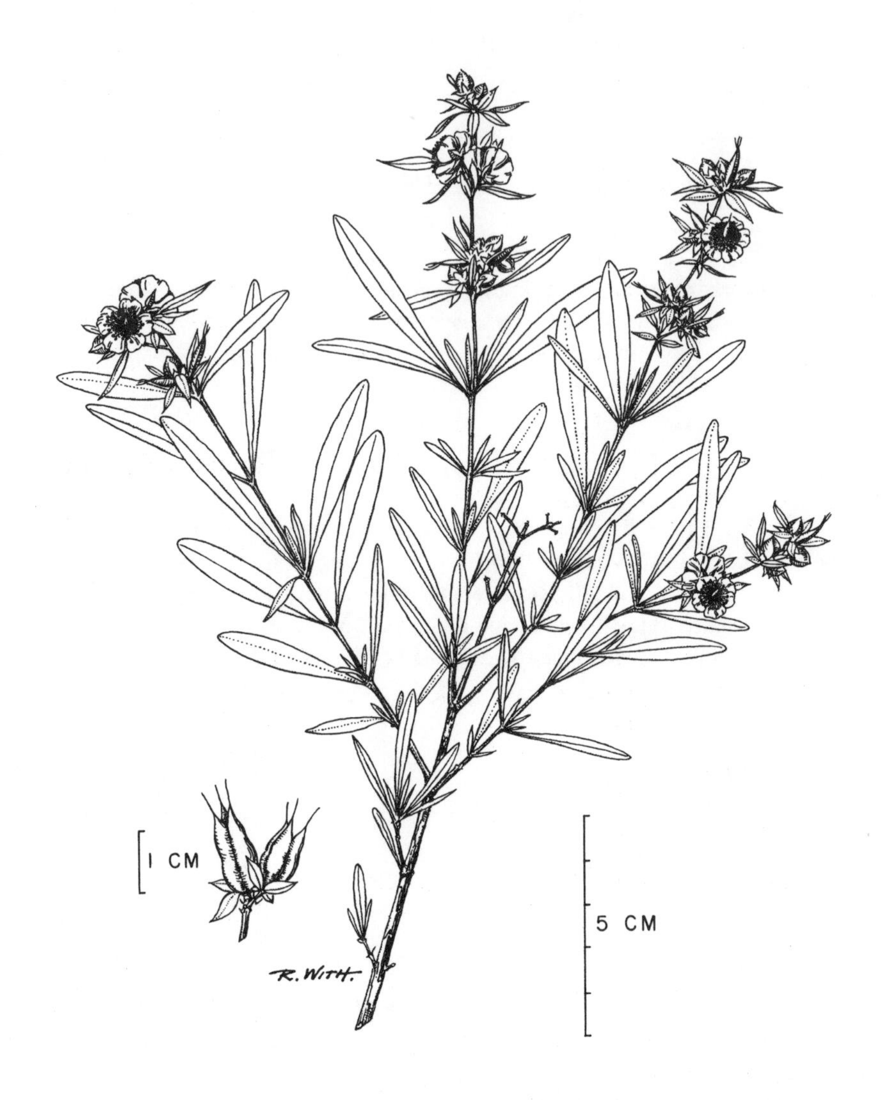

Hypericum prolificum

flat-topped clusters at the ends of the branches; each flower with 5 petals, numerous free slender stamens and a pistil with 5 locules; July and Aug. Fruit a dark brown ovoid pointed capsule with 5 styles united at first into a conspicuous beak, separating when the capsule splits; seeds cylindrical, tiny, and numerous; Aug. and Sept. (*kalmianum*—named in honour of Pehr Kalm, 1716-1779, a pupil of Linnaeus who also honoured him by the ericaceous genus *Kalmia*)

In sandy or rocky ground along the shores of lakes and rivers.

Common around the shores of Lake Huron and Georgian Bay; also locally along Lake Erie, the Niagara River, and the Ottawa River above Ottawa. (s.w. Que. and N.Y. to n. Mich., south to Ill. and Ohio)

Note The familiar herbaceous species of St. John's-wort *H. perforatum* L., which was introduced from Europe, has become an obnoxious weed in Ontario and elsewhere in North America by invading fields, meadows, and roadsides. (*perforatum*—perforated, in reference to the translucent glands of the leaves)

Field check Low bushy shrub; leaves opposite, sessile, narrowly oblong to linear, gland-dotted; tufts of smaller leaves in the axils of the principal ones; flowers yellow, showy, in small terminal clusters; fruit a pointed capsule with 5 united or separating styles.

Hypericum prolificum L. (*H. spathulatum* (Spach) Steud.) — Shrubby St. John's-wort

A medium-sized shrub with numerous erect branches growing to a height of 1.5-2 m. Branchlets glabrous and angled with wing-like ridges from the leaf bases; older stems grayish or brownish with peeling thin papery bark.

Leaves essentially similar to those of *H. kalmianum* but larger and with short petioles.

Flowers in small compact clusters at the ends of the branches and in the axils of the upper three or four pairs of leaves, similar to those of *H. kalmianum* except for a 3-chambered pistil; July and Aug. Fruit a brown lance-ovoid capsule with 3 slender styles united below, splitting when ripe into 3 sections, each tipped with a conspicuous style; seeds numerous, tiny, and cylindrical; Aug. and Sept. (*prolificum*—producing numerous offspring; *spathulatum*—spoon-shaped or spatulate, in reference to the shape of the narrow leaves)

In fields, on sandy plains, in woods, and along fencerows.

Local in Middlesex and Wellington counties; reported from the vicinity of Windsor and from Lambton County (without locality stated and no specimens seen). (N.Y. to Mich., south to La. and Ga.)

Note In Ontario, with the exception of the Ottawa Valley locations, Kalm's St. John's-wort follows the shorelines of the Great Lakes to the vicinity of Sault Ste. Marie, while shrubby St. John's-wort is mostly inland in its distribution.

Field check Opposite, gland-dotted, narrow, blunt-tipped leaves with tufts of smaller leaves in the axils; showy yellow flower clusters at the ends of the branches and also in the axils of upper pairs of leaves; capsule lance-ovoid, 3-chambered, with 3 persistent styles.

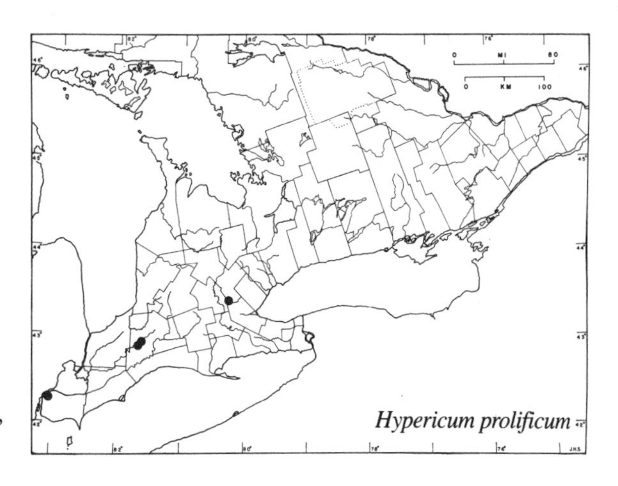

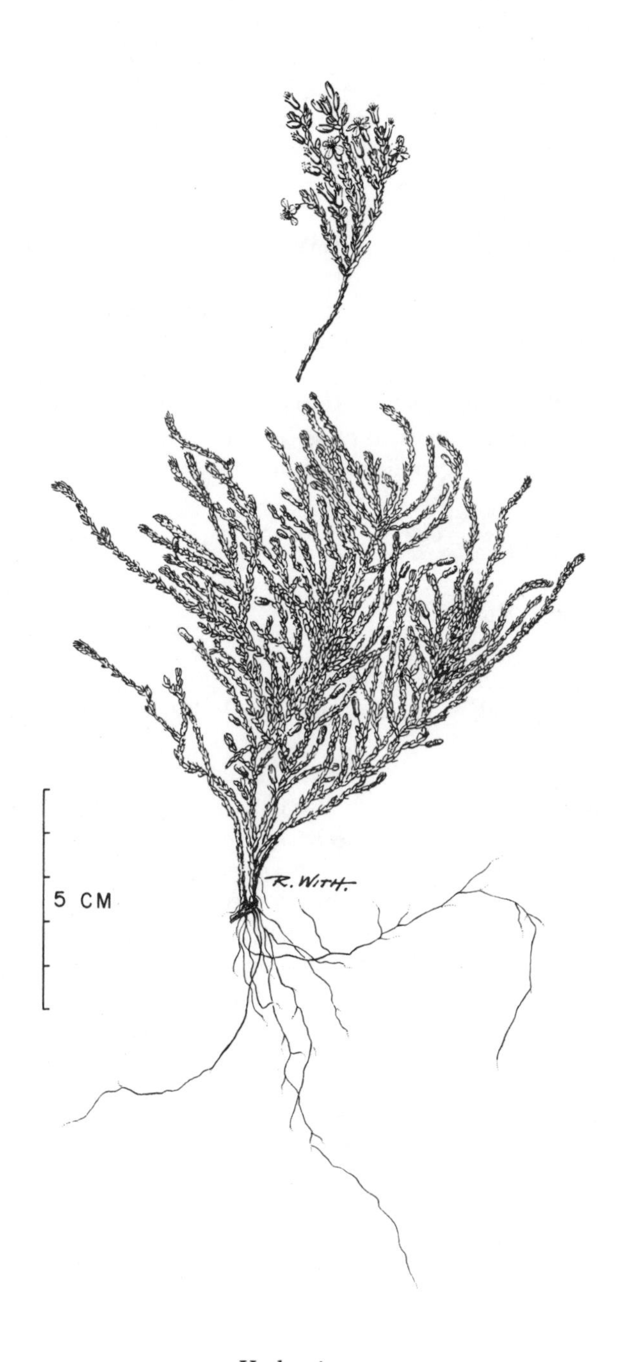

Hudsonia tomentosa

CISTACEAE – ROCKROSE FAMILY

Shrubs, subshrubs, and herbs with alternate or opposite entire simple leaves, sometimes revolute-margined and often pubescent. Flowers solitary or in cymose clusters, 5-parted; fruit a capsule. Seven or 8 genera and 175–200 species chiefly in warm temperate regions (especially the Mediterranean) and often in sunny sandy sites. The family name is based on *Cistus*, a genus that includes several handsome ornamentals.

Hudsonia L. – False heather

Bushy or much-branched and tufted perennials forming mats or low mounds, the stems clothed with alternate, sessile, persistent, scalelike downy leaves; flowers crowding along the upper part of the branchlets, small, yellow, and numerous enough to be rather showy. (*Hudsonia*—in honour of William Hudson, 1730–1793, an English botanist, apothecary, and author of *Flora Anglica*)

Hudsonia tomentosa Nutt. — Beach heath

A low and much-branched shrub, the erect or ascending tufts of branches usually less than 2 dm high, sometimes forming extensive mats. Branchlets grayish-green and hoary-pubescent, almost hidden by the closely appressed scalelike leaves; older stems at the base of the tufts of branches gray to reddish-brown.

Leaves alternate, simple, and evergreen, small and numerous, ovate-lanceolate to awl-shaped or scalelike, 2–4 mm long, densely gray-hairy, appressed and mostly overlapping each other, thereby clothing the stem.

Flowers 4–5 mm long, numerous, solitary at the ends of short leafy lateral branches, the 5 bright yellow petals surrounded by a hairy calyx; late June and July. Fruit a smooth, ovoid, few-seeded capsule contained within a persistent calyx. (*tomentosa*—densely woolly)

On sandy or silty beaches, on sand plains, or in sandy jack pine woods and clearings.

Mainly on northern shores of the Great Lakes drainage system; in the Rainy River District, along the eastern half of the north shore of Lake Superior, south of Lake Abitibi, and in the Ottawa Valley; also at Point Abino on Lake Erie. (P.E.I. and Lab. to Alta., south to Ind. and N.C.)

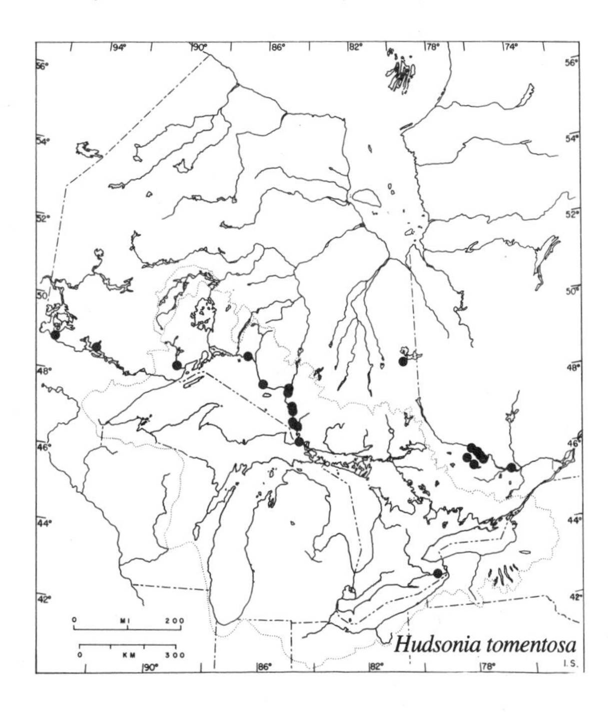

Note The beach heath is well adapted for growing in sand and on shifting dunes because of its long wiry tap root.

Field check Low, tufted or matted, evergreen shrub with minute grayish scalelike leaves clothing the numerous branches; flowers yellow, very small 5-parted, at the tips of the branches.

Daphne mezereum

THYMELAEACEAE – MEZEREUM FAMILY

About 50 genera and 500 species of both temperate and subtropical regions with the largest representation in Asia, Africa, and Australia. Mostly shrubs, some trees, woody vines, and herbs. Bark tough and fibrous. Leaves alternate or opposite, entire, deciduous or evergreen. Flowers in racemes or reduced to few-flowered heads or clusters, usually perfect and regular, the calyx tube 4- or 5-parted, petals small or lacking, stamens as many, or twice as many, as the calyx lobes, pistil simple or of several fused carpels, style single. Fruit a drupe, berry, or achene, rarely a capsule.

In two genera the bark is used for the making of paper: *Daphne cannabina* Wall., in Africa and India and *Edgeworthia tomentosa* (Thunb.) Nakai, one of several sources for rice paper, in China and Japan. The family name is based on an early generic name *Thymelaea* Mill., now included in *Daphne*. (*Thymelaea*—from the Greek roots meaning thyme and olive, in reference to the thymelike foliage and the small olivelike fruits of some species; *cannabina*—like hemp; *Edgeworthia*—in honour of M.P. Edgeworth, 1812-1881, English botanist in the East Indies; *tomentosa*—densely woolly)

Daphne L. – Mezereum, Daphne

About 70 species of shrubs of wide geographic distribution from Europe and northern Africa to temperate and subtropical Asia, Australia, and the Pacific region. Stems with tough fibrous bark, used in India and Africa to make paper. Leaves deciduous or evergreen. Flowers tubular, secreting nectar around the base of the ovary and visited by long-tongued insects. Several species are sold by nurseries as ornamentals for foundation planting, e.g., *D. mezereum* L., *D. cneorum* L., and a few cultivars. (*Daphne*—in Greek mythology the name of the nymph transformed, to the dismay of Apollo, into a laurel bush, hence also the common name in Greek for the laurel; *cneorum*—after *Cneorum*, a genus found in the Mediterranean and in Cuba)

Daphne mezereum L. — Daphne

A small, upright, and rather sparsely branched shrub seldom exceeding 1 m in height. Branchlets pale brown to grayish, pubescent at first with short crisp appressed hairs; older stems glabrous but with numerous scattered, dark, wartlike lenticels.

Leaves alternate, simple, and deciduous, sometimes in whorl-like clusters on short lateral branches; the blades oblong or oblanceolate, 3-8 cm long and 1-2 cm wide, thin, green above and gray-green below, smooth on both surfaces, the apex obtuse to acute, the base cuneate and tapering to a short petiole; margins entire and finely ciliate.

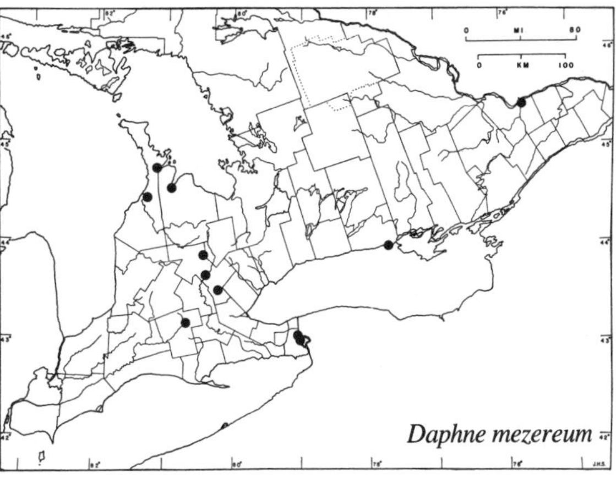

Flowers rose-purple or rarely white, very fragrant, appearing before the leaves in sessile lateral clusters of three or four; calyx tube 7-9 mm long, pubescent on the outside, the 4 calyx lobes petal-like, ovate, about 5 mm long; petals lacking; stamens 8, short, nearly sessile on the calyx tube,

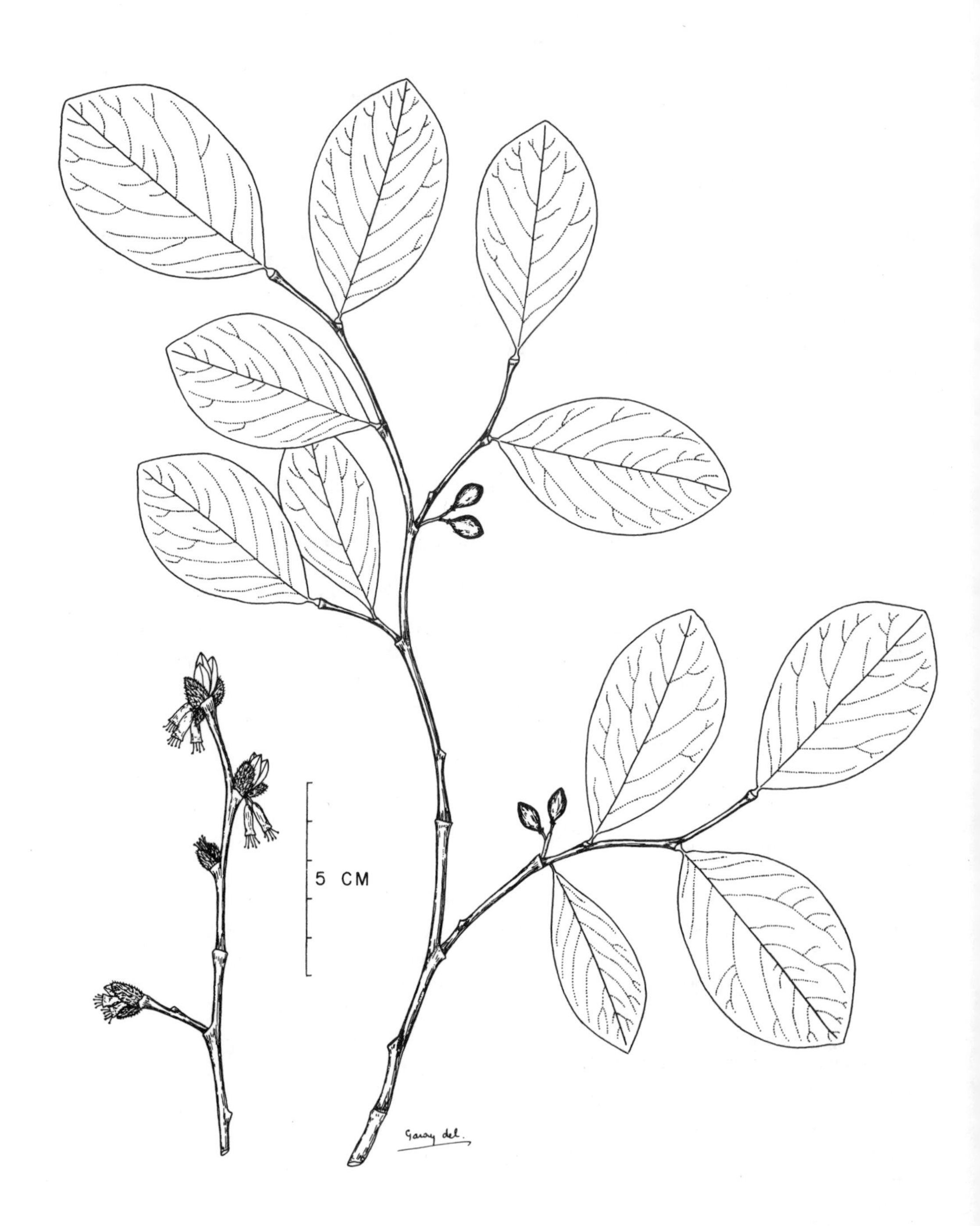

Dirca palustris

not exserted; style short with a large capitate stigma; late Apr. to May. Fruit a cluster of bright red, round-ovoid, berrylike drupes, yellow in the white-flowered form; Aug. and Sept. (*mezereum*—a modification of the Arabic name)

In open woods, especially in damp soil, often over limestone, along streams and river banks; a calciphile.

Escaped from cultivation in several localities across southern Ontario but well established in relatively few; known for many years from damp woods on the Dufferin Islands of the Niagara River near Niagara Falls; more recently discovered along a roadside north of Wiarton in the Bruce Peninsula; also at Ottawa, Presqu'ile Point on Lake Ontario, and along the Niagara Escarpment. (Nfld. to s. Ont., south to Ohio, N.S., and N. Eng.; native of Eurasia)

Note Dispersal of seeds is by birds which devour the berrylike fruits.

The attractive bright red fruits are a potential hazard wherever they are within reach of small children. They contain a glucoside and, if eaten, cause strong irritation and ulceration of the digestive tract which, in severe cases, may be fatal (Kingsbury, 1965, p. 35).

Field check Low shrub with fragrant, 4-parted, rosy-purple flowers in few-flowered sessile clusters along the stems in early spring; leaves thin, oblong, the margins ciliate; fruit a cluster of bright red drupes, in late summer.

Dirca L. — Leatherwood

A slow-growing shrub with the branching habit of a dwarf tree; branchlets flexible, swollen at the nodes; leaves alternate, simple, and deciduous; fruit a berrylike drupe. North American genus with one species in the east and a second, *D. occidentalis* Gray, in California. (*Dirca*—derived from *Dirce*, the name of the second wife of Lycus, in Greek mythology; *occidentalis*—western)

Dirca palustris L. — Leatherwood, Wicopy Moosewood

A low to medium-sized shrub or dwarf tree usually less than 2 m tall, the main trunk often divided close to the ground and developing a treelike system of erect or ascending branches. Branchlets stout, green at first, turning brown or grayish-brown, with conspicuous nodes and soft brittle wood, the bark fibrous and remarkably tough and pliable.

Leaves alternate, simple, and deciduous; blades up to 10 cm long and 7 cm wide, thin, smooth, and light green, elliptic to ovate or broadly oval, broadest about the middle, often somewhat rhombic or angular in outline with an obtuse or pointed apex and a rounded or tapered base; margins entire; petiole less than 3 mm long with an expanded dome-shaped base covering the bud for the following season.

Flowers perfect, pale yellow, and tubular, 6–9 mm long, in pendulous clusters of 2–5 as the leaves are unfolding, subtended by 2–4 very hairy dark brown enlarged bud scales which persist for several weeks; late Apr. and early May. Fruit a globose to ellipsoidal berrylike drupe 9–12 mm long, green at first then turning purplish-red, containing a single dark brown pit, and falling soon after ripening; June and July. (*palustris*—of swamps)

Usually in damp or shaded woods on rocky, sandy, or loamy soil; rarely on wooded sand dunes, in open jack pine woods or open rocky pastures.

Widespread throughout southern Ontario and northward to about 47° N on the east side of Lake Superior, reaching about the same northern limit as beech and sugar maple with which it is often associated; also known from a single site on a gravelly floodplain of the Whitefish River in O'Connor Township, Thunder Bay District, 48°20′N, 89°41′W. (N.S. and N.B. to Minn., south to La. and n. Fla.)

Note The common name leatherwood refers not to the wood but to the soft pliable bark which was once much used by Indians who cut it into long thin strips for thongs.

Field check Pliant branches with swollen nodes and tough fibrous bark; pale green entire alternate leaves; yellow tubular flowers in early spring.

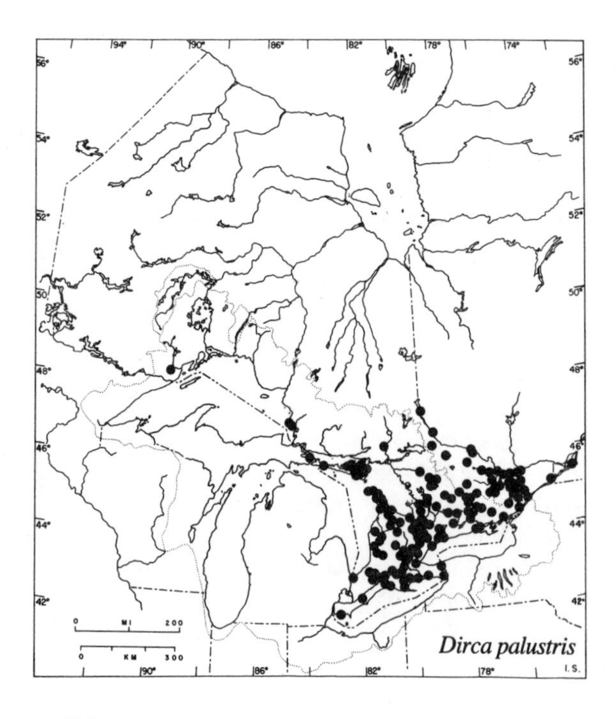

Dirca palustris

ELAEAGNACEAE — OLEASTER FAMILY

Shrubs and small trees with simple, alternate or opposite leaves; stems and leaves usually covered with silvery to brownish peltate scales or hairs. Flowers small, apetalous, solitary or clustered; fruit drupelike. Three genera and about 40 species, chiefly in the north temperate regions and often in steppes and coastal habitats. Some species in all three genera of this family fix atmospheric nitrogen in root nodules induced by the invasion of an actinomycete (bacterium).

Elaeagnus L. — Oleaster

Shrubs or small trees with alternate leaves in small lateral clusters on the twigs of the current year. The fruit of some species is edible but has a rather large stone. The Russian olive, *E. angustifolia* L., is frequently planted and is fairly hardy in all but the northern parts of Ontario. It has spines or thorns which are reduced leafless shoots. (*Elaeagnus* — from the Greek *elaea*, olive, and *agnos*, the chastetree, *Vitex agnus-castus* L.; *angustifolius* — narrow-leaved)

Elaeagnus commutata

Elaeagnus commutata Bernh. (*E. argentea* Nutt.)

Wolf-willow Silverberry

An upright much-branched shrub 0.5 – 3 m in height, spreading by stolons and forming thickets. Branchlets densely coated with rusty-brown scales; older stems pale brown to gray.

Leaves alternate, simple, and deciduous; blades 2 – 8 cm long and up to half as wide, ovate-lanceolate to elliptic or obovate, tapered to a blunt or acute tip, rounded or tapered at the base, gray to silvery-green above and silvery-white beneath, the lower surface often dotted with numerous tiny brown scales, especially along the midvein; margins entire or undulate; petioles usually less than 6 mm long.

Flowers heavily sweet-scented, tubular, and 4-parted, with somewhat spreading calyx lobes, pale to golden yellow within and silvery on the outside, 12 – 15 mm long, short-stalked, pendent, solitary or few from the axils of the leaves on the current year's branches but mostly close to the older wood; June and July. Fruit broadly ellipsoidal or round-ovoid and drupelike, 8 – 10 mm long, dry and mealy with a large striated pit, the outer surface densely coated with silvery scales so that the fruits look like silver berries; July and Aug. (*commutata* — changeable; *argentea* — silvery)

In sandy, gravelly, or clayey soils of shores and river banks; also on talus slopes, in clearings, and along the edges of thickets.

In Ontario found only north of 47° N from Lake Timiskaming and the north shore of Lake Superior to James Bay and Hudson Bay and west to Rainy River District. (Lower St. Lawrence River in Que. to Alaska, south to Utah, S. Dak., and Minn.)

Note The silvery sheen is caused by light reflection from the dead cells of the mature scales covering most parts of the plant. After the fruits have been boiled and are well soaked, the outer coat can be removed and the softened pits pierced for use as beads.

Field check Thicket-forming shrub of northern Ontario with alternate, entire, silvery-gray leaves; strongly scented flowers silvery white outside and yellow inside; silvery "berries".

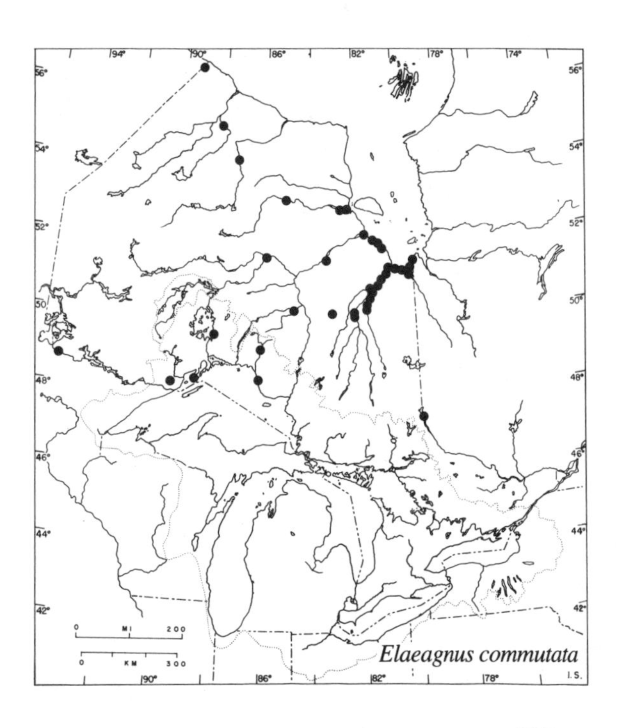

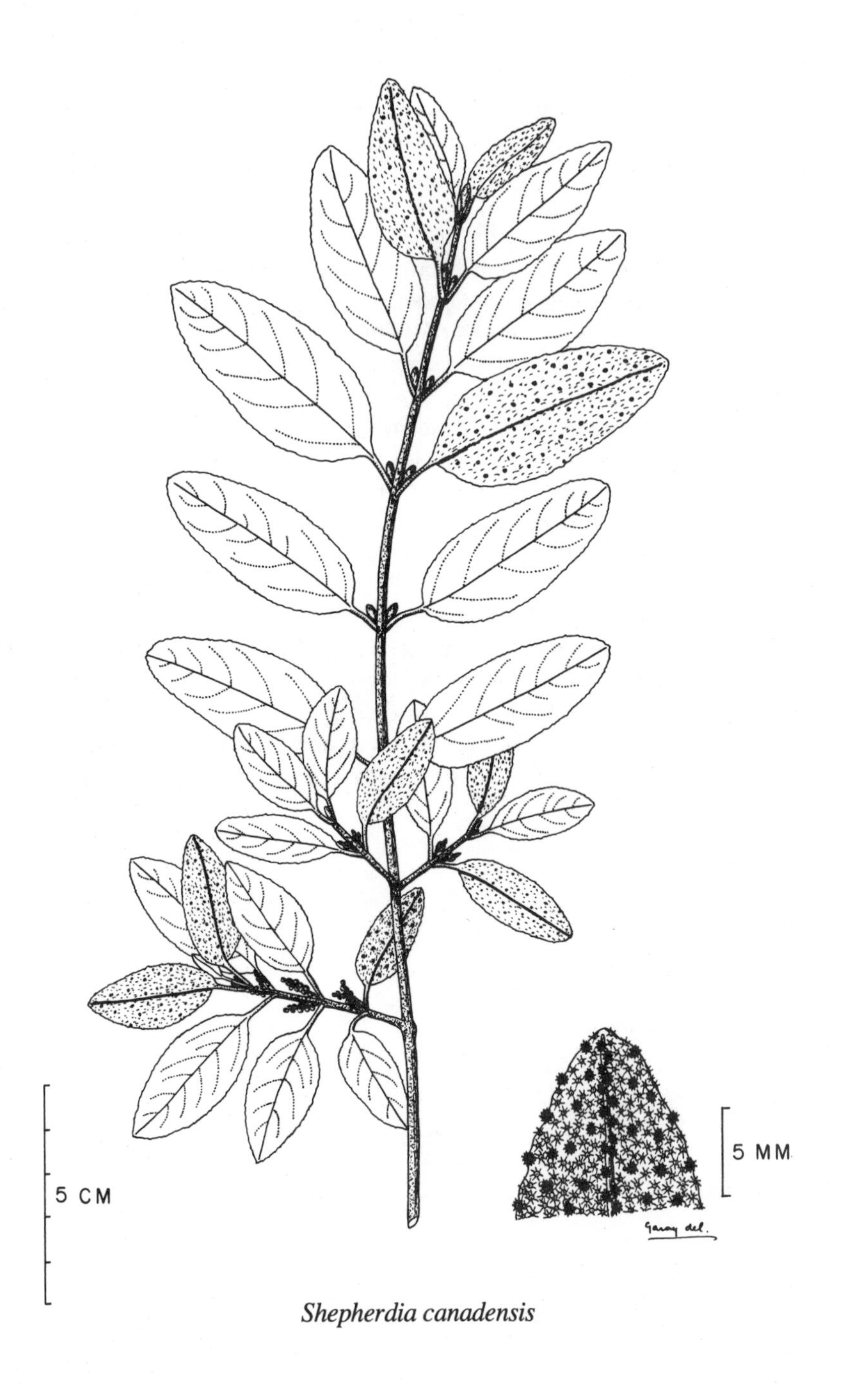

Shepherdia canadensis

Shepherdia Nutt. — Buffalo berry

Shrubs with opposite leaves and unisexual 4-parted flowers; floral tube with a ring of nectaries at the base of the calyx lobes; stamens 8; fruit berrylike. Three species in North America. (*Shepherdia*—named in honour of John Shepherd, 1764-1836, English botanist and horticulturist)

Shepherdia canadensis (L.) Nutt.

Buffalo berry
Soapberry

A low or medium-sized shrub, much-branched and spreading up to 2 m in height. Branchlets and buds bright rusty-brown, completely covered with a layer of fine white stellate hairs and numerous small brown overlapping circular scales; older bark dark brown to dark gray and minutely hairy.

Leaves opposite, simple, and deciduous; blades thickish, green or gray-green, and sparingly stellate-pubescent above, densely silvery stellate-pubescent beneath and dotted with numerous circular dark-centred brown scales; blades up to 5 cm in length and 3 cm in width, oval or elliptic with a rounded or obtuse apex and a rounded or tapered base; margins entire or minutely irregular due to the presence of marginal brown scales along the lower surface; veins impressed above and only the midvein conspicuous beneath; petiole about 1 cm long, grooved, and rusty-scurfy.

Flowers 3-5 mm across, greenish-yellow, unisexual, the male and female chiefly on separate plants, in dense clusters on short shoots at the nodes of the previous season's growth and opening before the leaves; late Apr. and early May. Fruit berrylike, consisting of an achene enclosed in the enlarged pulpy red or yellowish calyx tube, ovoid, slightly over 6 mm long; late June and July. (*canadensis*—Canadian)

In sandy, gravelly, or rocky soils; on shores, river banks, dry slopes, and in open rocky woods; occasionally in calcareous marshes.

Widespread throughout Ontario from Lake Erie to James Bay and Hudson Bay and from the St. Lawrence and Ottawa valleys to Rainy River; very common on limestone outcrops in the Bruce Peninsula and on Manitoulin Island. (Nfld. to Alaska, south to Oreg., n. Mex., Ohio, and N. Eng.)

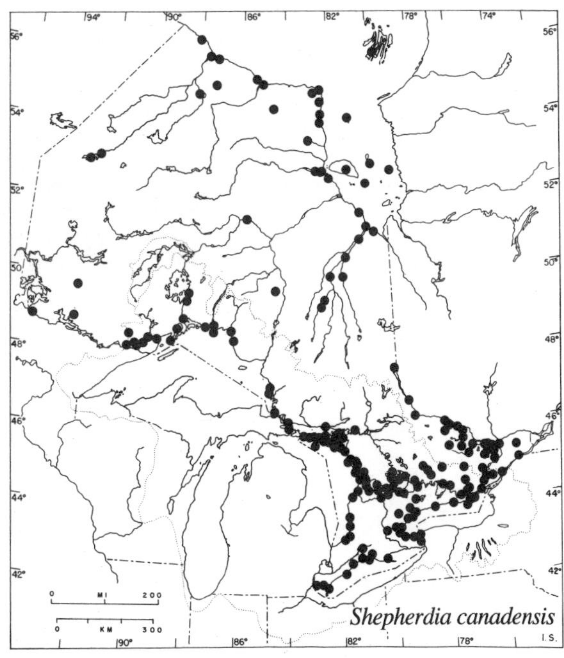

Shepherdia canadensis

Note The roots have bacteria-containing nodules that are capable of fixing free nitrogen from the atmosphere.

The fruit, mixed with sugar and water, can be whipped into a frothy, edible but somewhat bitter mixture. The bitter principle is from a substance called saponin, and the common name soapberry refers to the soaplike froth produced when the fruits are agitated in water.

Field check Low much-branched shrub with rusty-scurfy stems; opposite entire leaves with silvery brown-dotted lower surface; red or yellowish berrylike fruits.

Aralia hispida

ARALIACEAE – GINSENG FAMILY

Chiefly shrubs and trees, but also some woody vines and herbaceous perennials, with alternate simple or compound leaves, small flowers in umbels, heads, or compound inflorescences, and the fruit a drupe with one to several stones. Fifty to 60 genera, most abundant in the tropics of both hemispheres. Includes English ivy, *Hedera helix* L., a European woody vine climbing by aerial roots and often planted as a ground cover or against walls and buildings. Represented in Ontario by *Aralia* and *Oplopanax*. (*Hedera*—the Latin name for ivy; *helix*—twining or coiled)
The name of the family is based on the genus *Aralia*.

Aralia L.

A genus of about 30 species found in Indo-Malaysia, eastern Asia, Australia, and North America. (According to M.L. Fernald, the name *Aralia* was taken from the French-Canadian word *aralie* that accompanied specimens sent to Tournefort by the Quebec physician Michel Sarrasin de l'Etang, 1659–1734, and possibly this word originated with him. Linnaeus later adopted the name used by Tournefort. *Aralia*—derivation of name not obvious)

Aralia hispida Vent. — Bristly sarsaparilla

A low, leafy, and almost herbaceous perennial usually less than 1 m tall, with erect or ascending stems, woody and bristly from the base for about 1 dm. Young stems greenish to reddish and less abundantly spiny above; older stems brownish-gray, up to 1 cm in diameter.

Leaves alternate, pinnately compound, and deciduous, the largest ones clustered near the base of the stem; blades twice pinnate, with ovate-lanceolate sharply serrate leaflets 2–5 cm long or longer; petiole usually shorter than the blade and sometimes bristly.

Flowers creamy-white, 5-parted, small, and numerous, in umbels; June and July. Fruit dark purple-black, berrylike, 4–8 mm in diameter, with 1–5 bony stones; Aug. and Sept. (*hispida*—bristly)

In sandy, gravelly, and rocky soil, especially in clearings and along roadsides and railways.

Common from the St. Lawrence Valley west to Lake Superior and Lake-of-the-Woods, and northward to the upper part of the Albany River and Severn River drainage basins; rare south and west of the Canadian Shield; absent from the Boreal Forest and Barren Region. (Nfld. to Sask., south to Minn., and N.C.)

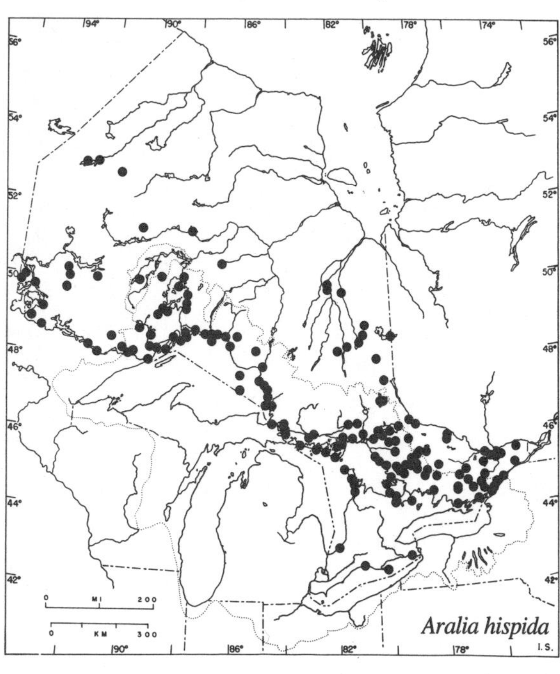

Field check Low, almost herbaceous perennial with woody stems bristly near the ground, twice-pinnate sharply serrate leaves, small creamy-white flowers, and black berries in stalked umbels.

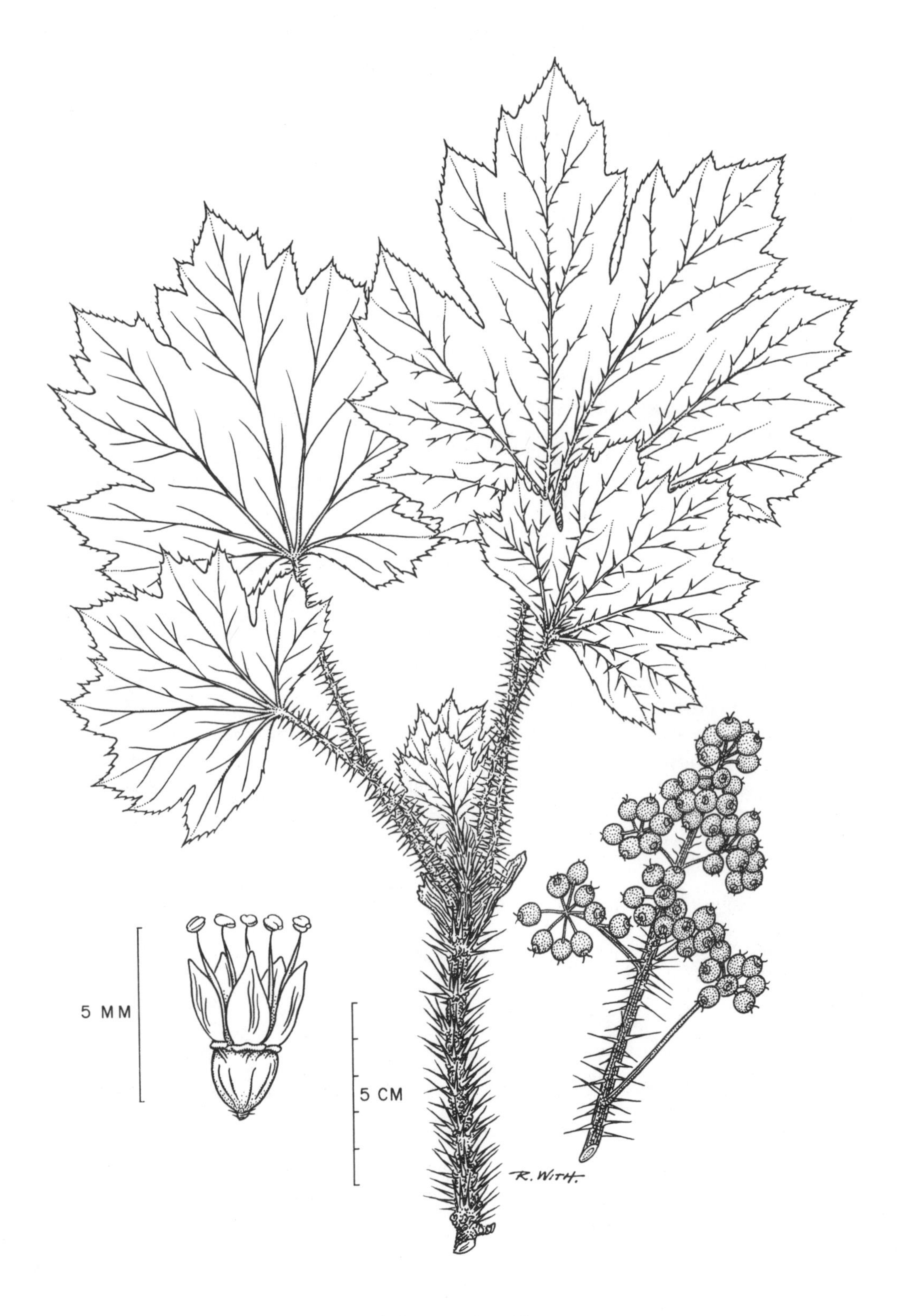

Oplopanax horridus

Oplopanax (T. & G.) Miq.

About 5 species of very prickly and sometimes arborescent shrubs native to eastern Asia and North America. (*Oplopanax*—from the Greek *hoplon*, a weapon, and *panax*, panacea, in reference to the all-healing powers ascribed to the roots of ginseng, a species in the genus *Panax*.

Oplopanax horridus (Sm.) Miq. (*Fatsia horrida* (Sm.) Benth. & Hook.; *Echinopanax horridus* (Sm.) Dcne. & Planch.) — Devil's-club

A sparsely branched straggling shrub less than 2 m tall, forming open colonies. Young stems dark brown at first, becoming paler with age, beset with formidable needle-sharp spines up to 1 cm long; older stems at the base of the plant rather ropelike and spongy with strong needle-sharp prickles and pale gray-brown papery bark.

Leaves very large, alternate, simple, and deciduous, the blade 2-4 dm across, nearly round, palmately lobed, cordate at the base, the lobes usually at least 5 in number, acuminate or cuspidate at the tip; dark green and smooth above, paler beneath with spines along the prominent veins; margins sharply toothed; petiole prominently spiny and often as long as the blade.

Flowers small, numerous, greenish-white, in stalked dense umbels along the axis of a cylindrical or narrowly pyramidal inflorescence 1-2 dm long; July. Fruit a bright red berrylike drupe, 4-6 mm long, slightly flattened and longer than wide; borne in showy pyramidal clusters; Aug. (*horridus*—shaggy; *Echinopanax*—from the Greek *echines*, hedgehog, hence prickly, and *panax*; *Fatsia*—a corruption of the Japanese name for a related plant)

In damp mixed woods and clearings of balsam woods.

The only location in Ontario is Porphyry Island, just east of the Sibley Peninsula in Thunder Bay District along the north shore of Lake Superior; also on Isle Royale and adjacent islands in Lake Superior which are part of Michigan, U.S.A. (s.w. Alaska and B.C., south to Calif., Oreg., Idaho, and Mont.; disjunct in the Lake Superior region)

Note The disjunct occurrence of this typically west coast shrub in Lake Superior has not been satisfactorily explained.

Devil's-club has the distinction of having caused a re-routing of the Canadian Pacific Railway through British Columbia since workers were practically unable to operate in some valleys where it abounds. It was the bane of early travellers and explorers in the West. An equally well-armed and formidable species occurs in the coniferous forests of eastern Asia.

Botanists have treated *Panax* and *Oplopanax* as masculine, feminine, and neuter in various manuals and floras. We are adopting the masculine ending following the arguments presented by Nicolson and Steyskal (1976).

Field check Straggling shrub with ropy stems and large toothed palmately lobed leaves beset with formidable needle-sharp spines; elongate clusters of flattish red "berries"; found on Porphyry Island, off the north shore of Lake Superior.

CORNACEAE – DOGWOOD FAMILY

Trees, shrubs, or rarely perennial herbs with mostly opposite, simple, entire or few-toothed petioled leaves without stipules and 4- or 5-parted regular flowers; fruit a drupe or berry with 1 or 2 stones. About 100 species in 7 or 12 genera (depending on whether *Cornus* is treated in a broad sense or split into 6 segregate genera), mainly in temperate regions but also in the mountains of the tropics. The only genus in North America is *Cornus s. lat.*, the genus from which the family derives its name.

Cornus L. – Dogwood

Deciduous shrubs, trees, and rarely herbs. Leaves with entire margins and with curved veins which run at an angle towards the margin but soon curve towards the tip of the leaf. Several species are used as ornamentals, including the handsome *C. florida* (flowering dogwood), which, however, is rarely successful in Ontario north of its native range. The fruits of the European *C. mas* L. (cornelian cherry) are edible and are used to make a preserve. At least two explanations have been given for the origin of the common name dogwood. Fernald (1950) suggested that it is a corruption from dagwood – from the old English dag, dagge, or dague – meaning a skewer or sharp-pointed instrument or weapon, hence dagger, since one of the European species (*C. sanguinea* L.) was used for the making of skewers by virture of its hard, smooth, fine-textured wood. Bailey (1935) stated that a decoction of the bark of *C. sanguinea* was used in England to wash mangy dogs. (*Cornus* — from the Latin *cornu*, a horn, perhaps in reference to the hardness of the wood)

Key to *Cornus*

a. Perennial herbs with slender woody rhizomes less than 5 mm in diameter; leaves in a whorl of 4 to 6 at the end of an erect stem less than 2 dm high; flower cluster solitary and terminal **_C. canadensis_**

a. Trees and shrubs with well-developed erect woody stems; leaves numerous, opposite or alternate; flower clusters several

- b. Stems with opposite branching; leaves all opposite and scattered along the stem or in one or two regular pairs near the ends of the branches
  - c. Flowers stalked, not surrounded by showy bracts; fruits globose
    - d. Pith of two-year-old branches white; stems greenish, reddish, or purplish
      - e. Stems greenish, reddish, or purplish; leaves lanceolate to ovate; fruits white or bluish-tinged **_C. stolonifera_**
      - e. Stems green, warty, usually dotted or streaked with purple; leaves broadly oval to nearly orbicular; fruits pale blue to greenish-white **_C. rugosa_**
    - d. Pith of two-year-old branches usually brown or at least a little darker than the wood, that of the current growth sometimes white; stems greenish or gray to brown
      - f. Flowers in an elongate round-topped cluster; fruits white, usually on bright red stalks **_C. racemosa_**
      - f. Flowers in a flat-topped or barely rounded cluster wider than long; fruits white or blue on green or purplish stalks
        - g. Leaves often drooping on arching petioles, smooth above; fruits blue or bluish-white on green stalks **_C. obliqua_**
        - g. Leaves not drooping, rough on the upper surface; fruits white on purplish stalks **_C. drummondii_**
  - c. Flowers sessile, surrounded by 4 large, white, petal-like bracts; fruits ellipsoid, shiny red, in stalked dense clusters **_C. florida_**
- b. Stems with alternate branching; leaves alternate or a few nearly opposite, often crowded (appearing whorled) near the ends of the branches **_C. alternifolia_**

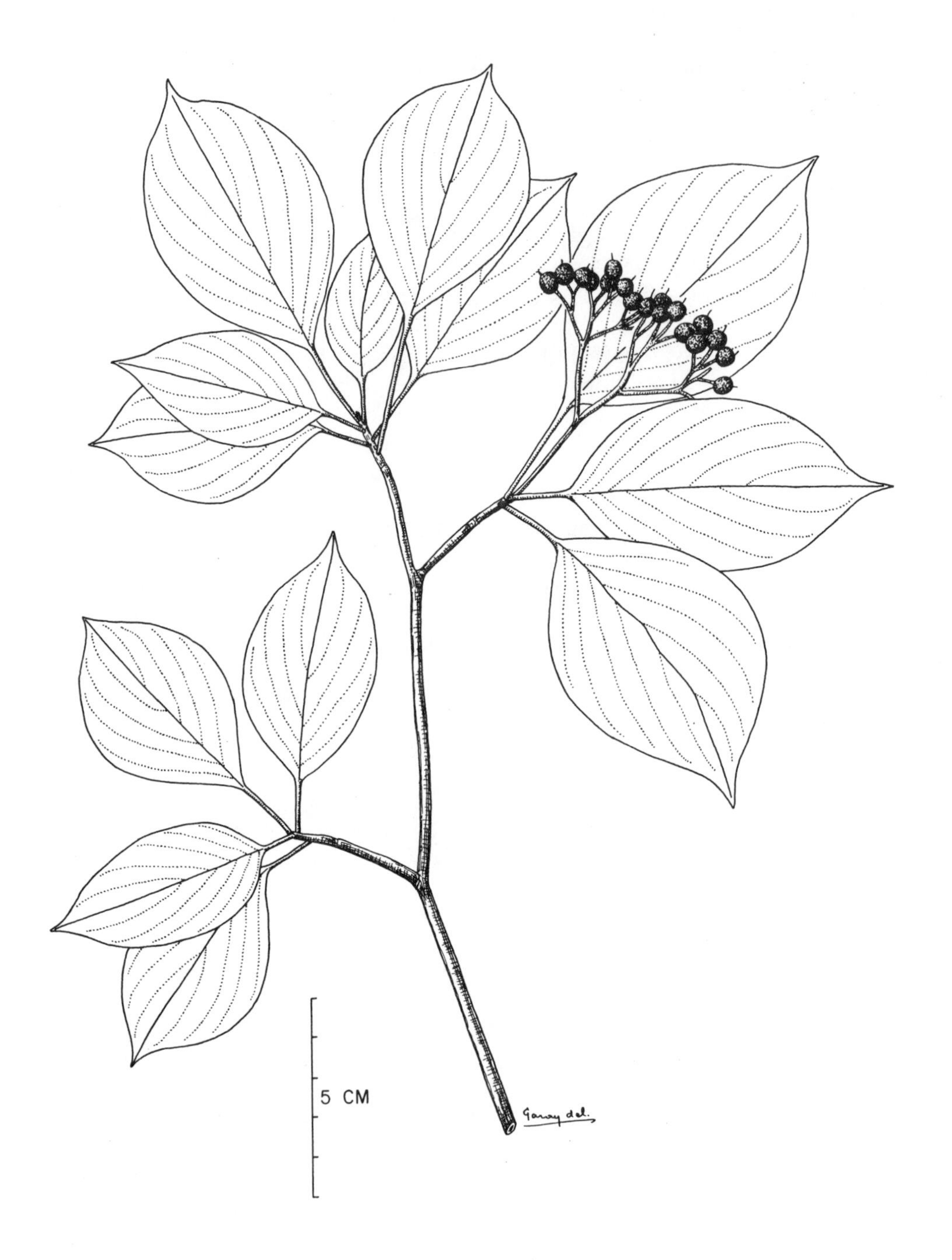

Cornus alternifolia

Cornus alternifolia L. fil.

Alternate-leaved dogwood
Green osier, Pagoda tree

A large shrub or small tree growing to a height of 4–6 m; when a tree, it is similar in form to sassafras with horizontal tiers of side branches which are wide-spreading with upcurved tips. Branchlets greenish-red to purple or brownish and glossy, with slender white pith, the alternate branches distinguishing this species from our other dogwoods.

Leaves simple, deciduous, and alternate, frequently so crowded at, and near, the ends of the branches as to appear opposite or whorled; blades rather thin, ovate or oval, 4–13 cm long and 2–7.5 cm wide, with a pointed tip and a rounded or narrowed base, dark green above, grayish and finely hairy beneath; margins entire; petiole slender and very short or up to 6 cm in length.

Flowers small, creamy-white, and numerous in large flat-topped clusters, opening about the middle of June. Fruit a stalked cluster of round, dark blue, berrylike drupes with a bloom, each drupe about 6 mm in diameter and on a red stalk; July and Aug. (*alternifolia*—with alternate leaves)

In thickets or open woods, on hillsides and ravine slopes.

Common in southern Ontario and becoming less frequent northward along the eastern shore of Lake Superior and west to Rainy River; northern limit near 49° N. (Nfld. to s. Man. and Minn., south to Mo. and Ga.)

Field check Large shrub or small tree with alternate branching and alternate leaves, these often crowded near the ends of the branches; leaves with entire margins and lateral veins curving towards the tip; dark blue "berries" on red stalks.

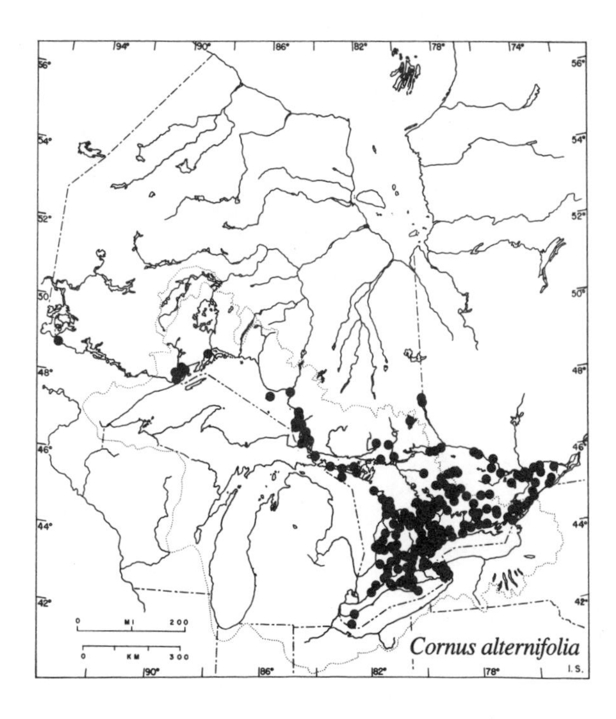

Cornus canadensis

Cornus canadensis L. **Bunchberry**

A low perennial with prostrate, slender, woody rhizomes bearing erect leafy stems usually less than 2 dm in height, sometimes forming patches or extensive colonies. Stems erect, slender, green, and herbaceous, somewhat ridged or striate, with several conspicuous nodes, each with a pair of papery scales or reduced foliage leaves.

Leaves simple, deciduous, and sessile or short-stalked, in a whorl-like cluster of 4–6 at the end of an erect stem, sometimes with an additional whorl lower down on the flowering stem; blades elliptic to broadly oval or obovate, 3–7 cm long, pointed at both ends, smooth, dark green above and paler beneath; margins entire; veins conspicuous and curved towards the tip of the leaf.

Flowers in a cluster above the terminal whorl of leaves at the end of a peduncle 1–4 cm long; individual flowers very small, 4-parted, yellowish-green to cream-coloured, the cluster surrounded by 4 white to pinkish-tinged petal-like bracts (modified leaves), the bracts broadly rounded, 1–2 cm long, abruptly tapered to a pointed tip, soon deciduous; May and June. Fruit a bright red berrylike drupe 5–8 mm in diameter, in clusters; July to Sept. (*canadensis*—Canadian)

On the forest floor in deciduous, mixed, and coniferous woods, frequently on hummocks, fallen logs, and stumps.

Throughout Ontario from Lake Erie northward to James Bay and from the St. Lawrence and Ottawa valleys to the Manitoba boundary. (s. Greenl. and Nfld. to Alaska, south to Calif. and N.J.; e. Asia)

Note A closely related and similar species, *C. suecica* L., occurs in Eurasia, Greenland, the eastern Canadian Arctic, Alaska, Yukon, northern British Columbia, and the Northwest Territories but has not been reported for Ontario. It differs chiefly in having small foliage leaves along the stem rather than small bracts. The two hybridize where their ranges overlap. (*suecica*—Swedish)

Field check Low herbaceous perennial with terminal whorls of rounded, entire-margined leaves; flowers in clusters surrounded by 4 conspicuous white petal-like bracts; clusters of red "berries".

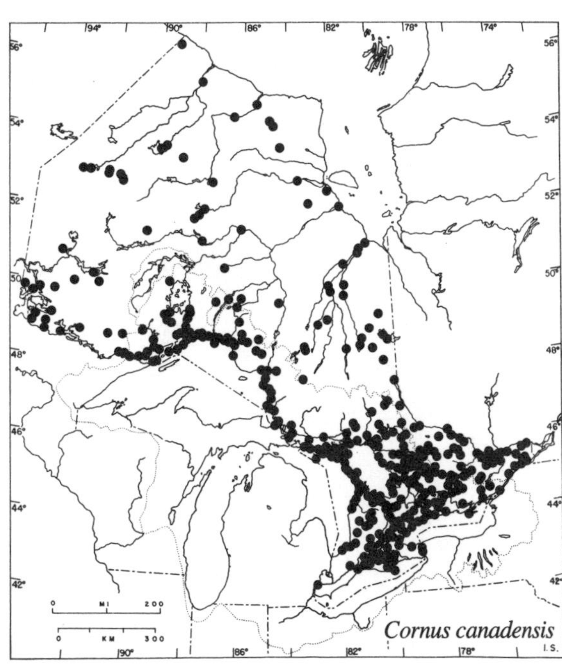

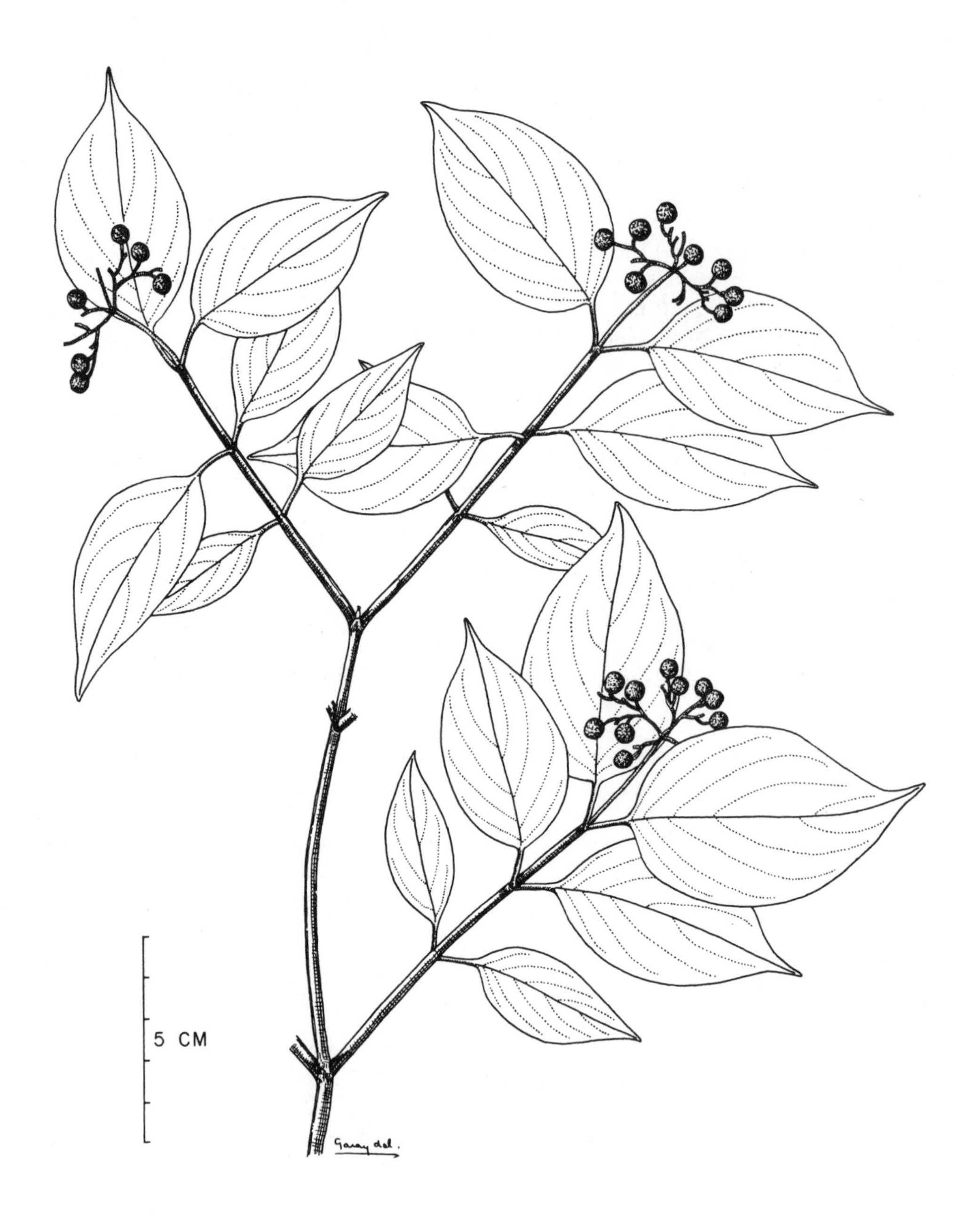

Cornus drummondii

Cornus drummondii C.A. Mey. (*C. asperifolia* in some manuals, but not Michx.) — Rough-leaved dogwood

An erect much-branched shrub up to 3-4 m in height, sometimes forming thickets. Branchlets reddish and minutely downy at first, becoming brownish to gray and smooth; pith slender, brown or rarely white.

Leaves simple, opposite and deciduous, the blades 5-9 cm long and 2.5-5 cm wide, lanceolate or narrowly oval to broadly ovate, long-tapering and sometimes slightly folded at the apex and rounded, tapered, or rarely cordate at the base, dark green and definitely roughened above, paler and finely woolly beneath; margins entire; petiole downy, 5-20 mm long.

Flowers small, creamy-white, and numerous, in loose flat clusters; June. Fruit a white berrylike drupe about 6 mm across, on a purplish-red stalk, in stalked clusters; Aug. (*drummondii*—named after T. Drummond, 1780-1853, the discoverer of this species)

In sandy or clayey soils, along the margins of woods, or near shores of lakes and streams.

Confined to the counties of Essex and Kent and the Regional Municipality of Haldimand-Norfolk. (s. Ont. to Neb., south to Tex. and Miss.)

Field check Shrub with brown pith and opposite taper-pointed leaves, rough above, paler and woolly beneath; leaf margins entire and lateral veins curved towards the tip of the leaf; white "berries" on purplish-red stalks.

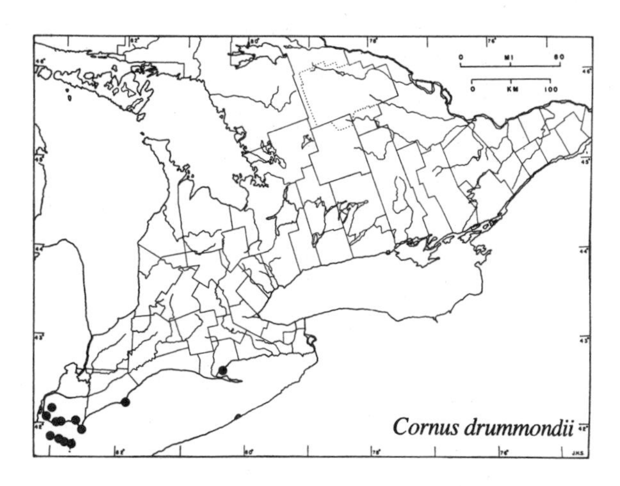

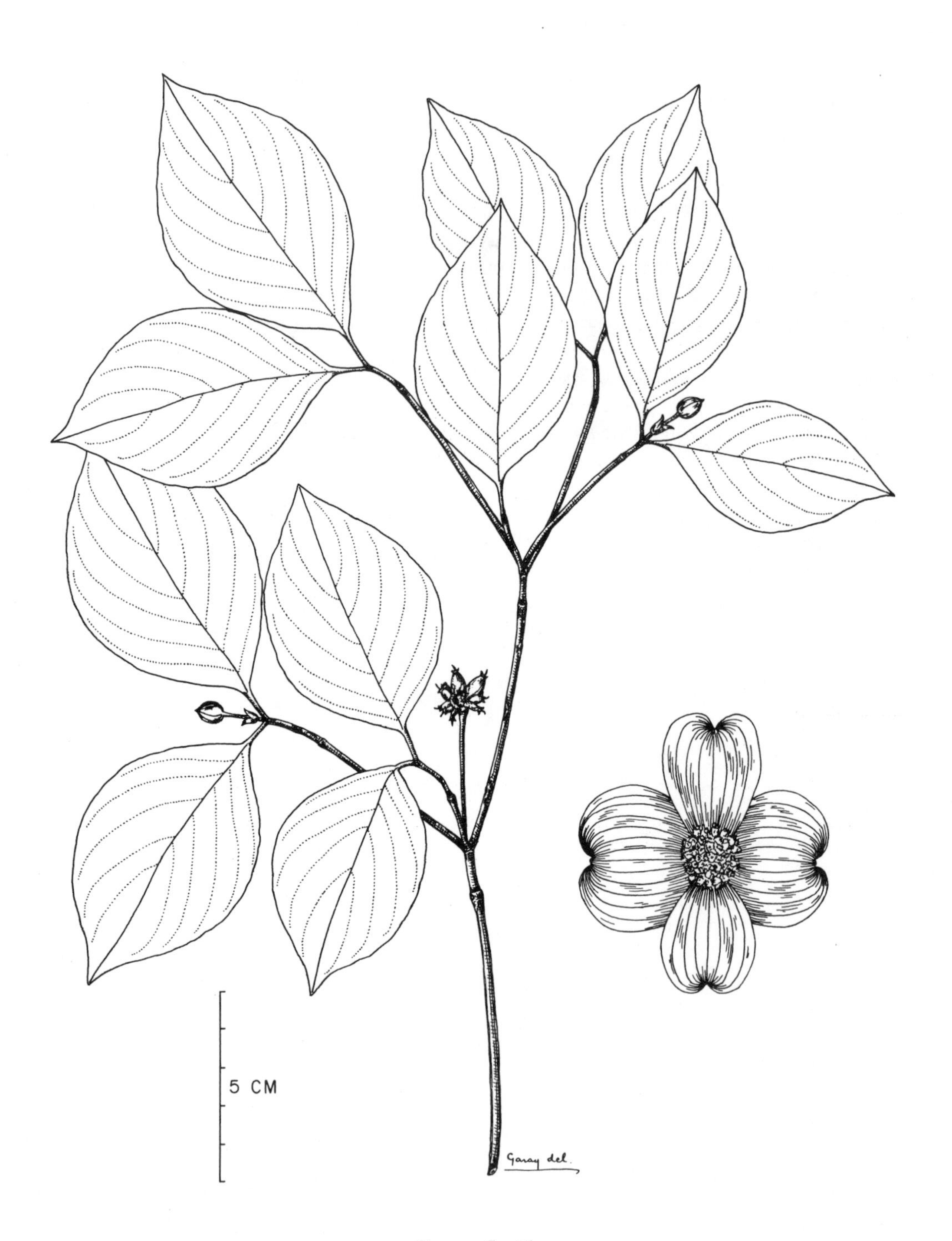

Cornus florida

Cornus florida L. — Flowering dogwood

An erect shrub up to 3 or 4 m in height or a small tree up to 9 m or more with a single trunk and branches more or less in horizontal whorls or tiers. Branchlets greenish, becoming reddish with a scattering of fine white hairs; bark on older stems and trunks dark reddish-brown to blackish, broken up into small angular or rounded scales; pith white.

Leaves simple, opposite, and deciduous, mostly in one or two pairs at the ends of the individual branches; blades ovate or elliptic to broadly oval, 5-15 cm long and 2.5-8 cm wide, rather thin, light to dark green above and pale grayish-green below, the tip pointed and the base rounded or tapered; margins entire; petiole slender and usually less than 2.5 cm long.

Flower buds stalked; flowers small, greenish-yellow or whitish, in dense clusters surrounded by 4 large and showy petal-like, white or pinkish, deciduous bracts (the expanded flower bud scales), the clusters thus simulating conspicuous and attractive single flowers; bracts broadly obovate to obcordate, 2-5 cm long, usually a little twisted, and with a red or purplish notch at the free end; opening as the leaves are unfolding; late May. Fruit an ovoid, shiny red, berrylike drupe about 10-12 mm long, beaked with tiny persistent calyx lobes, in prominently stalked tight clusters, each cluster usually containing several undeveloped fruits mixed with the mature ones; Aug. and Sept. (*florida*—flowering)

In acid soils, usually on the edge of sandy or wet woodland, in open woods, and on ravine slopes.

Confined in southern Ontario to the Deciduous Forest Region. (s.w. Maine and s. Ont. to Kans., south to Mex. and Fla.)

Note Flowering dogwood is one of the finest ornamentals among our native shrubs and trees but, unfortunately, the flower buds are killed in winter in areas outside its natural range. It is singularly attractive not only when flowering in the spring but also in the fall when the bright red clusters of fruits merge with the bronzy greens, reddish browns, and brilliant scarlet of its autumnal foliage.

Field check Shrub or small tree with opposite leaves, often with only one or two pairs at the ends of the branches; leaf margins entire and lateral veins curving towards the tip of the leaf; attractive 4-bracted flower clusters and stalked dense heads of shiny red ovoid berrylike fruits.

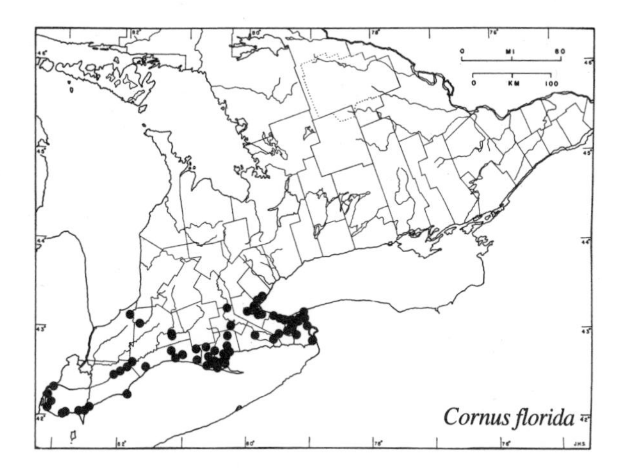

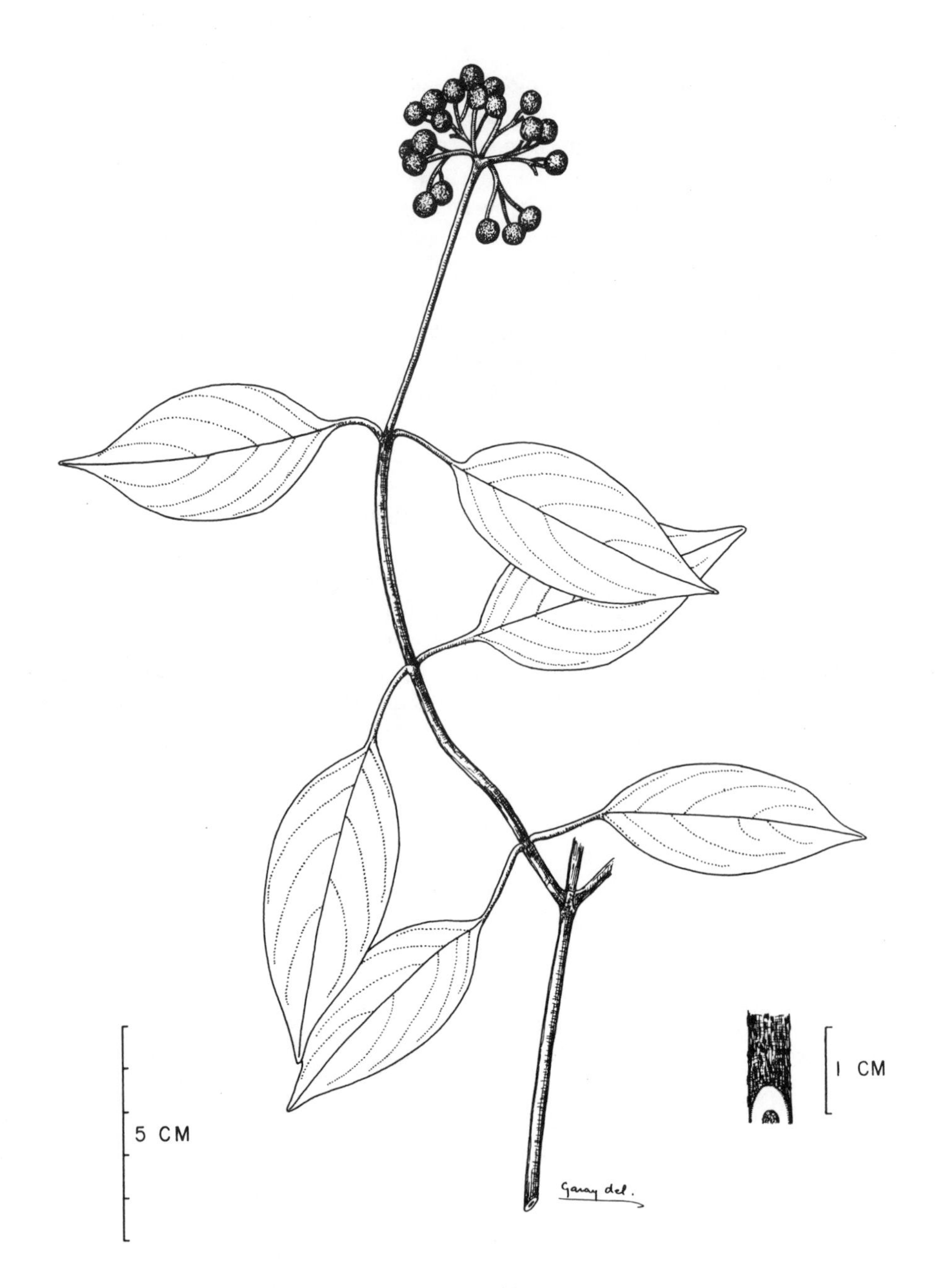

Cornus obliqua

Cornus obliqua Raf. (*C. amomum* ssp. *obliqua* (Raf.) Wilson) — Silky dogwood

An upright or spreading shrub up to 3 m in height with rather open branching. Branchlets grayish and finely hairy but later becoming smooth and reddish to purplish-brown; pith slender and brown.

Leaves simple, opposite, and deciduous; the blades 5–10 cm long and 1–5 cm wide, lanceolate-elliptic to oval, usually less than half as broad as long, tapered at the apex and gradually tapered or barely rounded at the base, dark green above, pale grayish-green, finely appressed-hairy, and microscopically papillose beneath; margins entire; petiole about 2 cm or less in length, often arching and causing the leaf to droop.

Flowers small and creamy-white in nearly flat-topped pubescent clusters; early July, the last of our dogwoods to bloom. Fruit a round, blue or bluish-white, berrylike drupe, 6–10 mm in diameter, in long-stalked clusters; late Aug. and Sept. (*obliqua*—sloping, the reference obscure; *amomum*-Latin name of a shrub)

In low damp ground along streams and in marshes, ditches, thickets, and open woods.

Widely distributed east and southwest of the Canadian Shield in southern Ontario; northward along the Bruce Peninsula to Manitoulin Island; also in the Ottawa Valley and the southern part of Nipissing District. (w. N.B. to N.Dak., south to Okla. and Ga.)

Field check Slender brown pith; narrow, long-pointed, often drooping opposite leaves on arching petioles; leaf margins entire and lateral veins curving towards the tip of the leaf; "berries" bluish-white.

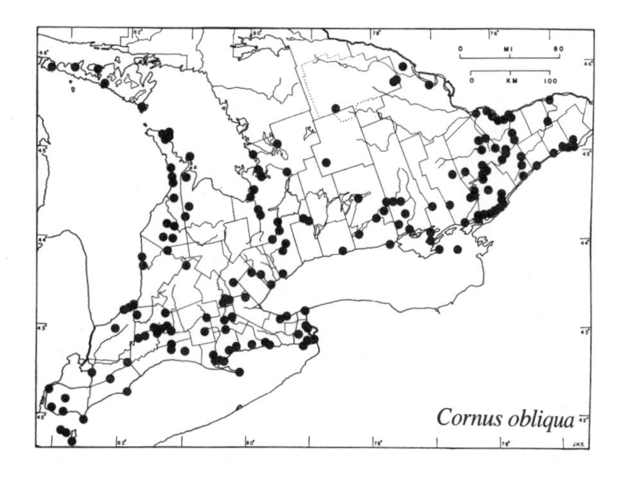

Cornus racemosa

An erect, much-branched shrub about 2.5 m tall, commonly forming compact thickets. Branchlets slender, smooth, and gray to light brown; pith slender, pale brown or white in current year's growth.

Leaves numerous, simple, opposite, and deciduous; the blades 5 - 10 cm long and 2 - 4 cm wide, dark green above, gray-green and minutely hairy below, elliptic or narrowly oval with a long tapered apex, the base tapered to a petiole 6 - 12 mm long.

Flowers small, creamy-white, and ill-scented, in loose elongate clusters nearly as high as broad; late June. Fruit a round, white, berrylike drupe less than 6 mm in diameter, usually on a bright red stalk and in red-stalked clusters; Aug. (*racemosa* — in racemes)

In thickets, moist soil along river banks, roadsides, and fencerows, on sandy slopes and limestone ridges.

Common in southern Ontario east and southwest of the Canadian Shield; in the Ottawa Valley northward to about 46° N; also along the Ontario-Minnesota boundary west of Lake Superior. (Maine to s.e. Man., south to Okla. and Md.)

Field check Densely leafy shrub with numerous gray branchlets; leaves opposite, with margins entire and lateral veins curving towards the tip of the leaf; elongate round-topped inflorescence; white "berries" on red stalks.

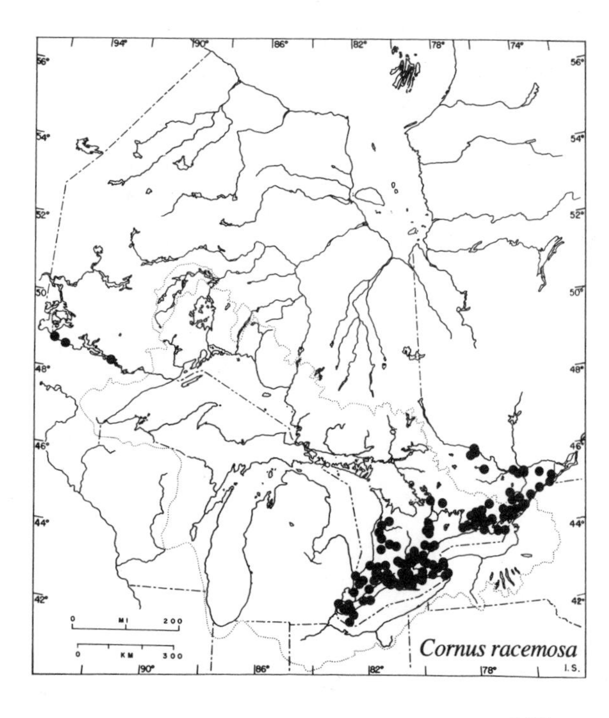

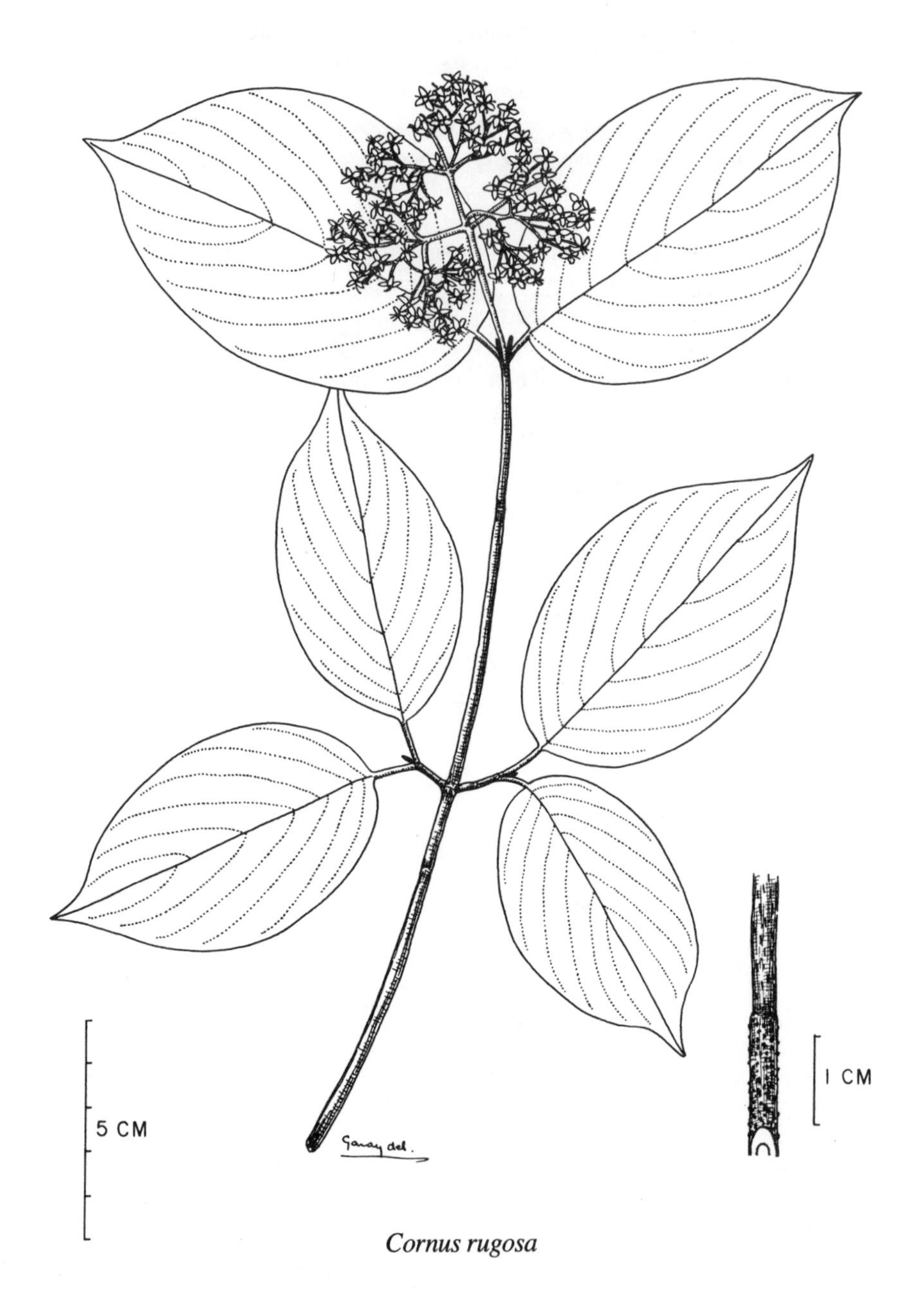

Cornus rugosa

Cornus rugosa Lam. — Round-leaved dogwood

A coarse erect shrub, sometimes treelike with one or a few dominant stems, growing to a height of about 3 m. Branchlets warty, pinkish to yellowish-green, dotted or streaked with purple or reddish-brown; older stems purplish, the pith large and white.

Leaves simple, opposite, and deciduous; blades 7–15 cm long and 5–12 cm wide, broadly oval to nearly round, abruptly pointed at the tip and broadly rounded at the base, pale to dark green and rough above, grayish below with a dense feltlike covering of short woolly hairs; margins entire; petiole 12–18 mm long.

Flowers small, white to creamy-white, in dense flat-topped clusters; late June. Fruit a round, pale blue to greenish-white, berrylike drupe, about 6 mm in diameter; Aug. (*rugosa*—wrinkled or rough)

In sandy, gravelly, or rocky soil, on limestone talus and ledges, in open woods, thickets, and on slopes of ravines.

Widely distributed in Ontario from Lake Erie northwestward to about 50° N at Lake-of-the-Woods and northeastward to about 47° 30′ N at Lake Timiskaming; common along the Niagara Escarpment and on Manitoulin Island. (N.S. to s.e. Man., south to Iowa and Va.)

Field check Stems green, dotted or streaked with purple; leaves opposite, broadly rounded, rough above and gray-hairy below; leaf margins entire and lateral veins curving towards the tip of the leaf; "berries" pale blue to greenish-white.

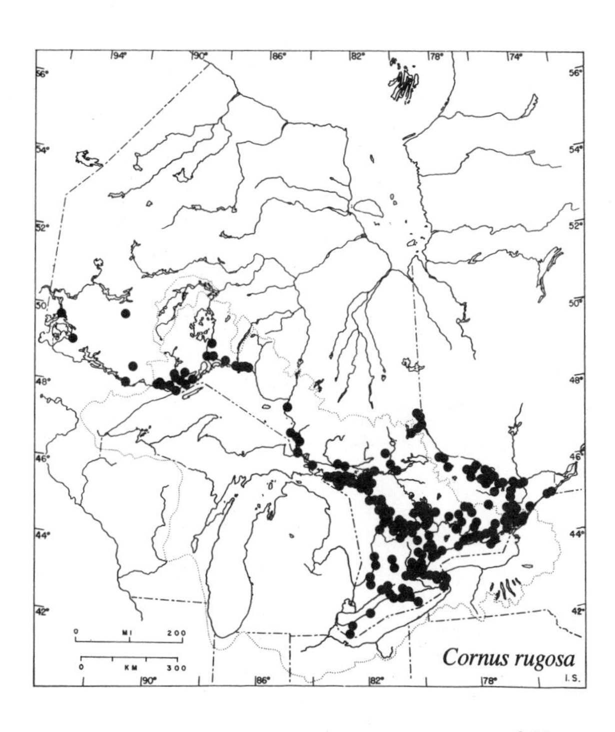

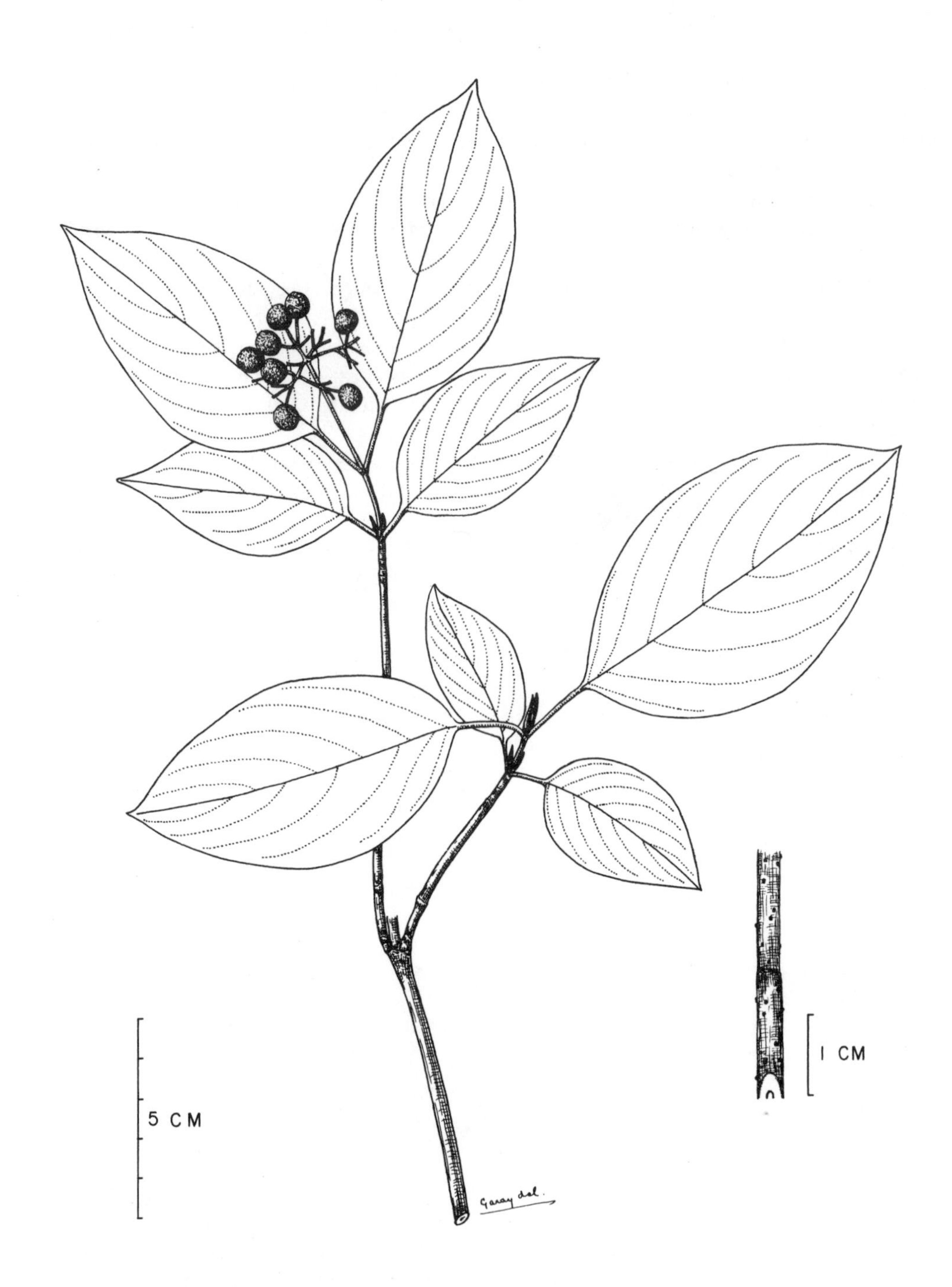

Cornus stolonifera

Cornus stolonifera Michx. (*C. alba* L. ssp. *stolonifera* (Michx.) Wang.) — Red osier

An erect, ascending or loosely spreading shrub with the branches sometimes prostrate and freely rooting, commonly forming thickets and reaching a height of about 2 or 3 m. Branchlets green and finely hairy, soon becoming smooth and purplish to bright red, the pith large and white.

Leaves simple, opposite, and deciduous; blades 5 - 15 cm long and 2.5 - 9 cm wide, dark green above, paler to whitened and finely to densely soft-hairy beneath, lanceolate-ovate to broadly oval, tapered or abruptly short-pointed at the tip and rounded or narrowed at the base; margins entire; petiole 0.6 - 2.5 cm long.

Flowers small, dull white, in flat-topped or slightly rounded clusters; late June. Fruit a round, berrylike drupe 6 mm or more in diameter, usually white or bluish-tinged, rarely lead-coloured; Aug. and Sept. (*alba*—white; *stolonifera*—bearing stolons)

In low damp ground, along shores, river flats, edges of marshes, in damp open woods and thickets, and along roadsides.

Common throughout southern Ontario, westward to the Manitoba boundary and northward to James Bay and Hudson Bay. (Nfld. to Alaska, south to Calif. and W. Va.)

Note The name *Cornus stolonifera* is still widely used for the red osier dogwood of eastern North America, but some botanists follow Wangerin's treatment (1910) in classifying the plants of our area as a subspecies of the Eurasian *C. alba* L., the latter then considered circumboreal.

Throughout the winter the bright red stems of red osier are a pleasing sight along roadsides and in open low ground.

Field check Stoloniferous shrub with reddish stems and large white pith; leaves opposite, the margins entire, and the lateral veins curving towards the tip of the leaf; white or bluish-tinged "berries" in clusters.

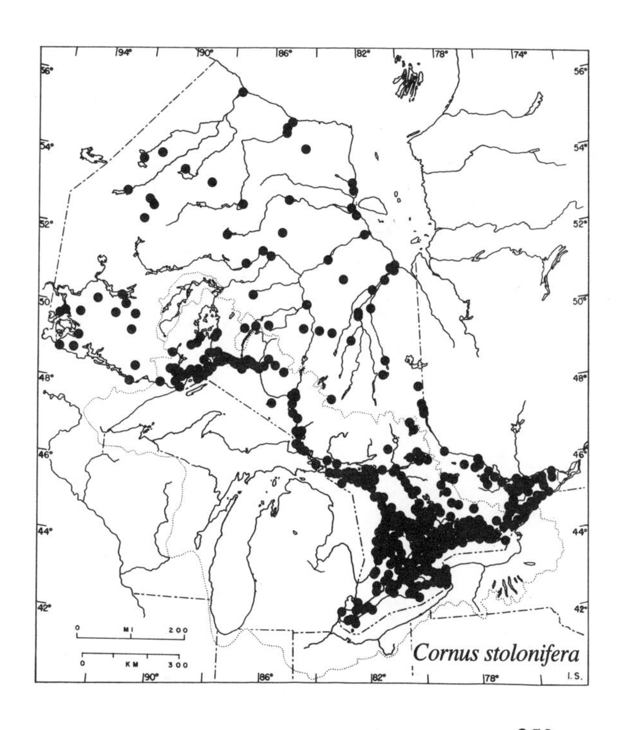

Cornus stolonifera

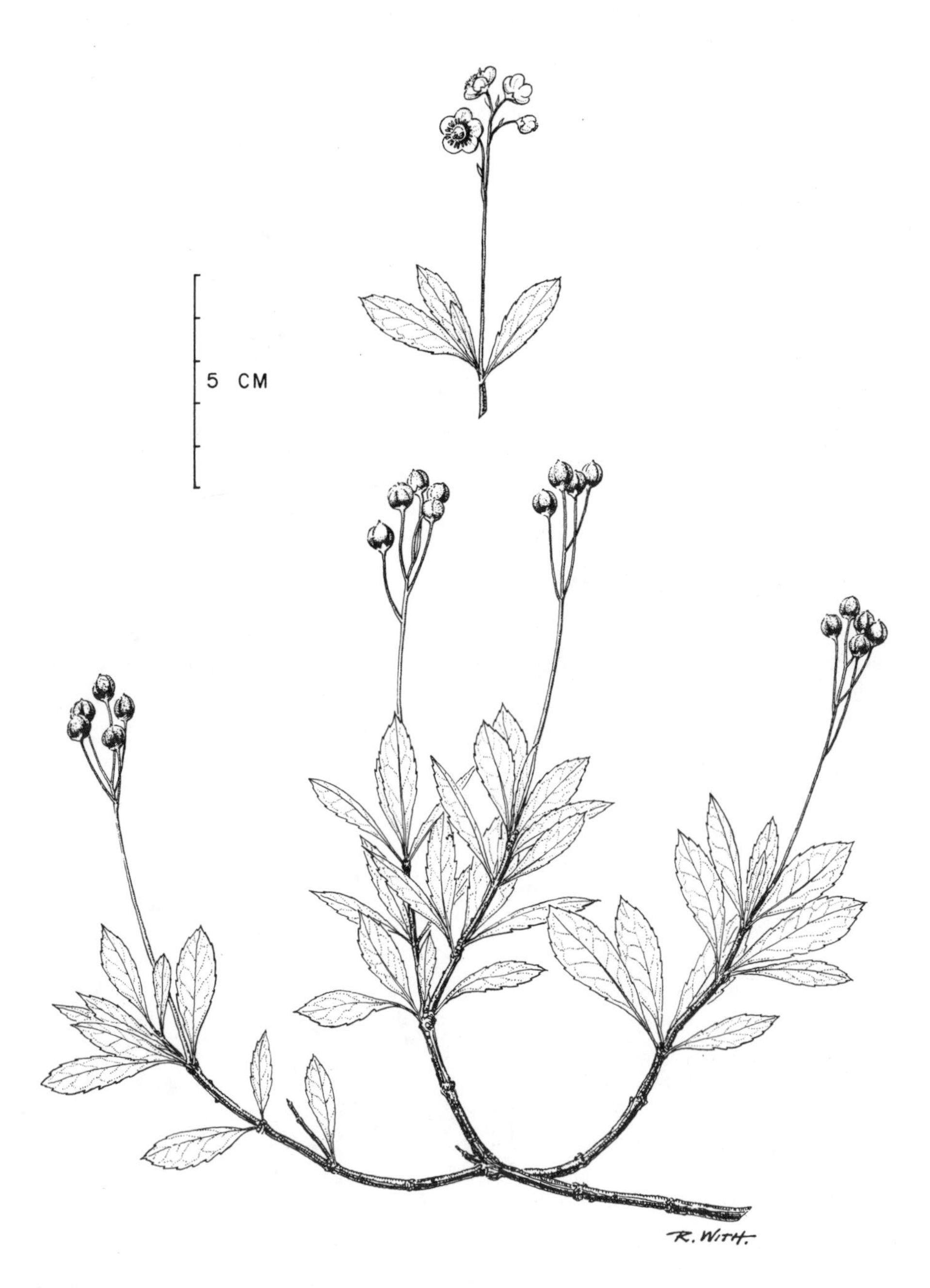

Chimaphila umbellata

PYROLACEAE — WINTERGREEN FAMILY

A small family of 3 genera and about 30 species, sometimes included in the *Ericaceae* as a subfamily. Mostly evergreen, slightly woody or herbaceous perennials of north temperate and arctic regions. Plants with well-developed rhizomes and alternate or whorled simple leaves lacking stipules. Flowers solitary or several in a terminal cyme or raceme, 4- or 5-parted, often with a nectariferous disc surrounding the pistil; anthers opening by apical pores. Fruit a capsule with innumerable small seeds.

Of the three genera *Chimaphila*, *Pyrola*, and *Moneses*, only the first is included here as warranting status as a half-shrubby perennial. The name of the family is based on the genus *Pyrola*.

Chimaphila Pursh — Pipsissewa, Wintergreen

About 8 species in northern Eurasia and North America. Low plants with creeping rhizomes and erect stems bearing thick, shiny, toothed leaves scattered or whorled. Flowers rosy-pink to white, terminating a long peduncle. (*Chimaphila*—from the Greek *cheima*, winter, and *philein*, to love, in allusion to the common name wintergreen)

Key to *Chimaphila*

a. Leaves elliptic-lanceolate (broadest at or below the middle), dark green above and variegated by a broad irregular white median stripe; veins not prominent; restricted to the shore of Lake Erie (Regional Municipality of Haldimand-Norfolk) ***C. maculata***

a. Leaves elliptic to oblanceolate (broadest at or beyond the middle), green on both sides, without white median stripe; veins impressed above and prominent beneath; widely distributed from Lake Erie to northwestern Ontario ***C. umbellata***

Chimaphila umbellata (L.) Bart. — Prince's-pine, Pipsissewa

A low, slightly woody perennial usually not more than 2.5 dm high. Stems slender, creeping at or just below ground level, freely rooting, giving rise to upright leafy and flowering branches singly or in groups; branchlets greenish or brownish, glabrous, and with fine longitudinal lines or ridges.

Leaves simple and evergreen, in terminal whorls and also scattered along the stems; blades 3–7 cm long and 1–2 cm wide, leathery, lustrous and dark green above, slightly paler beneath, veins impressed above and prominent below, elliptic to oblanceolate, the tip blunt or pointed, usually mucronate, the base tapered to a short grooved petiole; margins slightly inrolled and prominently toothed, especially near the tip where the teeth are often more crowded.

Flowers white to rose-pink, saucer-shaped, 10–15 mm wide, with 5 rounded, somewhat fleshy petals, 10 stamens and a broad central stigmatic disc; in umbel-like clusters of 3–10 on recurved or erect pedicels at the summit of stout peduncles that surpass the uppermost whorls of leaves; July and Aug. Fruit a depressed globose capsule 4–8 mm in diameter, usually erect when mature; Aug. and Sept. (*umbellata*—in umbels)

In dry, sandy, or rocky coniferous and mixed woods and clearings, usually in well-drained situations; on gravel terraces, limestone pavement, and jack pine barrens.

Widely distributed from Lake Erie to the upper part of the Severn River drainage basin and from the St. Lawrence and Ottawa valleys to Lake-of-the-Woods; northern limit at about 53° N. (Nfld. to B.C., south to Calif. and Ga.)

Note *C. umbellata* comprises a circumboreal complex in which several geographical varieties have been recognized: typical var. *umbellata* in Eurasia; var. *cisatlantica* Blake from Nfld. and Que. to e. Man., south to Minn. and Ga.; var. *occidentalis* (Rydb.) Blake from B.C. and s.e. Alaska to Calif. and Colo.; also n. Mich. (*cisatlantica*—on this side of the Atlantic, i.e., North American; *occidentalis*—western, i.e., in the western hemisphere)

A related species called spotted wintergreen (*C. maculata* (L.) Pursh) has been found near Turkey Point on Lake Erie (Regional Municipality of Haldimand-Norfolk). It differs in its more lanceolate leaves which have a conspicuous white median stripe. Records of a collection made at Hamilton (Regional Municipality of Hamilton-Wentworth) by W.Nicol in 1886 and another labelled "Baysville" (District-Municipality of Muskoka) made prior to 1904 have been excluded as there has been no confirmation of the existence of this rare plant in those two areas. (*maculata*—spotted)

Field check Low perennial with slender creeping rhizomes and thick, glossy, denticulate leaves in whorls; 5-parted pinkish flowers in umbel-like clusters on erect terminal peduncles.

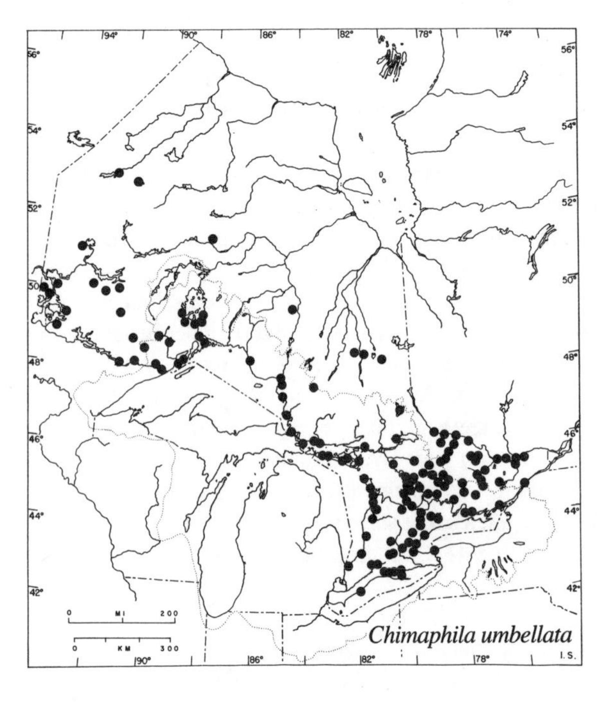

Chimaphila umbellata

ERICACEAE — HEATH FAMILY

A family of about 50 genera and over 1300 species widely distributed throughout the world but in the tropics confined to high elevations and in Australasia almost absent. Woody plants, including slender trailing or low undershrubs, large shrubs, and a few trees. Often conspicuous elements in acid peaty soils of moors and bogs.

Leaves without stipules, frequently evergreen, and mostly alternate, rarely opposite or whorled. Flowers regular or nearly so, usually perfect, the stamens as many, or twice as many, as the petals or corolla lobes and the anthers usually opening by terminal pores. Fruit a capsule, a berry, or drupelike. The family name is taken from the important genus *Erica* which has more than 500 species. (*Erica*—Greek *ereike*, heather)

Note The peak flowering periods of bog-inhabiting members of the heath family tend to be successive rather than synchronous, i.e., leatherleaf, followed by bog rosemary and bog laurel (exceptions in having considerable overlap), then velvet-leaf blueberry, Labrador tea, bog cranberry, and wintergreen. Two Ontario peat bogs near Guelph are currently sites for investigating various hypotheses that have been proposed (Reader, 1975, 1977) to explain this specialization, such as competition for pollinators and avoidance of the inevitable waste arising from the activities of non-specialized pollinating bees and, to a lesser extent, flies.

Andromeda L. — Andromeda

Low slender, evergreen shrubs with whitened or pale bluish-green foliage and small umbels of pink or white flowers. Two species, circumpolar in arctic and subarctic regions and extending southward mainly in bogs. (*Andromeda*-a mythical Ethiopian princess)

Key to *Andromeda*

a. Stems 3-6 dm high; leaves 2-5.5 cm long, whitened beneath with a close layer of short hairs; flowers 5-6 mm long in rather dense nodding clusters; pedicels less than 1 cm long ***A. glaucophylla***

a. Stems less than 2 dm high; leaves 1-3 cm long, whitened and glabrous beneath; flowers 6-7 mm long in erect open clusters; pedicels 1-3 cm long ***A. polifolia***

Andromeda glaucophylla

A low erect or trailing shrub with ascending branches, growing to a height of 3–6 dm. Branchlets round in section, brownish and glabrous; older stems gray to blackish.

Leaves alternate, simple, and evergreen, the young leaves often bluish-green; blades narrowly elliptic to linear, 2–5.5 cm long and 3–10 mm wide, firm and leathery, dark green above and whitened beneath with a close layer of short white hairs; margins entire and conspicuously inrolled, the tip sharp-pointed with a tiny spine (mucro), the base tapered to a very short stalk or the blade sessile.

Flowers in small drooping terminal clusters, white to pinkish, urn-shaped, about 5–6 mm long; May and June. Fruit a small roundish capsule less than 6 mm in diameter with a persistent style at the depressed and indented top, at first bluish but becoming brownish in age; fruits drooping when young but usually becoming erect as they mature; seeds light brown and numerous; late July and Aug. (*glaucophylla*—with bluish-green leaves)

In sphagnum bogs and spruce-tamarack swamps.

From the Ottawa and St. Lawrence valleys northwest to Lake-of-the-Woods and from the Niagara Peninsula north to James Bay and the Severn River drainage basin. (s.w. Greenl. and Nfld. to Man., south to Minn. and N.J.)

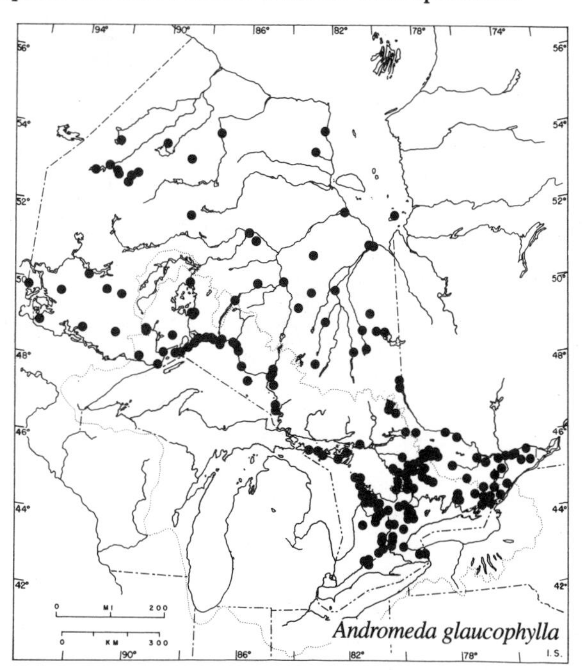

Note In the far north, a closely related plant called dwarf bog-rosemary (*Andromeda polifolia* L.) is known from half a dozen locations along, or close to, the Ontario shores of Hudson Bay and from two sites inland. It is a low shrub with prostrate, freely rooting, creeping stems bearing upright or ascending branches, usually less than 2 dm high, and smaller leaves (1–3 cm long) which are glabrous and glaucous beneath, their margins strongly revolute. This species is circumpolar. (Lab. to Alaska; Eurasia) (*Polifolia*—an old generic name)

Intermediates between these two species (with leaves finely pubescent beneath) have been found in the James Bay–Hudson Bay region and named *A.* × *jamesiana* Lepage. (*jamesiana*—for its discovery in the James Bay region)

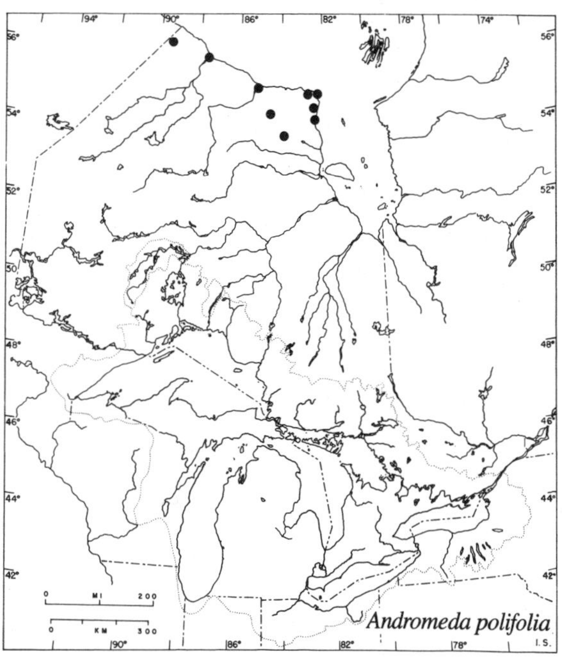

Field check Bog shrub; leaves alternate, evergreen, with inrolled margins and strongly whitened lower surface; young leaves often bluish; flowers urn-shaped, pinkish, in small drooping clusters; capsules with depressed indented tops, erect at maturity.

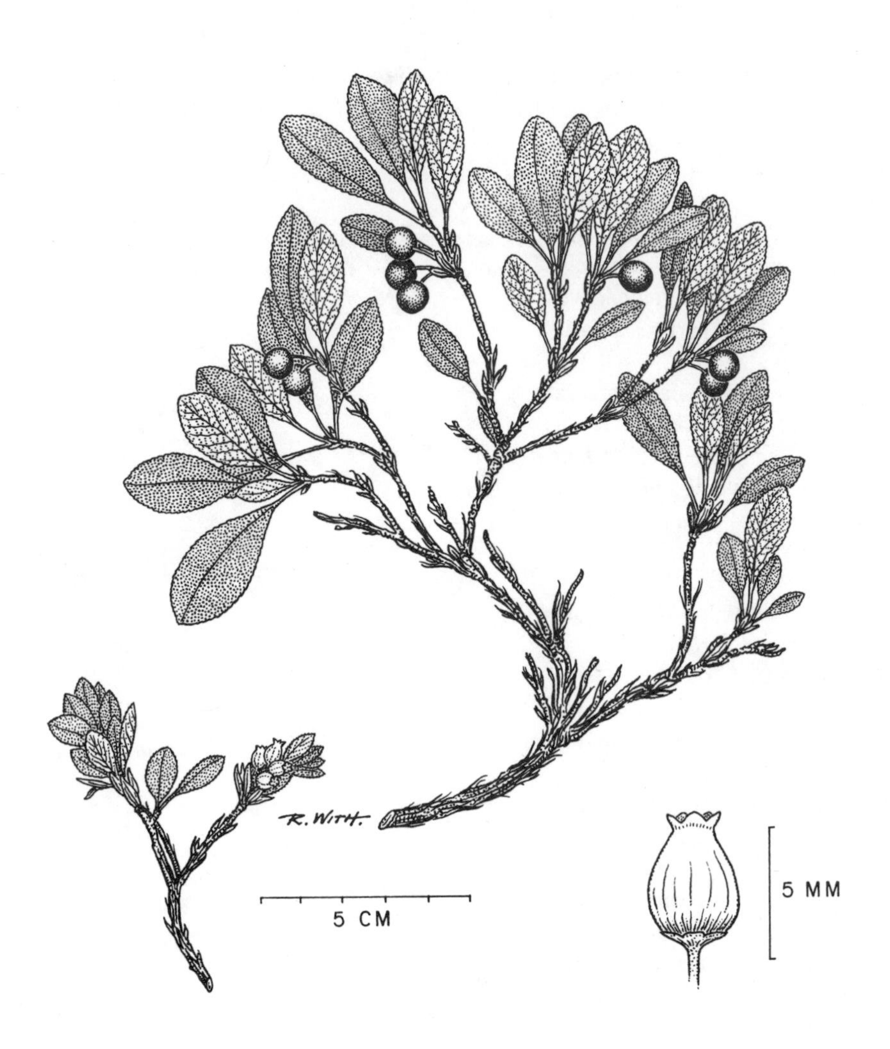

Arctostaphylos alpina

Arctostaphylos Adans. — Bearberry

A genus of 60–70 species primarily of western North America and Central America but with one species circumpolar in arctic regions. Trees, shrubs, and trailing woody plants with freely branching stems, alternate leaves, and white to pinkish flowers in scaly-bracted terminal clusters. (*Arctostaphylos*—from the Greek *arktos*, a bear, and *staphyle*, a bunch of grapes)

Key to *Arctostaphylos*

a. Leaves thin, persistent but withering, conspicuously veiny beneath and usually rugose above; margins finely crenate; fruit purplish-black or red, juicy; in Ontario found north of 52° N — ***A. alpina***

a. Leaves thick and leathery, evergreen, not conspicuously veiny; margins entire and somewhat inrolled; fruit red with dry mealy pulp; widely distributed — ***A. uva-ursi***

Arctostaphylos alpina (L.) Spreng. (including ssp. *rubra* (Rehd. & Wils.) Hult.; *A. rubra* (Rehd. & Wils.) Fern.) — Alpine bearberry

A prostrate matted or short-trailing shrub often much-branched and tufted, less than 2 dm high. Branchlets gray-brown to blackish, slender, brittle, and glabrous; older stems covered with loose, brown, papery bark.

Leaves alternate, simple, and deciduous or frequently withering and persisting in a bleached condition; young leaves often reddish tinged; blades of mature leaves thin, obovate or spatulate to oblanceolate, 2–6 cm. long, blunt to rounded or abruptly pointed at the apex, tapered at base; upper surface dull green, glabrous, often rugose-veiny, lower surface pale green, glabrous, with conspicuous reticulate venation; margins finely crenate; petiole slender, up to half as long as the blade.

Flowers in small clusters from terminal scaly buds, opening with the expanding leaves; corolla 4–5 mm long, urn-shaped, 5-parted, and white to pinkish; June and early July. Fruit a juicy berrylike drupe, 6–10 mm in diameter, purplish-black or red, edible but insipid; July and Aug. (*alpina*—alpine; *rubra*—red)

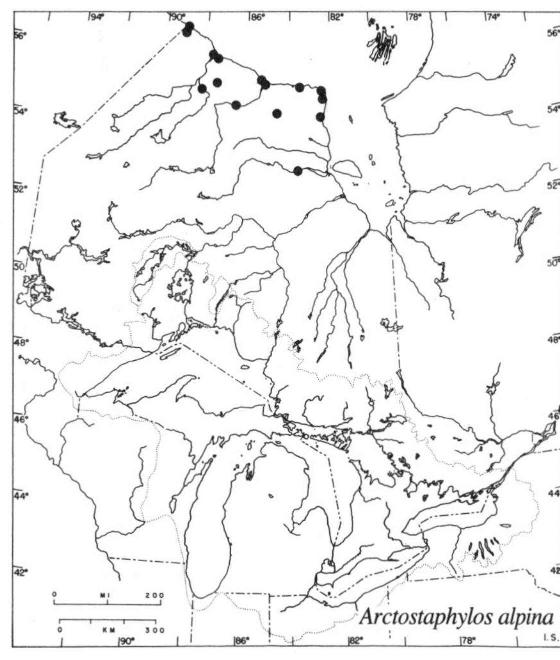

In tundra habitats: on sandy and gravelly beach ridges; in lichen heath and on calcareous till; on clay banks along rivers.

Along the shores of James Bay and Hudson Bay and inland along rivers and edges of lakes; southern limit about 53° N. (w. Greenl. and Nfld. to Alaska, south to B.C. and the higher mountains of Maine and N.H.; Eurasia)

Note Hultén (1968) has treated *A. rubra* as a subspecies of *A. alpina*, differing from

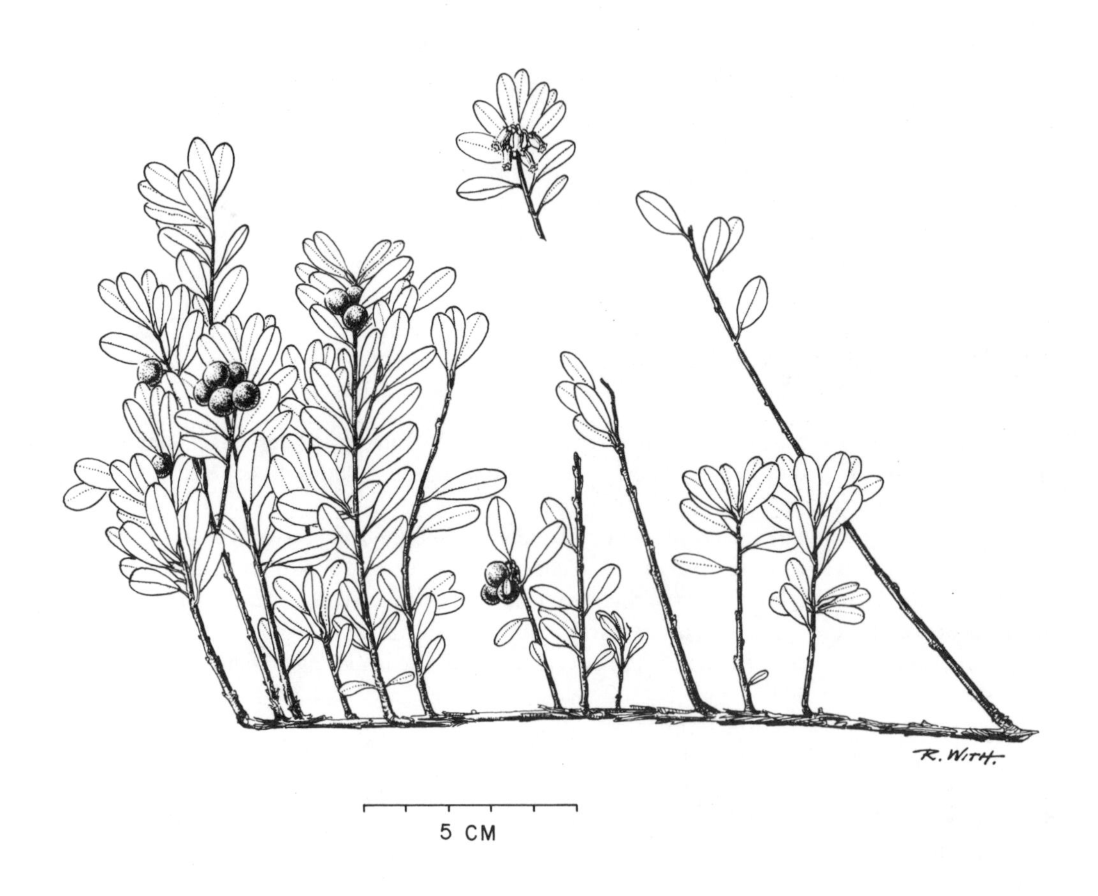

Arctostaphylos uva-ursi

typical ssp. *alpina* in its more ascending branchlets, long-petioled and less rugose thin leaves, and red fruits. Our description includes both subspecies.

Field check Prostrate shrub of tundra habitats; leaves thin, rugose, deciduous or withering and persisting; flowers opening with the expanding leaves; fruit berrylike, red or purple-black, juicy and edible but insipid.

Arctostaphylos uva-ursi (L.) Spreng. — Bearberry

A prostrate shrub with trailing stems up to several metres long and short upright or ascending branches 5-15 cm high. Branchlets reddish-brown, finely hairy, and sometimes glandular; older stems glabrous or hairy, reddish-brown to gray-black, with conspicuously peeling papery bark.

Leaves alternate, simple, and evergreen; blades oblong or oval, broadest at or above the middle, 1-3 cm long and 6-12 mm wide, firm and leathery, dark green above, paler and slightly hairy beneath, especially along the entire and slightly inrolled margins; apex blunt, rounded, or somewhat notched, base tapered to a short petiole.

Flowers small, white to pinkish or white with pink tips, in crowded terminal clusters on short branches; corolla 5-parted and urn-shaped or short-cylindrical, about 5-6 mm long; May and June. Fruit a reddish, dry, berrylike drupe about 8-10 mm in diameter with rather mealy and tasteless pulp; Aug. and Sept. (*uva-ursi*—grapes of the bear, hence bearberries)

In open, dry, sandy or rocky ground, especially along the shores of lakes and rivers; also in semi-open coniferous woods.

Widely distributed throughout Ontario from Lake Erie to Hudson Bay and from the St. Lawrence River to the Manitoba border. (w. Greenl. and Nfld. to Alaska, south to n. Calif. and Va.; Eurasia)

Note Several studies have recognized subspecies and varieties within *A. uva-ursi* but we have not attempted to distinguish among the specimens collected in Ontario. (See Packer and Denford, 1974.)

Field check Prostrate shrub with short ascending branches and oval to oblong evergreen alternate leaves; flowers small, pinkish, urn-shaped; fruit red, berrylike with dry, mealy pulp.

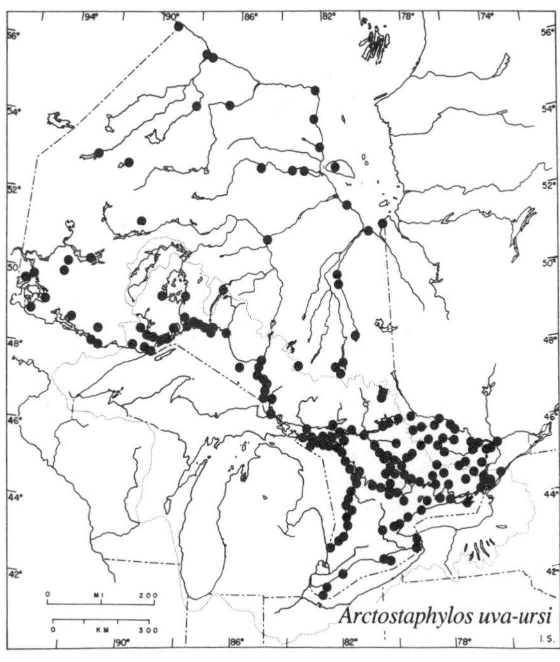

Chamaedaphne calyculata

Chamaedaphne Moench — Leatherleaf

The single species of leatherleaf is a scurfy-leaved evergreen shrub of bogs and shallow waters along lakeshores in the north temperate regions. Flowers pendent, solitary in the axils of reduced leaves along branches that are short-lived. (*Chamaedaphne*—from the Greek *chamae*, on the ground, and *daphne*, laurel)

Chamaedaphne calyculata (L.) Moench — Leatherleaf, Cassandra

A low, erect, much-branched shrub growing to a height of about 1 m. Branchlets brownish, minutely hairy, and covered with small, round, flat, scurfy scales; older stems grayish, the outer bark shredding and exposing smooth reddish inner bark.

Leaves alternate, simple, and evergreen, progressively smaller towards the ends of the flowering and fruiting branches; blades 1–4.5 cm long and 3–15 mm wide, oblong, oval, or elliptic, firm and leathery, green to brownish and glabrous above, pale brownish beneath with a close covering of tiny, white to brownish, round, flat, scurfy scales; margins entire to minutely crenulate; the tip blunt, rounded, or abruptly pointed, the base slightly rounded or tapered to a very short petiole.

Flowers whitish, 5-parted, urn-shaped or cylindrical, about 5–6 mm long, pendent in the axils of reduced leaves and forming one-sided leafy racemes at the ends of spreading branches; May and early June. Fruit a small, brownish, depressed round capsule less than 6 mm in diameter with a hairlike persistent style; capsules remaining on the branches for several years; seeds minute and numerous; July and Aug. (*calyculata*—with an outer calyx)

In peat bogs, around the shores of lakes, and along streams, often forming dense thickets.

From Lake Erie and the Ottawa and St. Lawrence valleys north to James Bay and the Severn River drainage basin and west to the Manitoba border. (Nfld. to Alaska, south to n.e. B.C., Iowa, and Ga.; Eurasia)

Field check Bog shrub; leaves alternate, evergreen, brownish and scaly beneath; flowers small, white, urn-shaped, pendent in the axils of reduced leaves at the ends of spreading branches.

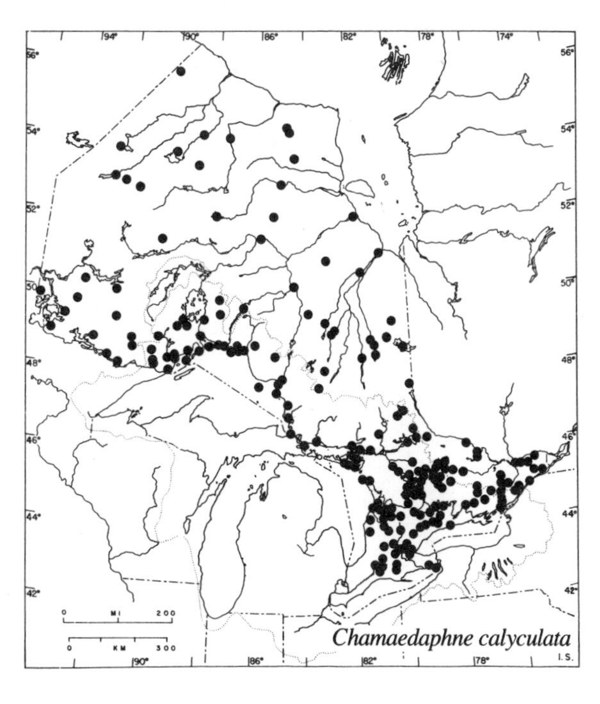

Chamaedaphne calyculata

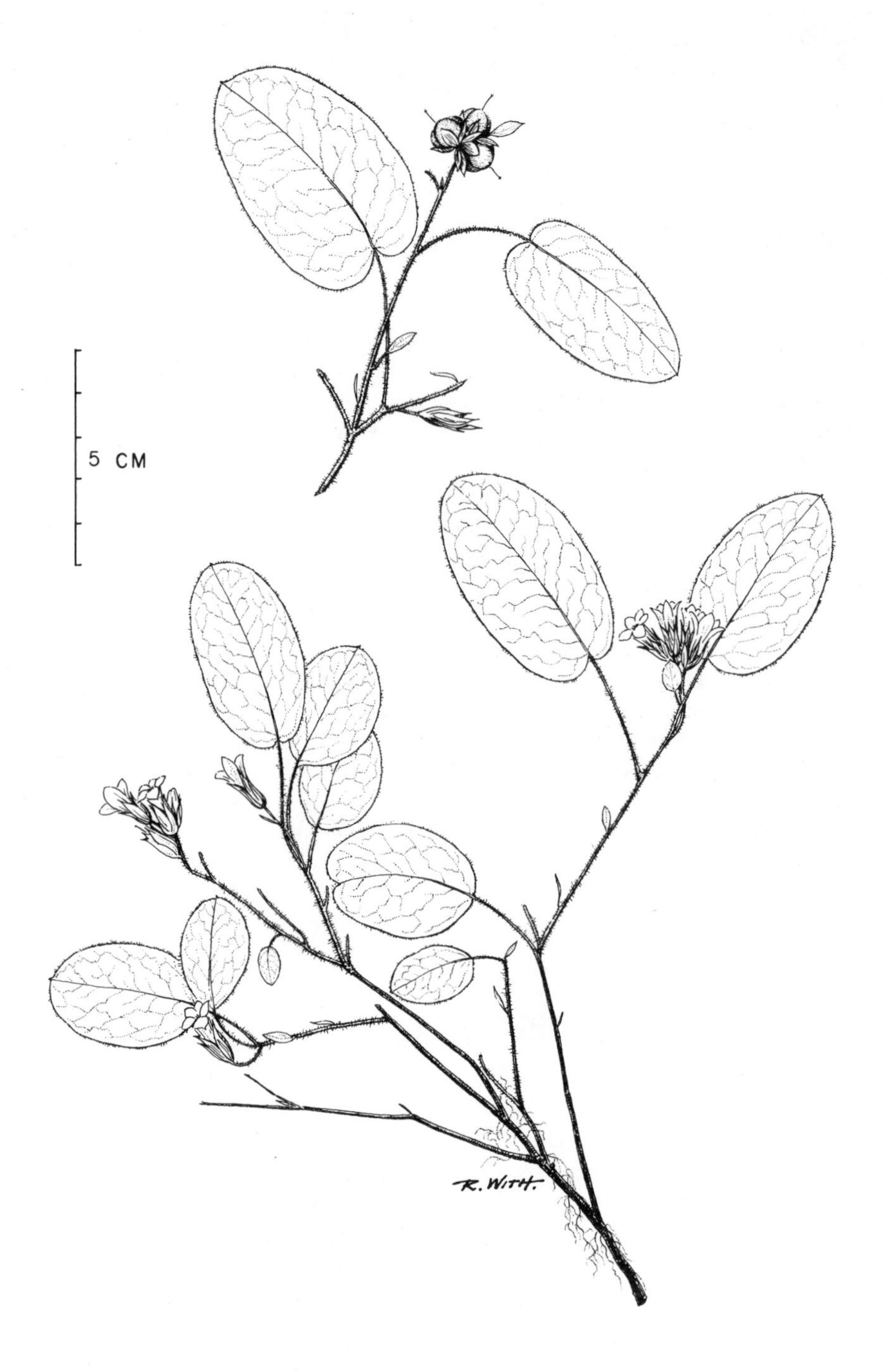

Epigaea repens

Epigaea L. — Trailing arbutus

A genus of 2 species, one in eastern North America and one in Japan. Prostrate, scarcely woody plants with evergreen leathery alternate leaves and small but spicy-aromatic pink to white flowers in crowded scaly-bracted terminal and axillary clusters. The roots lack root hairs and absorption is accomplished through a mycorrhizal relationship with certain fungi that live in association with the root tissues. (*Epigaea*—from the Greek *epi*, upon, and *gaea*, the earth, in reference to the trailing habit of growth)

Epigaea repens L. — Trailing arbutus

A prostrate evergreen creeper with sparingly branched, wiry or cordlike stems covered with bristly brown hairs and rooting at the nodes.

Leaves alternate, simple, and evergreen; blades up to 7 cm long, oval or broadly elliptical and somewhat leathery, apex blunt or pointed, the base rounded or heart-shaped; margins entire and ciliate with brownish hairs; petiole hairy and up to one-half the length of the blade.

Flowers spicy-fragrant, white to pink or rose-coloured; corolla tubular, about 12 mm long and with 5 spreading lobes; sepals 5, slender, nearly distinct, persistent; stamens 10 and stigma 5-lobed; the compact few-flowered clusters open in early spring; late Apr. and May. Fruit a small globular capsule, the 5 chambers containing numerous fine dark brown seeds which mature in late summer. (*repens*—creeping)

In sandy and rocky woods, often partially hidden by leaf litter.

Common across the central part of Ontario on the Canadian Shield and northward to about 50° N; local in bogs and sandy woods in southern Ontario. (Nfld. to s. Man., south to Iowa and Fla.)

Field check Prostrate or trailing evergreen plant with alternate oval leathery leaves and spicy-aromatic pinkish flowers in early spring.

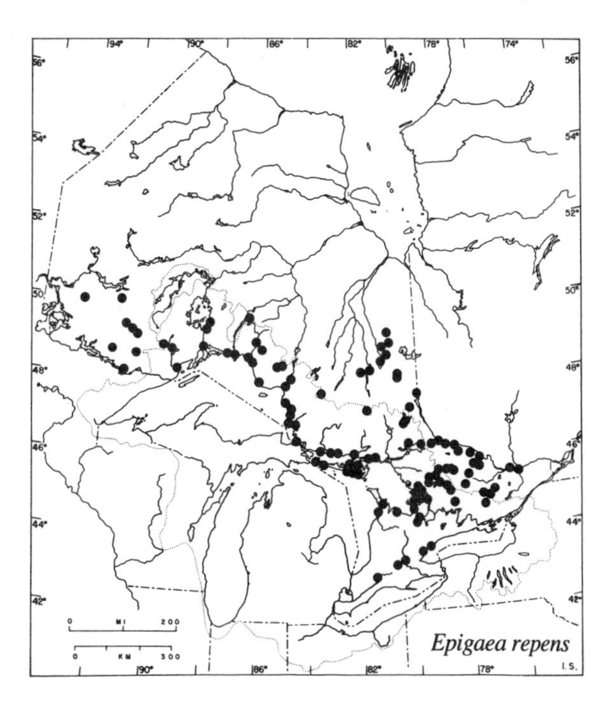

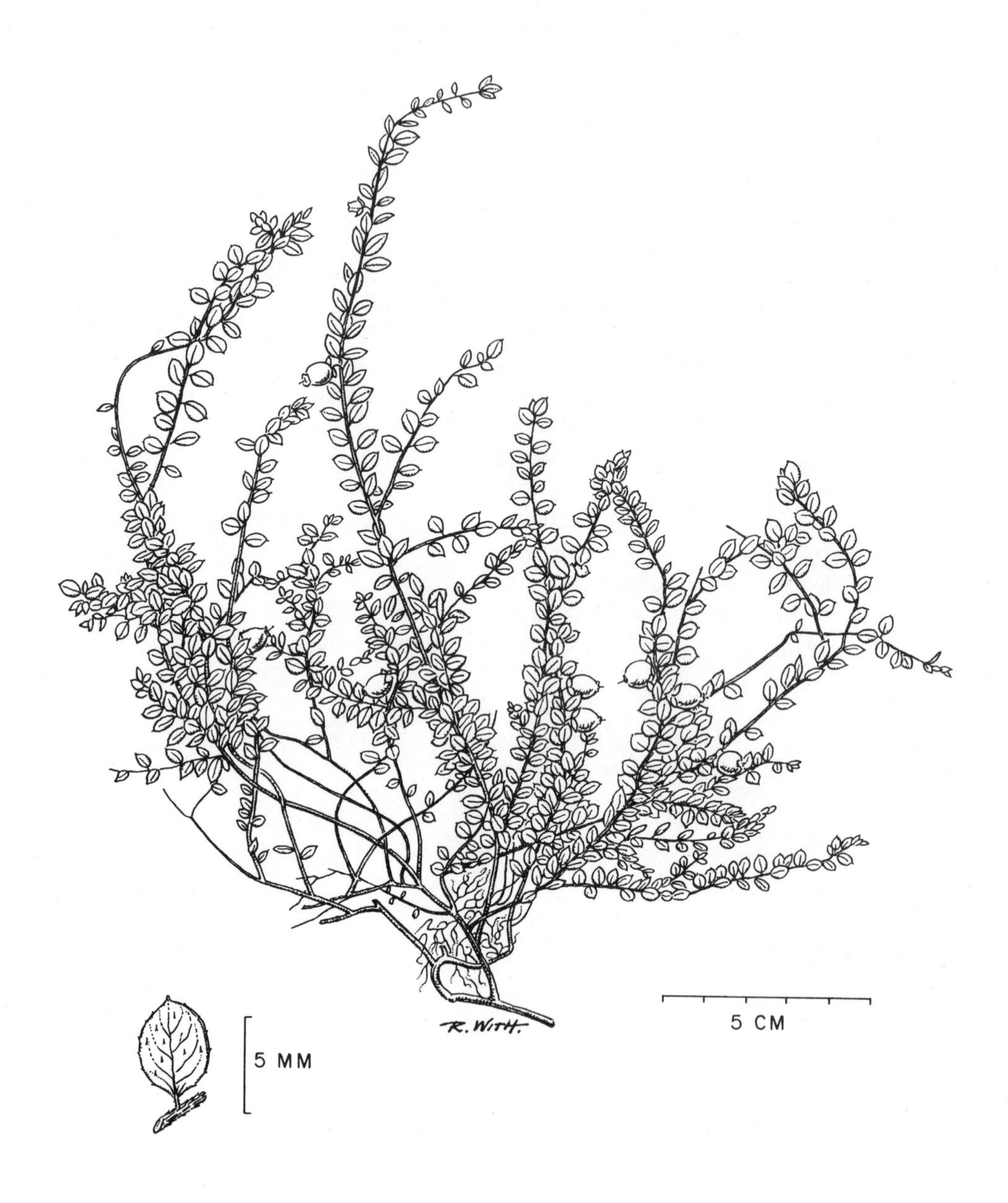

Gaultheria hispidula

Gaultheria Kalm ex L.

A genus of some 200 species, mostly around the Pacific Basin, only about half a dozen occurring in North America. Those in Ontario are almost herbaceous shrublets with alternate evergreen leaves, small white to pinkish axillary flowers on pedicels with 2 bractlets, and berrylike fruits. The fruit is really a capsule that is enclosed by the calyx, the latter becoming fleshy at maturity.

Wintergreen and snowberry are more closely related to eastern Asian species of *Gaultheria* than to those species of the genus that occur on the western coast of North America (Wood, 1972). (*Gaultheria*—commemorates an early botanist and court physician at Quebec, Jean-François Gaultier, 1708-1756)

Key to *Gaultheria*

a. Stems all creeping or trailing, covered with bristly brown hairs, and bearing numerous small leaves (2-10 mm long); flowers 4-parted; berries white, translucent **G. hispidula**

a. Stems of two kinds, horizontal at or below the ground level and erect, bearing a few leaves (1-5 cm long) crowded near the summit of the branches, not bristly-hairy; flowers 5-parted; berries red, opaque **G. procumbens**

Gaultheria hispidula (L.) Muhl. (*Chiogenes hispidula* (L.) T. & G.) — Snowberry

A prostrate evergreen plant with trailing, often matted slender stems covered with appressed brownish hairs.

Leaves small and numerous, alternate, simple, and evergreen, 2-10 mm long; blades firm, green to brownish-green above and much paler beneath, oval to elliptic, abruptly pointed at the tip and tapered at the base to an extremely short petiole; margins revolute and the lower surface with scattered, appressed, brown, bristlelike hairs.

Flowers few, nearly white, solitary in the axils of the leaves on short recurved pedicels subtended by a pair of bracts; corolla bell-shaped, deeply 4-parted, and about 2-3 mm long; May and early June. Fruit a translucent juicy white berry about 10 mm long, with a mild flavour of wintergreen; July and Aug. (*hispidula*—with fine rough hairs; *Chiogenes*—from the Greek *chion*, snow, and *genos*, offspring, referring to the snow-white berries)

In sphagnum bogs and moist coniferous woods, especially on moss-covered rocks and decaying logs or stumps.

Widely distributed in Ontario northward to James Bay and Hudson Bay; apparently

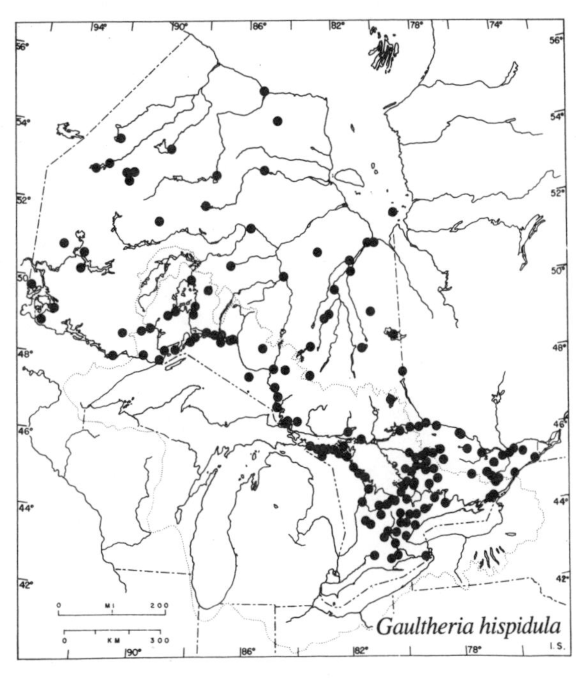

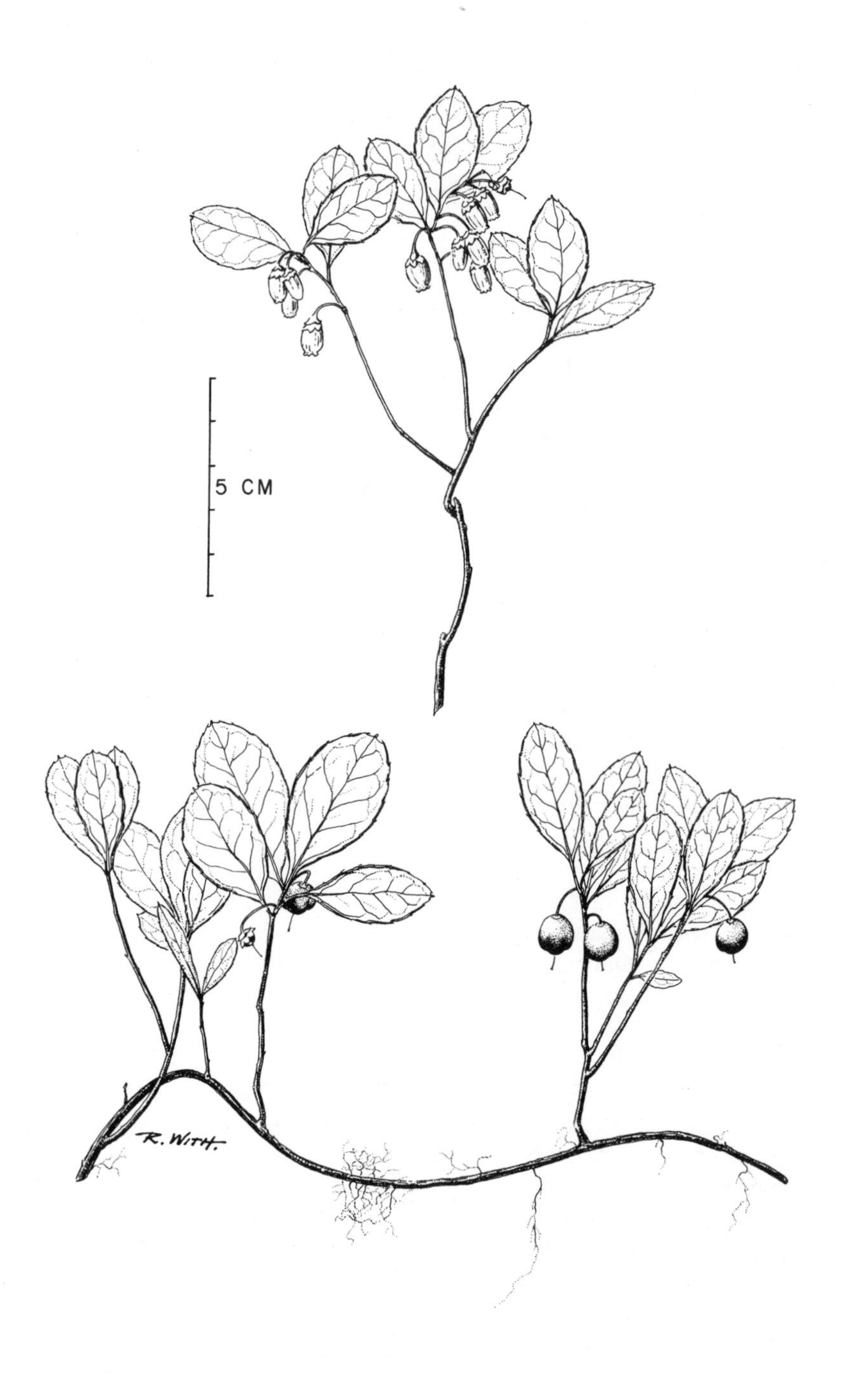

Gaultheria procumbens

rare in the Boreal Forest and Barren Region; not reported from the region at the western end of Lake Erie. (Nfld. to B.C., south to n. Idaho and N.C.)

Field check Creeping evergreen with numerous small oval leaves, bristly-hairy on the lower surface; flowers and fruits close to the stem; berries white, translucent.

Gaultheria procumbens L. — Wintergreen, Checkerberry Teaberry

A low evergreen plant of the forest floor with erect or ascending stems usually less than 1.5 dm high. Stem a slender woody branching rhizome below or at ground level, giving rise to leafy and flowering branches singly or in clumps; lower portion of aerial branches with a few remote, slender, reddish bracts or scales, the normal leaves clustered at the summit.

Leaves alternate, simple, and crowded near the top of the upright stem; blades oval, elliptical, or oblong, 1–5 cm long, leathery when mature, dark green and glossy above and paler beneath; margins somewhat revolute and obscurely toothed, the teeth incurved and bristle-tipped, the rounded tip of the leaf having a short projection (mucro); base of blade narrowed to a short, often reddish petiole.

Flowers solitary and nodding on curved pedicels in the axils of the leaves; corolla about 5–9 mm long, white, barrel-shaped, with 5 rounded lobes; June and early July. Fruit a capsule surrounded by the smooth fleshy calyx, the whole about 10 mm in diameter and resembling a small bright red cherry, edible with a distinctive wintergreen flavour; Aug. and Sept., often overwintering. (*procumbens*—lying on the ground)

In sandy and mossy woods and clearings.

Throughout central and southern Ontario except for the region at the western end of Lake Erie; northern limit at about 50° N. (Nfld. to s.e. Man., south to Minn. and Ga.)

Note The leaves and fruits have the characteristic flavour of wintergreen when chewed. Leaves may be steeped in boiling water to produce a pleasantly mild, aromatic tea. The wintergreen flavour is due to the presence of a glucoside that breaks down in water to produce methyl salicylate, which can also be synthesized chemically to produce the artificial wintergreen flavouring. The same glucoside occurs in the bark of the sweet birch (*Betula lenta* L.) and the yellow birch (*B. alleghaniensis* Britt.).

Wintergreen was reported from the Windsor area in 1914 but no specimen has been seen from anywhere in Essex County. It should be looked for in that area since it occurs nearby in Michigan and south of Lake Erie in Ohio.

Field check Low evergreen plant with oval, leathery, wintergreen-flavoured leaves clustered at the top of the stem; fruit a bright red "berry".

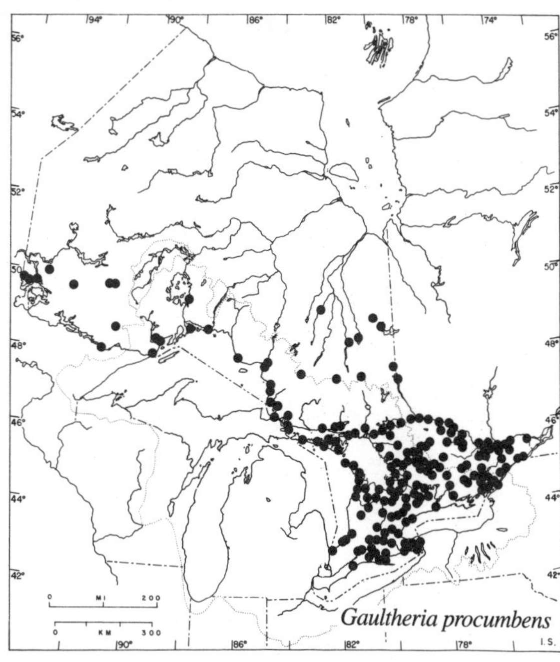

Gaylussacia baccata

Gaylussacia HBK — Huckleberry

There are about 50 species of this genus in North and South America, but only *G. baccata* occurs in Ontario. Huckleberries are freely branching shrubs that spread underground to form colonies. In habit they resemble blueberries (*Vaccinium* spp.), but their alternate deciduous leaves are often covered with shiny golden resinous dots, and their berries, though sweet and edible, are less desirable on account of the 10 hard nutlets they contain. The flowers are borne in lateral bracted racemes and the corolla varies from cream-coloured to orange or purplish-red. They are reported to be self-sterile and to be cross-pollinated by bees. (*Gaylussacia*—in honour of the famous French chemist and physicist Louis Joseph Gay-Lussac, 1778-1850)

Gaylussacia baccata (Wang.) K. Koch — Black huckleberry

An erect much-branched shrub growing to a height of 3-10 dm, rarely higher. Branchlets brownish and minutely hairy at first; older stems purplish-gray to blackish with small sections of the outer papery bark peeling off.

Leaves alternate, simple, and deciduous; blades oval to oblong, 2-5.5 cm long and 1-2.5 cm wide, firm, dark green above, paler beneath, both surfaces sprinkled with golden-yellow resinous dots (which are more noticeable below); margins entire and minutely hairy; tip rounded, blunt, or pointed, the base tapered to a short petiole.

Flowers small (about 6 mm long), in lateral, often one-sided racemes on gland-dotted pedicels, yellowish-orange to reddish, slenderly cylindrical; late May and June. Fruit an edible but seedy, reddish-purple to blackish, berrylike drupe; July and Aug. (*baccata*—bearing berries)

In acidic soils; in sandy or rocky woods and clearings, and in bogs.

From Lake Erie to Lake Nipissing and from the Ottawa and St. Lawrence valleys to Sault Ste. Marie; not reported from northern Ontario. (Nfld. to s. Algoma Dist. to Sask., south to La. and Ga.)

Field check A low much-branched shrub with ovate to elliptic, gland-dotted, alternate leaves; flowers orange to reddish, slenderly cylindrical; fruits purplish-black, berrylike, edible but seedy.

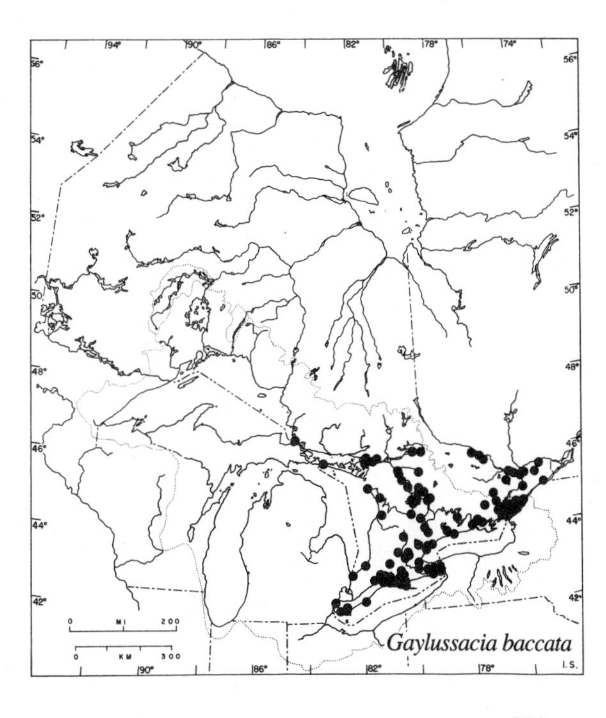

Kalmia angustifolia

Kalmia L. — Laurel

A small genus of 8 species found in North America and Cuba. Evergreen shrubs with naked buds, entire leathery leaves which may be alternate, opposite, or whorled, and showy 5-parted flowers in lateral or terminal clusters. The anthers are held in individual pockets in the corolla and the filaments are strongly arched so that when an insect lands on the corolla the anther is released and pollen is scattered over the visitor through the catapultlike action. (*Kalmia*—named by Linnaeus after one of his students, Pehr Kalm, 1716-1779, who travelled and collected in North America)

Pehr Kalm was the first botanist to study and collect extensively the Canadian flora, some of his specimens being grown in the first important botanic garden in northern Europe and others preserved in the Linnaean Herbarium, where they have been a major source of documentation on the flora of eastern North America (Taylor, 1976).

Key to *Kalmia*

a. Stems round in section; leaves stalked, broadly elliptic to oval (usually some more than 1.5 cm wide); flowers in spreading or drooping axillary (lateral) clusters **K. angustifolia**

a. Stems flattened (alternately at right angles in successive internodes); leaves sessile, narrowly oval or elliptic (usually less than 1.5 cm wide); flowers in erect terminal clusters **K. poliifolia**

Kalmia angustifolia L. — Sheep-laurel, Lambkill

A small erect shrub reaching a height of 6-10 dm. Branchlets round in section, at first brownish and minutely downy; older stems glabrous and grayish.

Leaves simple, evergreen, and opposite or in whorls of three; blades oval or elliptic, 1.5-5 cm long and 5-20 mm wide, firm or leathery, dark green above, nearly glabrous on both surfaces; margins entire and somewhat inrolled; the tip blunt or rounded and the base somewhat rounded or tapered to a distinct petiole 3-10 mm long.

Flowers 5-parted, deep pink, and showy, on long slender stalks in lateral clusters from the axils of the previous year's leaves; corolla saucer-shaped, 9-12 mm in diameter, and when first open, the stamens arching outwards with the anther of each held temporarily in a small pouch on the inner surface of the corolla; June and early July. Fruit a small, long-stalked, globose capsule less than 6 mm in diameter, containing numerous small seeds, the persistent style about as long as the capsule; clusters of capsules may persist for several years on the branches; late July and Aug. (*angustifolia*—with narrow leaves)

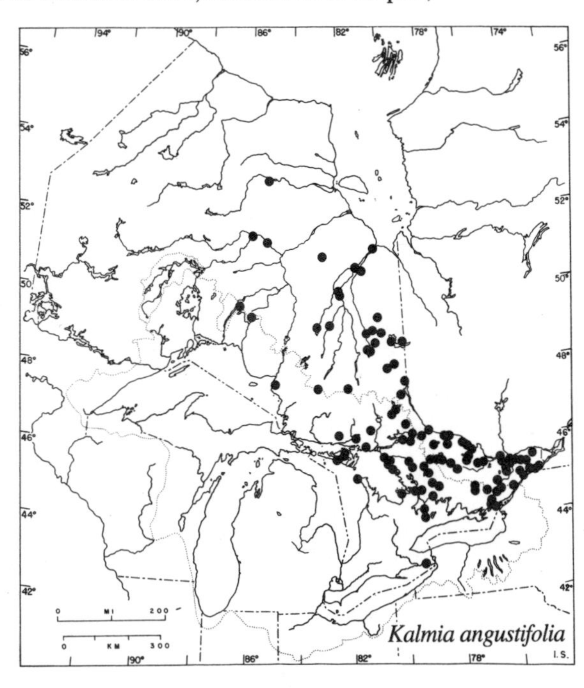

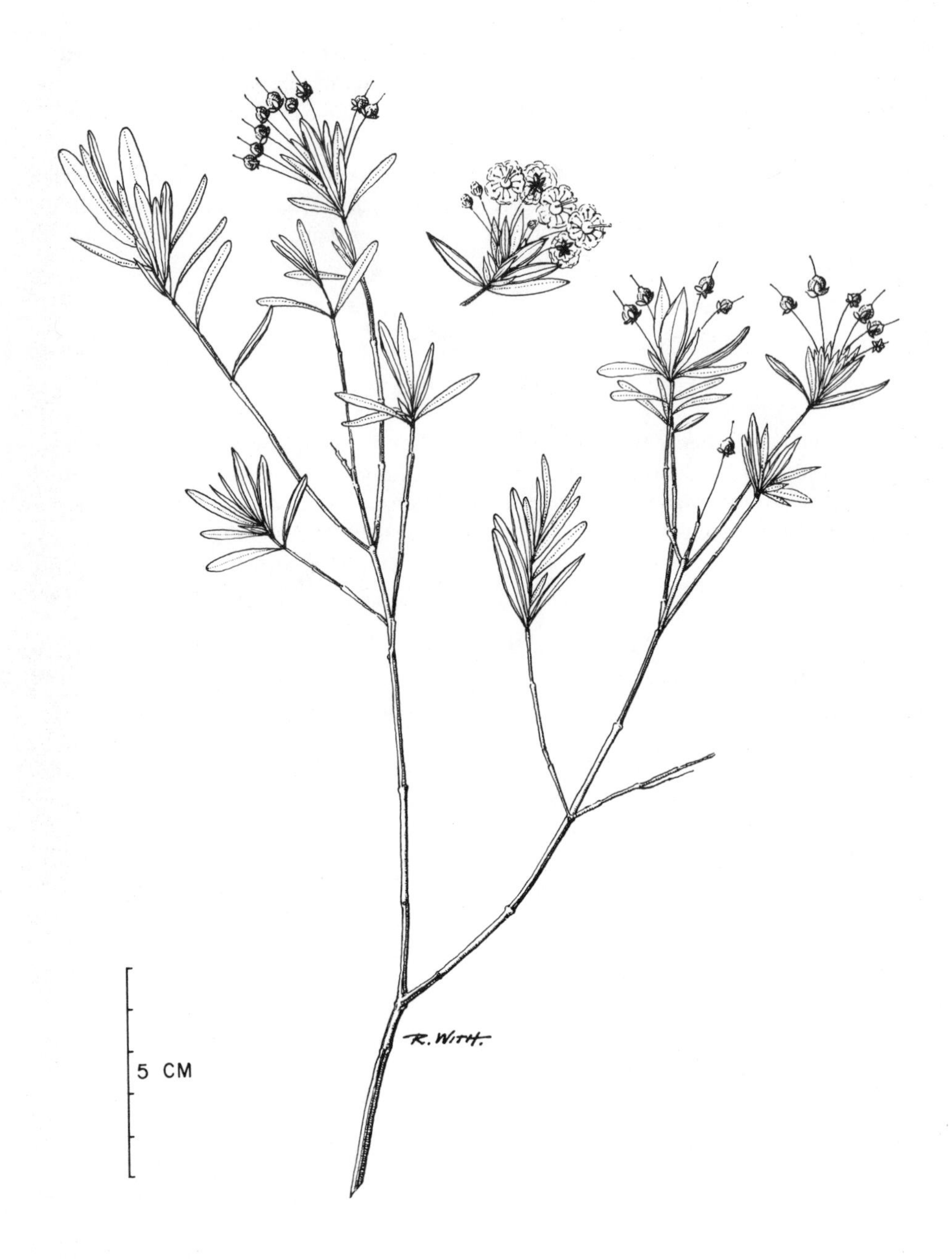

Kalmia poliifolia

In and around sphagnum bogs, in open jack pine woods, in moist coniferous woods, and in rocky or gravelly soil.

From the Ottawa and St. Lawrence valleys to Manitoulin Island and north to James Bay; absent from the Deciduous Forest Region except for the Wainfleet Bog in the Regional Municipality of Niagara; not reported in northern Ontario west of 87° W. (Nfld. to Ont., south to Mich. and Ga.)

Note The leaves of this species are poisonous to browsing sheep, hence the common name lambkill.

Field check Leaves evergreen, opposite, or in whorls of three; deep pink flowers in clusters from the leaf axils of the previous year's growth.

Kalmia poliifolia Wang. — Bog-laurel

A low straggling shrub growing to a height of about 6 dm. Branchlets with conspicuous nodes or joints, two-edged, and flattened alternately at right angles in each successive internode, glabrous, pale brown at first; older stems dark brown to blackish.

Leaves opposite, simple and evergreen; blades sessile, narrowly oval to elliptic, 1–4 cm long and 6–12 mm wide, blunt or pointed at the tip and rounded or tapered at the base, firm or leathery, dark green and glabrous above, conspicuously whitened beneath with a close covering of short white hairs; margins entire and inrolled; purple clavate glands along the midrib readily visible, especially on the lower surface of young leaves.

Flowers like those of *K. angustifolia* but somewhat paler, in terminal clusters on the branches of the current season's growth; May and June. Fruit a small globose capsule less than 6 mm in diameter, the style about as long as the capsule or longer; in stiff erect clusters, each capsule with numerous small seeds; July and Aug. (*poliifolia*—with leaves of *Polium*, a generic name given to a group of mints in which some species have strongly whitened leaves)

In swamps and sphagnum bogs, often with *Ledum* and *Chamaedaphne* in open spruce-tamarack muskeg.

Widely distributed from the Ottawa and upper St. Lawrence valleys to Hudson Bay, and from Lake Superior to Lake-of-the-Woods; local in several isolated bogs of the Deciduous Forest Region. (Nfld. to Yuk., south to Calif., Minn., and N.J.)

Field check Small bog shrub with two-edged stems; leaves evergreen, opposite, strongly whitened beneath; flowers pink, in erect terminal clusters.

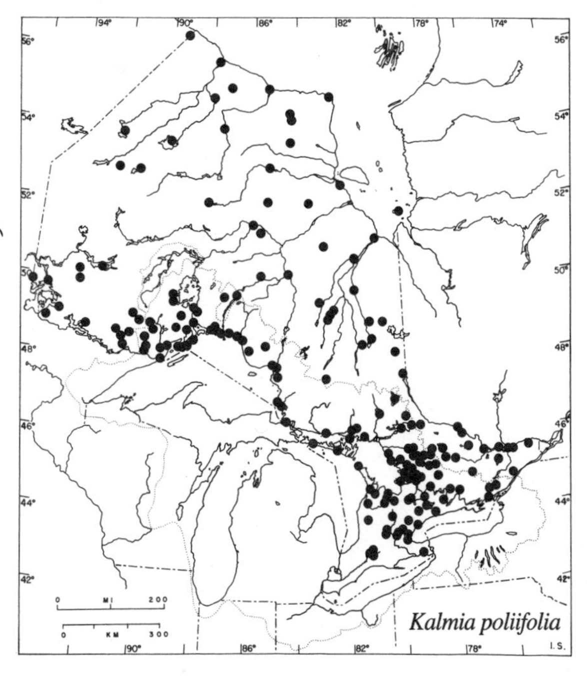

Kalmia poliifolia

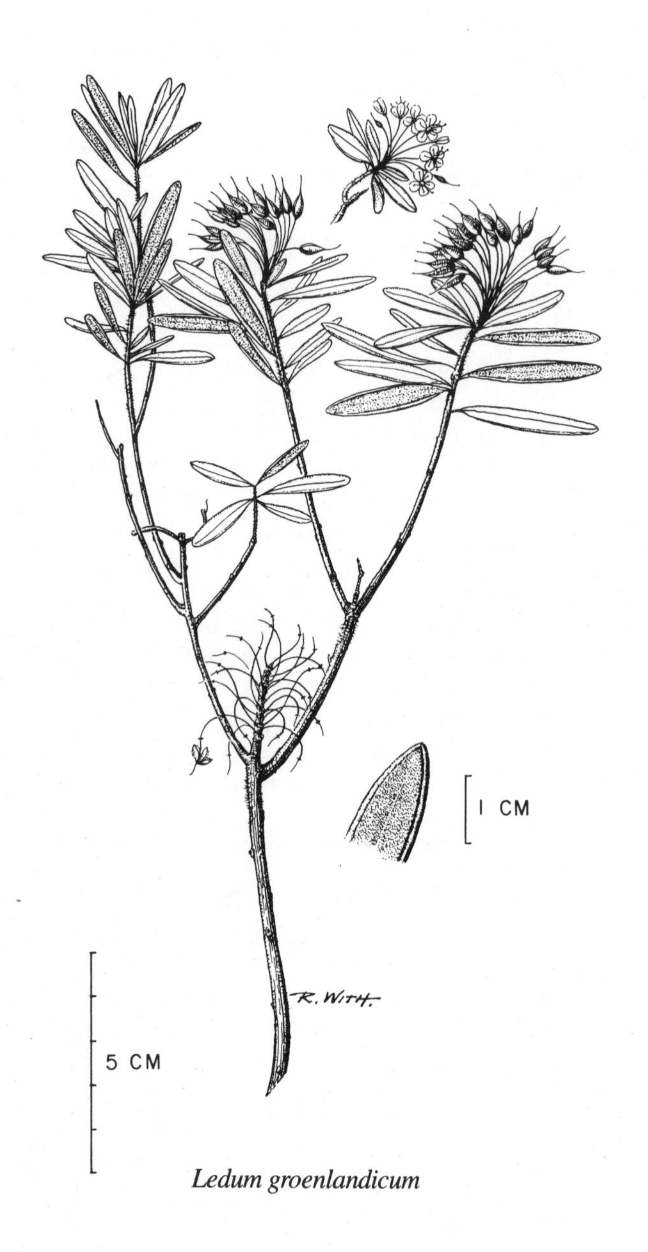

Ledum groenlandicum

Ledum L. — Labrador tea

A genus of 5–10 species found mostly in boreal and arctic regions. Low and freely branching shrubs with thick, leathery, alternate sessile or subsessile leaves with inrolled margins, the lower surface usually with a dense woolly covering of white to rusty-brown hairs. Flowers white, long-stalked, in crowded terminal clusters; petals distinct (not united). Fruit a brownish capsule in clusters which are often overtopped by the development of terminal leafy shoots. (*Ledum*—from the Greek *Ledon*, used originally for a species of *Cistus* (rockrose) which produces a sweet-smelling resin)

Key to *Ledum*

a. Upright shrub to about 1 m tall; leaves oval, oblong, or broadly linear (25–50 mm long, 5–20 mm wide), usually 3–4 times as long as wide; pedicel of mature capsule with gradual curvature; Lake Ontario to Hudson Bay — ***L. groenlandicum***

a. Low erect, decumbent, or trailing shrub less than 5 dm high; leaves narrowly oblong to linear (5–15 mm long, 1–3 mm wide), usually 5–8 times as long as wide; pedicel of mature capsule sharply curved near its summit; shores of Hudson Bay and northern part of James Bay. — ***L. decumbens***

Ledum groenlandicum Oeder — Labrador tea

A low evergreen shrub growing to a height of about 1 m. Branchlets woolly with curly brown hairs; older stems becoming glabrous, grayish to purplish or reddish-brown.

Leaves alternate, simple, and evergreen, fragrant when crushed; blades 2.5–5 cm long and 5–20 mm wide, usually 3–4 times as long as wide, firm, thick, and leathery, oval, oblong, or broadly linear, rounded or tapered at the base to a short stalk; the upper surface dark green and without hairs but somewhat wrinkled, with an impressed midvein; margins rolled inward on the underside, the lower surface covered with a dense coating of whitish or light brown to rusty-coloured curly hairs.

Flowers with distinct (not united) petals, creamy-white, small (less than 6 mm long) but borne on slender stalks in crowded roundish showy clusters at the ends of the branches; late May, June, and early July. Fruit a capsule 5–6 mm long, tipped with a hairlike style, splitting from the bottom upwards and releasing numerous fine seeds, the empty capsules persisting for several years, their pedicels with gradual uniform curvature; late July and Aug. (*groenlandicum*—of Greenland)

In and around sphagnum bogs, in swamps and wet woods.

Widely distributed from Lake Ontario to Hudson Bay and from the Ottawa and St. Lawrence valleys to Lake-of-the-Woods;

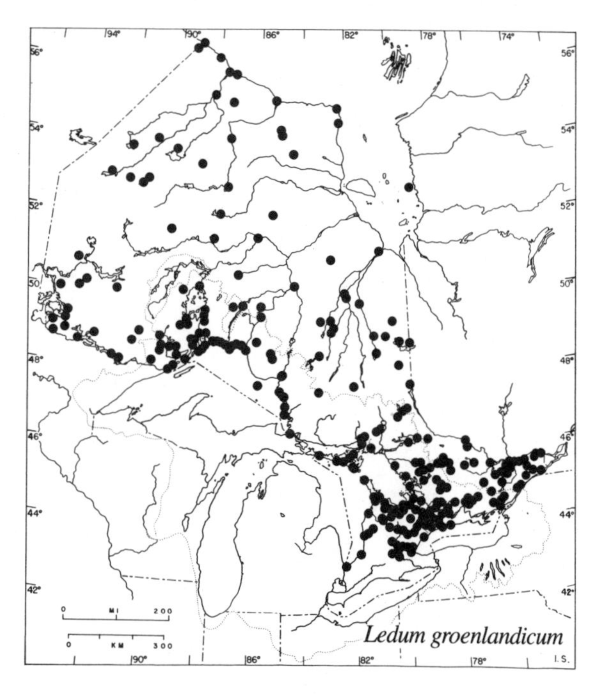

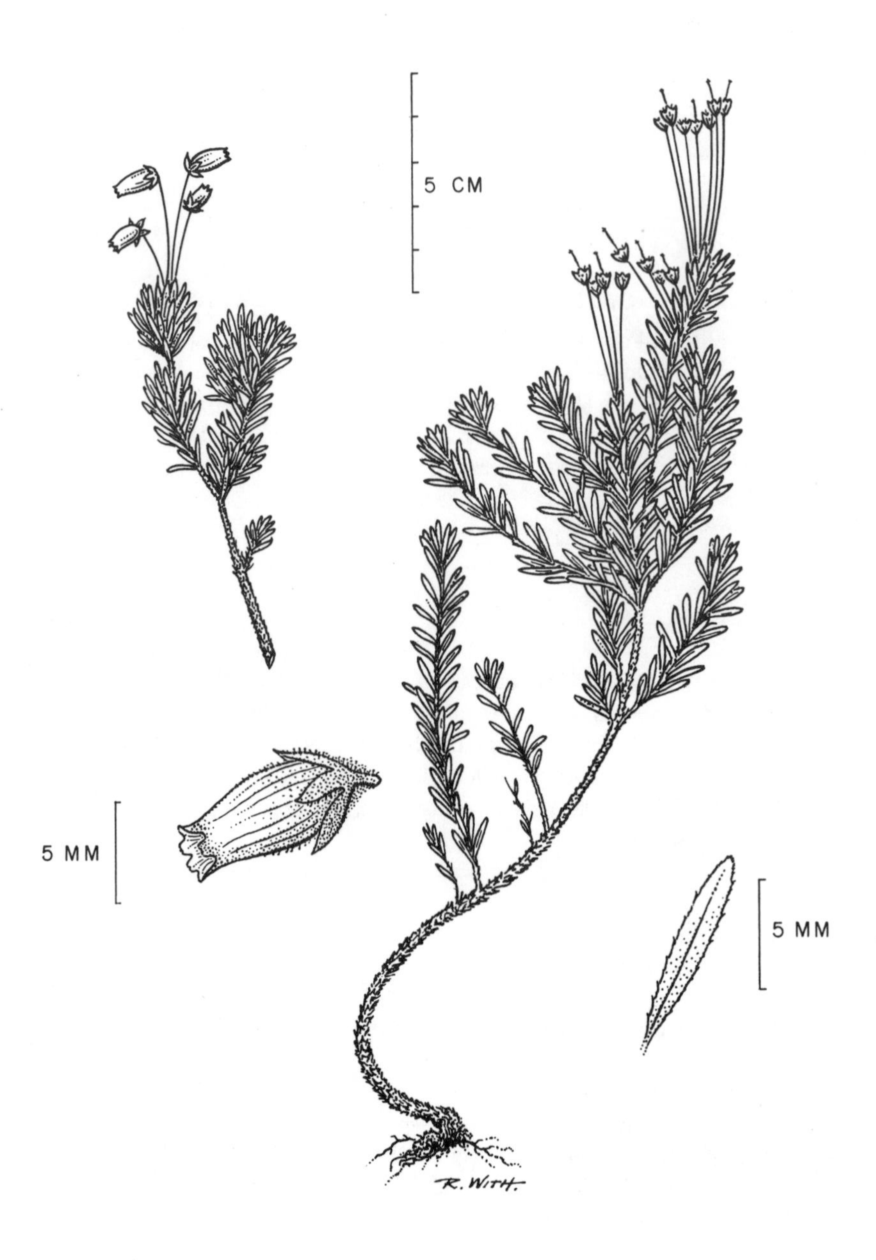

Phyllodoce caerulea

absent from the region immediately north of Lake Erie. (s. Greenl. and Nfld. to Alaska, south to Oreg. and N.J.)

Note A related species called dwarf Labrador tea (*L. decumbens* (Ait.) Lodd.) is a low, erect trailing or decumbent evergreen shrub less than 5 dm high, with small, narrow, almost linear leaves up to 1.5 cm long and 1–3 mm wide, usually 5–8 times as long as wide; pedicels of mature capsules sharply curved or bent near the summit. (*decumbens*—lying along the ground)

This species occurs in the arctic tundra along the shores of Hudson Bay and the northern part of James Bay. (w. Greenl. and Lab. to Alaska; e. Eurasia)

Field check Low evergreen bog shrub; leaves leathery with inrolled margins and a thick layer of rusty-brown hairs beneath, aromatic when crushed.

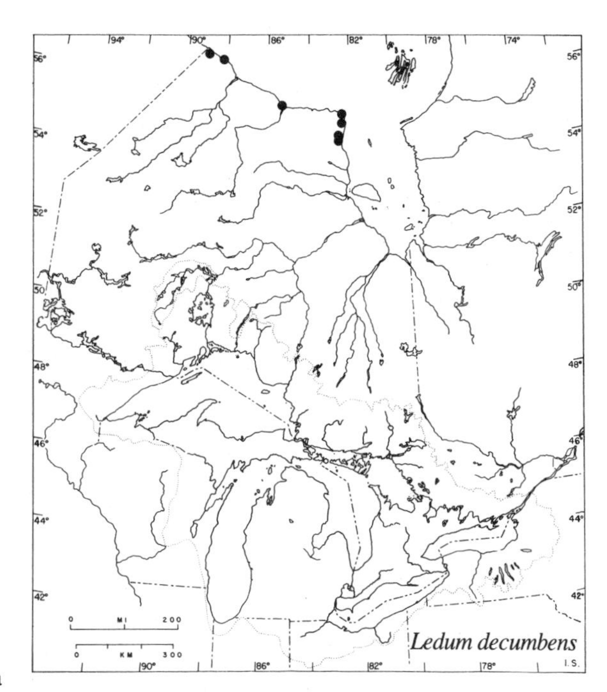

Phyllodoce Salisb. — Phyllodoce

Much-branched dwarf shrubs with crowded, linear, blunt, serrulate evergreen leaves. Flowers bisexual, 4–6-parted. Fruit a many-seeded capsule. About 7 species, circumpolar in the northern hemisphere. (*Phyllodoce*—the name of a sea nymph in Roman mythology)

A single species occurs in Ontario.

Phyllodoce caerulea (L.) Bab. — Phyllodoce

Low much-branched heathlike shrub usually less than 2 dm high. Stems reddish-brown to gray, the younger portions bearing the numerous crowded leaves, older stems bare but roughened with persistent flattened, peglike leaf bases; stem woody, 2–5 mm thick at base.

Leaves simple, alternate, and evergreen, closely spiralled around the stem; blade linear, 4–10 mm long, blunt at the tip; the margins minutely roughened or appearing serrulate with a row of small stalked reddish glands; green on both sides, the midrib depressed and on the lower surface obscured by a broad band of short, curled, whitish hairs; petiole very short with a constriction where it meets the swollen peglike base.

Flowers nodding on slender reddish-glandular pedicels 1–2 cm long from the axils of the uppermost leaves, solitary or in small umbel-like clusters; sepals 5, about 4 mm long, lanceolate, and glandular-pubescent; corolla purple, 5-lobed, cylindrical to urceolate, about 1 cm long; stamens 10, included; style straight with a capitate stigma, usually persistent in early fruit after the corolla has fallen off; July. Fruit a nearly globose many-seeded capsule barely exceeding the persistent sepals, on erect pedicels 3–4 cm long; August. (*caerulea*—sky-blue, but inappropriate if in reference to the colour of the fresh flowers which are purple and only become blue in dried specimens)

On beach ridges in spruce-lichen woodland; also to be expected on tundra, peaty or rocky slopes, in lichen mats and snow-bed depressions.

A single site known in Ontario on the southwestern coast of Hudson Bay near the mouth of the Sutton River (55° 7′ N, 83° 52′ 35″ W), 4 August 1978, R.Sims 2397A (TRT No.208294. The original specimen label gave the longitude erroneously as 85° 52′ 35″ N). (Greenl. and Nfld. to Alaska, south to N.S., Gaspé, Que., Hudson Bay, and mountains of Maine and N.H.; Iceland; Eurasia)

Note In view of the distribution of this species in the arctic tundra of North America and particularly on the east and west sides of Hudson Bay, it is surprising that it was not found earlier in various parts of the coastal tundra strip along the Hudson Bay portion of Ontario or the northwestern part of James Bay. Other species of shrubs of similar range that may be expected in Ontario are *Cassiope hypnoides* (L.) D.Don, *Diapensia lapponica* L., *Loiseleuria procumbens* (L.) Desv. (alpine-azalea), *Salix herbacea* L., and *Salix uva-ursi* Pursh. (bearberry willow).

Field check Low heathlike shrub with crowded, blunt, linear, glandular-serrulate evergreen leaves and nodding purplish tubular flowers on slender reddish-glandular pedicels in terminal few-flowered clusters; fruit an erect-stalked many-seeded capsule.

Rhododendron L. — Rhododendron

A large genus of 600 or more species chiefly of north temperate regions and with the greatest diversity in eastern Asia from the Himalayas through Malaysia to southern China and Japan. Shrubs and small trees usually with thick leathery leaves (deciduous in azaleas but evergreen in most rhododendrons) and large scaly winter buds. The flowers are large and showy, in terminal clusters, with the stamens conspicuously exserted; fruit a capsule.

Many species of azalea and rhododendron are cultivated for their large handsome flowers. (*Rhododendron*—from the Greek *rhodon*, a rose, and *dendron*, a tree)

Key to *Rhododendron*

a. Evergreen shrub less than 3 dm high; branches prostrate and often matted, or erect and mound-forming; leaves entire, leathery, scurfy beneath; flowers at the ends of leafy branches (June-July); arctic (near shores of Hudson Bay) ***R. lapponicum***

a. Deciduous shrub up to 1 m high; branches erect or ascending; leaves thin, ciliate, pubescent beneath (not scurfy); flowers at ends of naked branches appearing before or with the new leaves (May); southern (near Ottawa River) ***R. canadense***

Rhododendron canadense

Rhododendron canadense (L.) Torr. **Rhodora**

A low shrub about 1 m or less in height with stiffly ascending branches. Young stems pale brown to grayish, older stems dark gray to blackish. The persistent terminal fruiting clusters of the previous season are soon bypassed by the elongating leafy shoots of the current year.

Leaves alternate, simple, and deciduous, the young ones pubescent with pale to rusty-coloured hairs; mature blades elliptic to narrowly oblong, 2 – 5 cm long, the tip acute to blunt and abruptly short-pointed, the base wedge-shaped; dark green above and somewhat paler beneath, permanently finely pubescent on the midrib and veins of the lower surface; margins ciliate, permanently revolute; petioles 1 – 3 mm long.

Flowers showy, rose-purple, on short glandular pedicels, in terminal umbel-like clusters appearing before or with the expanding leaves, corolla 2 – 3 cm long, split nearly to the short tubular base into 3 parts, the two lower ones oblong and spreading, the upper one shallowly 3-lobed; stamens 10, about as long as the corolla; style exceeding the stamens, capitate; May. Fruit a cylindrical capsule, 7 – 15 mm long, glandular-puberulent, and somewhat glaucous, at first capped by the threadlike styles, later splitting lengthwise into its 5 sections; July to Aug. The pedicels are slender in the flowering stage but elongate slightly and become conspicuously thickened as the fruit matures. The capsules eventually fall off and leave a tight cluster of short stout pedicels, each with a cup-shaped persistent calyx at the tip. (*canadense* — Canadian)

In peat bogs; also in damp thickets, on dry heaths, acid barrens, and rocky slopes.

Known in Ontario only from the Alfred peat bog (45° 29′ N, 74° 52′ W) and the Newington Bog (45° 07′ N, 74° 58′ W), the latter site discovered by P. M. Catling in 1983 and not shown on the map on p. 393; also found nearby on the opposite side of the Ottawa River and farther east along both sides of the St. Lawrence River. (Nfld., s. Que., and e. Ont., south to e. Pa. and n. N.J.)

Note This species differs from other rhododendrons by its deciduous habit and from azaleas by its deeply split corolla and 10, rather than 5, stamens. It is sometimes treated as a separate genus, *Rhodora* L. The name *Rhodora* was chosen as a short title for the journal of the New England Botanical Club, which began publication in January, 1899.

Field check Low shrub with stiffly ascending branches and elliptic-oblong deciduous leaves; flowers showy, rose-purple, in May, before or with the expanding leaves; fruit a stalked capsule; extreme southeastern corner of Ontario.

Rhododendron lapponicum

Rhododendron lapponicum (L.) Wahl. — Lapland rosebay

A freely branching dwarf shrub, prostrate and matted or forming low moundlike bushes up to 3 dm high. Branchlets thickish, gray-brown, and scurfy.

Leaves alternate, simple, and evergreen, somewhat crowded at the ends of the branches; blades 1–2 cm long, thick and leathery, elliptic or oval to oblanceolate or narrowly obovate, rounded to blunt and minutely pointed at the tip, tapered to the base, the upper surface dull dark olive-green, the lower surface rusty-brown, deeply pitted and densely covered on both sides with resin dots; margins entire and slightly revolute; petiole very short.

Flowers very aromatic, numerous, in umbel-like clusters at the ends of the branches; corolla bell-shaped, deep purple, 1–2 cm broad; stamens 5–10, filaments purple; stamens and style well exserted; calyx and pedicels densely scurfy; June and July. Fruit an erect, stalked, ovoid-cylindrical capsule 3–6 mm long, at first tipped by the slender style which is soon lost when the capsule splits into its 5 sections; July and Aug. (*lapponicum*—of Lapland)

On sandy, gravelly, or calcareous soil, beach ridges, and mossy tundra.

Chiefly on the coast and off-shore islands of Hudson Bay and inland around large lakes (e.g., Hawley Lake); also on islands in James Bay (N.W.T.) and along the east coast of James Bay and Hudson Bay. (e. Greenl. and Nfld. to Alaska, south to n. B.C., Banff, Alta., where local only, Gaspé Pen., Que., and the high mountains of N.Y. and N. Eng.; rare in cent. Wis.; Scandinavia; Siberia)

Field check Dwarf matted or mound-forming shrub with scurfy branchlets and small evergreen leaves dull dark green above and brownish-scurfy beneath; flowers showy, deep purple, aromatic; fruit a stalked capsule; northern part of Hudson Bay drainage basin and islands in James Bay.

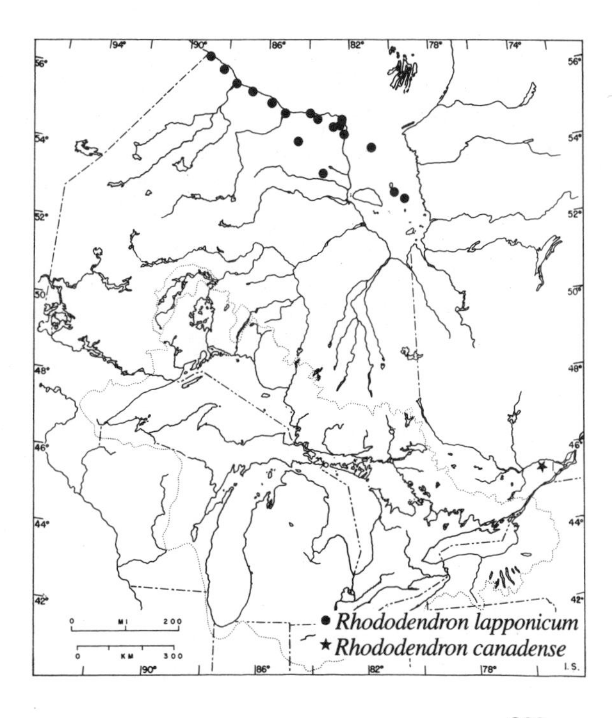

Vaccinium L.

About 150 species in the north temperate regions and also on some mountains in the tropics. Deciduous or evergreen shrubs and small trees with alternate, simple, sessile or short-stalked leaves. Flowers 4- or 5-parted, solitary in the leaf axils or in lateral or terminal clusters; ovary inferior; fruit a many-seeded red, blue, or black berry. The genus is commercially important for the edible fruits of blueberry, cranberry, and mountain cranberry. The native species are plants of sandy, rocky, or peaty (acid) soils and vary in habit from slender prostrate vines and low depressed or mat-forming bushes to upright and openly branched shrubs over a metre in height or taller (2–3 m) in high-bush blueberry. (*Vaccinium*—the Latin name for the blueberry)

Hybridization is known to occur in the blueberries, particularly between *V. angustifolium* and *V. myrtilloides* and less frequently between either of these species and *V. pallidum*, producing intermediates difficult to identify. The variation in *V. corymbosum* is reported to be due to its supposed origin from hybrids involving southern species beyond our range. In southern Ontario the commonest species of *Vaccinium* are low sweet blueberry (*V. angustifolium*), velvet-leaf blueberry (*V. myrtilloides*), and high-bush blueberry (*V. corymbosum*).

The genus *Vaccinium* is a heterogeneous group and is more easily understood when divided into its natural subgenera, five of which are represented in Ontario. These subgenera are sometimes treated as distinct genera, particularly the true cranberries (*Oxycoccos*).

Field checks

Blueberries: Stems warty, leaves deciduous, elliptic to oval or obovate, 2.5–9 cm long; flowers narrowly bell-shaped to cylindric, 5-parted, in clusters; anthers not awned; berries blue to black, edible.

Bilberries: Northern; leaves deciduous, spatulate, elliptic, broadly oval or nearly round, less than 2.5 cm long; flowers narrowly urn-shaped or broadly bell-shaped, 4- or 5-parted, borne singly or in groups of two or three in leaf axils; anthers awned; berries blue to blue-black, mostly sweet.

Mountain Cranberries: See the field check for *V. vitis-idaea* var. *minus*, which is our only member of this group.

Deerberries: See the field check for *V. stamineum*, which is the only species of this group in our area.

Cranberries: Creeping or trailing evergreen vines with small elliptic to narrowly triangular leathery leaves less than 15 mm long; flowers pink, 4-parted, borne on slender pedicels with a pair of bractlets or small leaves; petal segments reflexed; stamens exserted; berries red, sour.

Key to *Vaccinium*

a. Leaves deciduous; berries blue to blue-black or rarely greenish
  b. Mature leaves 2.5–9 cm long; flowers numerous in terminal or lateral clusters
    c. Low shrubs (usually less than 1 m high)
      d. Stems not warty; leaves 2.5–9 cm long; flowers and fruit on long slender stalks; corolla open, bell-shaped, with core of conspicuously exserted stamens (Deerberries) ***V. stamineum***
      d. Stems warty; leaves 2.5–5 cm long; flowers and fruit on short stalks; corolla narrowly bell-shaped or cylindric, stamens not exserted (Blueberries)
        e. Leaves finely toothed with bristle-tipped teeth
          f. Teeth numerous and closely spaced; erect low shrubs with spreading branches ***V. angustifolium***
          f. Teeth sparse and margins often finely hairy; stiffly branched upright shrub ***V. pallidum***
        e. Leaves entire, with or without fine hairs
          g. Stems and leaves densely velvety-hairy with short stiff hairs ***V. myrtilloides***
          g. Stems and leaves glabrous or sparsely hairy ***V. pallidum***
    c. Tall shrubs up to 2 m or more in height (High-bush blueberries) ***V. corymbosum***
  b. Mature leaves 6–25 mm long; flowers solitary in the leaf axils or 2 or 3 from scaly axillary buds (Bilberries)
    h. Margins entire or with only a few teeth near the base
      i. Leaves thin; flowers 5-parted, solitary in the lower leaf axils ***V. ovalifolium***
      i. Leaves firm to leathery; flowers 4-parted, borne singly or 2 or 3 from scaly axillary buds ***V. uliginosum* ssp. *pubescens***
    h. Margins minutely toothed all around ***V. caespitosum***
a. Leaves evergreen; berries red
  j. Lower leaf surface black-dotted with bristlelike glands (Mountain cranberries) ***V. vitis-idaea* var. *minus***
  j. Lower leaf surface glandless, pale or whitened (Cranberries)
    k. Leaves oblong-elliptic, blunt, paler beneath; pedicels with 2 small leaflike bracts ***V. macrocarpon***
    k. Leaves narrowly elliptic, pointed, strongly whitened beneath; pedicels with 2 tiny, usually coloured, slender bractlets ***V. oxycoccos***

Vaccinium angustifolium

Vaccinium angustifolium Ait. — Low sweet blueberry

An erect low shrub usually less than 6 dm high, with numerous spreading or ascending branches; stoloniferous and forming large patches. Branchlets greenish-brown, minutely warty, glabrous or finely hairy in lines running down from the nodes; older stems reddish-brown to blackish, glabrous, with ridged and flaky bark.

Leaves alternate, simple, and deciduous; blades oval to elliptic, 2.5 – 4.5 cm long and 6 – 15 mm wide, tapered or pointed at both ends, thin, bright green and glabrous on both surfaces or with a few hairs along the veins; margins minutely serrate with bristle-tipped teeth; petiole very short.

Flowers in terminal or lateral crowded clusters before or with the unfolding leaves; corolla narrowly bell-shaped, white or pale pink, 5-parted, less than 6 mm long; May and early June. Fruit a sweet, edible, blue berry with a bloom, 6 – 12 mm in diameter; June to Aug. (*angustifolium* — with narrow leaves)

In dry, sandy, or rocky clearings and open woods, and along roadsides; also in and around sphagnum bogs.

Widely distributed from Lake Erie northward to the Lake Superior region, becoming less frequent north of 50° N (see Vander Kloet, 1978). (Nfld. to Man., south to Iowa and W. Va.)

Note A polymorphic species in which outcrossing is obligatory, pollination is by insects, and seed dispersal mainly by birds and mammals. According to Vander Kloet (1976), no seedlings have been established in eastern Ontario for forty years and a decrease in the populations there is predicted.

Field check Low colonial shrub with minutely warty stems; leaves glabrous or hairy on the veins, finely serrate with bristle-tipped teeth; berries blue, sweet.

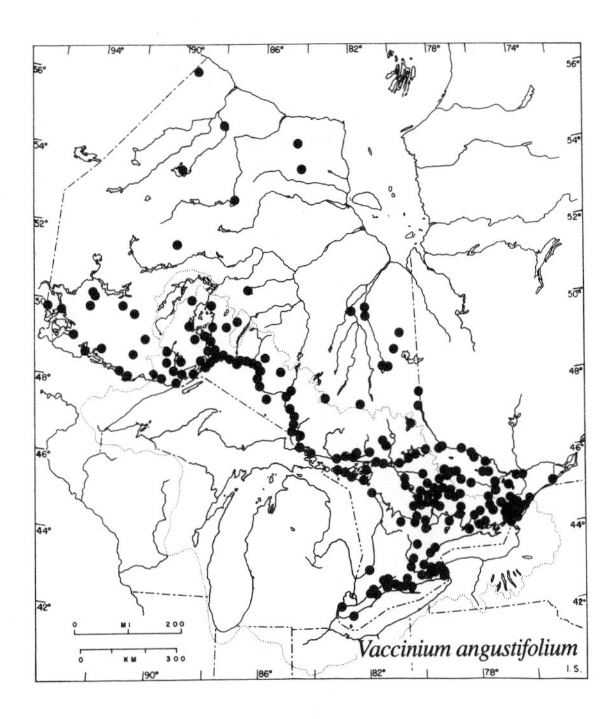

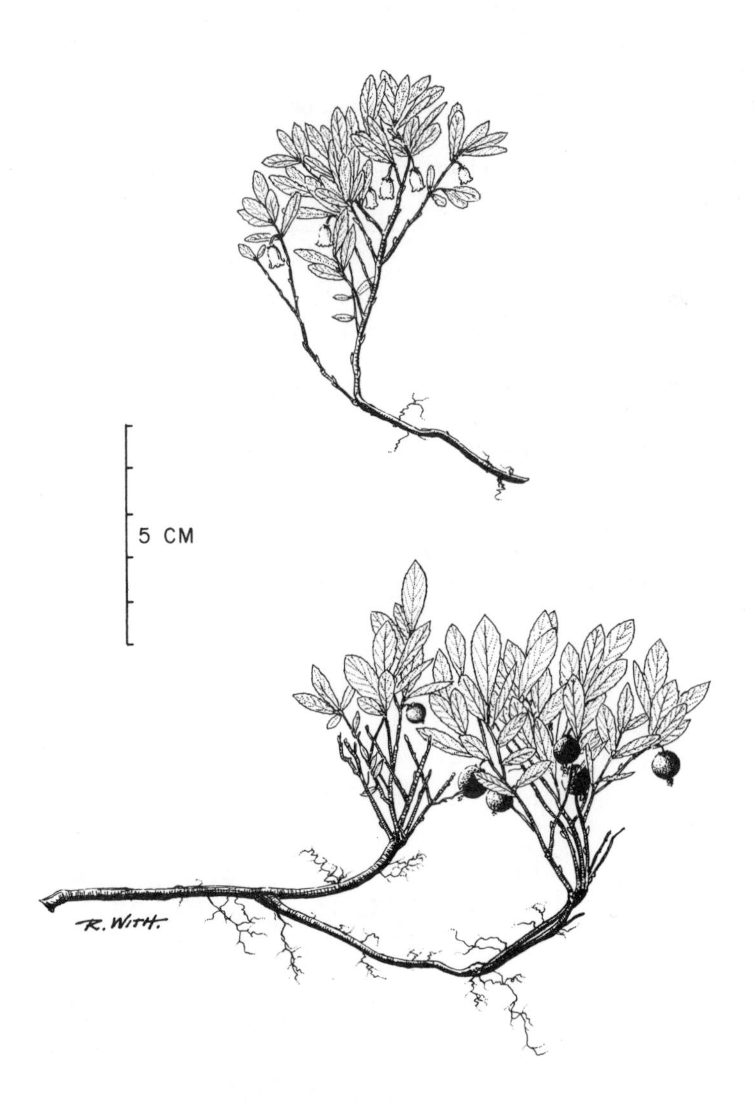

Vaccinium caespitosum

Vaccinium caespitosum Michx. — Dwarf bilberry

A low, tufted, much-branched shrub less than 3 dm high, spreading by stolons to form mats or patches. Branchlets green to brownish, minutely hairy, somewhat angled or ridged; older stems more round in section, with close gray-brown bark.

Leaves alternate, simple, and deciduous; blades 1 - 2.5 cm long and 3 - 12 mm wide, thin, obovate or spatulate (wider above the middle), abruptly wedge-shaped towards the apex with a rounded or pointed tip and gradually tapered to a slender wedge-shaped (cuneate) base, uniformly green and glabrous on both surfaces; margins finely serrate with bristle-tipped teeth; petiole very short or the blade sessile.

Flowers pink to reddish, less than 6 mm long, narrowly urn-shaped with 5 petals united almost to the tip, solitary in the axils of the lower leaves; June and early July. Fruit a sweet blue to blue-black berry with a bloom, about 5 - 6 mm in diameter; late July and Aug. (*caespitosum* — forming cushions)

In rocky, gravelly, or sandy clearings and in thickets in the coniferous forest.

Along the north shore of Lake Superior and northward to the upper drainage basins of the Moose and Albany rivers; northern limit at about 53° N. (Nfld. to Alaska, south to Calif. and n. N. Eng.)

Field check Low mat-forming northern shrub; leaves small (less than 2.5 cm long), thin, broadest well beyond the middle, finely serrate; flowers pink, 5-parted, solitary in the lower leaf axils; berries sweet, blue with a bloom.

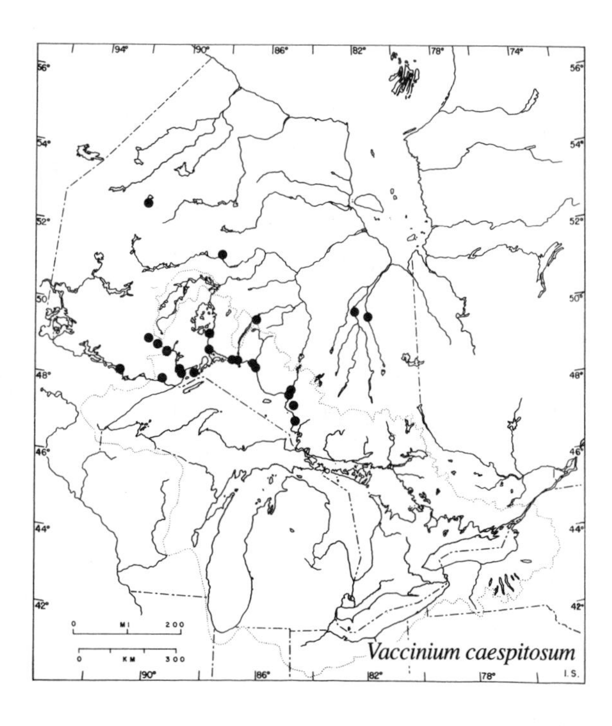

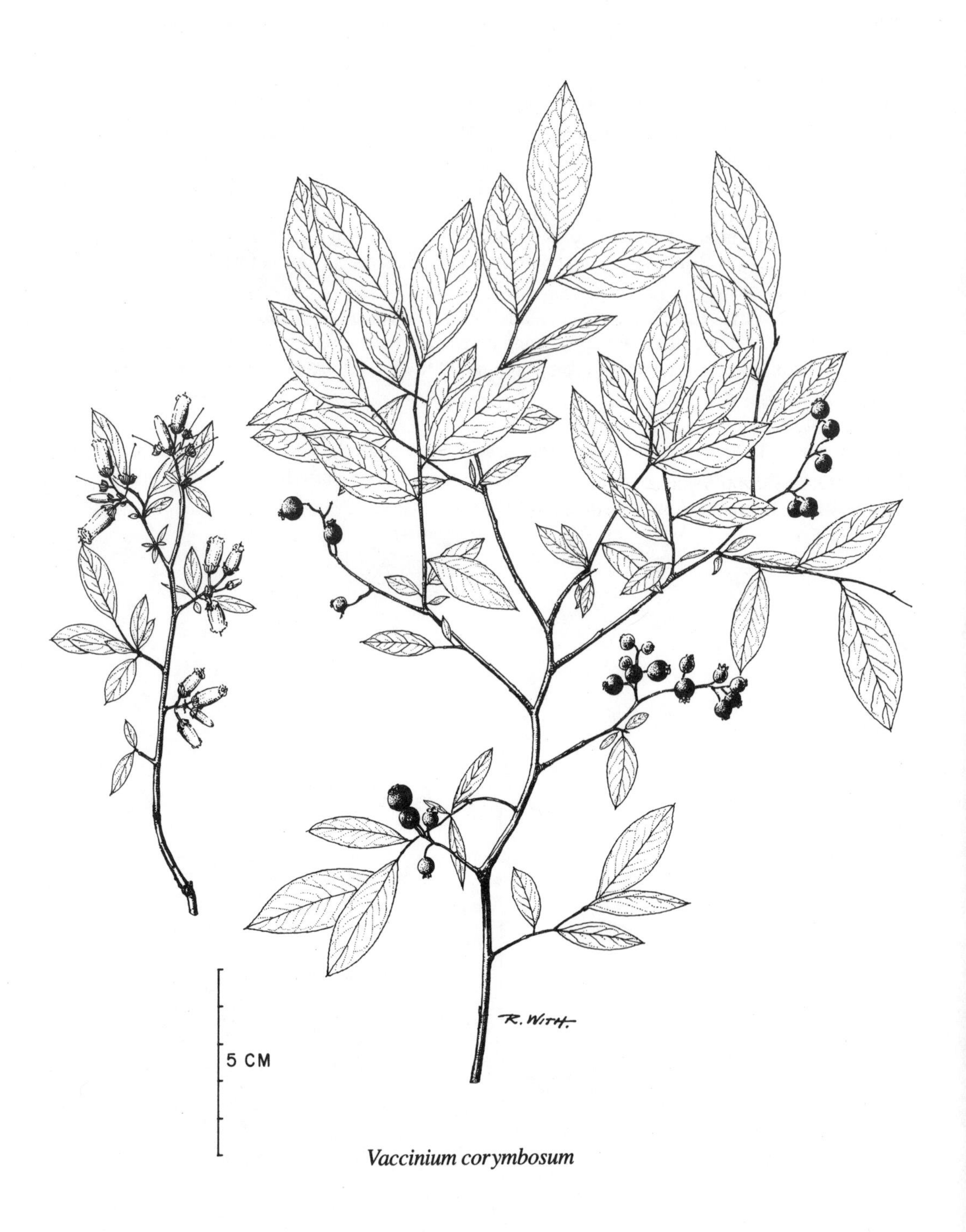

Vaccinium corymbosum

Vaccinium corymbosum L. — High-bush blueberry

A medium-sized shrub reaching a height of 2 - 3 m. Branchlets green to brownish and minutely hairy in lines running down from the nodes; older stems reddish-brown to blackish, somewhat ridged and warty.

Leaves alternate, simple, and deciduous; blades oval to ovate or elliptic, 3.5 - 7.5 cm long and 1.5 - 3 cm wide, blunt to pointed and often bristle-tipped at the apex, tapered or slightly rounded to the base; dark green and glabrous above or with fine hairs along the main veins, green or slightly paler beneath and hairy along the veins or over the entire surface; margins entire, ciliate, or with slender gland-tipped teeth; petioles very short.

Flowers rather densely clustered at the ends of the branches; corolla broadly urn-shaped or cylindrical, about 8 - 10 mm long, 5-parted, white or pink-tinged; the largest-flowered of our blueberries, opening as the leaves unfold or rarely when the leaves are half grown; late May and June. Fruit a sweet, juicy, blue or blue-black berry with a bloom, about 8 - 10 mm in diameter; July and Aug. (*corymbosum* — in corymbs, referring to the arrangement of flowers in the clusters)

In low woods, at the edges of swamps or ponds, and rarely in open sandy clearings.

In the Deciduous Forest Region, in Northumberland County, and in eastern Ontario between the Ottawa and St. Lawrence rivers. (N.S. to Wis., south to Ark. and n. Fla.)

Note Our description includes several varieties of high-bush blueberry, which have been named in some manuals, as well as the closely related black high-bush blueberry, *V. atrococcum* (Gray) Heller. (*atrococcum* — black-berried)

Field check Our tallest blueberry (up to 2.5 - 3 m); in swamps or wet woods; leaves 3.5 - 7.5 cm long; berries sweet and juicy, blue-black with a bloom.

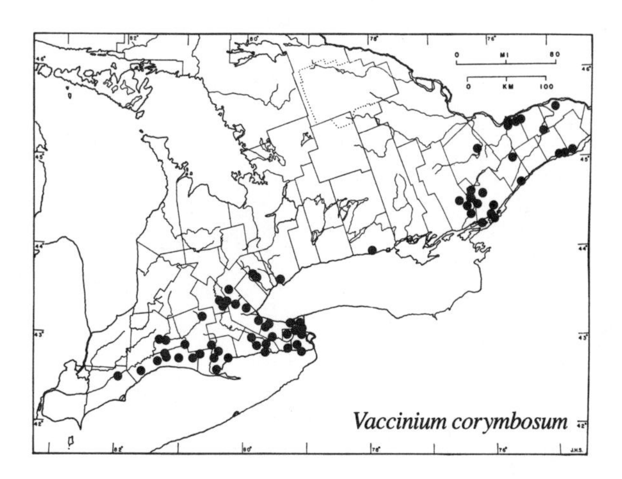

Vaccinium corymbosum

Vaccinium macrocarpon

Vaccinium macrocarpon Ait. — American cranberry, Large cranberry

A prostrate, trailing, slender evergreen vine with wiry or cordlike stems, usually much branched and intertwining, often 1 m or more in length, the upright or ascending leafy or flowering branches less than 2 dm high. Branchlets light brown to reddish-brown and minutely pubescent; older stems with a papery outer layer peeling off in shreds and exposing the smooth dark bark beneath.

Leaves alternate, simple, and evergreen; blades elliptic or oblong-elliptic, leathery, dark green above and pale beneath, 5 - 15 mm long and 2 - 5 mm wide; margins entire and somewhat revolute, the tip of the blade rounded or blunt, and the base abruptly contracted to a very short petiole.

Flowers solitary or 2 - 6 in a cluster from the axils of the lower reduced leaves of the erect or ascending leafy branches; corolla divided into 4 lanceolate pale pink segments (6 - 10 mm long) which are separate almost to the base and spreading or reflexed; stamens 8, exserted, the whole giving the appearance of a tiny shooting star (*Dodecatheon* sp., a member of the primrose family); pedicel slender, pubescent, with 2 small, green, leaflike bracts above the middle; July. Fruit an edible, but rather acid, red berry 1 - 2 cm in diameter, globose to ellipsoid, obovoid, or rarely pyriform, frequently remaining on the vine over winter; late Aug. to Oct. (*macrocarpon* — from the Greek *makros*, long, and *karpos*, fruit)

In swamps and open bogs, and on wet shores of ponds and streams.

From Lake Erie to the eastern shore of Lake Superior in the west and to the Ottawa River in the east; also in the Rainy River District; apparently reaching its northern limit near 49° N. (Nfld. to w. Ont. and Minn., south to Ill., Ark., and N.C.)

Note Wild and cultivated plants of this species are the main source of the commercial cranberries in eastern North America. When sweetened, they make excellent jellies, preserves, and sauces.

Field check Creeping evergreen vine with small, blunt, nearly sessile leathery leaves; pedicels with 2 small leaflike bracts above the middle; berries red, sour.

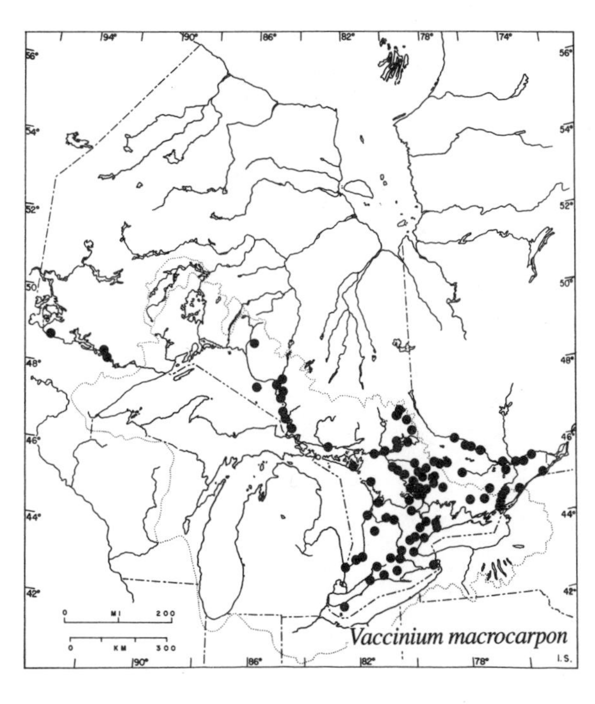

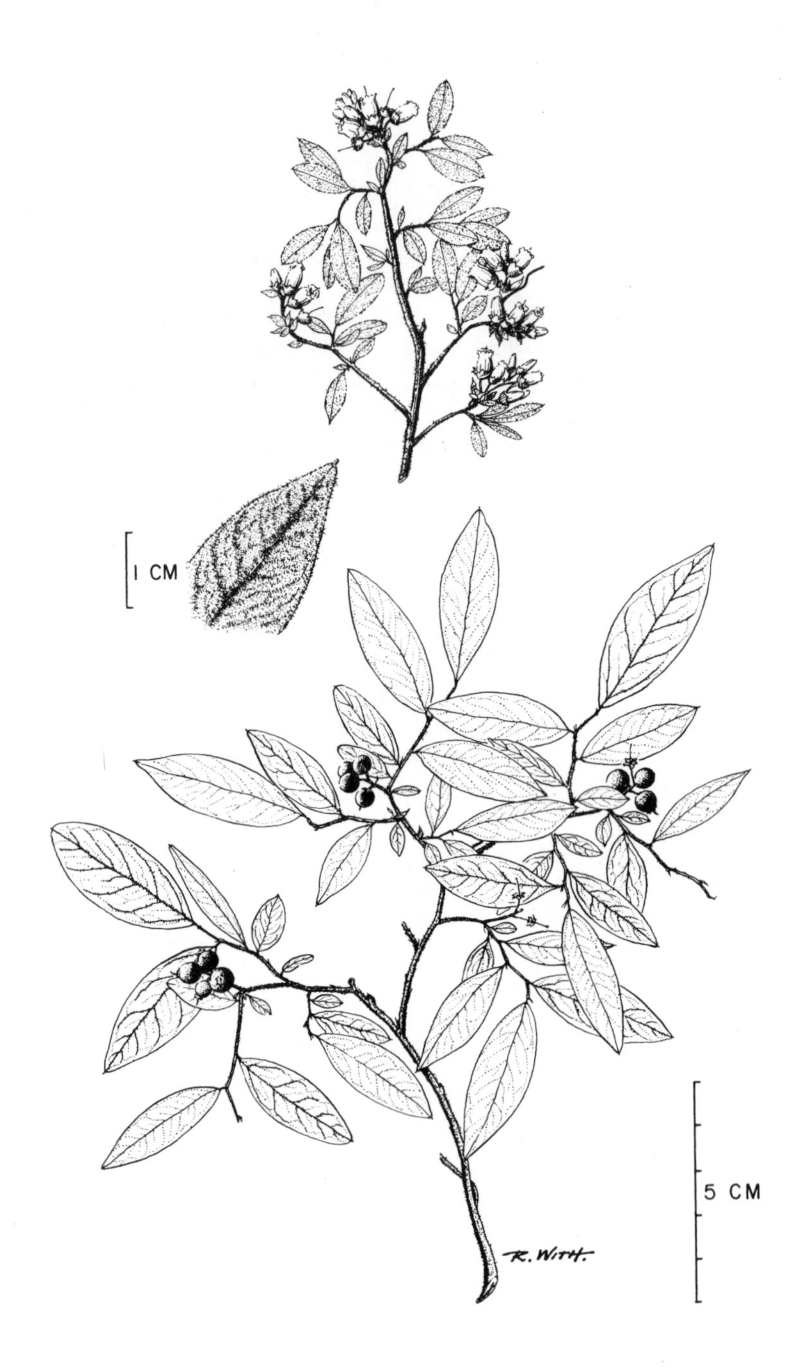

Vaccinium myrtilloides

Vaccinium myrtilloides Michx. (*V. canadense* Kalm) — Velvet-leaf blueberry

A low shrub with spreading or ascending branches, reaching a height of 3 – 6 dm. Young branches greenish-brown, densely velvety-hairy with a coating of stiffish, white hairs; older stems reddish-brown to blackish with crowded warty lenticels, the finely hairy outer bark finally peeling off.

Leaves alternate, simple, and deciduous; blades oval to elliptic, 2.5 – 5 cm long and 1 – 2.5 cm wide, tapered or pointed at both ends or somewhat rounded at the base, thin, dark green above, paler beneath, downy on both surfaces or becoming glabrous above except along the veins; margins entire and finely hairy; petioles very short and hairy.

Flowers in crowded clusters at the ends of short leafy branches, opening as the leaves are expanding; corolla bell-shaped or short-cylindrical, 5-parted, less than 6 mm long, greenish-white to creamy-coloured, and pink or purple-tinged; late May and June. Fruit an edible but sour blue berry, usually with a heavy bloom, 6 – 9 mm in diameter; July to Sept. (*myrtilloides* — resembling *myrtillus*, in reference to *Vaccinium myrtillus*, a European blueberry; *canadense* — Canadian)

In dry or moist, sandy or rocky clearings and open woods; also in sphagnum bogs and swamps.

Widely distributed throughout Ontario except for the northernmost areas near Hudson Bay. (Nfld. to B.C., south to Mont. and the mountains of Va.)

Field check Low shrub with velvety-hairy warty stems; colonial; leaves downy, at least below, the margins finely hairy; berries sour, blue with a bloom.

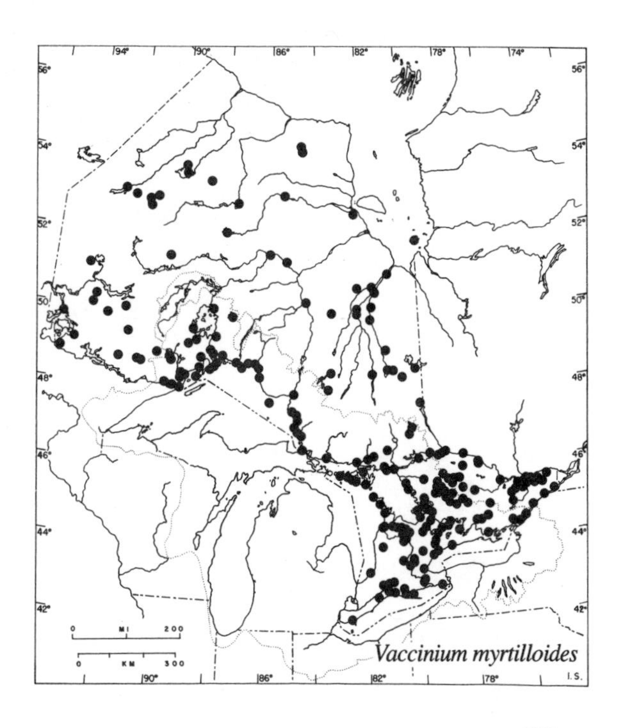

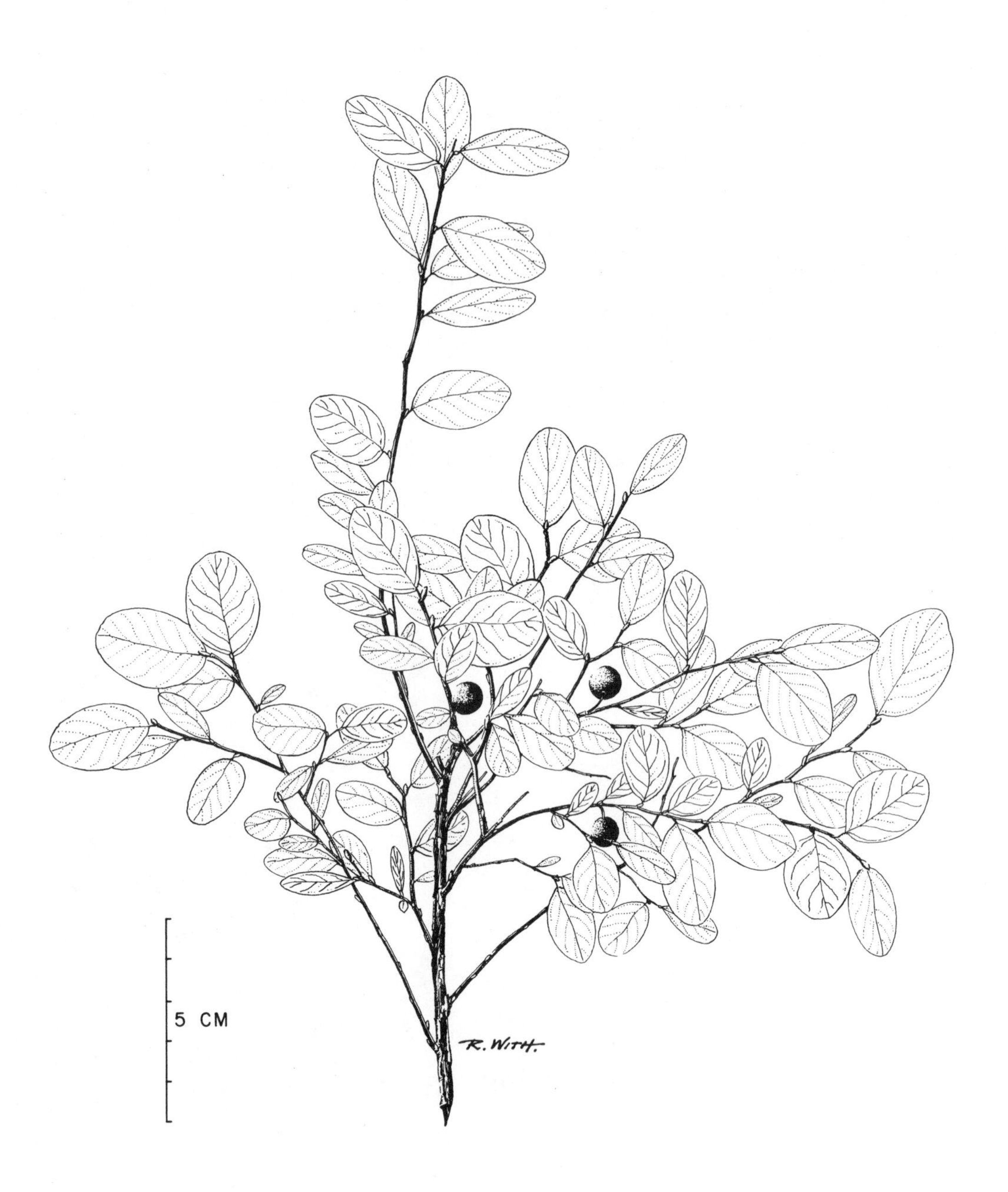

Vaccinium ovalifolium

Vaccinium ovalifolium Sm. — Oval-leaved bilberry

A straggling shrub reaching a height of 8–12 dm. Branchlets at first brownish and glabrous, with sharp longitudinal ridges or conspicuously angled; older stems purplish-gray to blackish with strongly flaking outer bark.

Leaves alternate, simple, and deciduous; blade thin, broadly oval to nearly round, 1–3 cm long and 0.5–1.8 cm wide, blunt or rounded at the tip and often with a small abrupt point (mucro), rounded or broadly wedge-shaped at the base, dull green above, paler beneath, glabrous on both surfaces; margins entire or with a few small, widely spaced, gland-tipped teeth near the base; petioles about 1–2 mm long.

Flowers solitary on short stalks from the axils of the lower leaves on the current year's growth; corolla about 6 mm long, 5-parted, broadly bell-shaped and pinkish, opening when the leaves are only half grown or sometimes before the leaves appear; June. Fruit a dark blue to blue-black berry with a bloom, 6–9 mm in diameter, often with a disagreeable flavour; July and Aug. (*ovalifolium*—with oval-shaped leaves)

In rocky mixed woods and along shores of lakes.

Along the eastern shore of Lake Superior from Michipicoten Harbour and Michipicoten Island south to the vicinity of Sault Ste. Marie. (Nfld., N.S., and Que.; Ont. and n. Mich.; Alaska to Oreg. and Mont.; e. Asia)

Field check Straggling shrub with angled branchlets; leaves oval to roundish, glabrous, entire or with a few small teeth below the middle; flowers pinkish, 5-parted, solitary in lower leaf axils on current year's growth; berries blue-black with a bloom; restricted to east shore of Lake Superior.

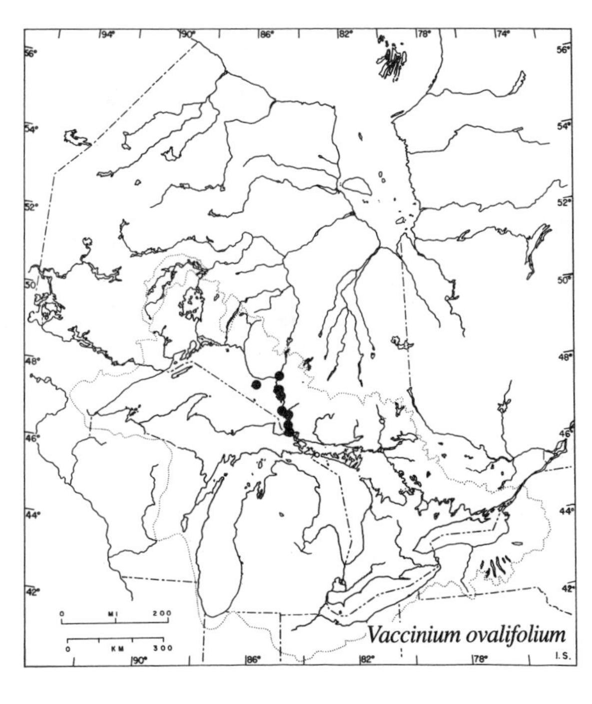

Vaccinium ovalifolium

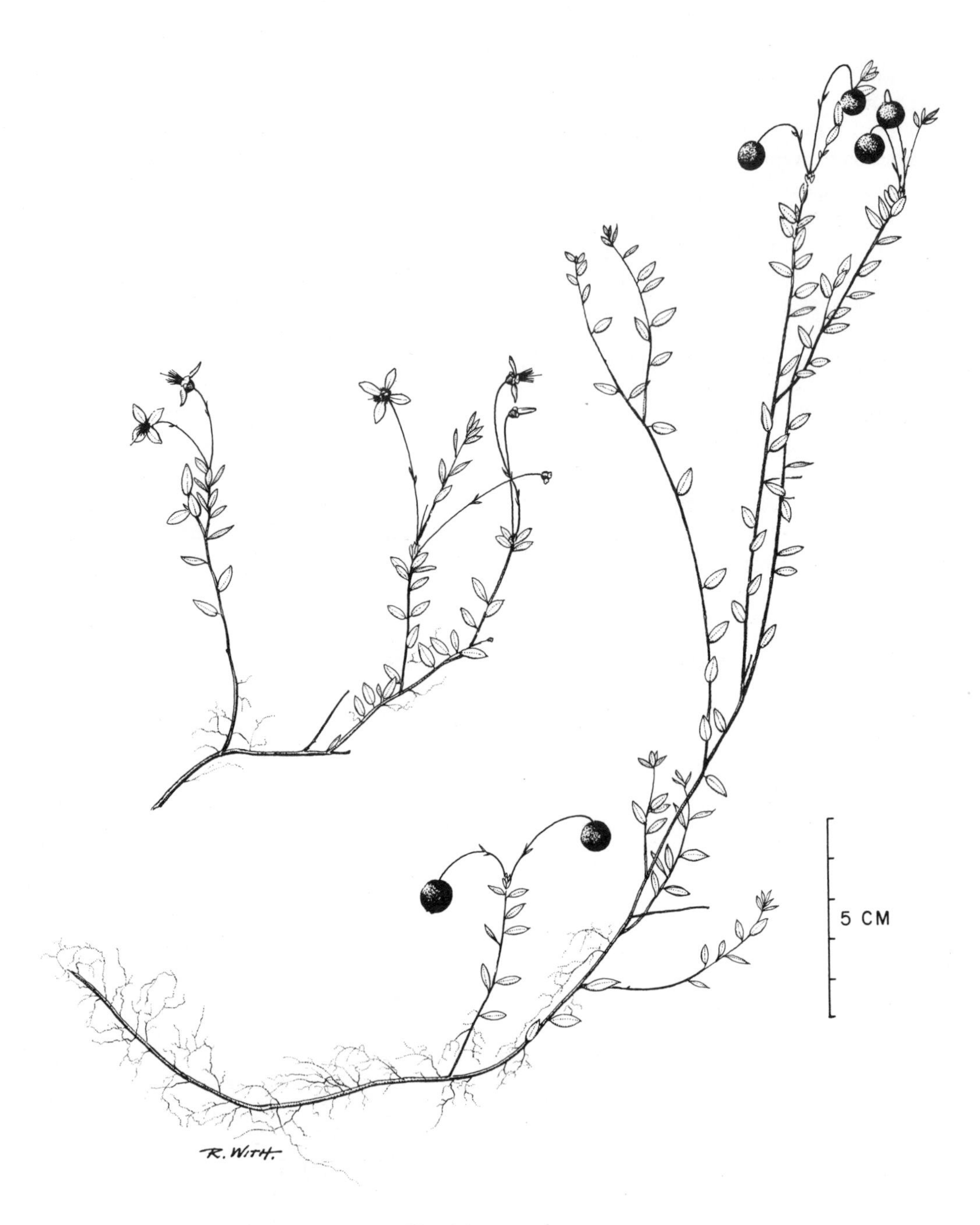

Vaccinium oxycoccos

Vaccinium oxycoccos L. — Small cranberry

A creeping or trailing prostrate vine with very slender wiry stems often rooting at the nodes and growing to half a metre or more in length, the erect or ascending leafy and flowering branches usually less than 2 dm high. Branchlets light brown to reddish-brown, minutely pubescent, the outer bark of older stems peeling in pale shreds or strips and exposing the smooth dark inner bark.

Leaves alternate, simple, and evergreen; blades 2 - 10 mm long, 1 - 3 mm wide, elliptic-ovate or narrowly triangular, strongly whitened beneath, the tip acute or blunt and the base rounded; margins entire and strongly revolute; petiole very short.

Flowers solitary or in clusters of 2 - 6 at the ends of the leafy branches; corolla with 4 lanceolate, pale pink, recurved segments 5 - 6 mm long; stamens 8, prominently exserted; pedicel slender, erect, smooth or slightly pubescent and with 2 mostly reddish-coloured, scalelike bractlets at or below the middle; late June and July. Fruit a globose berry, 5 - 15 mm in diameter, at first pale and speckled but becoming reddish when ripe, edible but sour; Aug. to Oct. (*oxycoccos* — sour berry, an old generic name of Greek origin still used by botanists desiring to separate cranberries from blueberries, deerberries, and other subgenera of *Vaccinium*)

In peat bogs, on wet, acid or sour soils, and on tundra.

Throughout Ontario except for the region at the western end of Lake Erie. (Nfld. to Alaska, south to Oreg., Mich., and N.C.; Eurasia)

Field check Creeping evergreen vine with small, mostly pointed and strongly revolute, nearly sessile leaves whitened beneath; pedicels with 2 slender reddish-coloured bractlets at or below the middle; berries red, sour.

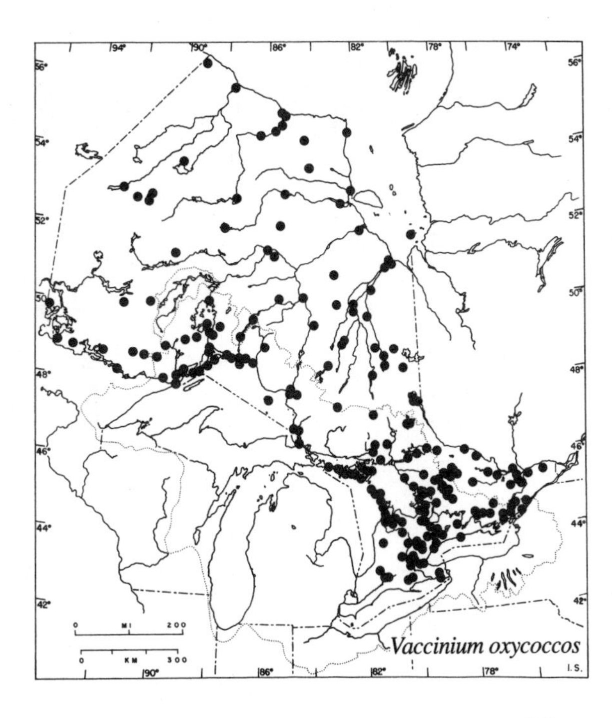

Vaccinium oxycoccos

Vaccinium pallidum

Vaccinium pallidum Ait. (*V. vacillans* Torr.) — Dryland blueberry

A low shrub with stiff erect stems and short ascending branches, usually less than 1 m high. Branchlets yellowish-green to brownish, glabrous or minutely hairy in lines running down from the nodes; older stems brownish and glabrous, somewhat ridged or angled, the bark wrinkled or closely warty, with numerous tiny lenticels.

Leaves alternate, simple, and deciduous; blades 2.5 - 5 cm long and 1 - 2.5 cm wide, thickish, broadly oval to obovate or elliptic, pointed or rounded at the apex and often with a minute sharp point (mucro), rounded or tapered at the base, dull green above, paler beneath, minutely hairy at first on both surfaces, becoming glabrous at maturity except along the veins and margins; margins entire or minutely toothed; petiole very short and glabrous or minutely hairy.

Flowers at the ends of short side branches, opening before the leaves are fully expanded; corolla 5-parted and narrowly bell-shaped or cylindrical, about 6 mm long, creamy-white to buff-coloured or greenish, often tinged with red; late May and early June. Fruit a sweet, edible, dark blue berry with a slight bloom, 6 - 9 mm in diameter; late July to Sept. (*pallidum*—pale; *vacillans*—swaying)

In dry sandy open woods and on hillsides, especially under oak and pine.

In the Deciduous Forest Region from Windsor to the Niagara River, Hamilton, and Toronto; also on islands in the upper St. Lawrence River. (N.H. to s.e. Wis., south to e. Okla. and Ga.)

Note Although the name *V. vacillans* has been frequently used for this shrub, it must be replaced by the earlier name proposed by Aiton.

Field check Low stiffly branched shrub of dry sandy habitats; leaves thickish, from one-third to one-half as broad as long; flowers whitish to greenish, often tinged with red; berries dark blue with a slight bloom, sweet, and edible.

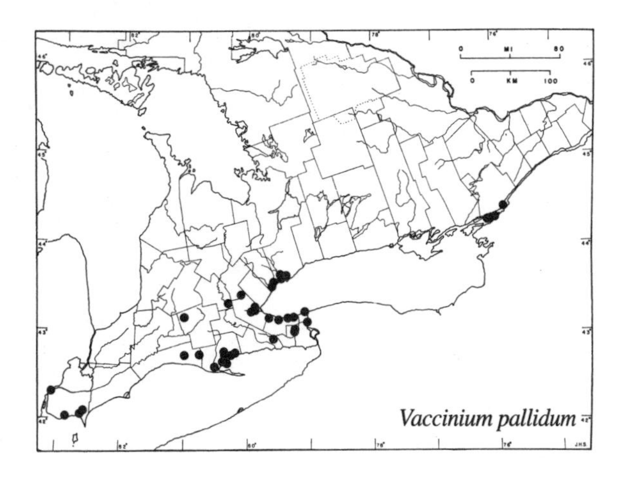

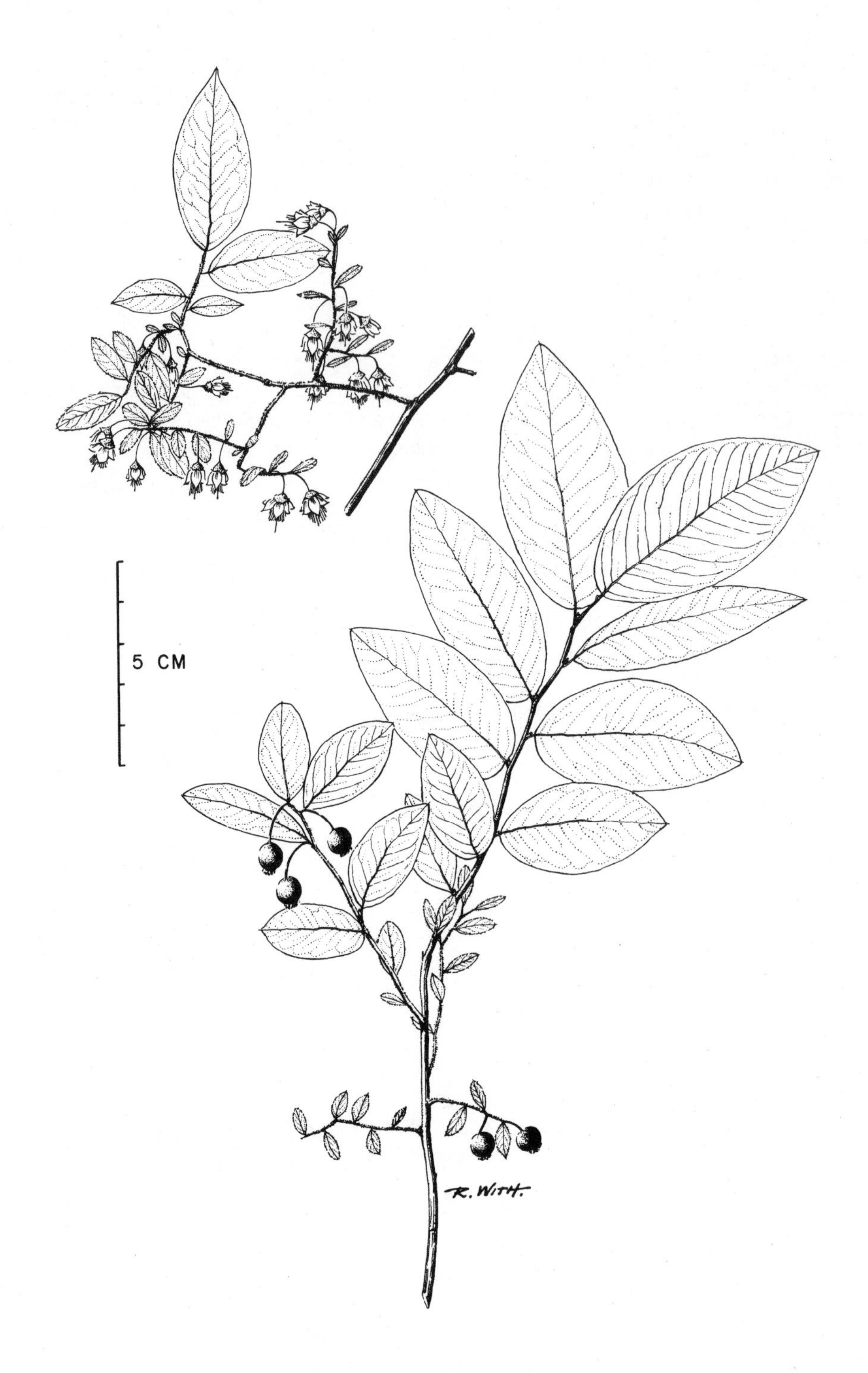

Vaccinium stamineum

Vaccinium stamineum L.

Deerberry
Squaw huckleberry

A low, slender shrub with spreading branches, seldom exceeding 1 m in height. Branchlets minutely hairy and gray to brownish, becoming glabrous and gray to blackish with the outer bark peeling off in thin papery shreds.

Leaves alternate, simple, and deciduous; blades 2.5–9 cm long and 0.5–4 cm wide, oval to elliptic, broadest at or above the middle, blunt or pointed at the tip and often with a minute sharp point (mucro), rounded to broadly wedge-shaped at the base, firm, dark green above, much paler or conspicuously whitened beneath, at first minutely hairy on both surfaces but becoming glabrous above except along the midrib; margins entire and finely hairy; petiole less than 3 mm long and densely hairy.

Flowers numerous in loose clusters on short divergent branches, gracefully pendent on long slender stalks from the axils of conspicuous leafy bracts resembling the smallest leaves; corolla shallowly open, bell-shaped, about 6–9 mm in diameter, with 5 white widely spreading lobes; stamens 10, in a fringelike ring around the spinelike style; late May and June. Fruit a scarcely edible, yellowish-green to bluish, round, juicy berry, 6–9 mm in diameter, with a few soft seeds, falling as soon as ripe; July and early Aug. (*stamineum*—with stamens, in reference to the prominent stamens)

In dry rocky woods, thickets, or clearings.

Restricted to the Niagara River region and to the Thousand Islands area in the St. Lawrence River; previously reported from London and Hamilton but not recently collected in those areas. (Mass. to Ind., south to Mo., La., and Fla.)

Note This species is in danger of extinction in the Niagara River site; one of the stands in the Thousand Islands is within the boundaries of the St. Lawrence Islands National Park and efforts are being made to protect the plants there.

Field check Low slender shrub; leaves thin, entire, whitened beneath, broadest at or above the middle; flowers shallowly open, bell-shaped, numerous in leafy-bracted clusters; stamens conspicuous in a ring around the spinelike style; berries yellowish-green to bluish; restricted to a few sites in the Niagara River and Thousand Islands regions.

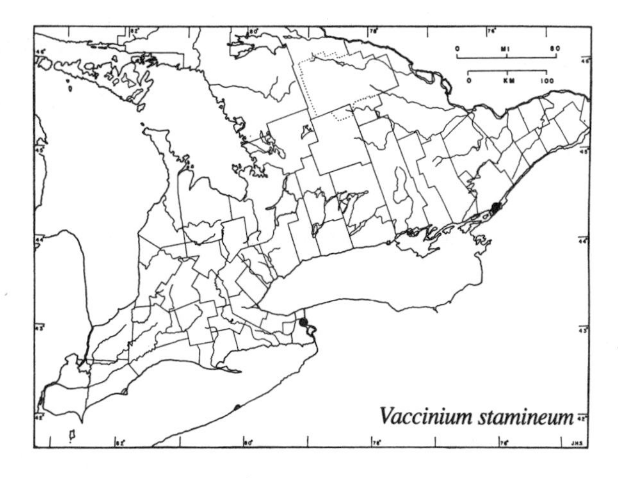

Vaccinium uliginosum ssp. *pubescens*

Vaccinium uliginosum L. ssp. *pubescens* (Wormsk. ex Hornem.) Young — Bog bilberry

A low shrub with many stiff erect or spreading branches, rarely exceeding 6 dm in height. Branchlets reddish-brown and glabrous, the thin whitish epidermal layer peeling off the older stems and exposing grayish-brown to purplish-black bark.

Leaves alternate, simple, and deciduous; blades rather firm to leathery, narrowly elliptic to nearly round, 5 - 25 mm long and 5 - 15 mm wide, pointed, rounded, or indented at the tip and tapered at the base, dark green and glabrous above, paler and sparsely short-hairy beneath, conspicuously veiny, especially beneath; margins entire and slightly inrolled; petioles very short or the blade sessile.

Flowers usually 4-parted, rarely 5-parted, narrowly urn-shaped, 5 - 6 mm long, white or pale pink, solitary or in small clusters; late June and early July. Fruit a sweet, blue to blue-black berry with a bloom, about 4 - 6 mm in diameter; late July and Aug. (*uliginosum*—of wet or marshy places; *pubescens*—pubescent)

In sphagnum bogs and along the rocky or sandy shores of lakes and rivers.

From the north shore of Lake Superior to James Bay and Hudson Bay. (Greenl. and Nfld. to Alaska, south to Ont. and N.Y.)

Note Our description includes the alpine bilberry *V. uliginosum* var. *alpinum* Bigel. which is sometimes treated as a distinct species, *V. gaultherioides* Bigel. The alpine bilberry is a smaller plant with very small roundish leaves and is alpine and high-arctic in distribution. Shores of James Bay and Hudson Bay, e.g., Cape Henrietta Maria. It is to be looked for also along the north shore of Lake Superior and on the adjacent islands. (*alpinum*—alpine; *gaultherioides*—resembling *Gaultheria*)

Field check Low, stiff, much-branched northern shrub; leaves firm, less than 2.5 cm long, veiny, with entire inrolled margins; flowers mostly 4-parted, solitary or in small clusters; berries sweet, blue or blue-black with a bloom.

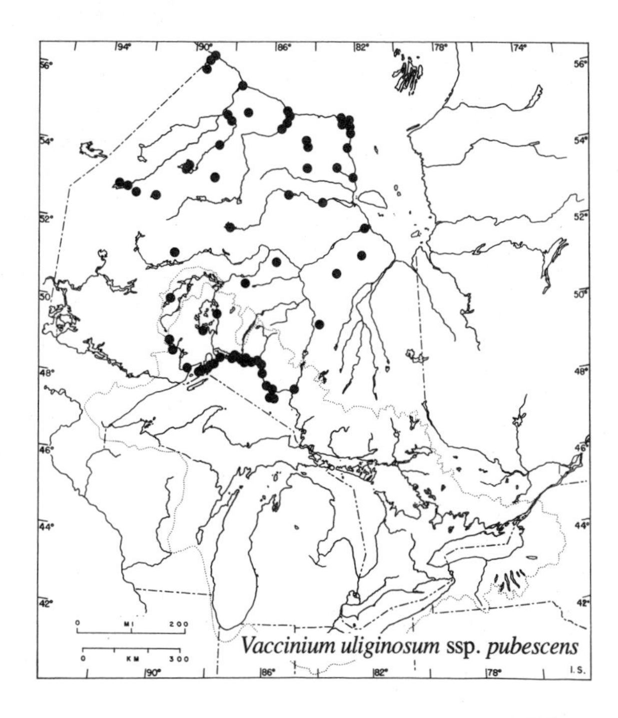

Vaccinium uliginosum ssp. *pubescens*

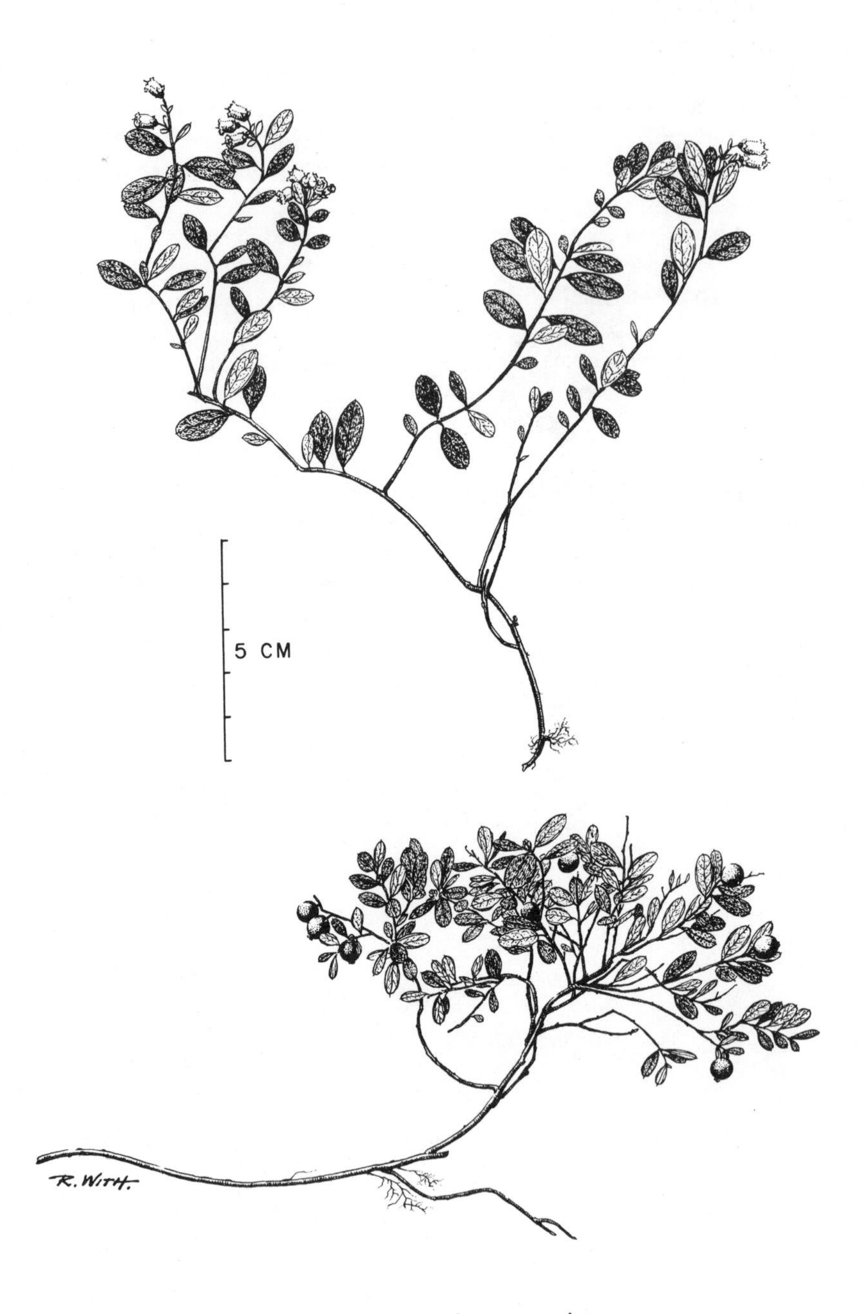

Vaccinium vitis-idaea var. *minus*

Vaccinium vitis-idaea L. var. *minus* Lodd. — Mountain cranberry

A prostrate, trailing, evergreen dwarf shrub with erect, slender branching stems, usually less than 15 cm high, often forming mats. Branchlets greenish-brown to reddish, nearly glabrous or with short crisp curly white hairs; older stems brown to blackish, glabrous or with a few crispy hairs, the outer bark peeling off.

Leaves alternate, simple, and evergreen; blades firm and leathery, oval to elliptic, 0.5 – 2 cm long and 3 – 15 mm wide, blunt, rounded, or indented at the tip, tapered or broadly wedge-shaped at the base, dark green, glossy, and glabrous above, much paler and black-dotted with dark bristlelike glands beneath; margins entire or wavy and conspicuously revolute, with curly hairs near the base; petioles 1 – 2 mm long and crispy-hairy.

Flowers in small one-sided clusters from scaly buds at the ends of the branches, each flower on a short glandular pedicel which has two small bracts near its base; calyx 4-parted, with glandular margins; corolla white, pinkish or reddish, 4-parted, open, bell-shaped, the style exserted; June and July. Fruit a sour and slightly bitter dark red berry about 8 – 10 mm in diameter, persistent over winter and more palatable the following spring. (*vitis-idaea*—grape of Mt. Ida; *minus*—small)

In rocky or sandy clearings, on moss-covered boulders and stumps, or in sphagnum bogs and muskeg.

From the north shore of Lake Superior and adjacent islands to James Bay and Hudson Bay. (w. Greenl. to Alaska, south to B.C., the Great Lakes, Que., and N. Eng.; Iceland; Eurasia)

Note *V. vitis-idaea* is circumboreal in distribution, the North American plants differing slightly from the European and usually treated as a geographical variety (var. *minus*) or as a subspecies (ssp. *minus* (Lodd.) Hult.). In Europe it has various common names, including cowberry, lingen, lingberry, and red whortleberry.

Both in Europe and wherever it is available in North America, the berries of this species are eaten raw or cooked in a manner similar to our use of commercial cranberries.

Field check Dwarf mat-forming northern shrub; leaves glossy and evergreen, less than 2.5 cm long, with scattered, dark, bristly glands on the lower surface; flowers 4-parted, in terminal clusters; berries dark red, sour.

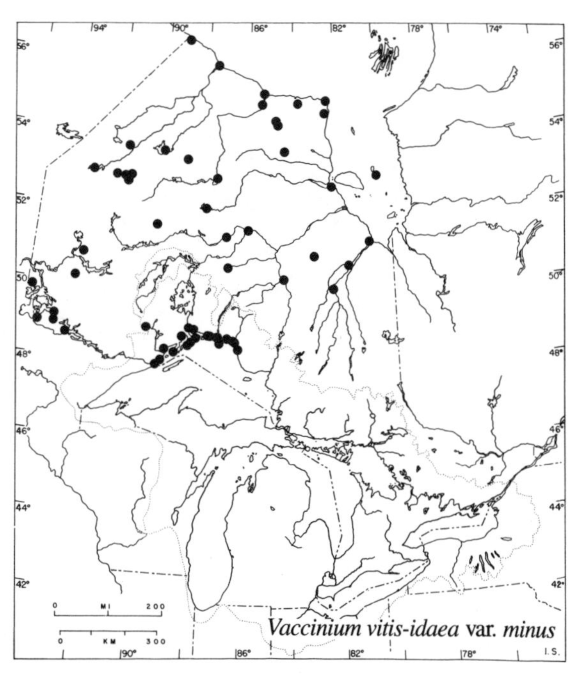

Vaccinium vitis-idaea var. *minus*

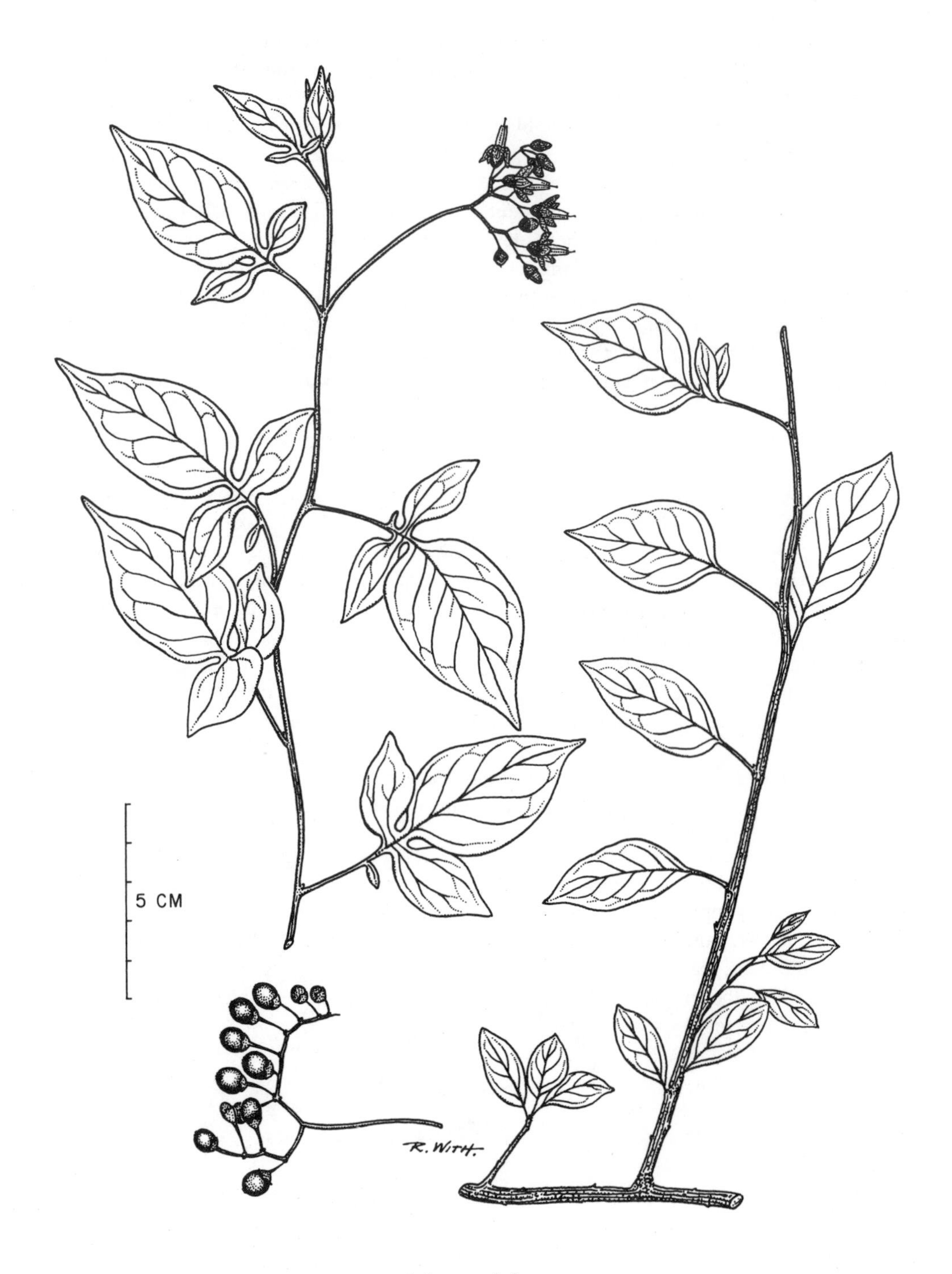

Solanum dulcamara

SOLANACEAE – NIGHTSHADE FAMILY

A large, world-wide, and primarily tropical family of 80–90 genera and 2000–3000 species, with a few members in the temperate parts of North America and Eurasia. Herbs, shrubs (rarely climbing), and small trees. Leaves usually alternate, simple, lobed, or compound, deciduous, lacking stipules. Flowers in cymes or solitary, perfect, mostly regular and 5-parted on bractless pedicels; ovary superior. Fruit a many-seeded berry or capsule.

Pollination is by insects, some flowers with long tubular corollas being visited by moths. Several genera include species of economic importance, e.g., *Solanum* (potato, tomato, egg-plant), *Capsicum* (red pepper), and *Nicotiana* (tobacco). (*Capsicum*—possibly from the Latin *capsa*, a chest, in reference to the form of the fruit; *Nicotiana*—in honour of Jean Nicot, 1530–1600, a French consul to Lisbon who introduced tobacco to the royal court of Portugal)

The name of the family is based on the genus *Solanum*, of which we have included here one species, a vine with somewhat woody stems and persistent woody bases.

Solanum L. – Nightshade

A genus of nearly 2000 species in tropical and temperate regions. Herbs, shrubs, vines, or small trees.

The calyx and corolla are rounded or wheel-shaped and the corolla is plaited in the bud; the stamens have short filaments and their anthers converge in a group closely surrounding the style. The fruit is a berry usually with 2 locules. (*Solanum*—a Latin name which is unexplained)

Solanum dulcamara L.

Bittersweet nightshade
Climbing nightshade

A perennial climbing or twining vine which may reach a length of 2–4 m. Branchlets green to brownish and herbaceous, smooth or minutely downy; older stems near the base of the plant woody and persistent, brown to gray, and up to 1 cm or more in diameter.

Leaves alternate, simple, and deciduous; the blade elliptic to lance-oval or broadly ovate with or without one or two pairs of lobes at the base, the unlobed leaves usually characteristic of the lower part of the stem, the lobed ones on the upper floriferous portion; margins entire; petiole short or up to nearly the length of the blade, both blade and petiole glabrous to minutely downy.

Flowers 1–1.5 cm across, somewhat saucer-shaped, with 5 spreading or reflexed blue-purple petal lobes and a central column of yellow stamens with a slender protruding style; in terminal or lateral, somewhat forking or cymose, open-branched stalked inflorescences; blooming throughout the season; June to Aug. or later. Fruit a many-seeded, juicy, red berry with a thin translucent skin. (*dulcamara*—from the Latin *dulcis*, sweet, and *amarus*, bitter)

In fields, fencerows, waste places, thickets, and clearings, usually near habitation but often appearing indigenous.

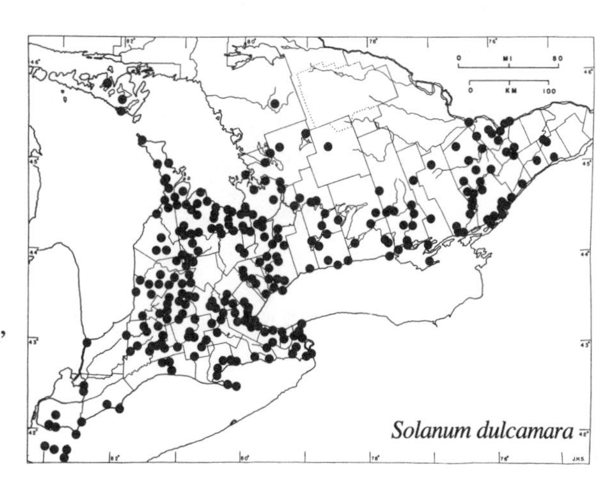

Solanum dulcamara

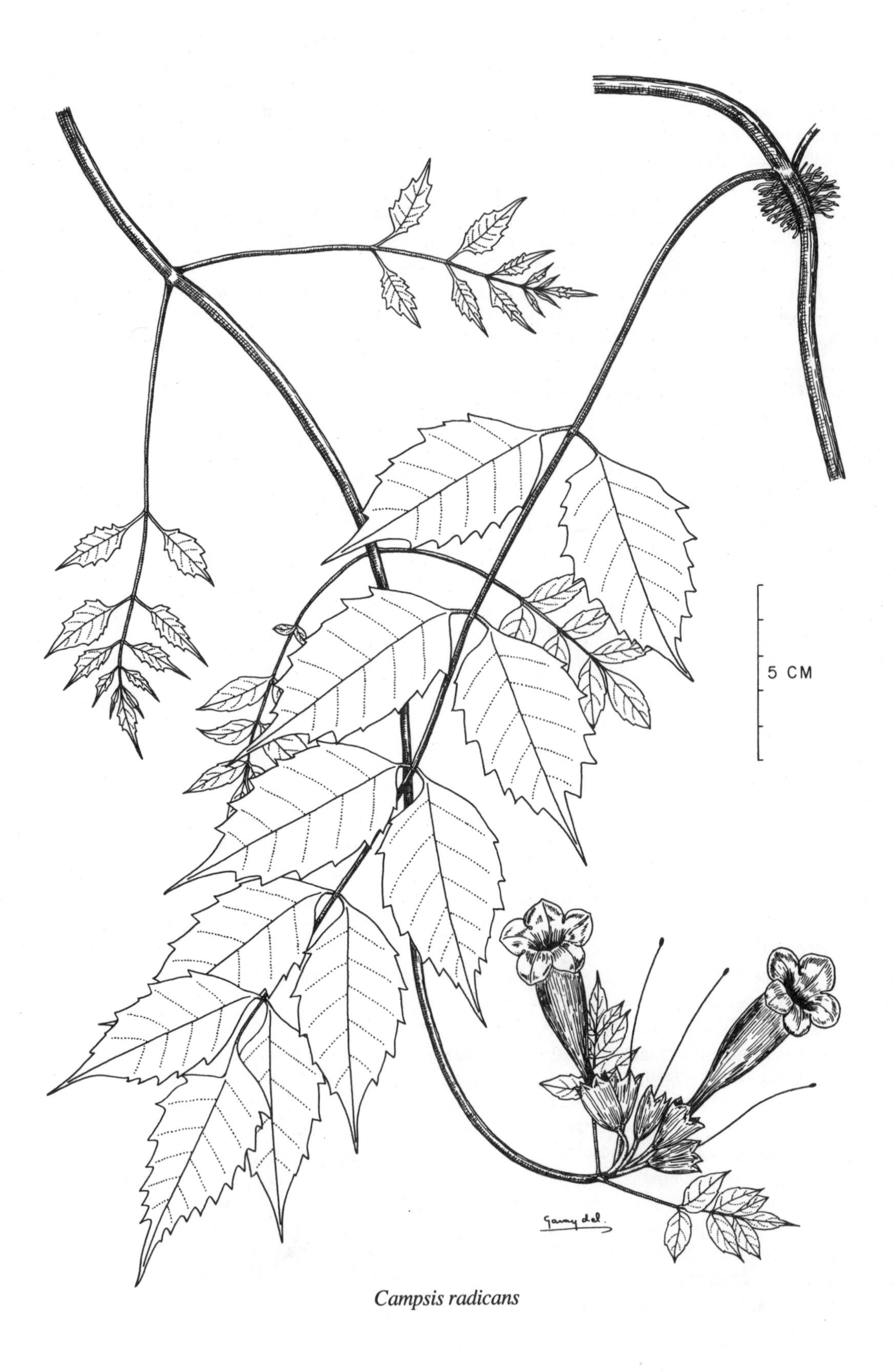

Campsis radicans

Introduced from Eurasia and naturalized throughout southern Ontario from Lake Erie to Manitoulin Island and from Lake Ontario to the Ottawa district; northern limit near 46° N; a report from the vicinity of Sault Ste. Marie needs confirmation. (widely introduced in southern Canada and the U.S.A.)

Note The specific name *dulcamara* and the derived common name bittersweet nightshade are based on a report that the roots, if chewed, taste bitter at first then sweet. This could be a dangerous experimental test since this plant belongs to the Nightshade Family in which so many plants contain toxic substances. The berries are considered to be toxic although some people have eaten a few without serious results.

Bittersweet is the common name for *Celastrus scandens* L., the tough, woody, twining vine with persistent orange-red capsules, and it is considered preferable to restrict its use to that plant.

Field check Perennial climbing vine with simple and basally lobed alternate leaves, purple and yellow flowers, and juicy translucent red berries.

BIGNONIACEAE – TRUMPET CREEPER FAMILY

Trees, shrubs, or lianas with opposite, usually compound but sometimes simple leaves without stipules. Flowers with 5-lobed funnel-shaped or bell-shaped irregular or bilabiate corolla which is deciduous. Fruit a capsule with winged flattened seeds. Chiefly tropical with 100-120 genera and 600-800 species. The tropical lianas show great variation in method of climbing; some are stem-twiners, others root-climbers, and still others climb by means of tendrils with or without adhesive discs or hooks. Our only representative is in the genus *Campsis*.

The family name is based on *Bignonia*, the genus of the cross-vine. (*Bignonia*—named for Abbé Jean-Paul Bignon, 1662-1743, court librarian at Paris, a friend of Tournefort)

Campsis Lour. – Trumpet creeper

Two species of woody vines, trailing or high-climbing by means of aerial roots along the stem, with pinnately compound leaves and large orange-red trumpet-shaped flowers on very short stalks. Both species have a unique system of nectaries (see Elias & Gelband, 1975). In addition to the nectar-secreting ring in the flower, there are four sets of nectaries outside the flower, comprising minute glands on the petioles, over the developing fruit, on the calyx, and on the corolla lobes. They begin functioning at different times and attract ants, some species displaying the "ant-guard" relationship known previously only in tropical or subtropical plants. The copious nectar secreted from the ring encircling the ovary differs in composition from the secretions of the extra-floral nectaries and attracts humming birds and bees. (*Campsis*—Greek *kampsis*, a bend or curve, in reference to the incurved stamens)

Campsis radicans (L.) Seem. — Trumpet creeper

A vigorous scrambling or high-climbing woody vine adhering to the trunks of trees by means of aerial roots and reaching a length of 5 m or more. Branchlets green or reddish, smooth or minutely downy at first, with a fringe of hairs across the stem at the base of each pair of leaves.

Leaves large, opposite, pinnately compound, and deciduous, each leaf long-petioled with 7 - 13 (usually 9 or 11) stalked, ovate or ovate-lanceolate leaflets 4 - 8 cm long and 1 - 4 cm wide; leaflets coarsely toothed, smooth and bright green above, a little paler and smooth or hairy along the veins beneath.

Flowers orange-red, trumpet-shaped, 5 - 8 cm long, with 5 spreading and slightly irregular lobes, on short stalks in conspicuous crowded clusters at the ends of gracefully curved branches; July and early Aug. Fruit a thick pod, 10 - 15 cm long, with a longitudinal ridge on each side; seeds winged, attached in several rows to the prominent central partition; late Aug. to Oct. (*radicans* — rooting, in reference to the aerial roots)

In sandy soil of clearings and edges of woods, climbing over conifers and up the trunks of deciduous trees; also escaping to roadsides and fencerows.

Native only along the Lake Erie shore of Essex County and on the Erie Islands near Point Pelee; formerly in the Chatham area, Kent County; frequently planted as an ornamental vine as far north as Georgian Bay but barely hardy in the Ottawa region. (N.J. to s. Ont. and Ill., south to Tex. and Fla.)

Note The trumpet creeper is a handsome ornamental vine that attaches itself easily to brick, masonry, or wooden sides of buildings. However, its vigorous growth habit can cause problems if the vine is allowed to spread onto shingled roof tops. It can also be grown on a tall stump or post as an accent point in a medium-sized or large garden. In Ontario this is the only vine pollinated to a large extent by birds.

The other species in this genus, *C. grandiflora* Loisel., is native in China and Japan. (*grandiflora* — with large flowers)

Field check Woody climber with aerial roots; opposite pinnately compound leaves; large orange-red trumpet-shaped flowers in clusters at the ends of curving branches; long thick pods producing winged seeds.

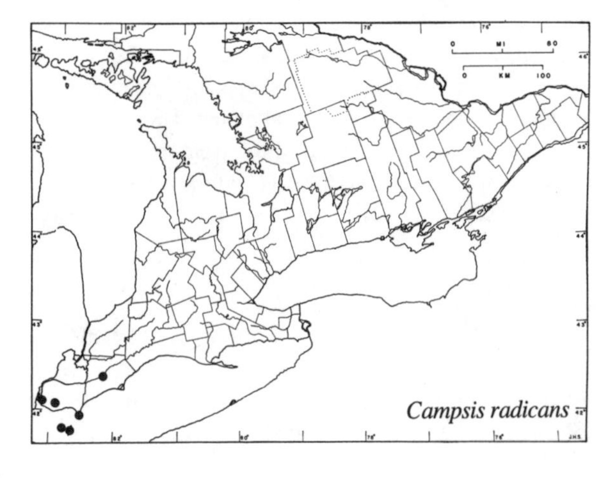

RUBIACEAE — MADDER FAMILY

One of the largest of flowering plant families with 400-500 genera and 6000-7000 species, most abundant in the tropics but with some representatives in temperate and even in arctic regions. Trees, shrubs, and herbs with opposite or whorled leaves; the stipules may be fused to form a sheath around the stem or they may become leaflike to produce, by various degrees of splitting or fusion, a whorl of 4, 6, 8, or 10 leaves at the node. Inflorescence usually a much-branched cyme or cymose panicle but sometimes the flowers solitary or few in a reduced cluster. Flowers perfect, regular, 4- or 5-parted, sometimes of two different shapes; petals united, borne above the inferior ovary. Fruit a capsule, berry, or twin 1-seeded nutlets which split apart at maturity. Insect pollination is common and some woody tropical members are inhabited by ants.

Several genera are of economic importance, including *Coffea* and *Cinchona*, which are sources of alkaloids (coffee and quinine, respectively), and *Gardenia*, which has a number of species cultivated for their handsome fragrant flowers. The family name is based on the genus *Rubia*. (*Rubia*—from the Latin *ruber*, red, in reference to the red dye obtained from the roots of *Rubia tinctorum* L.; *Coffea*—an Arabian name for the beverage; *Cinchona*—after Countess Cinchon, wife of a Spanish Viceroy in Peru who, in 1638, was cured of a fever by the use of bark of a Peruvian tree; *Gardenia*—in honour of Alexander Garden, M.D., a resident of Charleston, South Carolina, who corresponded with Linnaeus; *tinctorum*—with properties of a dye)

Cephalanthus L. — Buttonbush

About 15 species in temperate and tropical America, Asia, and Africa. Shrubs or small trees with numerous small white flowers borne in densely packed and long-stalked heads. (*Cephalanthus*—from the Greek *kephale*, head, and *anthos*, flower)

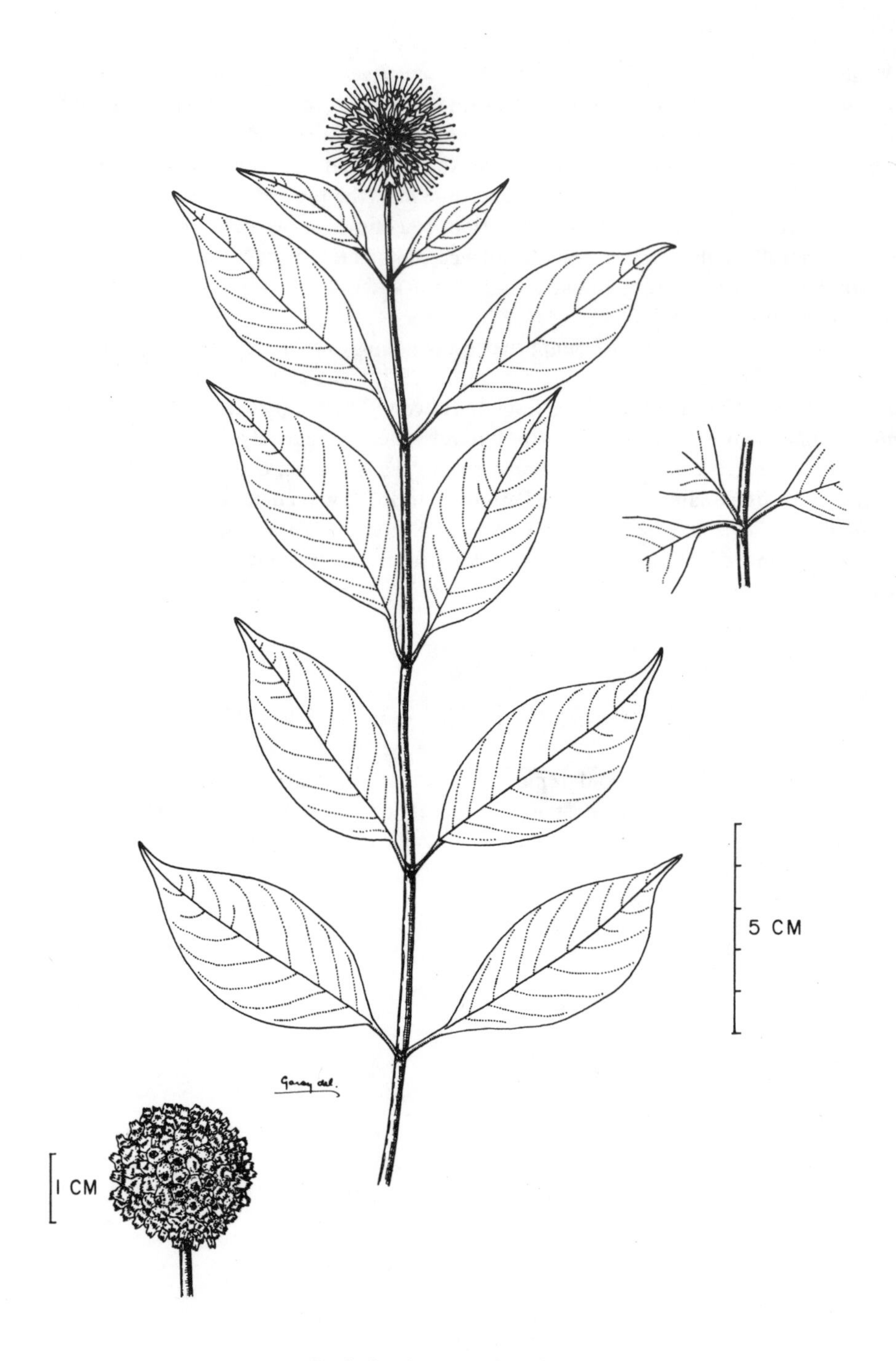

Cephalanthus occidentalis

Cephalanthus occidentalis L. **Buttonbush**

A large spreading shrub usually with several stems from the ground or more rarely treelike, up to 3 m in height. Branchlets green at first, later turning brown, with scattered pale lenticels; older stems gray-brown to purplish-gray, with light brown pith and smooth or furrowed bark.

Leaves opposite or whorled in threes or rarely in fours and deciduous; blades up to 18 cm long and 7.5 cm wide, thickish, bright green and glossy above and usually paler or softly hairy beneath, elliptic-lanceolate to broadly ovate, the apex acute, acuminate, or obtuse and the base rounded or tapered; margins entire or slightly wavy; petioles stout, grooved, and up to 2 cm long, with a pair of short, triangular, sharp-pointed stipules at the base.

Flowers small, perfect, creamy-white, tubular, and 4-parted, with a prominently exserted swollen-tipped style, borne in large numbers (100–200) in tightly packed spherical heads measuring 2–4 cm across, the numerous threadlike styles forming a complete halo; heads solitary or in dichasial groups of two or more at the ends of the branches, or on long stout stalks from the axils of upper leaves; mid-July. Fruit a spherical head of numerous brown cone-shaped nutlets with compressed sides and a persistent terminal 4-toothed remnant of the calyx; Sept. and Oct. (*occidentalis*—western, i.e., found in the western hemisphere)

Usually in damp habitats where the roots are water-logged for at least the early part of the season: along streams, bordering ponds, bogs, or marshes, and in ditches.

Widely distributed in southern Ontario, northward to 45° N and beyond to the Magnetawan River on the west and the Barron River on the east; absent or rare in the intermediate upland area which includes most of Nipissing district. (N.S. to Minn., southwest to Calif. and south to Mex. and Fla.; West Indies)

Field check Shrub of wet situations; opposite or whorled entire leaves; dense round heads of white flowers and hard spherical balls of dry brown fruits.

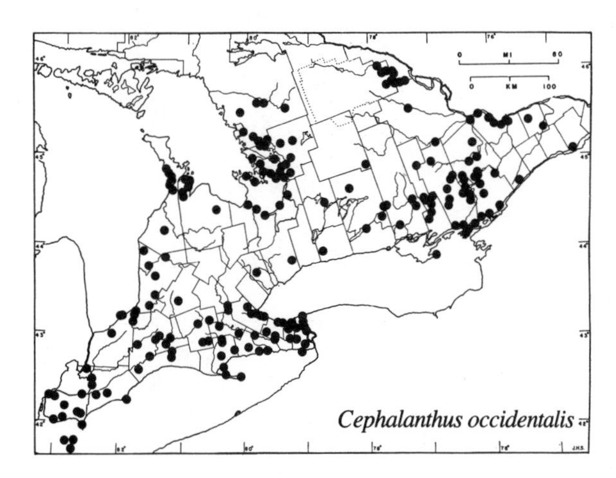

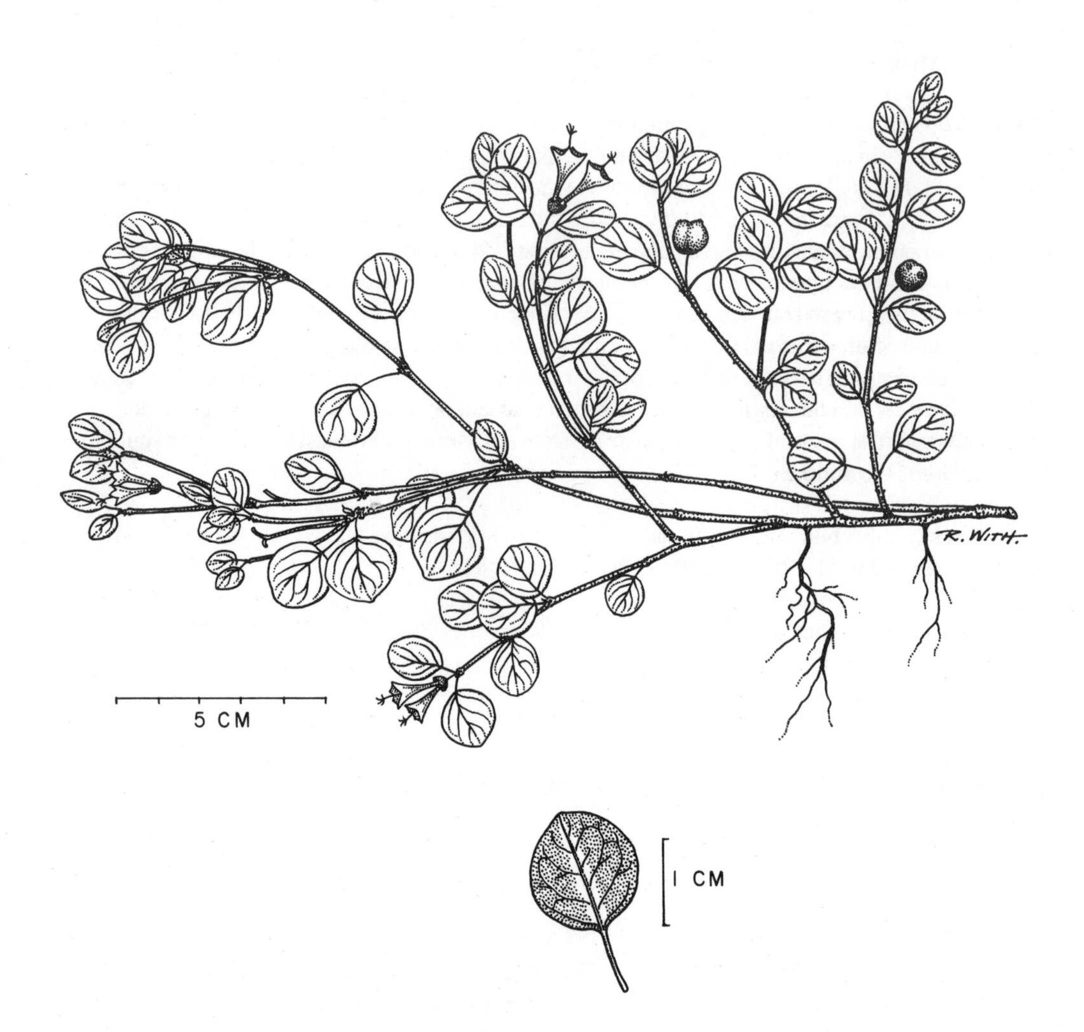

Mitchella repens

Mitchella L. — Partridgeberry

Only two species are known, one in North America and the other in northeastern Asia. Prostrate creeping plants with slender woody or wiry stems bearing small, ovate, opposite evergreen leaves and minute stipules. Flowers in pairs, the corolla funnel-shaped with spreading lobes, densely bearded on the inside. Fruit a berrylike double drupe. (*Mitchella*—named by Linnaeus in honour of a botanical correspondent in Virginia, Dr. John Mitchell, 1676–1768)

Mitchella repens L. — Partridgeberry, Two-eyed berry, Running box

A small trailing evergreen vine usually less than 5 dm long, with slender wiry persistent stems; rather sparsely branched except near the base of the plant; rooting at the nodes.

Leaves opposite, simple, and evergreen; blades round-ovate, 1–2.5 cm long and about as wide, blunt at the tip and rounded to truncate at the base, smooth and dark green with a broad pale midrib and often variegated with white lines on the upper surface; petioles often as long as the blade, usually with two lines of minute hairs.

Flowers fragrant, in terminal or axillary pairs; corolla white to purplish-tinged, 10–15 mm long, tubular with 4, occasionally 3 or 5–8, spreading to reflexed lobes densely bearded on the inside surface; June and July. Fruit a bright red berrylike double drupe, the ovaries of the two flowers uniting into one structure crowned by the persistent calyx teeth of the two flowers of the pair and containing 8 bony nutlets; fruit remaining on the plant over the winter; Aug. and Sept. (*repens*—creeping)

Part of the ground cover in dry or moist soil of mixed woods, cedar and pine forests.

Widespread throughout southern and central Ontario, west to the eastern end of Lake Superior; northern limit at about 47° N. (Nfld. to Minn., south to Tex. and Fla.)

Note The flowers are heterostylous, those on some plants having short stamens concealed within the tube, the styles long and protruding, an arrangement called "pin", while on other plants the stamens are exserted and the styles included, an arrangement called "thrum". This device, which promotes cross-pollination, is further reinforced by failure of pollen to be functional on the stigma of the same flower (Keegan et al., 1979). As plants of both kinds are not always growing close enough together for cross-pollination to occur, some patches of partridgeberry may be sterile.

The berries are edible but rather tasteless. A form with white instead of red berries is occasionally found.

Field check Small prostrate vine with slender wiry stems; stalked, round-ovate, entire opposite leaves; pairs of 4-parted, white or purplish-tinged, tubular fragrant flowers; scarlet overwintering berries.

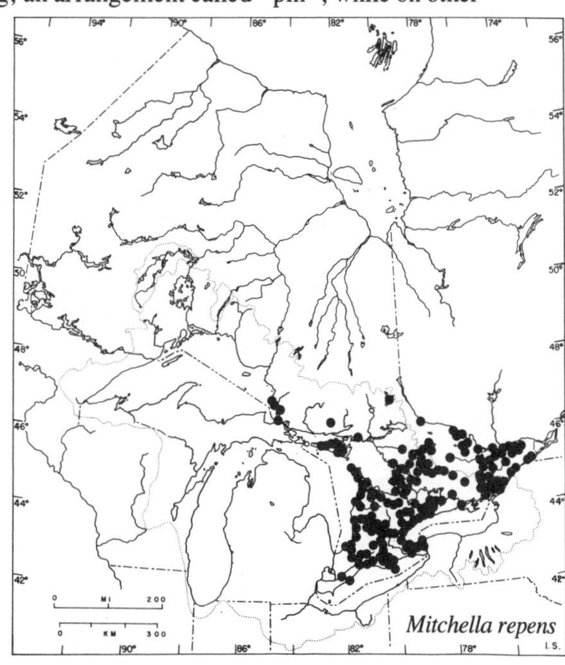

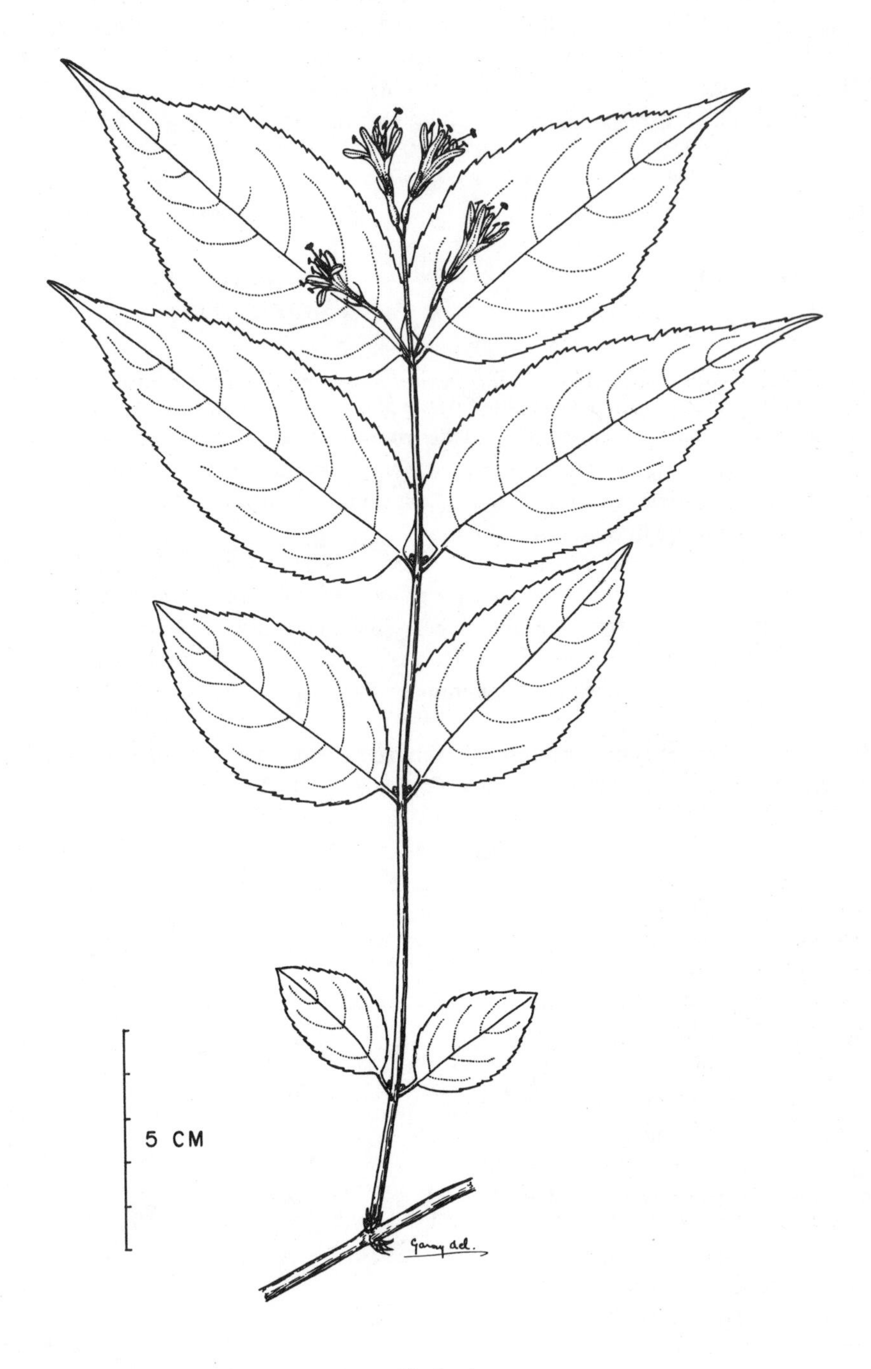

Diervilla lonicera

CAPRIFOLIACEAE – HONEYSUCKLE FAMILY

Shrubs, small trees, vines, and rarely herbs with opposite, mostly simple or lobed leaves (compound in *Sambucus*). Flowers with inferior ovary and 2 - 5(- 8) united carpels. Fruit a berry, drupe, achene, or capsule. About 15 genera and 500 species chiefly in the north temperate zone but some occur in mountain regions of the tropics. Includes a number of ornamental shrubs and vines, such as *Weigela, Kolkwitzia, Viburnum, Lonicera*. (*Weigela*—after Christian Ehrenfried von Weigel, 1748 - 1831; *Kolkwitzia*—in honour of Richard Kolkwitz, 1873 - 1932, Professor of Botany at Berlin)

Diervilla Mill. – Bush honeysuckle

Three species of shrubs confined to North America. Winter buds with several pairs of pointed bud scales. Flowers yellow but changing to orange or red after anthesis; pollination by insects of the Order *Hymenoptera*. (*Diervilla*—named for the French surgeon Dr. N. Dièreville who travelled in eastern Canada in 1699 - 1700 and introduced the shrub into France)

Ornamentals with showy clusters of crimson to pinkish tubular flowers are sold by nurseries under the name of the closely related genus *Weigela*.

Diervilla lonicera Mill. — Bush honeysuckle

A low upright shrub usually less than 1 m high. Branchlets green or reddish, often with two lines of minute hairs running lengthwise along the stem; older twigs brownish to gray.

Leaves opposite, simple, and deciduous; blades 5 - 13 cm long and 1.5 - 6 cm wide, oblong-ovate to lanceolate, with a long tapering and sometimes curving tip and a rounded or wedge-shaped, often asymmetrical base, dark green and usually smooth above, or hairy on the midrib, pale green and smooth to densely short-hairy below; margins sharply serrate and with a fringe of short hairs; petioles 3 - 12 mm long. Plants with the leaves densely hairy beneath, *D. lonicera* var. *hypomalaca* Fern., are more common northwestward.

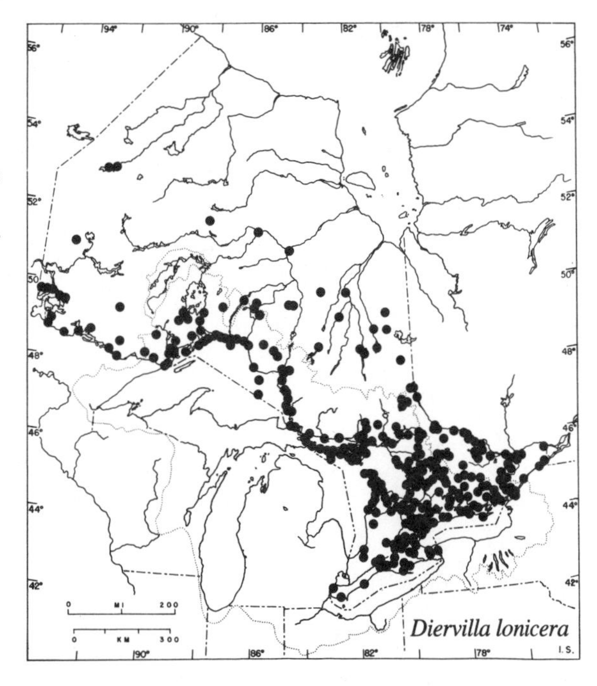

Diervilla lonicera

Flowers narrowly tubular, perfect, 5-parted, in short-stalked clusters of 2 - 6 (often 3) at the ends of the branches and in the axils of the leaves along the stem; petals pale yellow at first, turning orange or brownish-red; June and July. Fruit a slender brown capsule, the free end pointed and tipped by a threadlike beak with calyx lobes bristlelike and persistent, solitary or in small clusters; July to Sept. (*lonicera*—after *Lonicera*, the generic name of the honeysuckle; *hypomalaca*—from the Greek *hypo*, below, and *malakos*, soft, referring to the soft hairy coating on the lower surface of the leaves in the variety)

Usually in dry soil in sandy or rocky woods and thickets, on cliffs, ridges, sand dunes, and hillsides, and in open pastures.

Common in southern Ontario and north

Linnaea borealis ssp. *longiflora*

to 50° N with a few scattered locations in the upper drainage basins of the Albany and Severn rivers; absent from the Boreal Forest and Barren Region. (Nfld. to e. Sask., south to Iowa and N.C.)

Note A study in Michigan (Schoen, 1977) indicates that the flowers of *Diervilla lonicera* are protogynous, self-incompatible, and adapted for pollination by bumblebees and hawkmoths.

Field check Low shrub; opposite serrate tapered leaves; clusters of yellow-red tubular flowers and pointed slender-beaked capsules.

Linnaea Gronov. — Twinflower

Slender trailing perennial vine with small evergreen leaves. *Linnaea* is a monotypic genus in the northern hemisphere, and the species *L. borealis*, which occurs in Eurasia as the typical subspecies *borealis*, is separated in America as subspecies *longiflora* (Torr.) Hult. (*Linnaea*—named in honour of Carolus Linnaeus, 1707-1778, the famous Swedish naturalist with whom the typical subspecies of this plant was a special favourite)

Linnaea borealis L. ssp. *longiflora* (Torr.) Hult. — Twinflower

A prostrate creeping or trailing evergreen vine often reaching a length of 2 m or more, with numerous short, erect, leafy side branches less than 1 dm high. Branchlets slender and wiry, green to reddish-brown and finely hairy; older stems woody but rarely exceeding 2-3 mm in diameter.

Leaves opposite, simple, and evergreen; the blade oval, rounded, or obovate, 1-2 cm long, tip blunt and base abruptly tapered to a short petiole, upper surface of the blade and margins with straight, bristlelike hairs; margins revolute with a few blunt teeth beyond the middle; petiole with crinkly hairs shorter than those of the blade.

Flowers small, perfect, pinkish, bell-shaped, and delicately fragrant, in pairs (rarely 3 or more) at the summit of a Y-shaped stalk 5-10 cm long bearing 2 tiny bracts at the fork, 2 bractlets at the ends of the branches and, finally, each flower subtended by a pair of minute bracts; peduncle, its branches, and all the bracts and bractlets glandular-hairy; June to Aug. Fruit a tiny dry 1-seeded capsule enclosed by the persistent calyx; Aug. to Sept. (*borealis*—northern; *longiflora*—long-flowered)

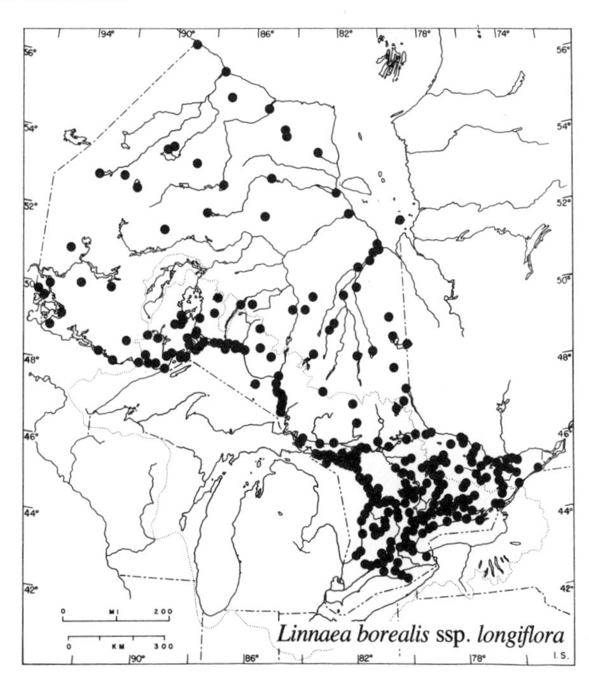

Linnaea borealis ssp. *longiflora*

On the forest floor and in clearings of coniferous and mixed woods; on hummocks in sphagnum bogs and in cedar woods; on rotten logs, moss-covered boulders, talus slopes, and cliff bases; on the banks of rivers and lakes and along the edges of woods.

Widely distributed from Lake Erie north

to Hudson Bay and from the upper part of the St. Lawrence River west to Lake-of-the-Woods; absent in the area at the western end of Lake Erie. (Greenl. to Alaska, south to Calif. and Md.; Eurasia)

Note Twinflower is pollinated by insects. The capsules separate from their stalks below the sticky glandular bracts and are dispersed by animals to which they adhere.

The American subspecies has also been known as *L. borealis* ssp. *americana* (Forbes) Hultén and as var. *americana* (Forbes) Rehder. (*Americana*—American)

Field check Slender trailing evergreen vine with small, roundish, and sparingly blunt-toothed opposite leaves; pairs of delicately fragrant pinkish bell-shaped flowers on erect Y-shaped glandular-hairy stalks.

Lonicera L. — Honeysuckle

Erect and twining shrubs or vines with opposite, simple, entire deciduous (or rarely evergreen) leaves, some of which may be perfoliate. Flowers showy and fragrant, those with long tubular corollas pollinated by long-tongued insects such as hawkmoths, or by humming birds. Insects of the Orders *Diptera* and *Hymenoptera* pollinate species with short or more open corollas but some bees are known to bite a hole at the base of the corolla to obtain the nectar, thereby leaving their role as a pollinator unfulfilled.

Although honeysuckle flowers are perfect, they are functionally unisexual, thus promoting cross-pollination. When the flowers first open, the stamens are uppermost and ready to shed pollen while the style with its immature stigma curves downward. After the pollen is shed, the style curves upward bringing the now receptive stigma uppermost while the empty stamens curve downward. Although dependent on weather conditions, the transposition of these organs usually takes place in about three days. Changes in flower colour are correlated with stages of pollination and fertilization. Dispersal is due in part to the regurgitation of undigested seeds by birds which eat the juicy berries. (*Lonicera*—in honour of the sixteenth century German herbalist and physician Johann Lonitzer, 1499-1569)

Key to *Lonicera*

a. Erect shrubs; leaves all distinct; flowers in axillary pairs
- b. Pith of branches white, solid; flowers yellowish
  - c. Bark gray-brown, shredding; flowers and fruits on long stalks (1 cm or more); fruits red, purple, or purple-black
    - d. Leaves acuminate; the 4 bracts of the flower stalks broad and leaflike, the inner ones enclosing the purple-black fruits — ***L. involucrata***
    - d. Leaves blunt or rounded; bracts of the flower stalks none or small and inconspicuous
      - e. Leaves smooth, the margins and short petioles ciliate; berries ovoid, distinct, and widely divergent — ***L. canadensis***
      - e. Leaves short-hairy beneath, the margins not ciliate; berries round, partly or nearly united, not divergent — ***L. oblongifolia***
  - c. Bark reddish-brown, peeling in conspicuous papery layers; flowers and fruits on short stalks (1 cm or less); fruits blue — ***L. villosa***
- b. Pith of branches brown with a central cavity; flowers pink or white — ***L. tatarica***

a. Trailing or twining vines, rarely shrublike; lower leaves distinct, upper 1-4 pairs united across the stem to form discs; flowers in terminal whorled clusters or in interrupted spikes from the terminal discs
  - f. Leaves smooth above, whitened or bluish-white and smooth or hairy beneath, margins not ciliate — ***L. dioica***
  - f. Leaves hairy on both surfaces, pale green beneath, margins ciliate — ***L. hirsuta***

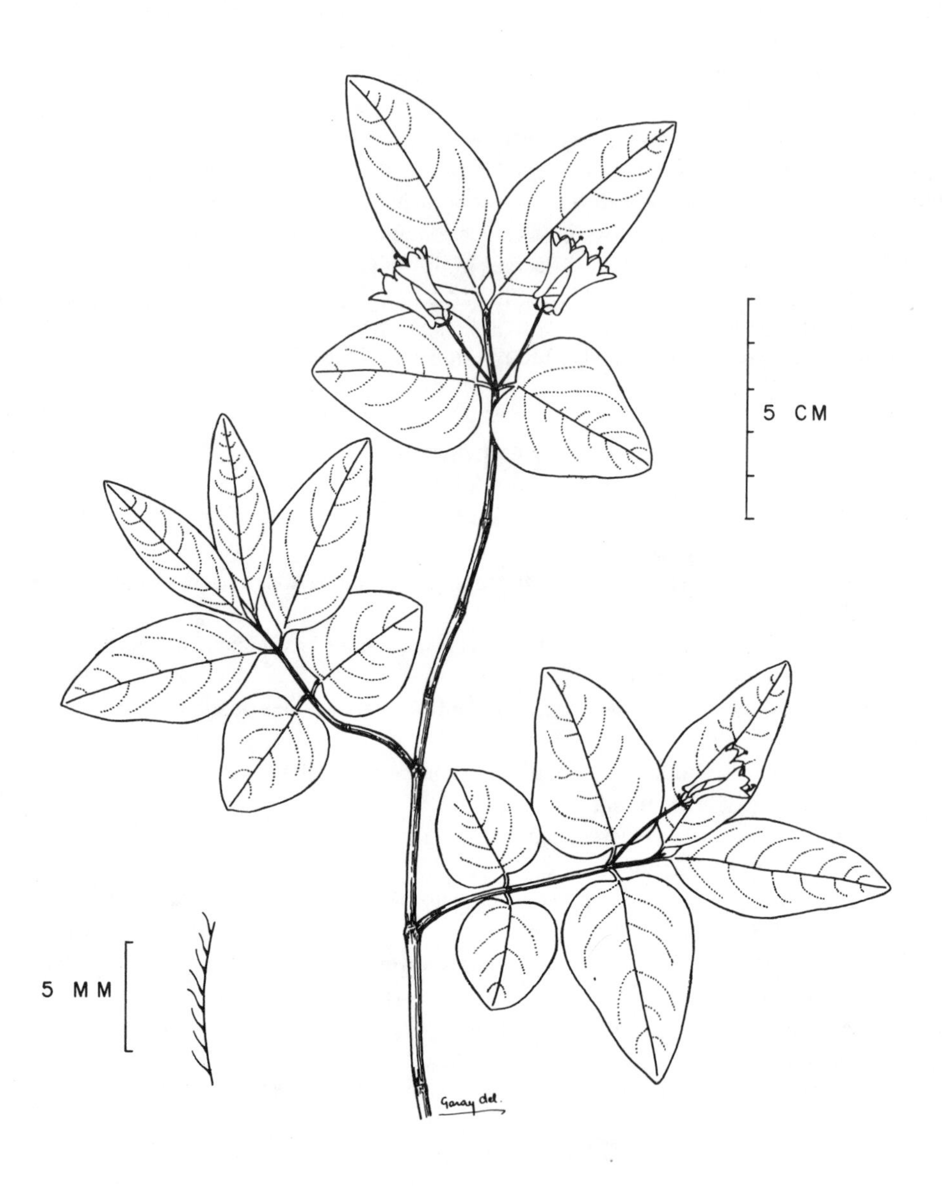

Lonicera canadensis

Lonicera canadensis Bartr. — Fly honeysuckle

An erect and loosely branched or straggling shrub up to 1.5 m in height. Branchlets green or purplish and smooth, later becoming gray to brownish; bark on older twigs shreddy with fine threadlike portions coming loose.

Leaves opposite, simple, and deciduous; blades 2.5–9.5 cm long and 2.5–5 cm wide, thin, ovate-oblong to elliptic, acute or blunt at the tip, rounded, shallowly heart-shaped, or tapered at the base, bright green and smooth above, a little paler beneath; margins and the short (less than 1 cm) petiole ciliate.

Flowers pale yellow, tubular to funnel-shaped, 12–18 mm long, with short flaring lobes and an asymmetrical swelling at the base; in pairs on long slender stalks from the axils of the leaves; opening as the leaves are still expanding; May and June. Fruits in long-stalked divergent pairs of 3–4-seeded ovoid red berries; late June and July. (*canadensis*—Canadian)

Usually in damp or shaded ground in woods or thickets, on ravine slopes, and around swamps and bogs.

Common throughout southern Ontario and from Lake Abitibi and the upper part of the Moose River drainage basin west to Lake-of-the-Woods; northern limit near 50° N. (Que. to w. Ont., south to Iowa and N.C.)

Field check Straggling shrub with broadly elliptic opposite leaves; leaf margins and petioles ciliate; pairs of yellow flowers on long slender stalks; elongate, divergent, red twin berries.

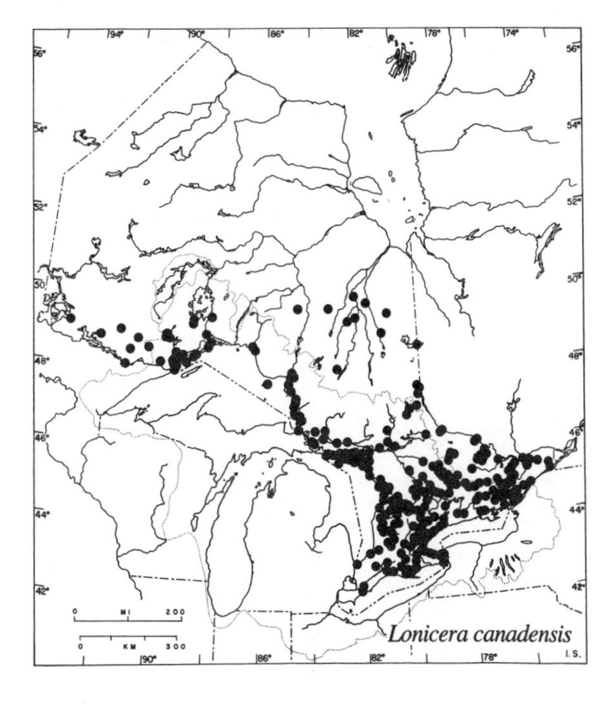

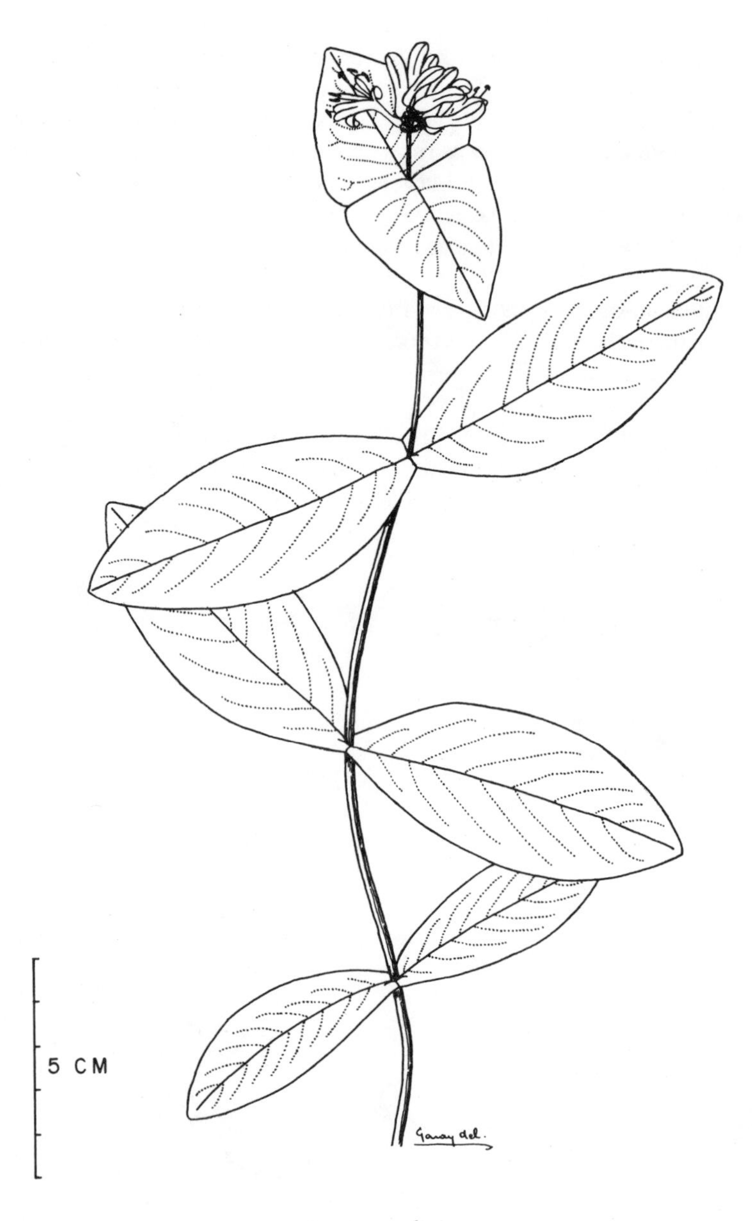

Lonicera dioica

Lonicera dioica L. — Glaucous honeysuckle

A semi-erect shrub or twining woody vine growing to a length of 3 m or more. Branchlets smooth and green or purplish at first; older stems turning brown to gray; bark shredding and peeling off.

Leaves opposite, simple, and deciduous; blades sessile or the lower ones distinctly stalked with a petiole about 1 cm long, the upper 1 - 4 pairs of leaves usually connate and forming saucerlike discs; blades of lower leaves 4 - 12 cm long and 1 - 6 cm wide, rounded or blunt at the tip, dark green and smooth above, strongly whitened to bluish-white and smooth or hairy beneath; margins entire.

Flowers greenish-yellow to orange to purplish as they fade, tubular to funnel-shaped, 12 - 18 mm long with 2 spreading lips; in stalked clusters from the centre of the terminal leafy discs; late May to early July. Fruit an orange-red berry; in terminal clusters subtended by the leafy discs; July and Aug. (*dioica*—dioecious or in two households, hardly appropriate for a species with perfect flowers)

In dry or moist habitats; open woods, thickets, rocky ridges, slopes, shores and fencerows.

Common in southern Ontario, especially in the calcareous areas east and southwest of the Canadian Shield and in northern Ontario from the north shore of Lake Superior to Lake-of-the-Woods and James Bay, and the upper Severn River drainage basin. (N. Eng. and Que. to B.C., south to Okla. and Ga.)

Note The above description includes specimens that have been identified as *L. dioica* var. *glaucescens* (Rydb.) Butters, a variety which may be distinguished by its fewer connate leaves (often only one pair), larger leaves, smaller flowers, pubescence on the lower surface of the leaves, and sometimes glands on the ovary. However, study has shown that these characters intergrade when large samples of the population are examined. (*glaucescens*—becoming glaucous, i.e., strongly whitened or somewhat bluish-green)

Field check Straggling shrub or twining climber; leaves opposite, whitened beneath, the 1 - 4 upper pairs joined across the stem; orange-red berries in terminal clusters subtended by leafy discs.

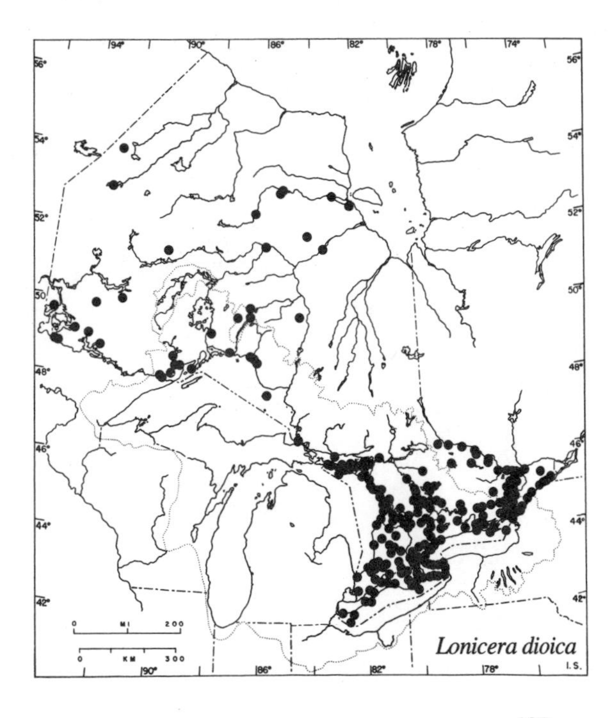

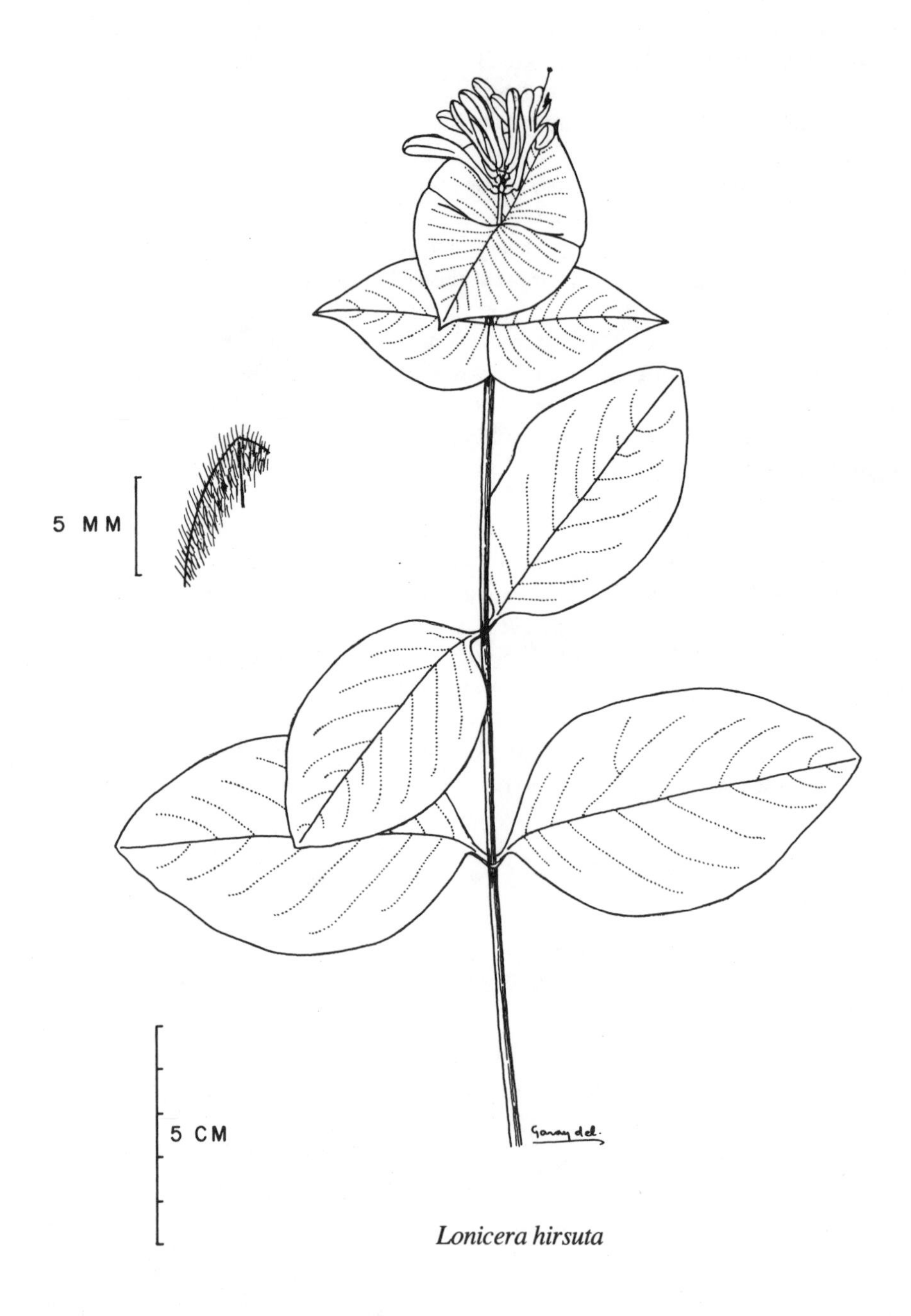

Lonicera hirsuta

Lonicera hirsuta Eat. — Hairy honeysuckle

A trailing or high-climbing woody vine growing to a height of about 3 m. Branchlets green or purplish, glandular-hairy, often spotted with purplish-brown, becoming brown to gray with conspicuous shredding bark.

Leaves opposite, simple, and deciduous; blades sessile or with a petiole seldom exceeding 1 cm in length, the upper 1 - 2 pairs connate and forming saucerlike discs usually with pointed tips; blades of the lower leaves ovate, oval, or broadly elliptic, obtuse to acute at the tip and rounded or tapered at the base, 5 - 13 cm long and 2.5 - 9 cm wide, dull green above with appressed silky hairs; margins ciliate with a fringe of silky hairs that shine in transmitted light, a character distinguishing the hairy honeysuckle from the otherwise similar glaucous honeysuckle.

Flowers yellow to orange, narrowly tubular with gradually flaring lobes, 15 - 25 mm long, in a single stalked cluster or in several whorls forming interrupted spikes from the centre of the terminal leafy discs; late June to early Aug. Fruit an orange-red several-seeded berry, in a single terminal cluster or in separate clusters subtended by leafy discs; Aug. to Oct. (*hirsuta*—stiffly hairy)

In dry or moist habitats; sandy, gravelly, or rocky woods, thickets, slopes and clearings; also in swampy or mossy woods.

Occasional throughout most of southern Ontario except the easternmost and westernmost parts; not reported beyond 52° N. and extending into the Boreal Forest and Barren Region only along the Moose River drainage system. (Que. and Ont., south to Nebr. and Pa.)

Field check Straggling shrub or twining climber; hairy opposite leaves with ciliate margins; upper 1 - 2 pairs of leaves connate; orange-red berries in stalked clusters.

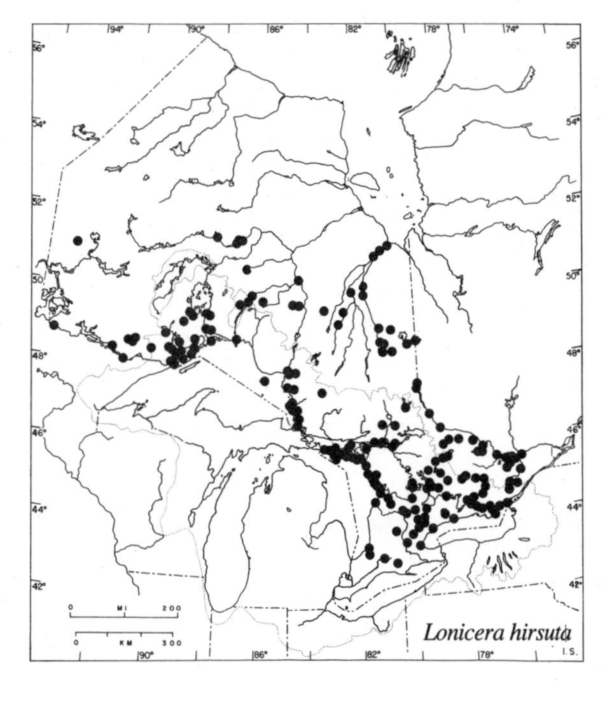

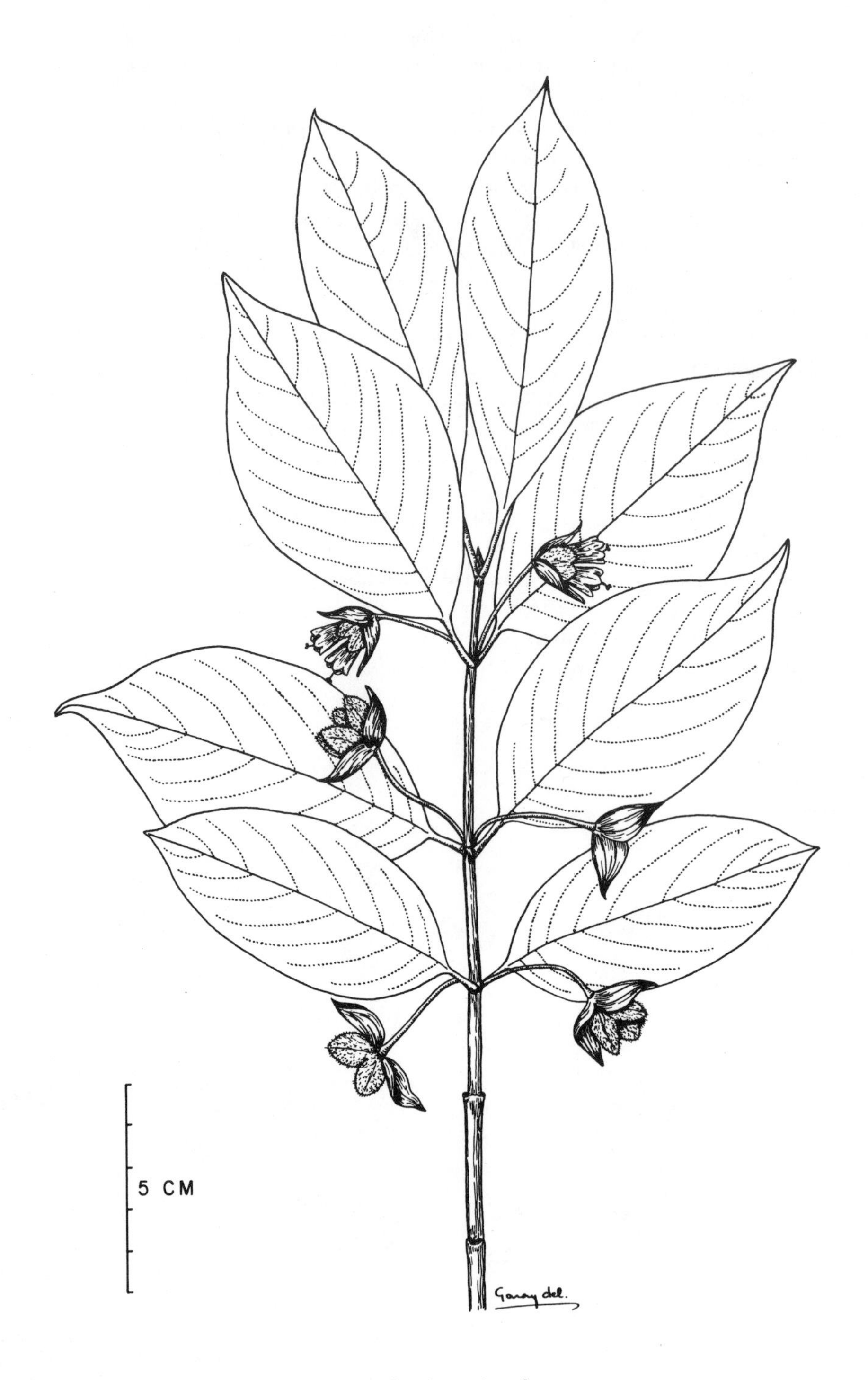

Lonicera involucrata

Lonicera involucrata (Richards.) Banks

Bracted honeysuckle
Northern honeysuckle
Fly honeysuckle

An upright shrub with erect ascending branches, 2–3 m tall. Branchlets greenish to purplish, smooth or minutely hairy; older stems slightly angled or four-sided with solid white pith and gray-brown bark which becomes shreddy.

Leaves opposite, simple, and deciduous; blades 5–15 cm long and 2.5–7.5 cm wide, elliptic, oval or obovate with an acute or abruptly pointed (acuminate) tip and a rounded or tapered base, dark green and nearly smooth above, paler and hairy beneath, especially along the conspicuous veins; margins fringed with white hairs; petioles about 1 cm long.

Flowers about 1 cm long, in pairs at the ends of long stout stalks from the axils of the leaves, each pair of flowers subtended by 2 large, pointed, glandular-hairy, rounded, green to purple bracts and 2 broad smaller bractlets; corolla yellow tinged with red, glandular-hairy, narrowly tubular, with straight or barely spreading rounded lobes; June and July. Fruit a pair of shiny, purple-black berries 6–9 mm across, covered by the enlarged bracts and bractlets, the outer pair of which is finally reflexed; July to Sept. (*involucrata*—with an involucre, in reference to the prominent bracts that accompany the flowers and fruits)

In cool moist habitats: in swampy woods and thickets, in bogs, and along streams and lakeshores.

On Michipicoten Island and along the north shore of Lake Superior, northward to James Bay and Hudson Bay; not recorded south of 47° N in Ontario. (Que. to s.e. Alaska, south to Calif., n. Mex., and Wis.)

Note The common name fly honeysuckle is ascribed to at least four shrubby honeysuckles: *L. oblongifolia, L. canadensis, L. villosa,* and *L. involucrata*. It would be preferable to adopt bracted honeysuckle for the last-named species.

Field check Large pointed opposite leaves; purple-bracted yellow and red tubular flowers on long stalks from the leaf axils; glossy purple-black berries almost hidden by the bracts.

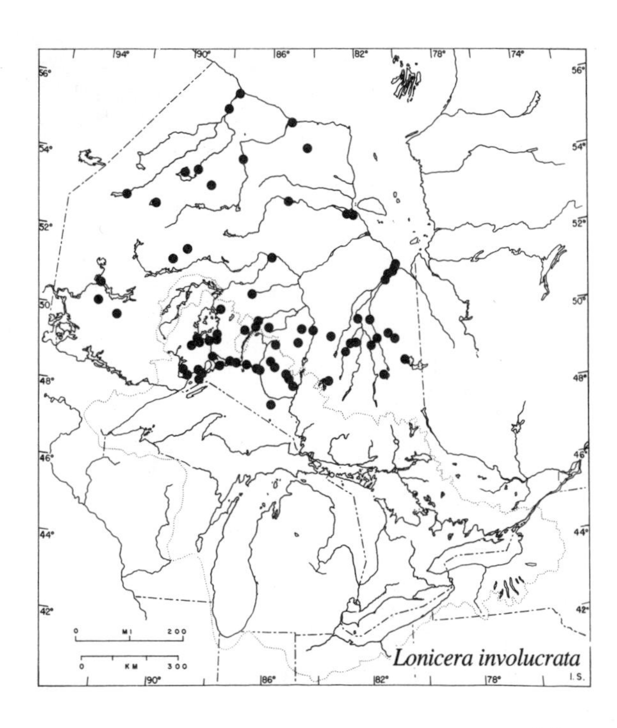

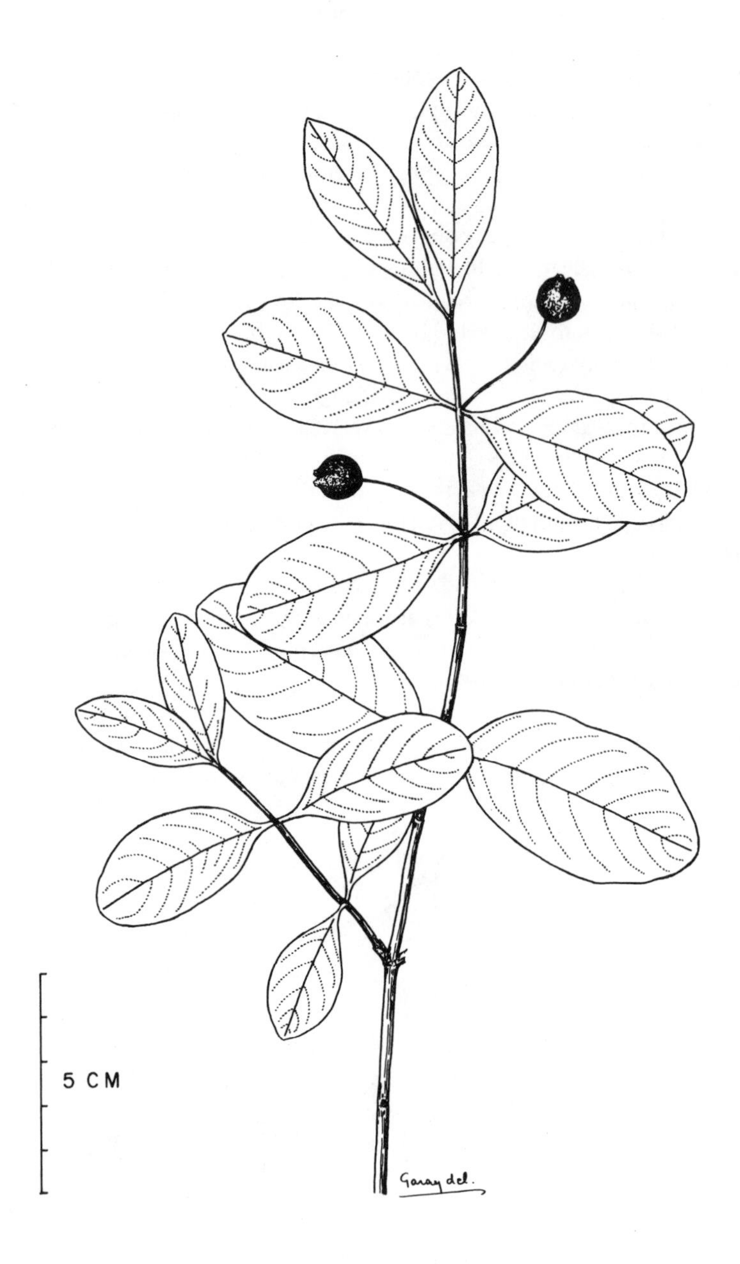

Lonicera oblongifolia

Lonicera oblongifolia (Goldie) Hook. — Swamp fly honeysuckle

A small erect shrub with ascending branches, commonly forming thickets and growing to a height of 1–1.5 m. Branchlets green to purplish, smooth or minutely hairy; older branches brownish-gray to blackish with solid pith and shredding bark.

Leaves opposite, simple, and deciduous, sessile or with a petiole less than 3 mm long; blades 2.5–9 cm long and 1–4 cm wide, rather thick, oblong or elliptic to narrowly obovate, rounded or blunt at the tip and tapered at the base, green above and paler beneath, downy when young but becoming smooth in age; margins entire, not ciliate.

Flowers yellowish-white, narrowly tubular with 2 spreading lips, about 12 mm long, borne in pairs at the ends of slender stalks up to 4 cm long from the axils of the leaves; ovaries of each pair of flowers more or less united; June and July. Fruit a long-stalked pair of orange-red to red or purple several-seeded berries usually united below for at least half their length but sometimes distinct; July and Aug. (*oblongifolia*—with oblong leaves)

Usually in cool moist habitats: marshes, swamps, and bogs or cedar woods; and in jack pine woods, often over calcareous rock.

Common in the limestone areas east and southwest of the Canadian Shield, on the Bruce Peninsula, Manitoulin Island, and northward to the Lake Superior drainage basin and James Bay, reaching a northern limit at about 53° N; rare in the Deciduous Forest Region and in the Boreal Forest and Barren Region. (N.S. to Sask., south to Minn. and Pa.)

Field check Low upright shrub forming thickets; leaves opposite, oblong; flowers and fruits in pairs on long stalks.

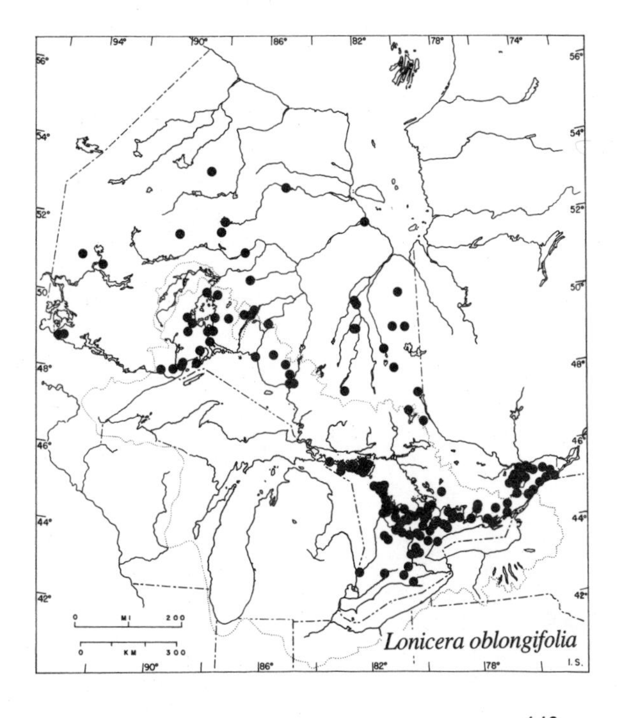

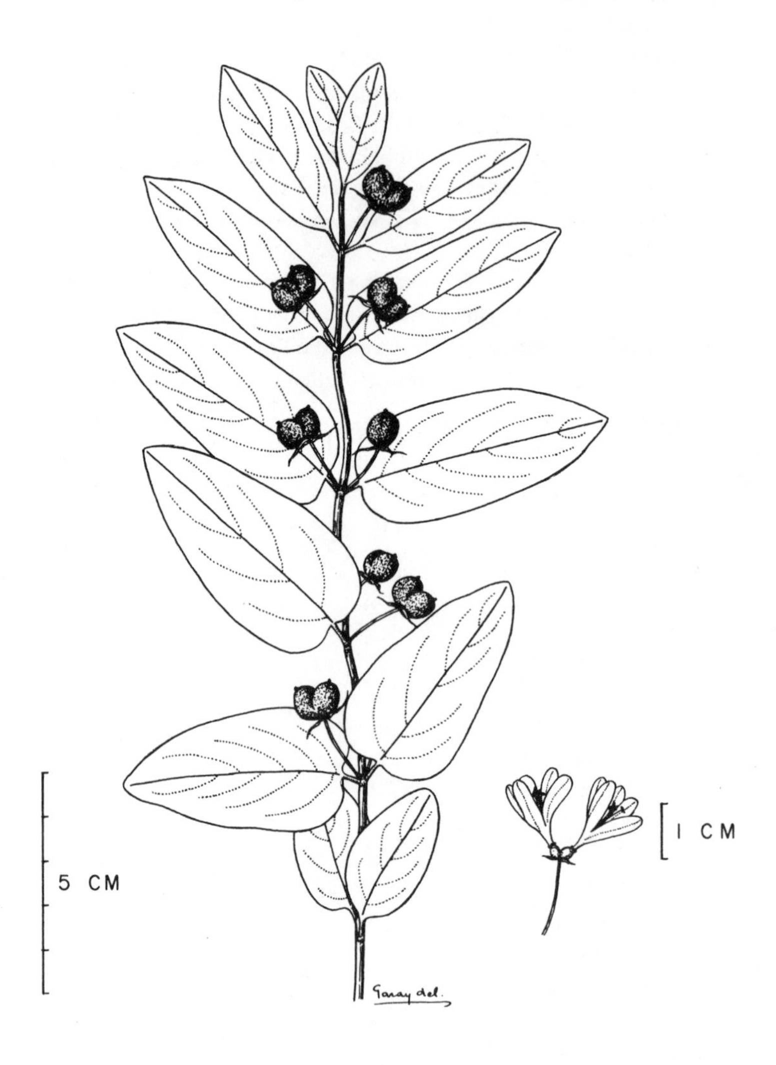

Lonicera tatarica

Lonicera tatarica L. — Tartarian honeysuckle

A large erect shrub reaching a height of 3 m or more; branchlets slender, smooth, and green, later turning brown to brownish-gray, the pith brown with a hollow core.

Leaves opposite, simple, and deciduous; blades 2-6 cm long and 1-4 cm wide, thin, green and glabrous on both sides, oblong to oval, pointed, rounded, or blunt at the tip and rounded, truncate, or sometimes a little heart-shaped at the base; margins entire; petioles usually less than 6 mm long, glabrous or with a few hairs.

Flowers numerous, in pairs on slender stalks from the axils of the leaves, pink or white, with a short slender tube and broadly flaring lobes, the ovaries of each pair of flowers separate or slightly united at the base; May and June. Fruit a pair of red, orange, or yellow berries partly united at the base; June to Aug. (*tatarica*—of Tartary)

In open woods and thickets, along lakeshores, edges of cliffs, and roadsides.

Commonly planted as an ornamental shrub and a frequent escape in the region east and southwest of the Canadian Shield in southern Ontario; also found at a few widely separated locations in northern Ontario to at least 50° N. (native of Eurasia)

Note Although *L. tatarica* is the honeysuckle most frequently encountered as an escape in fields, wasteland, and semi-natural areas, the following species have also been reported:

(1) *L. maackii* (Rupr.) Maxim., a species of northeastern Asia, has been introduced into gardens in the vicinity of Hamilton and has now become established as a wild plant in that region. It is an upright and spreading shrub to 5 m and may be distinguished from *L. tatarica* by its long-pointed (acuminate) leaf blades (4-9 cm long), which are narrowed at the base, and by its pure white flowers (turning yellow in age) borne on peduncles generally less than 6 mm long, shorter than the petioles of the subtending leaves. (*maackii*—in honour of the Russian botanist Richard Maack, 1825-1886)

(2) and (3) *L. morrowii* Gray, a native of Japan, and *L. xylosteum* L. from Eurasia (European fly honeysuckle) may be distinguished from *L. tatarica* by the fact that the lower surface of the leaves and the outside of the yellowish-white corolla are pubescent. In *L. morrowii* the leaves are usually broadest at or below the middle and the bracts are about as long as the ovaries. In *L. xylosteum* the leaves are usually broadest above the middle with the apex more or less acuminate and the bracts are much shorter than the ovaries. Both species have been collected as escapes in southern Ontario. (*morrowii*—in honour of its discoverer James Morrow, 1820-1865; *xylosteum*—an old generic name for a plant called boneweed, from the Greek *xylon*, wood, and *osteon*, bone)

(4) *L.* × *bella* Zab., a hybrid between *L. tatarica* and *L. morrowii*, which displays characters intermediate between those of its parent species, has been found several times in southern Ontario as an escape. (*bella*—beautiful)

Field check Coarse shrub with hollow branches and brown pith; smooth, oval, blunt opposite leaves; numerous long-stalked pink flowers in pairs; twin berries partly united at the base.

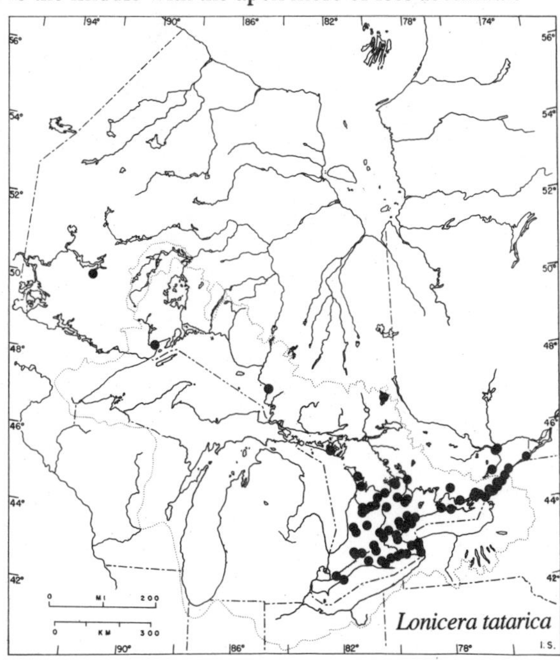

Lonicera tatarica

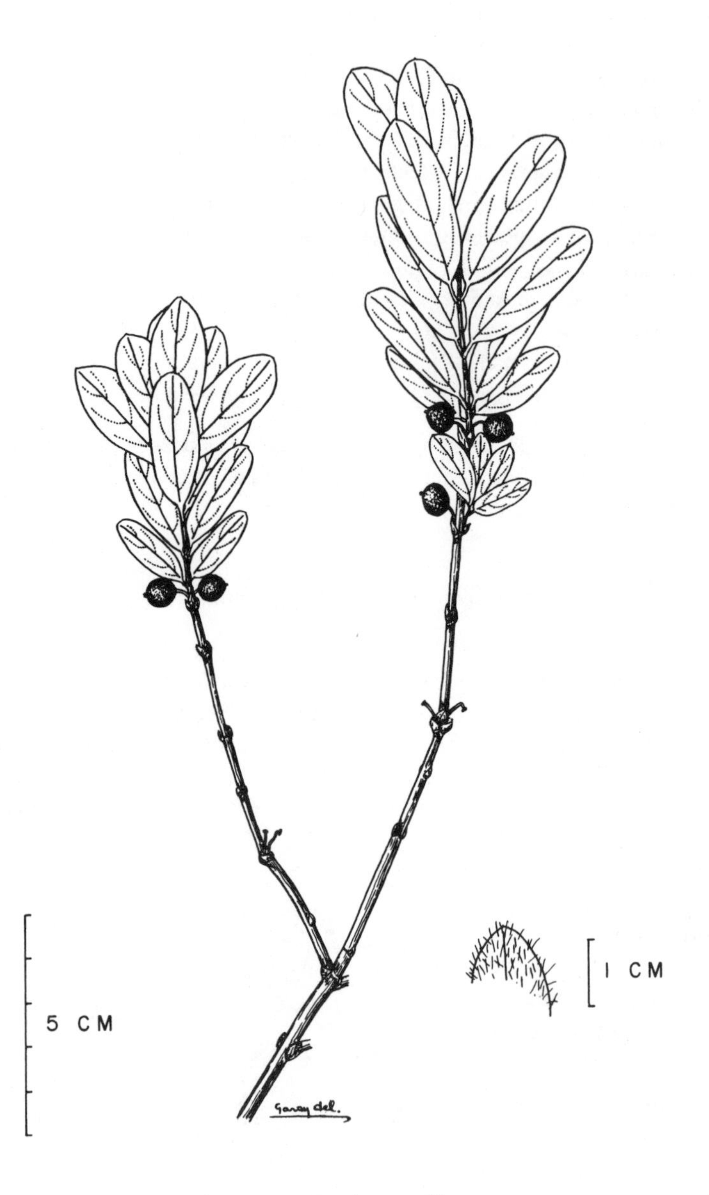

Lonicera villosa

Lonicera villosa (Michx.) R. & S. — Mountain fly honeysuckle

A low erect or ascending shrub with stiff branches, usually less than 1 m high, occasionally forming thickets. Branchlets purplish-red, villous with scattered long soft hairs, especially at the nodes, bark of older twigs reddish-brown to gray, the outer papery layers soon splitting and peeling to expose reddish-brown inner layers.

Leaves opposite, simple, and deciduous, sessile or on very short petioles (less than 3 mm long); blades rather firm, 2.5 – 6 cm long and 1 – 3 cm wide, elliptic to oblong to oblong-oblanceolate, rounded or blunt at the tip, sometimes mucronate, rounded or tapered at the base, dark green and with appressed hairs above, paler and hairy especially along the veins beneath; margins ciliate and often revolute.

Flowers yellowish, tubular to funnel-shaped, in pairs at the ends of short hairy stalks from the axils of the lower leaves, each pair of flowers subtended by 2 narrow, scalelike, persistent bracts longer than the ovary; May and June. Fruit dark blue, edible, appearing like a single berry but consisting of 2 fused ovaries (from the pair of flowers) surrounded by a blue fleshy narrow-mouthed cup derived from connate bractlets; June to Sept. (*villosa* - hairy)

In sphagnum or tamarack bogs, swamps, damp thickets, and clearings; on wet shores and jack pine plains.

Rather rare in the upland area east of Lake Huron and south of Georgian Bay and in the Ottawa - St. Lawrence lowlands; more common from Lake Nipissing to James Bay, along the north shore of Lake Superior and west to Lake-of-the-Woods; northern limit at about 54° N; absent from the Deciduous Forest Region. (Nfld. to Man., south to Minn. and N. Eng.)

Note Several varieties have been distinguished on the basis of variations in pubescence of the leaves and branchlets. Three of these have been reported from Ontario (see Fernald, 1950, p.1333).

Both *L. villosa* and the western North American *L. cauriana* Fern. are closely related to the Eurasian *L. coerulea* L. and are treated by some botanists as varieties of the latter, thereby recognizing a wide-ranging species distinct from other species of *Lonicera* but variable within itself. (*coerulea* — sky-blue; *cauriana* — from *Caurus*, the northwest wind)

Field check Low upright shrub with peeling papery bark; young stems hairy; leaves opposite, elliptic to oblong, sessile or short-stalked; flowers in pairs on short stalks with slender bracts; fruit a dark blue edible berry.

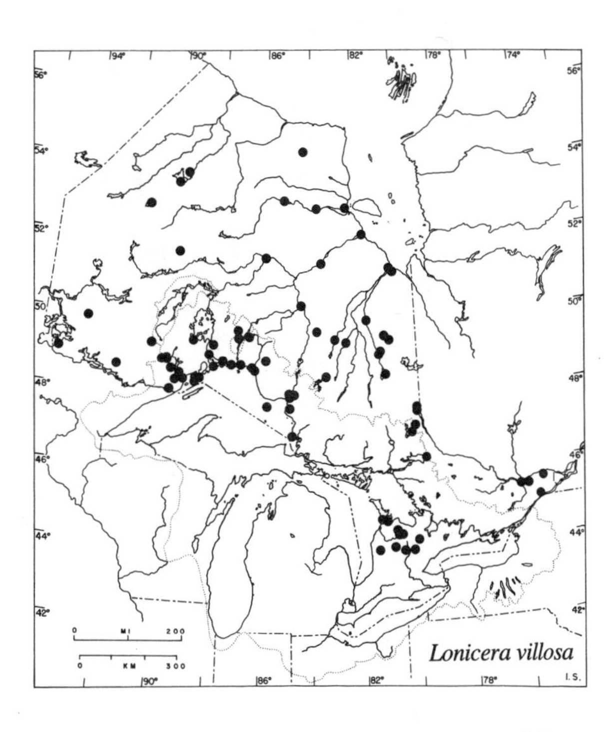

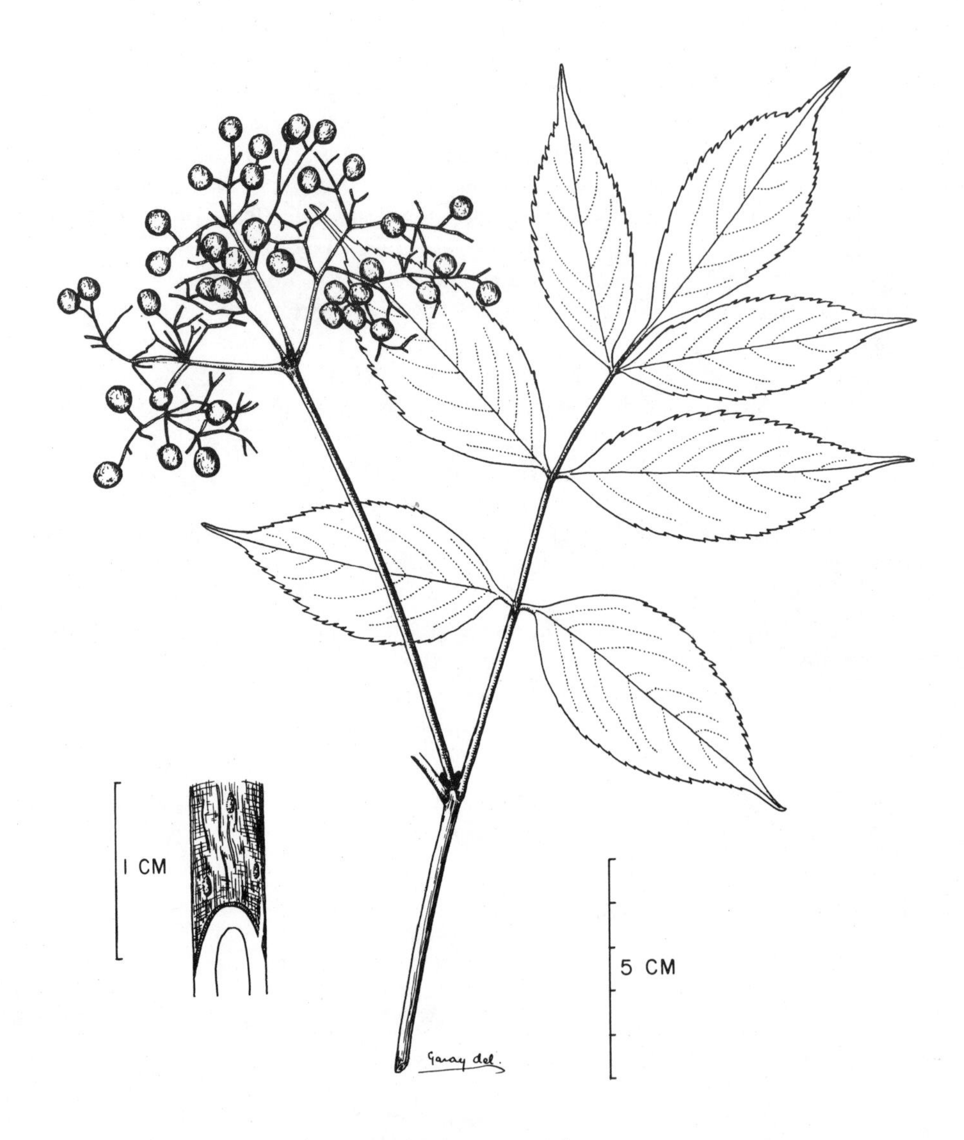

Sambucus canadensis

Sambucus L. — Elderberry

Shrubs or small trees; bark with conspicuous warty lenticels; stems with large pith; foliage and flowers often with heavy rank odour; unique in the Honeysuckle Family in having compound leaves. The flowers are perfect and probably pollinated by insects, although nectaries are lacking and little is known about specific pollinators. Several species are used as ornamentals because of the large showy clusters of flowers and red or purple-black fruits. About 40 species of cosmopolitan distribution. (*Sambucus*—from the Greek *sambuke*, the name of an ancient musical instrument supposed to have been made from the wood of the European elder, *S. nigra* L.; *nigra*—black)

Key to *Sambucus*

a. Stems (two years old and older) with large white pith; leaflets 5-11 (usually 7); flowers opening after the leaves have developed (mid-July), in broad, flat, or slightly rounded clusters; fruits purple-black, edible ..... *S. canadensis*

a. Stems with large orange to red-brown pith; leaflets 5-7 (usually 5); flowers opening with the unfolding leaves (mid-May to June), borne in elongate, pyramidal or rounded clusters which are longer than broad; berries red, inedible ..... *S. pubens*

Sambucus canadensis L.

Elderberry
American elder

An erect, soft-stemmed or barely woody and somewhat stoloniferous shrub growing to a height of 3 m. Branchlets yellowish-gray and smooth or nearly so; older stems stout, with warty gray-brown bark and large white pith.

Leaves large, opposite, pinnately compound, and deciduous, with a petiole 2.5-5 cm long; leaflets 5-11 (usually 7), sessile or short-stalked, elliptic, 5-15 cm long and 2.5-5.5 cm wide, sharply serrate, abruptly and conspicuously sharp-pointed at the tip and rounded or tapered, frequently asymmetrical at the base, bright green above, paler and mostly smooth or a little hairy along the veins beneath, the terminal leaflet often broader than the lateral ones and some of the leaflets (usually the lower pairs) deeply parted or divided into two or three segments; stipules and stipels sometimes present; foliage and twigs ill-scented when bruised.

Flowers white, 5-parted, small, and numerous, in flat or slightly rounded, long-stalked, compound, terminal broad cymes 10-18 cm across, heavily-scented when open in July, blooming late when the red-berried elder already has ripe or nearly ripe fruit. Fruit a round, juicy, purple-black, edible, berry-like drupe, less than 6 mm across, in large clusters; Aug. and Sept. (*canadensis*—Canadian)

In low ground, swamps, thickets, edges of woods, roadsides and fencerows.

Common in southern Ontario and north to Lake Nipissing and to the North Channel of Lake Huron; northern limit near 46° 30′ N; a report from the vicinity of Sault Ste. Marie awaits confirmation. (N.S. to s.e. Man., south to Okla. and Ga.)

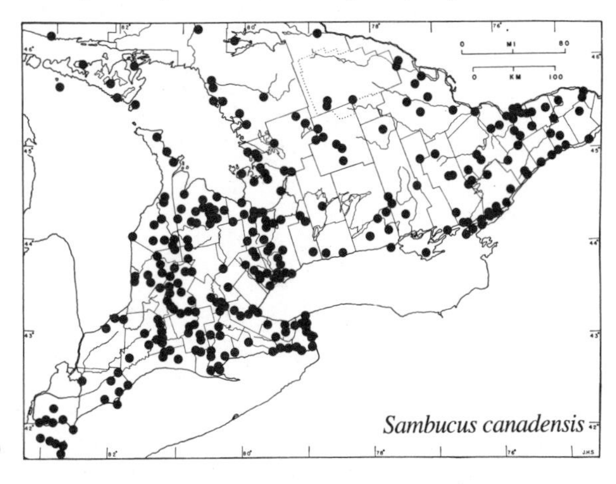

Sambucus canadensis

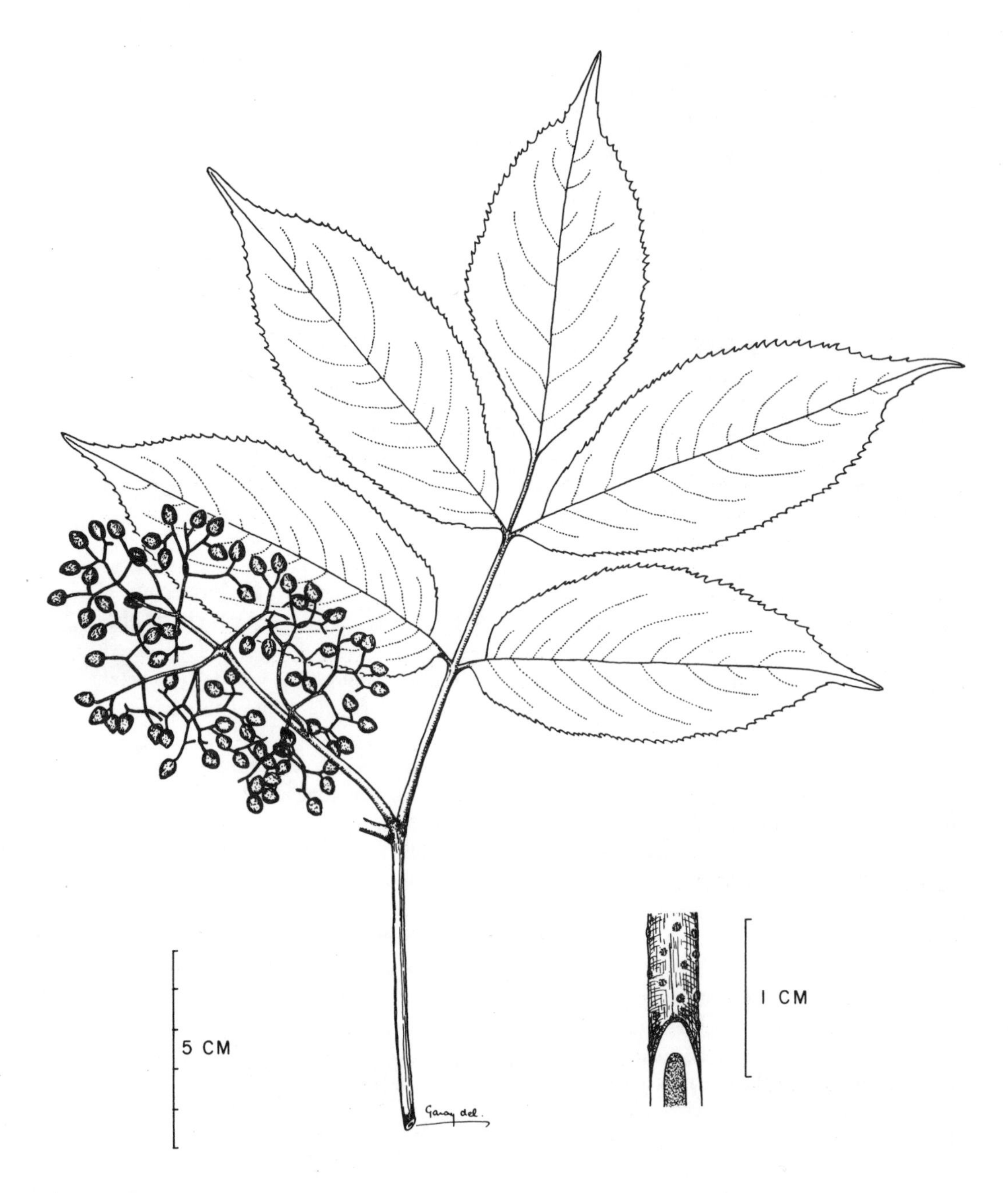

Sambucus pubens

Note In Colonial times the berries were often used in the preparation of soups, pies, jelly, and wine; the flowers dipped in batter and fried; the stems, after removal of the pith, served for pipe stems, spiles for tapping maple trees, whistles, and peashooters.

The European species, *S. nigra* L., is similar to ours and its fruit is widely used for pies, jelly, and wine.

Field check White pith; large opposite pinnately compound leaves usually with 7 sessile or short-stalked leaflets; flowers in broad, flattish clusters in July; purple-black edible "berries" in late summer.

Sambucus pubens Michx. (*S. racemosa* L. ssp. *pubens* (Michx.) Hult.)

Elderberry Red-berried elder

An erect soft-stemmed or barely woody shrub growing to a height of about 4 m. Branchlets yellow-brown and hairy, becoming thick, with warty gray-brown bark and large orange to reddish-brown pith.

Leaves large, opposite, pinnately compound, and deciduous; leaflets 5–7 (usually 5), ovate-lanceolate, 5–13 cm long and 2.5–5.5 cm wide, abruptly and conspicuously sharp-pointed at the apex and rounded or somewhat cordate at the usually asymmetrical base, green above, paler and either smooth or downy beneath, short-stalked or prominently stalked, with or without stipules and stipels; margins sharply serrate; petioles 2.5–5 cm long; foliage and branchlets ill-scented when bruised.

Flowers 5-parted, small, numerous, white, and ill-scented, in terminal, elongate, rounded or pyramidal compound clusters 5–13 cm long, usually longer than broad, the purplish buds opening with the developing leaves about five or six weeks before those of the American elder; May and June (from May 15 in southern Ontario to June 30 in the northern part of the province). Fruit a round, inedible, red, berrylike drupe less than 6 mm across, in elongate clusters; July and Aug. (*pubens*—luxuriant; *racemosa*—in racemes)

In thickets, ravines, open woods, and clearings, along roadsides and edges of woods, and in fence rows.

Common throughout southern Ontario and around the north shore of Lake Superior, becoming infrequent northward to about 51° N along the Moose River. (Nfld. to Man., south to Tenn. and N. Eng.)

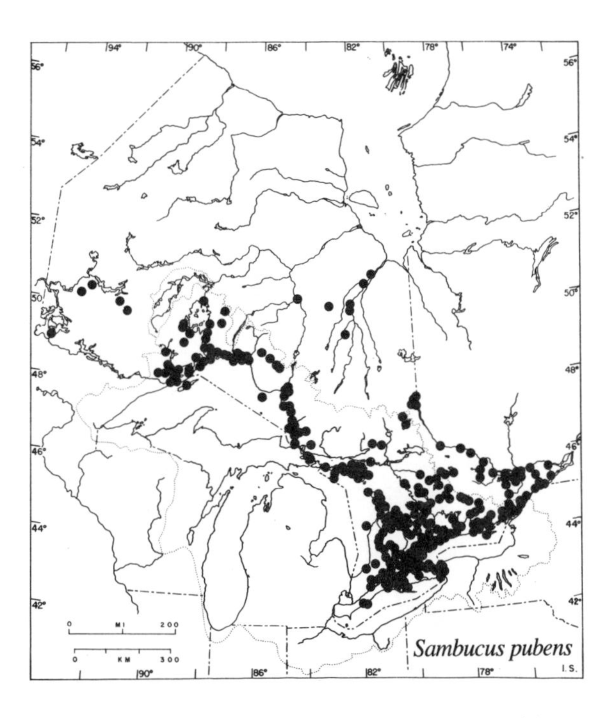

Note Red elderberry is one of few shrubs tolerant of smelter fumes (see Hutchinson, 1976). Hultén has treated the North American red elderberry as part of a circumpolar complex, *S. racemosa* ssp. *pubens*, embracing our eastern taxon as well as western taxa with pyramidal inflorescences and red or black fruit.

Field check Brown pith; large opposite pinnately compound leaves usually with 5 stalked leaflets; ill-scented flowers in elongate clusters in May and June with the developing leaves; red "berries" in midsummer.

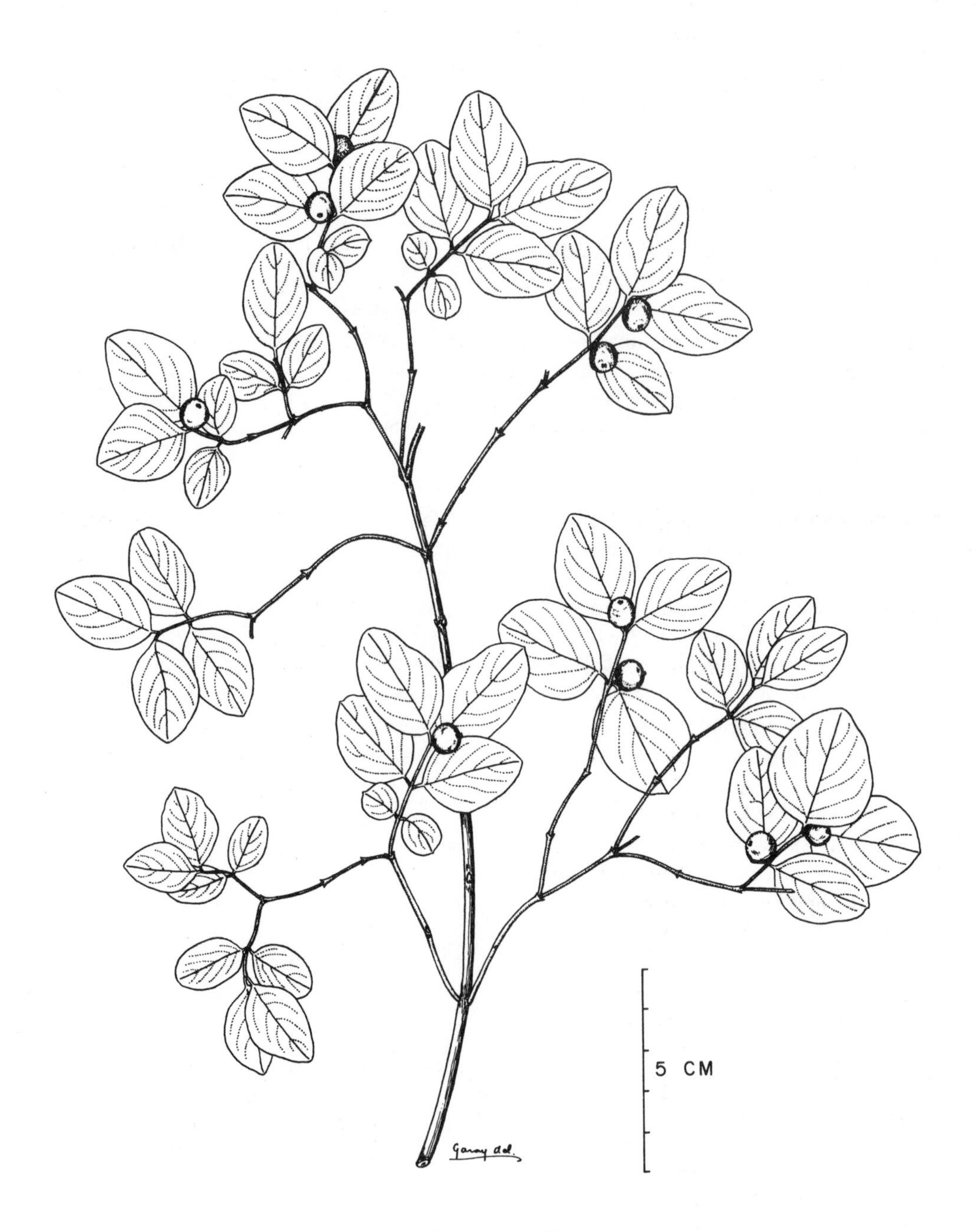

Symphoricarpos albus

Symphoricarpos Duham. — Snowberry

Low bushy shrubs with opposite, entire to crenate or lobed leaves and terminal or axillary clusters of small pendent flowers which are pollinated chiefly by bees and wasps. Seventeen species in North America and one in central China. (*Symphoricarpos*—from the Greek *symphorein*, to bear together, and *karpos*, fruit, in allusion to the clustered berries)

Key to *Symphoricarpos*

a. Leaves rather thin, the margins mostly entire (or wavy-toothed to lobed on young vigorous shoots); flowers short-stalked, solitary or in short spikelike clusters of 2 – 5; corolla about 6 mm long, the style not exserted; berries white with a dark end spot *S. albus*

a. Leaves firm or thickish, the margins entire, wavy-toothed, or lobed; flowers sessile in elongate crowded clusters of 2 – 10; corolla about 9 mm long, the style exserted; berries dull white and soon becoming discoloured and blackish *S. occidentalis*

Symphoricarpos albus (L.) Blake (*S. racemosus* Michx. in some manuals.) — Snowberry

A small ascending or spreading shrub usually less than 1 m high, forming low thickets. Branchlets slender, smooth or minutely hairy, light brown at first, turning purplish to gray and darker in age; bark becoming shreddy to fibrous; pith small and brown with a hollow central core.

Leaves opposite, simple, sessile, and deciduous; blades rather thin, elliptic-oblong or rhombic-ovate to nearly orbicular, 1 – 5 cm long a and 1 – 3 cm wide, blunt or rounded and sometimes minutely pointed at the tip, rounded or tapered at the base, dark green and smooth or with a few appressed hairs above, paler and smooth to minutely downy or whitened beneath; margins entire and minutely hairy. Leaves on vigorous young shoots larger and often wavy-toothed or lobed.

Flowers small, perfect, 5-parted, about 6 mm long; the corolla pink and white, tubular or narrowly bell-shaped with short lobes, hairy on the inside; stamens and style not projecting from the corolla tube; in short clusters or spikelike racemes of 1 – 5 at the ends of the branches and in the axils of the upper leaves; June and July. Fruit a round, white, spongy, berrylike drupe 6 – 12 mm across, with a dark spot at the free end, solitary or a few together, persistent through the winter; Aug. and Sept. (*albus*—white; *racemosus*—in racemes)

In sandy or rocky open ground, in thickets, or on well-drained talus slopes and ridges.

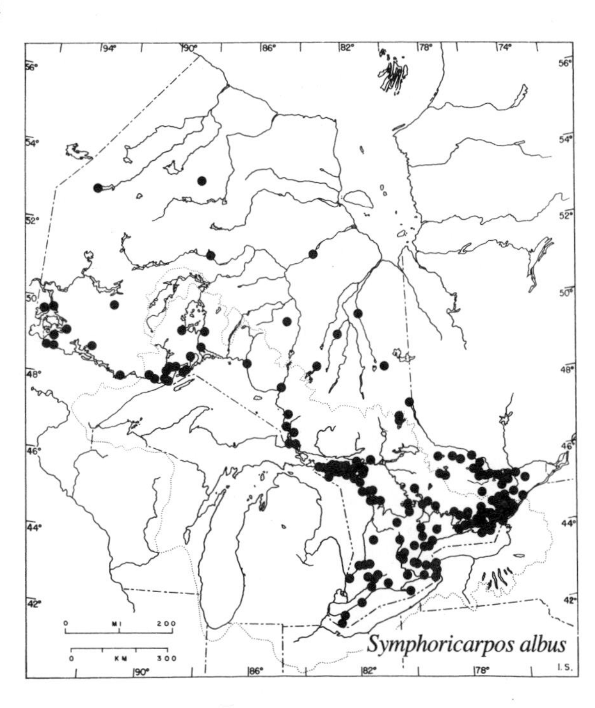

Symphoricarpos albus

Symphoricarpos occidentalis

Common in southern Ontario and from Thunder Bay to Lake-of-the-Woods but becoming rare northward to a limit near 53° N. (Que. to B.C., south to Colo. and Va.)

Note A variety of this species from the Pacific slope, *S. albus* var. *laevigatus* (Fern.) Blake, is cultivated and sometimes escapes to roadsides and waste places. It can be distinguished by its taller and coarser habit (up to 2 m), smooth leaves more frequently lobed and wavy-toothed, more numerous flowers in elongate, mostly terminal clusters, and larger fruits (up to 18 mm across). (*laevigatus*—smooth)

Field check Low spreading shrub; stems with hollow pith; leaves oblong to ovate, opposite, and sessile; small pink flowers in few-flowered clusters; white "berries" with a dark end spot.

Symphoricarpos occidentalis Hook. — Wolfberry, Buckbrush

A small shrub up to 1 m in height, stiffly erect and freely-branching, spreading by stolons, and commonly forming dense low thickets. Branchlets slender, reddish-brown, and minutely hairy, turning gray with age; bark soon becoming shreddy; pith small and brown.

Leaves opposite, simple, and deciduous; blades elliptic to oval, 2–7.5 cm long and 1.5–5.5 cm wide, thick and firm, from nearly sessile to stalked with a petiole almost 1 cm long, the apex blunt or rounded, frequently mucronate, the base rounded or tapered, smooth and dark green above, paler and smooth or hairy along the veins beneath; margins entire, wavy-toothed, or lobed, minutely hairy, and somewhat revolute.

Flowers small, perfect, 5-parted, pale pink, and bell-shaped with spreading lobes, densely hairy on the inside; corolla about 9 mm long, the stamens and style exserted; in crowded elongate clusters of 2–10 from the ends of the branches and in the axils of the leaves; July and Aug. Fruit a round, dull greenish-white, spongy, berrylike drupe containing small 1-seeded pits; in rather crowded clusters and soon becoming discoloured and blackish; Aug. and Sept. (*occidentalis*—western)

In sandy or rocky soil, in dry open fields and along railway embankments; usually in disturbed ground or clearings.

Apparently introduced from the West and established in a dozen or more localities across southern Ontario, along the north shore of Lake Superior, and around Lake-of-the-Woods. (w. Ont. to B.C., south to n. Wash., N.Mex., and Mo.; naturalized eastward to Que., Pa., and N. Eng.)

Field check Thicket-forming shrub with nearly sessile, oval to elliptic opposite leaves; small pink flowers in crowded clusters; dull white berries soon turning blackish.

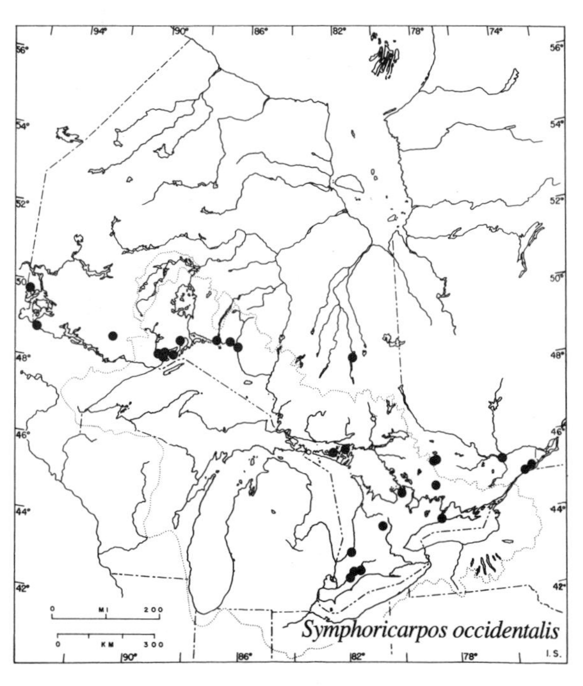

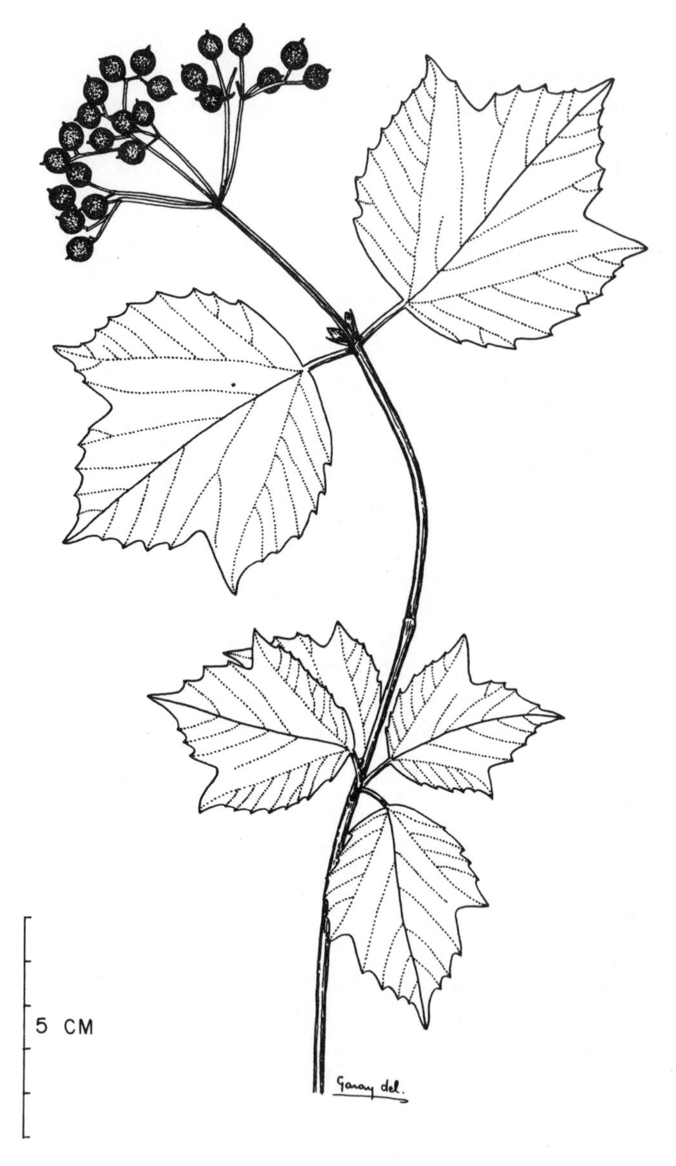

Viburnum acerifolium

Viburnum L. — Arrow-wood, Viburnum

Erect or decumbent shrubs and small trees with simple entire toothed or palmately lobed leaves and numerous white or pink flowers in compound cymes, sometimes the outer florets enlarged and sterile. Fruit a drupe with a large central stone. Pollination is mainly by insects of the Orders *Hymenoptera* and *Diptera*. Many species are in cultivation as ornamentals. Several of our native species are attractive both in flower and in fruit and exhibit interesting changes of colour during the ripening of the fruits. Roughly 150–200 species, chiefly of temperate regions. (*Viburnum*—classical Latin name, possibly given to *V. lantana* L., the wayfaring tree; an alternative derivation has been suggested from the Latin *viere*, to weave together, in reference to the pliable branchlets of some members of the genus)

Key to *Viburnum*

a. Leaves 3-lobed, palmately 3–5 nerved from the base
- b. Leaves soft-downy beneath with clustered hairs and tiny brown to blackish resinous dots; fruit blue ***V. acerifolium***
- b. Leaves smooth or with scattered single hairs and no resinous dots beneath; fruit orange-red
  - c. Stipules slender with thickened tips; flowers numerous, of two kinds, the outer enlarged and showy, on leafy shoots from terminal buds ***V. trilobum***
  - c. Stipules none; flowers few, all alike, on 2-leaved short branches from lateral buds ***V. edule***

a. Leaves without lobes, pinnately nerved
- d. Erect shrubs with small leaves (2.5–10 cm long); flowers all alike
  - e. Leaves entire, finely or irregularly toothed but not coarsely dentate; cymes short-stalked or sessile
    - f. Leaves finely and sharply serrate; winter buds gray; cymes sessile ***V. lentago***
    - f. Leaves entire or irregularly crenulate or wavy-toothed; winter buds golden brown; cymes short-stalked ***V. cassinoides***
  - e. Leaves coarsely toothed; cymes long-stalked
    - g. Leaves usually hairy beneath, sessile or short-stalked with bristlelike stipules; teeth 4–11 on each side ***V. rafinesquianum***
    - g. Leaves smooth or hairy on the veins and in the vein axils beneath, clearly stalked, without evident stipules; teeth 9–21 on each side ***V. recognitum***
- d. Sprawling shrubs with large leaves (10–21 cm long); flowers of two kinds, the outer enlarged and showy ***V. alnifolium***

Viburnum acerifolium L. — Maple-leaved viburnum

A low shrub with slender ascending branches, usually less than 2 m in height. Branchlets green and smooth or minutely hairy; older stems reddish or purplish-gray.

Leaves opposite, simple, and deciduous; blades 3-lobed and palmately veined, the two spreading lateral lobes separated from the median lobe by shallow clefts; blades 6–12 cm long and about as wide, nearly round to ovate or ovate-lanceolate, sharp-pointed at the tip and rounded or cordate at the base, dark green and sparsely hairy above, paler and softly downy beneath with clustered hairs and numerous minute brown or blackish resinous dots scattered over the lower surface; margins coarsely dentate; petioles downy, 1–3 cm long, with a pair of

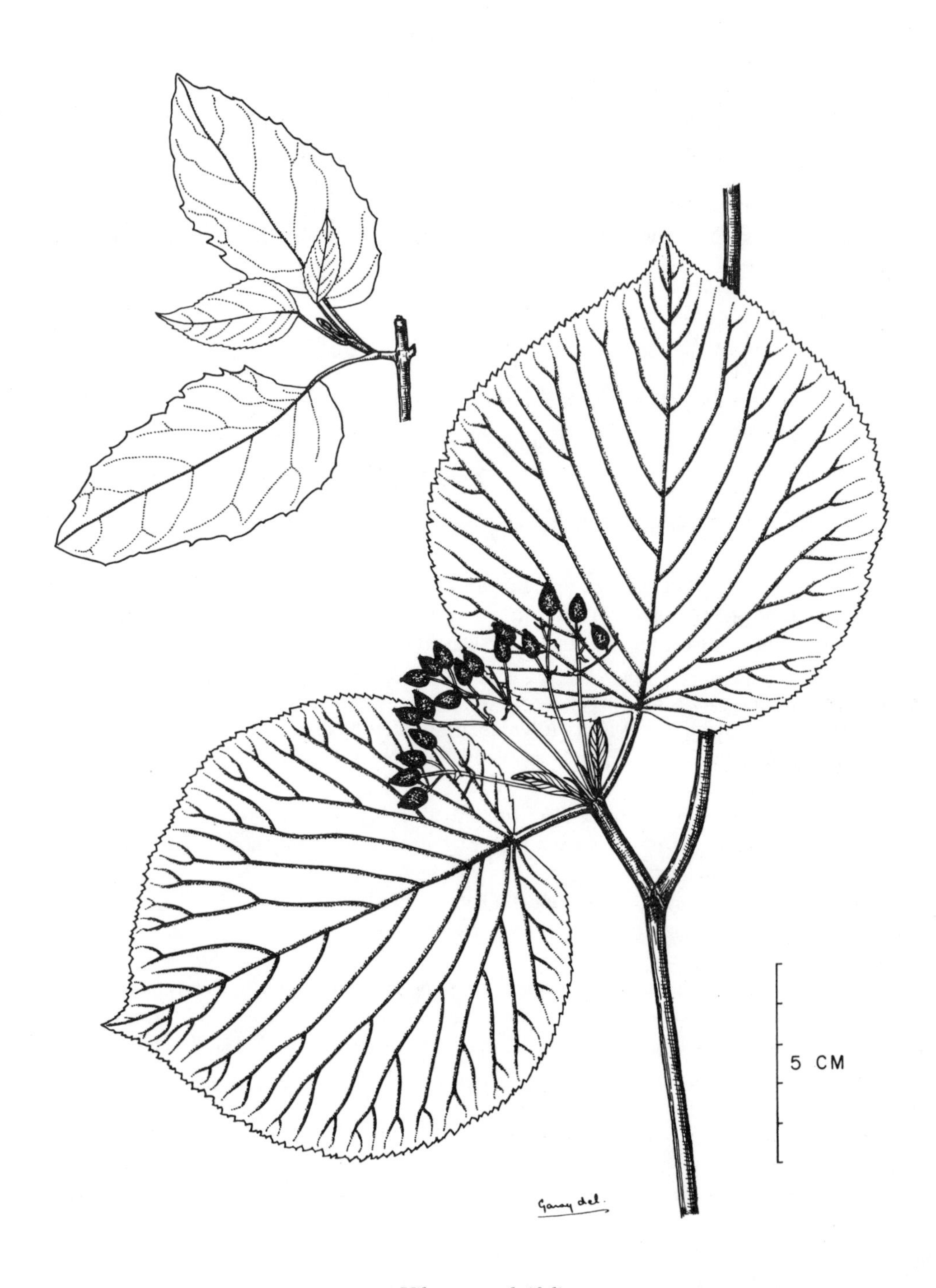

Viburnum alnifolium

bristle-tipped stipules; leaves at the ends of some of the branches have poorly developed lateral lobes.

Flowers creamy-white, small, numerous, in long-stalked terminal clusters 2.5–9 cm across; mid-June. Fruit a round to ovoid berrylike drupe, at first green, then red turning to dark blue or purple-black; in open clusters; Sept. and Oct. (*acerifolium*—with leaves like *Acer*, the genus of maples)

In dry or moist sandy, rocky, or clayey soil of open woods, thickets, ravine slopes, and hillsides.

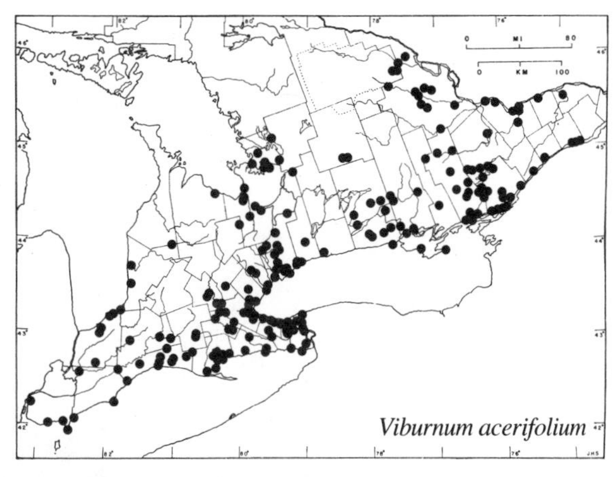

Common just north of Lake Erie and Lake Ontario, rare farther northward to about 46° N in the east and 45° N in the west; apparently absent from the Bruce Peninsula. (Que. and N.Eng. to Minn., south to Tenn. and Ga.)

Field check Maplelike opposite leaves downy and minutely resin-dotted beneath; small creamy-white flowers in stalked terminal clusters; dark blue to purple-black "berries".

Viburnum alnifolium Marsh. — Hobble-bush

A low, spreading or sprawling shrub less than 2 m in height with the branches often prostrate and rooting at the nodes and tips. Branchlets covered with a dense light cinnamon-brown scurfy pubescence; older stems purplish-brown and smooth or sometimes a little ridged and warty; winter buds large, naked (protected by rudimentary foliage leaves rather than by scales), scurfy-covered, and conspicuous from late summer to the following spring.

Leaves opposite, simple, and deciduous, large and coarse, characteristically arranged in distant pairs along the stem and on its short branches in a horizontal steplike series; blades 10–20 cm long and 7–18 cm wide, broadly oval to almost round, dark green above, paler beneath, covered when young with a light brown scurfy pubescence which persists on the prominent veins on the lower surface, tip abruptly pointed, base rounded or heart-shaped; margins closely toothed; petioles scurfy-hairy, 1–6 cm long, bearing a pair of stipules with free bristlelike tips, the stipules later deciduous. Leaves on sprout growth are often thin, narrowly ovate, coarsely toothed and smooth on both surfaces (shown at upper left in the illustration).

Flowers white and of two kinds, in short-stalked, saucer-shaped clusters up to 13 cm across; marginal flowers with enlarged flattened corollas, sterile, and surrounding the more numerous but less conspicuous fertile ones; late May and early June. Fruit a small, ovoid, berrylike drupe

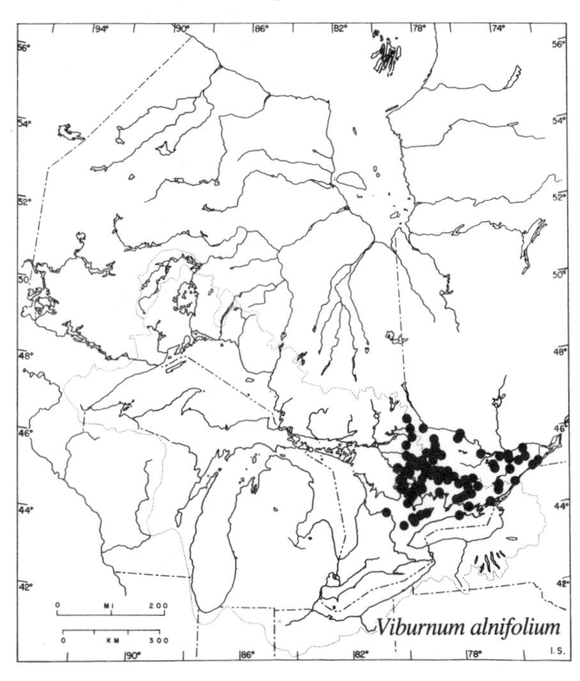

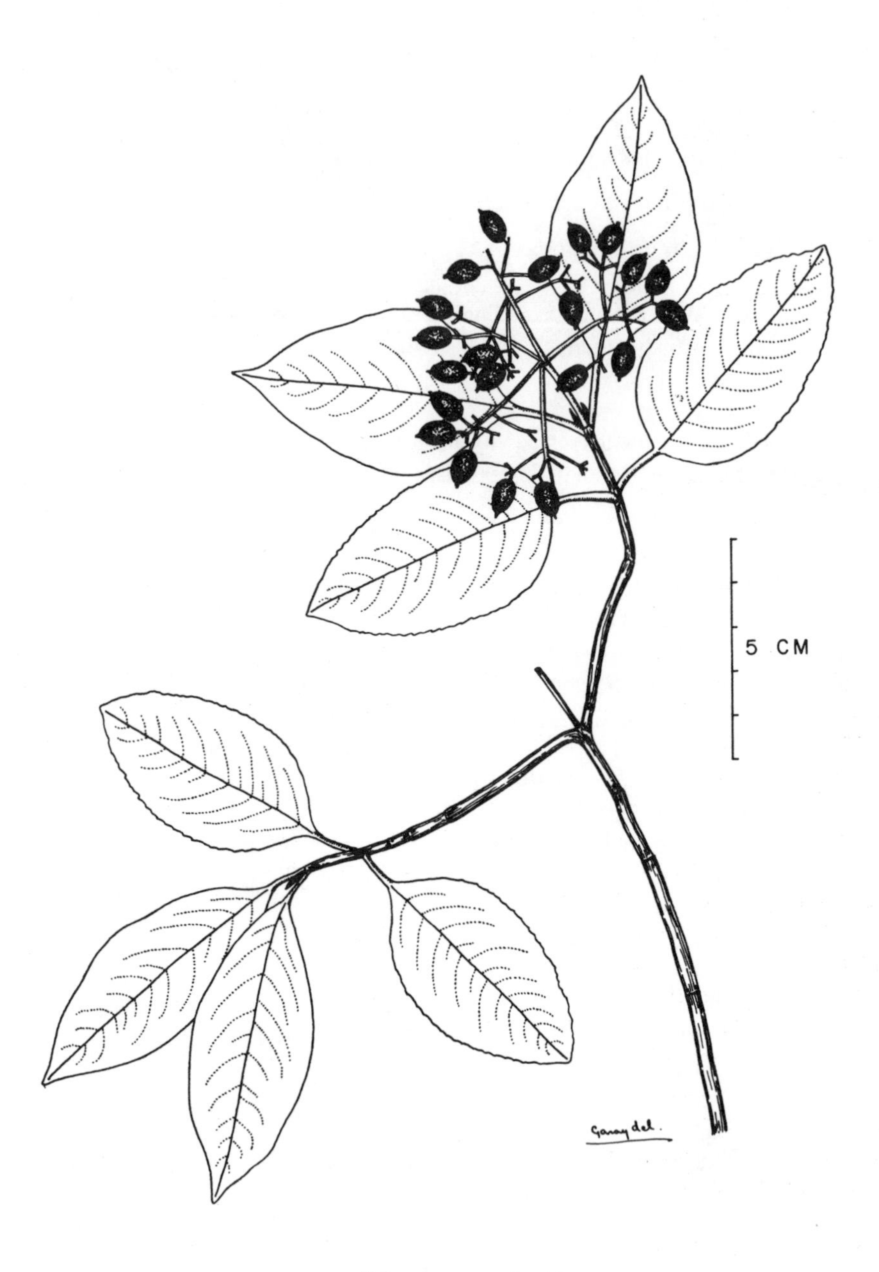

Viburnum cassinoides

about 9 mm long, changing colour from green to crimson to purple-black as it matures; Aug. and Sept. (*alnifolium*—with leaves like *Alnus*, the alder)

In damp woods, cool shaded ravines, and thickets.

Common on the Canadian Shield in southern Ontario, less common to the southwest of the Shield; absent from the Deciduous Forest Region and north of 47° N. (N.S. to e. cent. Ont., south to Tenn. and Ga.)

Note A related species, *V. lantana* L., the European wayfaring tree, is sometimes planted in southern Ontario as an ornamental and may be found occasionally as an escape along roadsides and in semi-natural surroundings. It somewhat resembles *V. alnifolium* but its leaves are usually longer than wide, its habit is erect and more dense, all of its flowers are perfect, and it prefers more open habitats. (*lantana*—flexible)

Field check Sprawling habit; large, round, veiny, toothed, opposite leaves; showy marginal flowers bordering a saucer-shaped cluster; purple-black fruits; large scaleless scurfy winter buds.

Viburnum cassinoides L.

Wild raisin
Withe rod

An erect, rather stiffly branched shrub up to 4 or 5 m in height, spreading at the top. Branchlets often scurfy at first, becoming smooth; older twigs purplish, somewhat warty and ridged; winter buds narrow, covered by a single pair of golden-brown scurfy scales.

Leaves opposite, simple, and deciduous; blades 4.5-9.5 cm long and 2.5-5 cm wide, oval, oblong, or oblanceolate to rhombic-ovate, with an abruptly short-acuminate blunt tip and a narrowed or rounded base, dull dark green above, paler and sometimes scurfy beneath; margins entire, crenate or wavy-toothed; main vein pale above and brown-hairy beneath; petioles grooved, 0.5-2 cm long.

Flowers creamy-white, ill-scented, all alike, and perfect, in rather flat, terminal short-stalked clusters 5-10 cm across; late June. Fruit a nearly round or ellipsoid drupe 6-9 mm long, whitish-yellow at first, then quickly turning colour through pink to bright blue to blue-black, with a bloom; Aug. and Sept. (*cassinoides*—like *cassine*, in reference to *Ilex cassine*, a kind of holly)

In moist or acid soils, damp sandy banks, thickets, low woods, and swamps and around the edges of bogs.

Common in all but the western part of southern Ontario but becoming rare north of Lake Timagami and Lake Abitibi; westward to the eastern end of Lake Superior; northern limit at about 49° N in the east. (Nfld. to Ont., south to Ind. and Ala.)

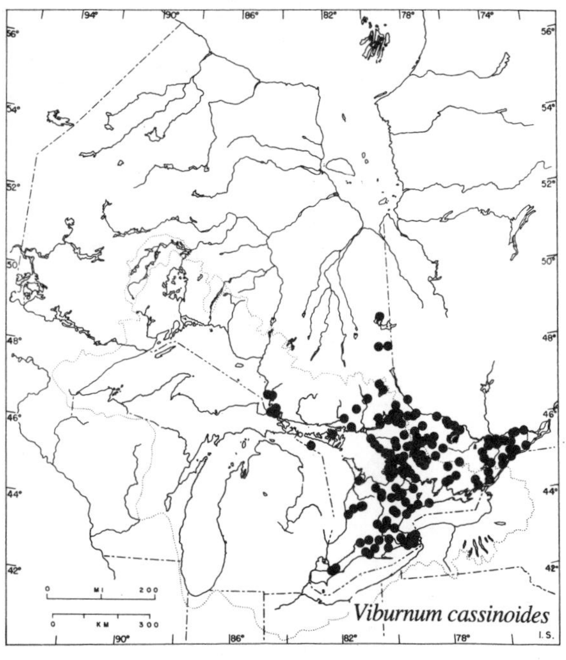

Viburnum cassinoides

Field check Entire, crenate or wavy-toothed, opposite leaves; short-stalked, rather flat clusters of ill-scented white flowers; open clusters of blue fruits; narrow golden-brown winter buds.

Viburnum edule

Viburnum edule (Michx.) Raf.

Mooseberry, Pembina Squashberry, Low-bush cranberry

An erect or straggling shrub usually not higher than 2 m. Branchlets smooth, purplish-brown, often angled or with longitudinal ridges; bark on older branches gray to brown.

Leaves opposite, simple, and deciduous; blades usually 3-lobed with two shallow clefts separating the median lobe from the divergent lateral lobes; leaves at the ends of the main branches often unlobed or with poorly developed lobes; blades 4–12 cm long, 2.5–12 cm wide, roundish to ovate or ovate-lanceolate, sharp-pointed at the tip, rounded, cordate, or tapered at the base, dark green and smooth above, paler beneath and hairy along the veins and in their axils; margins dentate; petioles smooth, 0.5–4 cm long, without stipules but commonly with several stalked glands along the edge of the leaf near its junction with the petiole.

Flowers creamy-white, small, in comparatively few-flowered stalked open clusters about 2.5 cm across, usually borne on short 2-leaved branches from lateral buds on the wood of the previous season; June and early July. Fruit an ovoid or round berrylike drupe 6–12 mm long, at first yellow, later orange or red; pits large, flat, and ovate; July and Aug. (*edule*—edible)

In damp woods, swampy clearings, and bogs and along lakeshores and stream banks.

Widely distributed in northern Ontario from the north shore of Lake Superior to Hudson Bay and James Bay; rare south of 48° N and absent in southern Ontario. (Nfld. to Alaska, south to Oreg. and N. Eng.)

Note This is the most northerly of our viburnums. A delicious jelly can be made from the ripe fruit.

Field check Three-lobed opposite leaves, hairy on the veins beneath; few white flowers and orange-red fruits on 2-leaved side branches from previous season's growth; northern Ontario.

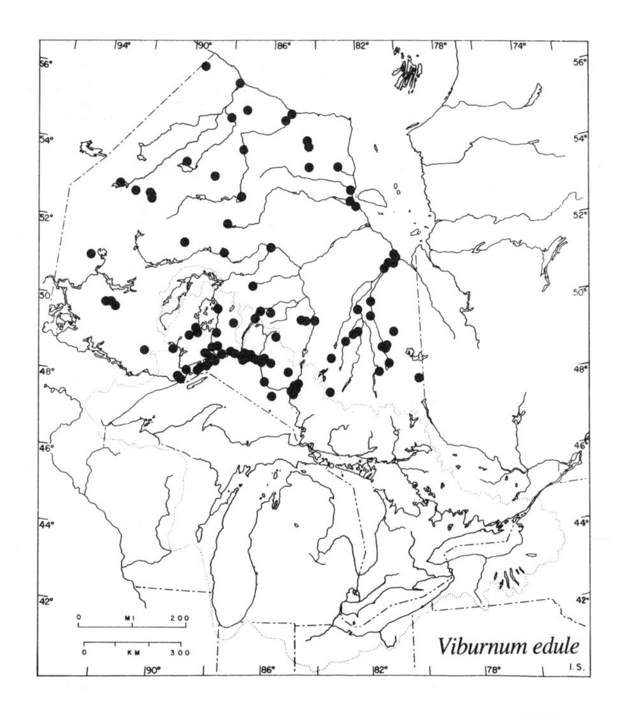

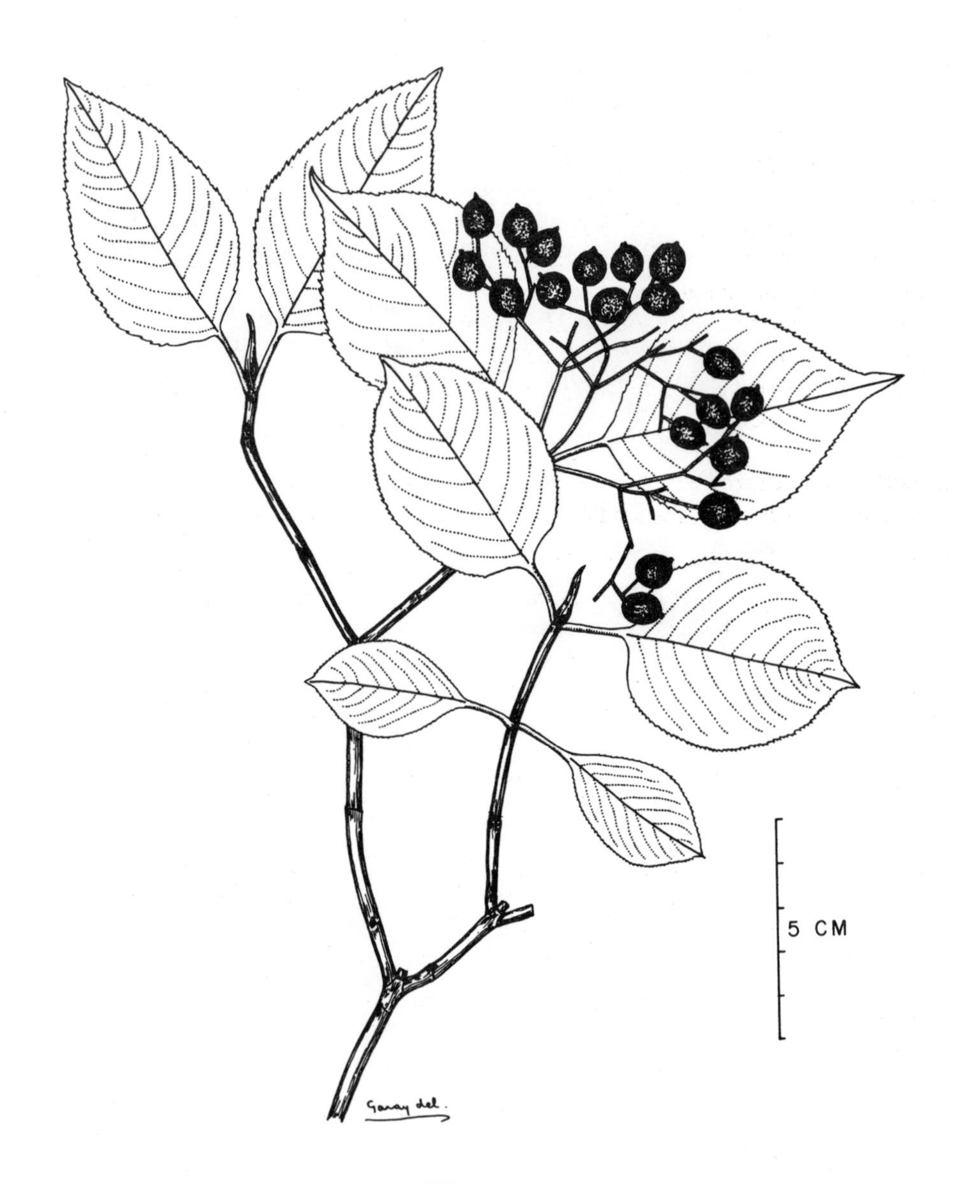

Viburnum lentago

Viburnum lentago L. Nannyberry

A large shrub, sometimes treelike, with a spreading top, growing to a height of 6 m or more, occasionally thicket-forming. Branchlets slender, brownish, and slightly scurfy, later becoming purplish-brown to grayish and smooth or a little ridged; winter buds gray, long, and narrow, the terminal flower buds swollen at the base.

Leaves opposite, simple, and deciduous; blades 5 - 10 cm long and 2.5 - 7 cm wide, elliptic-lanceolate to ovate, tapered or abruptly pointed with a well-marked tip, rounded or tapered at the base, dark green above, a little paler beneath, both surfaces generally smooth or a little scurfy on the veins beneath; margins finely and sharply serrate with pale, often gland-tipped incurved teeth; petioles 0.5 - 2.5 cm long, grooved, with winglike margins.

Flowers creamy-white, sweet-scented, all alike, and perfect, in terminal sessile clusters 5 - 10 cm across; late May and early June. Fruit a nearly round to ellipsoid drupe up to 12 mm long, blue-black with a bloom, in open clusters; pit large and flattish; the pulp sweet and edible; Aug. to Oct. (*lentago* — an old generic name)

In swamps and marshes; along shores and the edges of low woods and thickets.

Common in southern Ontario, especially in the limestone areas east, south, and west of the Canadian Shield; also in the region between Lake Superior and Lake-of-the-Woods. (s.w. Que. to s.e. Sask., south to Colo. and Ga.)

Field check Large shrub with finely toothed opposite leaves; petioles winged, grooved, and wavy-margined; sweet-scented white flowers in sessile clusters; open cluster of blue fruits; elongate and pointed gray winter buds; terminal flower buds swollen at the base.

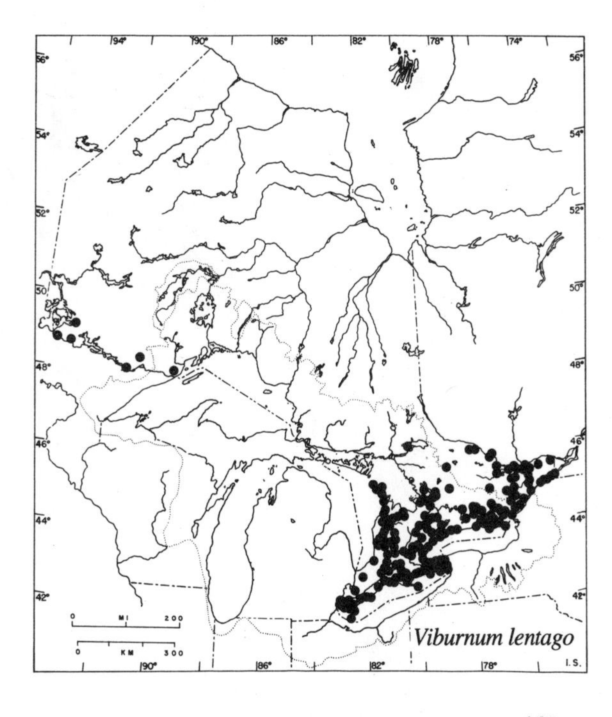

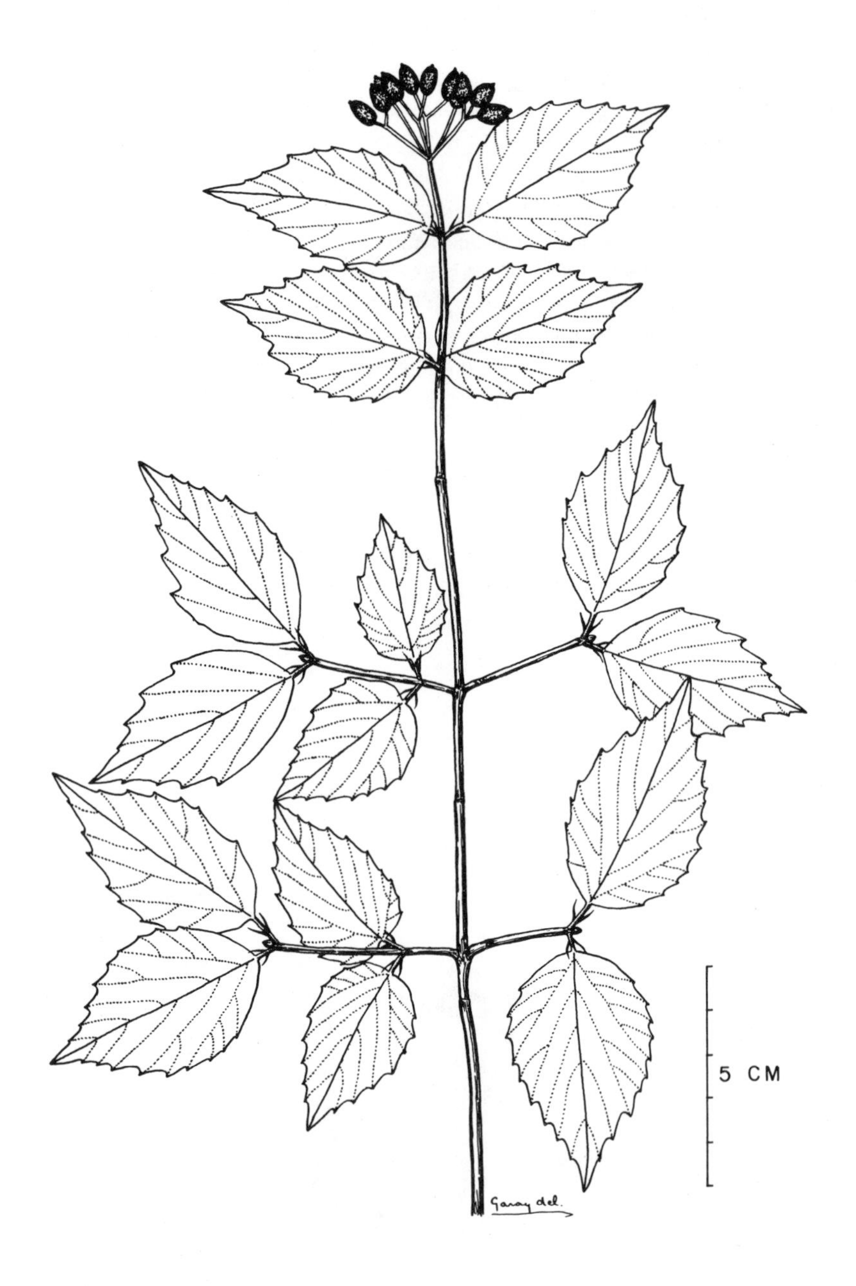

Viburnum rafinesquianum

Viburnum rafinesquianum Schult. — Downy arrow-wood

A small shrub, erect or spreading, usually less than 1.5 m high. Branchlets yellowish-brown, smooth or a little downy; older branches smooth, purplish to brownish-gray.

Leaves opposite, simple, and deciduous; blades 3.5–9 cm long and 1.5–5.5 cm wide, ovate to oblong-lanceolate, acute or long-tapering at the tip, rounded or heart-shaped at the base; smooth and dark green above, paler and smooth or downy, at least along the prominent veins, beneath; margins coarsely dentate with 4–11 teeth on each side; petioles grooved, short, rarely 1 cm long, with a pair of bristlelike stipules at the base which are usually longer than the petiole.

Flowers creamy-white, perfect, in terminal stalked clusters 2.5–7.5 cm across; late May and early June. Fruit an ellipsoidal dark purple drupe 6–9 mm long, in open clusters; Aug. and Sept. (*rafinesquianum*—in honour of Constantine Samuel Rafinesque-Schmaltz, 1783–1840, a brilliant pioneer naturalist, famous for his numerous contributions to natural history and for his eccentricities)

In dry thickets, open woods, hillsides, and river banks or on calcareous slopes and ledges.

In southern Ontario chiefly south of the Canadian Shield, from the east end of Lake Ontario to Georgian Bay, the Bruce Peninsula, and Manitoulin Island, and north to the Ottawa River; from the west end of Lake Ontario throughout the Deciduous Forest Region; also in the region between Lake Superior and Lake-of-the-Woods. (s.w. Que. to Man., south to Mo., Ark., and Ga.)

Field check Low shrub of dry places; coarsely toothed, mostly stipulate, sessile or short-stalked opposite leaves; open clusters of purple-black fruits.

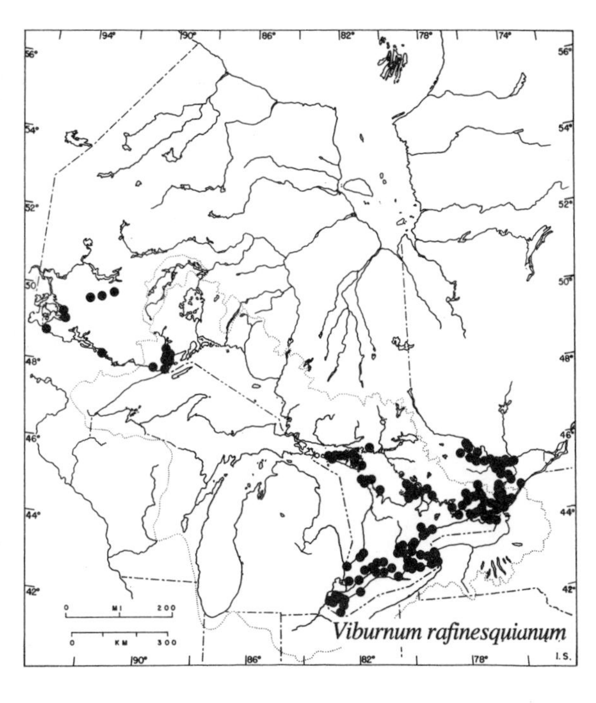

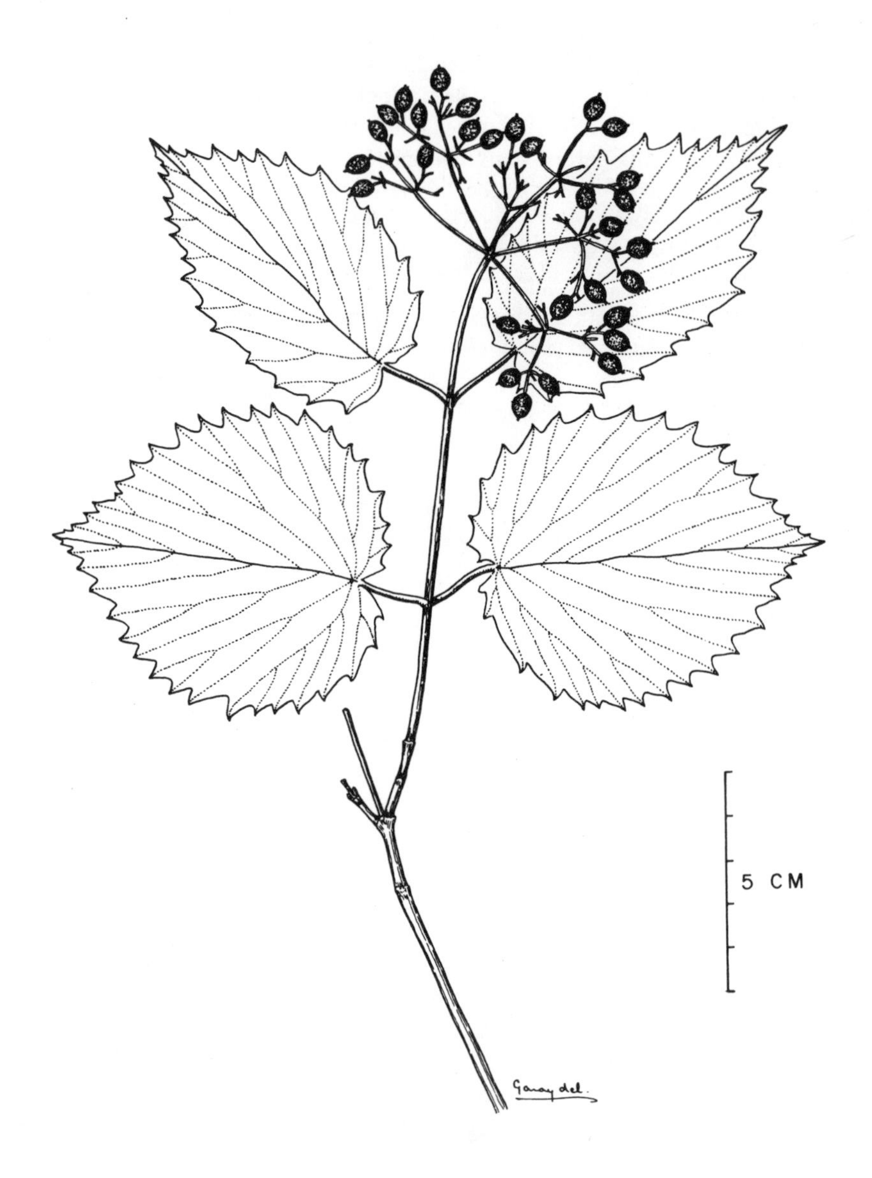

Viburnum recognitum

Viburnum recognitum Fern. (*V. dentatum* L. var. *lucidum* Ait.) — Southern arrow-wood

An upright or slightly spreading shrub growing to a height of 3 m or more. Branchlets smooth, brownish at first, later becoming purplish-gray, bluntly angled and with scattered corky warts.

Leaves opposite, simple, and deciduous; blades 3.5 - 10 cm long and 3 - 7.5 cm wide, ovate, broadly oval, or nearly round, acute or long-tapered at the tip, rounded or heart-shaped at the base, smooth and bright green above, paler and smooth or hairy beneath, especially along the prominent veins and in their axils where the hairs are tufted; margins coarsely dentate with 9 - 21 teeth on each side; petioles grooved, 1 - 3 cm long; stipules usually lacking.

Flowers creamy-white, ill-scented, perfect, in stout-stalked terminal clusters 5 - 9 cm across; mid-June. Fruit a round to ovoid, dark blue or blue-black drupe 6 - 9 mm long, in open clusters; Aug. and Sept. (*recognitum*—recognized or restudied; *dentatum*—toothed; *lucidum*—shining)

Generally in low wet situations: swampy woods and thickets.

Occasional in two disjunct parts of southern Ontario; the Deciduous Forest Region and the region between the Ottawa and St. Lawrence rivers. (N.B. to Ont. and Mich., south to n. Ohio and S.C.)

Field check Tall shrub of wet places; coarsely toothed and stalked opposite leaves lacking stipules; open clusters of blue-black fruits.

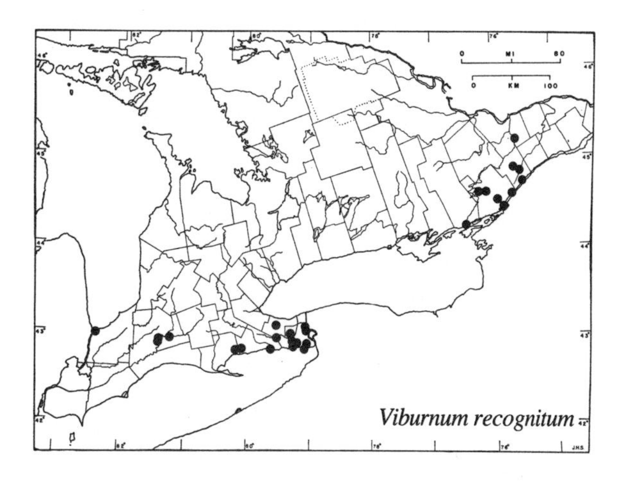

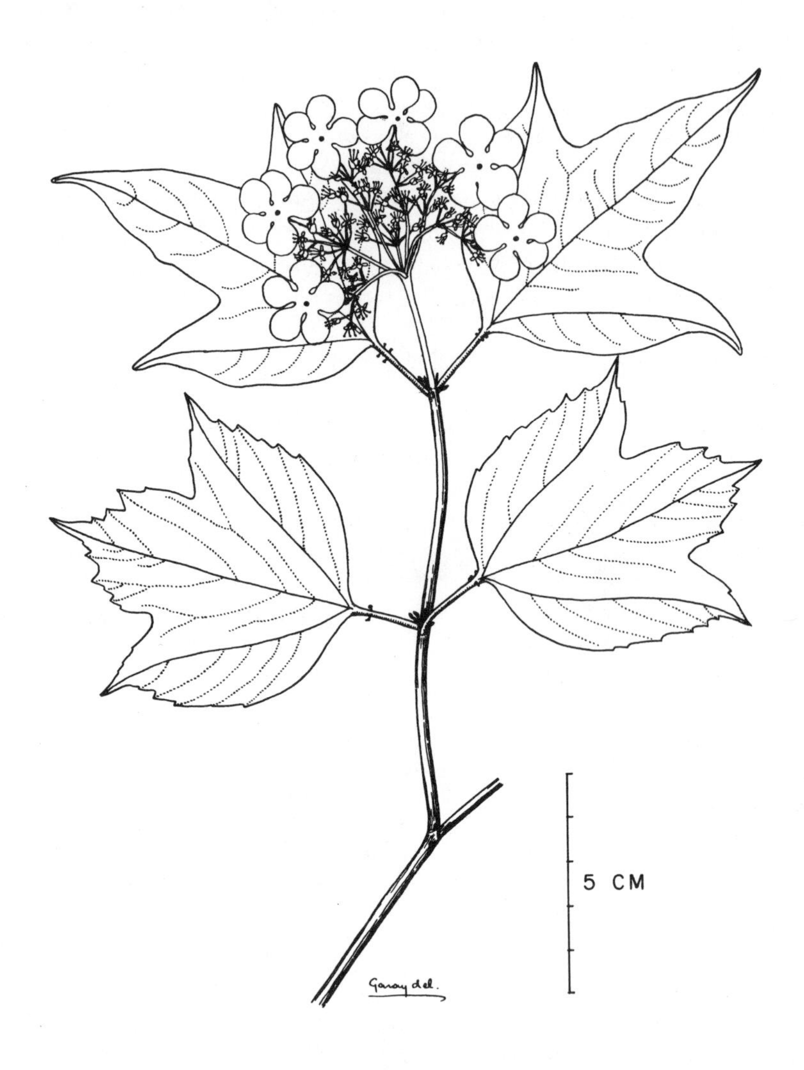

Viburnum trilobum

Viburnum trilobum Marsh. (*V. opulus* L. var. *americanum* Ait.) — High-bush cranberry

A coarse upright shrub growing to a height of 3 or 4 m. Branchlets smooth and gray to brownish-gray.

Leaves opposite, simple, and deciduous; blades broadly wedge-shaped, 5 – 11 cm long and about as wide, sharply 3-lobed with a deep cleft separating the median lobe from each strongly spreading lateral lobe, sharp-pointed at the tip, rounded or a little heart-shaped at the base, dark green and smooth above, paler and smooth or thinly hairy below; margins entire or toothed; petioles 1 – 4 cm long, conspicuously grooved, with a pair of slender thick-tipped stipules at the base and several small club-shaped glands near the junction with the blade; as in *V. edule* and *V. acerifolium*, leaves at the ends of some of the branches may be unlobed or with poorly developed lobes.

Flowers white, in showy, flat-topped, stalked terminal clusters up to 15 cm across; outer flowers sterile, with expanded flat corollas, surrounding the smaller, more numerous fertile ones; June and July. Fruit a round or ellipsoid, orange to red, juicy, berrylike drupe; in open clusters; Aug. and Sept. (*trilobum*—three-lobed; *opulus*—an old generic name; *americanum*—American)

In damp soil around swamps and bogs and along streams or in low cool open woods and thickets.

Common in southern Ontario, north to James Bay and northwest to Lake Superior and Lake-of-the-Woods; northern limit at about 52° N. (Nfld. to B.C., south to Wash. and N. Eng.)

Note The fruit can be used to make a delicious jelly, especially tasty with meat.

V. opulus is the European guelder rose, a species often planted as an ornamental in southern Ontario and occasionally found as an escape along roadsides and in woods. It resembles *V. trilobum* very closely but differs in having bristle-tipped stipules and the glands at the junction of the petiole and blade are concave with saucerlike discs. The fruits of *V. opulus* often remain on the bush all winter, and although they look as if they should be attractive to birds, they are apparently scorned until other food is gone, and even then they are eaten reluctantly.

Some botanists consider the differences between the Eurasian and the American high-bush cranberry to be worthy of recognition only at the varietal level, as in the synonym given.

Field check Tall shrub of moist places; 3-lobed opposite leaves; showy marginal flowers around a flat cluster of inconspicuous fertile flowers; open clusters of orange-red berrylike fruits.

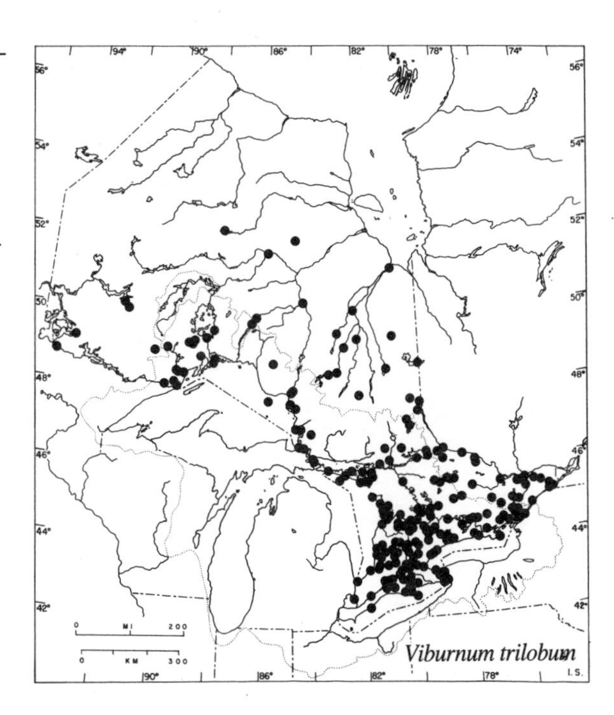

LITERATURE CITED

ARGUS, G.W.

1964 Preliminary reports on the flora of Wisconsin, no. 51: Salicaceae; the genus *Salix*—the willows. Transactions of the Wisconsin Academy of Sciences, Arts & Letters, 53 : 217–272.

1980 The typification and identity of *Salix eriocephala* Michx. (Salicaceae). Brittonia 32 : 170–177.

ARGUS, G.W. and D.J. WHITE

1977 The rare vascular plants of Ontario/Les plantes vasculaires rares de l'Ontario. Syllogeus no. 14, 1-63/1-66.

BRAUN, E.L.

1961 The woody plants of Ohio. Columbus, Ohio State University Press. 362 pp.

CODY, W.J.

1962 *Sorbaria sorbifolia* (L.) A.Br., False Spiraea, persisting and spreading after cultivation in Canada. Canadian Field-Naturalist 76 : 104–107.

DALTON, D.A. and A.W. NAYLOR

1975 Studies in nitrogen fixation by *Alnus crispa*. American Journal of Botany 62 : 76–80.

DALY, G.T.

1966 Nitrogen fixation by nodulated *Alnus rugosa*. Canadian Journal of Botany 44 : 1607–1621.

DEL TREDICI, P.

1977 The buried seeds of *Comptonia peregrina*, the Sweet Fern. Bulletin of the Torrey Botanical Club 104 : 270–275.

DORN, R.D.

1975 A systematic study of *Salix* section Cordatae in North America. Canadian Journal of Botany 53 : 1491–1522.

EINSET, J.

1951 Apomixis in American polyploid blackberries. American Journal of Botany 38 : 768–772.

ELIAS, T.S. and H. GELBAND

1975 Nectar: its production and functions in Trumpet Creeper. Science 189 : 289–291.

FERNALD, M.L.

1950 Gray's manual of botany. 8th ed. New York, Van Nostrand. 1632 pp.

FURLOW, J.J.

1979 The systematics of the American species of *Alnus* (Betulaceae). Rhodora 81 : 1–121, 151–248.

GILLIS, W.T.

1971 The systematics and ecology of Poison-ivy and the Poison-oaks (*Toxicodendron*, Anacardiaceae). Rhodora 73 : 72–159, 161–237, 370–443, 465–540.

GLEASON, H.A.

1952 Illustrated flora of the Northeastern United States and adjacent Canada. New York, New York Botanical Garden. 3 vols.

GLEASON, H.A. and A. CRONQUIST

1963 Manual of vascular plants of Northeastern United States and adjacent Canada. New York, Van Nostrand. 810 pp.

HARDIN, J.W.

1973 The enigmatic Chokeberries (*Aronia*, Rosaceae). Bulletin of the Torrey Botanical Club 100 : 178–184.

HITCHCOCK, C.L., A. CRONQUIST, M. OWNBEY, and J.W. THOMPSON

1955-69 Vascular plants of the Pacific Northwest. University of Washington Publication in Biology 17 : 5 vols.

HODGDON, A.R. and F.L. STEELE

1962 Glandularity in *Rubus allegheniensis* Porter. Rhodora 64 : 161–168.

1966 *Rubus* subgenus *Eubatus* in New England. Rhodora 68 : 474–513.

HULTÉN, E.

1968 Flora of Alaska and neighboring territories. Stanford, Stanford University Press. 1008 pp.

HUTCHINSON, T.C.

1976 The impact of pollution. In Morton, J.K., ed., Man's impact on the Canadian flora. Canadian Botanical Association Bulletin, Supplement to vol. 9, no. 1 : 25–29.

KEEGAN, C.R., R.H. VOSS, and K.S. BAWA

1979 Heterostyly in *Mitchella repens* (Rubiaceae). Rhodora 81 : 567–573.

KINGSBURY, J.M.

1965 Deadly harvest; a guide to common poisonous plants. London, Allen & Unwin. 128 pp.

LEPAGE, E.

1976 Les bouleaux arbustifs du Canada et de l'Alaska. Le Naturaliste canadien 103 : 215–233.

LI, HUI-LIN.

1952 Floristic relationships between eastern Asia and eastern North America. Transactions of the American Philosophical Society. New Ser. 42 : 371–429.

MCKAY, S.M.

1973 A biosystematic study of the genus *Amelanchier* in Ontario. (M. Sc. Thesis, University of Toronto.)

MORRIS, M.,D.E. EVELEIGH, S.C. RIGGS, and W.N. TIFFNEY, JR.

1974 Nitrogen fixation in the Bayberry (*Myrica pensylvanica*) and its role in coastal succession. American Journal of Botany 61 : 867–870.

MUNIYAMMA, M. and J.B. PHIPPS

1979 Cytological proof of apomixis in *Crataegus* (Rosaceae). American Journal of Botany 66 : 149–155.

NICHOLSON, D.H. and G.C. STEYSKAL

1976 The masculine gender of the generic name *Styrax* Linnaeus (Styracaceae). Taxon 25 : 581–587. (Includes reference to *Panax*.)

PACKER, J.G. and K.E. DENFORD

1974 A contribution to the taxonomy of *Arctostaphylos uva-ursi*. Canadian Journal of Botany 52 : 743–753.

PHIPPS, J.B. and M. MUNIYAMMA

1980 A taxonomic revision of *Crataegus* (Rosaceae) in Ontario. Canadian Journal of Botany 58 : 1621–1699.

PRINGLE, J.S.

1971 Taxonomy and distribution of *Clematis*, Sect. *Atragene* (Ranunculaceae), in North America. Brittonia 23 : 361–393.

READER, R.J.

1975 Competitive relationships of some bog ericads for major insect pollinators. Canadian Journal of Botany 53 : 1300–1305.

1977 Bog ericad flowers; self-incompatibility and relative attractiveness to bees. Canadian Journal of Botany 55 : 2279–2287.

RODRIGUEZ-BARRUECO, C.

1969 The occurrence of nitrogen-fixing nodules on non-leguminous plants. Botanical Journal of The Linnaean Society 62 : 77–84.

ROWE, J.S.

1972 Forest regions of Canada. Ottawa, Canadian Forest Service. Pub. no. 1300. 172 pp.

SCHOEN, D.J.

1977 Floral biology of *Diervilla lonicera* (Caprifoliaceae). Bulletin of the Torrey Botanical Club 104 : 234–240.

SCHRAMM, J.R.
1966 Plant colonization studies on black wastes from anthracite mining in Pennsylvania. Transactions of the American Philosophical Society 56 : 1–194.

SOPER, J. H. and E.G. VOSS.
1964 Black Crowberry in the Lake Superior Region. Michigan Botanist 3 : 35–38.

STEELE, F.L. and A.R. HODGDON
1963 Hybridization of *Rubus hispidus* and *R. setosus*. Rhodora 65 : 262–270.
1970 Hybrids in *Rubus* subgenus *Eubatus* in New England. Rhodora 72 : 240–250.

VANDER KLOET, S.P.
1976 A comparison of the dispersal and seedling establishment of *Vaccinium angustifolium* (the Lowbush Blueberry) in Leeds County, Ontario, and Pictou County, Nova Scotia. Canadian Field-Naturalist 90 : 176–180.
1978 Systematics, distribution and nomenclature of the polymorphic *Vaccinium angustifolium*. Rhodora 80 : 358–376.

WAGNER, JR., W.H.
1974 Dwarf Hackberry (Ulmaceae: *Celtis tenuifolia*) in the Great Lakes Region. Michigan Botanist 13 : 73–99.

WANGERIN, W.
1910 Cornaceae. Das Pflanzenreich IV.229 : 1–106.

WOOD, C.E.
1972 Morphology and phytogeography: the classical approach to the study of disjunctions. Annals of the Missouri Botanical Garden 59 : 107–124.

Addendum

BREITUNG, A.J.
1952 Key to the Genus *Rubus* of the Ottawa Valley. Canadian Field-Naturalist 66 : 108–110.

BOTANICAL AUTHORS

A.Br. Braun, Alexander Carl Heinrich 1805–1877 Germany Professor of Botany, Director of Berlin Botanical Garden

Adans. Adanson, Michel 1727–1806 France Botanist and explorer, author of *Familles des plantes*

Ahles, Harry E. 1924– United States Botanist, North Carolina and Massachusetts

Ait. Aiton, William 1731–1793 England Botanist, royal gardener at Kew, author of *Hortus Kewensis*

Anderss. Andersson, Nils Johan 1821–1880 Sweden Student of *Salix*

Ashe, William Willard 1872–1932 United States

B.Ehrh. Ehrhart, Johann Balthazar 1700–1756 Germany

Bab. Babington, Charles Cardale 1808–1895 England

Bailey, Liberty Hyde 1858–1954 United States Authority on horticulture, student of *Rubus*

Ball, Carleton Roy 1873–1958 United States Student of *Salix*

Banks, (Sir) Joseph 1743 - 1820 England Naturalist, traveller, president of The Royal Society

Barratt, Joseph 1796–1882 United States Geologist, Connecticut and Pennsylvania

Bart. Barton, Benjamin Smith 1766–1815 United States

Bartr. Bartram, William 1739–1823 United States Botanist, Philadelphia

Batsch, August Johann Georg Karl 1761–1802 Germany

Bebb, Michael Schuck 1883–1895 United States Student of *Salix*

Benth. Bentham, George 1800–1884 England Co-author (with J.D. Hooker) of *Genera Plantarum*, president of The Linnaean Society

Bernh. Bernhardi, Johann Jacob 1774–1850 Germany

Bigel. Bigelow, Jacob 1787–1879 United States Professor of Botany at Boston, student of *Rubus*

Blake, Sydney Fay 1892–1959 United States

Blanchard, William Henry 1850–1922 United States

Blume, Carl Ludwig von 1796–1862 Netherlands

Böcher, Tyge Wittrock 1909– Denmark Student of arctic flora, co-author (with K. Holmen and K. Jakobsen) of *Flora of Greenland*

Boivin, Joseph Robert Bernard 1916– Canada Student of the Canadian flora

Borkh. Borkhausen, Moriz Balthasar 1760–1806 Germany

Britt. Britton, Nathaniel Lord 1859–1934 United States Director of New York Botanical Garden, co-author (with A. Brown) of *Illustrated Flora of the Northern United States and Adjacent Canada*

Butt. Butters, Frederick King 1878–1945 United States Professor of Botany, Minnesota

C.A.Mey. Meyer, Carl Anton Andrejewicz von 1795 - 1855 U.S.S.R.

Carr. Carrière, Elie Abel 1818–1896 France Horticulturist

Cham. Chamisso, Adalbert von 1781–1838 Germany Botanist aboard Russian expedition to Alaska, 1816–1817

Clausen, Robert Theodore 1911– United States

Cockerell, Theodore Dru Alison 1866–1948 United States

Coult. Coulter, John Merle 1851–1928 United States

Crantz, Heinrich Johann Nepomuk von 1722–1799 Austria Physician and botanist; Professor at the University of Vienna

Cronq. Cronquist, Arthur John 1919- United States Student of of Compositae, Curator at New York Botanical Garden Herbarium, co-author (with Gleason) of the 1950 revision of Britton and Brown's *Illustrated Flora of the Northeastern United States and Adjacent Canada*

DC. de Candolle, Augustin Pyramus 1778–1841 Switzerland

Decne. Decaisne, Joseph 1807–1882 France Horticulturist

Desf. Desfontaines, René Louiche 1750–1883 France Student of trees and shrubs suitable for cultivation

Desv. Desvaux, Augustin Nicaise 1784–1856 France

Don, David 1799–1841 England Professor at King's College, Librarian of The Linnaean Society

Dore, William G. 1912- Canada Student of Gramineae and botanist at Biosystematics Research Institute, Ottawa

Dorn, Robert D. 1942- United States Student of *Salix* and local floras in northwestern U.S.A.

Duham. Duhamel du Monceau, Henri Louis 1700–1781 France

Dunal, Michel Félix 1789-1852 France Botanist at the University of Montpellier

Du Roi, Johann Philippe 1741–1785 Germany

Eat. Eaton, Amos 1776–1842 United States Author of *Manual of Botany for the Northern States*

Eggl. Eggleston, Willard Webster 1863–1935 United States Student of flora of New England (especially Vermont) and of *Crataegus*

Ehrh. Ehrhart, Jakob Friedrich 1742–1795 Switzerland German botanist and pupil of Linnaeus

Ell. Elliot, Stephen 1771–1830 United States Botanist in South Carolina, author of *A Sketch of the Botany of South Carolina and Georgia*

Engler, Heinrich Gustav Adolf 1884–1930 Germany Director of the Botanical Garden and Museum at Berlin

Fern. Fernald, Merritt Lyndon 1873–1950 United States Student of flora of eastern North America, co-author (with Robinson) of the 7th edition of Gray's *A Manual of Botany* and author of the 1950 revision of this manual

Fluegge, Johann 1775–1816 Germany Student of *Salix*

Focke, Wilhelm Olbers 1834–1922 Switzerland Student of *Rubus*

Forbes, James 1773–1861 England Gardener and horticulturist

Fritsch, Karl 1864–1934 Austria

Gaertn. Gaertner, Philipp Gottfried 1754–1825 Germany Apothecary

Gillis, William Thomas 1933–1979 United States Student of Anacardiaceae, especially toxic members, Assistant Professor at Michigan State University

Gleason, Henry Allan 1882–1975 United States Co-author (with Cronquist) of 1950 revision of Britton and Brown's *Illustrated Flora of the Northeastern United States and Adjacent Canada*

Goldie, John 1793–1886 Scotland Botanical collector in eastern North America in 1819

Grauer, Sebastien 1758–1820 Germany

Gray, Asa 1810–1888 United States Professor at Harvard University, author of *Manual of the Botany of the Northern United States*, co-author (with John Torrey) of *A Flora of North America*

Greene, Edward Lee 1843–1915 United States Professor of Botany at the University of California

Gronov. Gronovius, Johannes Fredericus 1690–1762 Netherlands Dutch botanist and friend of Linnaeus

Hagerup, Olaf 1889–1961 Denmark Keeper of the herbarium, Copenhagen

Hara, Hiroshi 1911- Japan Taxonomist and phytogeographer, student of floras of Japan and eastern North America

HBK Humboldt, Bonpland, and Kunth

Humboldt, Friedrich Wilhelm Heinrich Alexander von 1769–1859 Germany Naturalist and explorer

Bonpland, Aimé Jacques Alexandre 1773–1858 France Botanical explorer on expedition to South America

Kunth, Carl Sigismund 1788–1850 Germany Professor at Berlin; wrote up expeditions of Humboldt and Bonpland

Heller, Amos Arthur 1867–1944 United States Botanical collector; editor of the periodical *Muhlenbergia*

Hitchc. Hitchcock, Charles Leo 1902– United States Taxonomist and authority on the flora of the Pacific Northwest

Hook. Hooker, (Sir) William Jackson 1785–1865 England Student of arctic flora, horticulturist, prolific author and editor

Hornem. Hornemann, Jens Wilken 1770–1841 Denmark

Hult. Hultén, Oskar Eric Gunnar 1894–1980 Sweden Student of circumpolar arctic flora

Hyland, Fay 1900– United States Student of woody plants of the state of Maine

Jacq. Jacquin, Nikolaus Joseph von 1727–1817 Austria

Kalm, Pehr 1716–1779 Sweden Botanical collector in North America in 1747–1749, pupil of Linnaeus

Kerner, Johann Simon von 1755–1830 Germany

Kirchner, Georg 1837–1885 Germany

Knerr, Ellsworth Brownell 1861–1942 United States

Koch, Karl Heinrich Emil 1809–1879 Germany Professor of Botany at Berlin, dendrologist, explorer

Kuntze, Carl Ernst Otto 1843–1907 Germany

L. Linnaeus, Carolus (original name, Carl von Linné) 1707–1778 Sweden Naturalist, explorer, founder of the binomial system of nomenclature for plants and animals

L.fil. Linné, filius, Carl von 1741–1783 Sweden Son of Carolus Linnaeus

Lam. Lamarck, Jean Baptiste Antoine Pierre Monnet de 1744–1829 France Author of floras of France, co-author (with Poiret) of *Encyclopédie méthodique*

Lange, Johan Martin Christian 1818–1898 Denmark

Lepage, (Abbé) Ernest 1905–1981 Canada Student of arctic and subarctic North American flora

Lévl. Léveillé, (Abbé) August Abel Hector 1863–1918 France

L'Hér. L'Héritier de Brutelle, Charles Louis 1746–1800 France Botanist and magistrate

Lindl. Lindley, John 1799–1865 England Student of Orchidaceae and British flora

Link, Johann Heinrich Friedrich 1767–1851 Germany Founder of the Berlin-Dahlem Herbarium

Lodd. Loddiges, Conrad 1738–1826 England Nurseryman, horticulturist

Loisel. Loiseleur-Deslongchamps, Jean Louis Auguste 1774–1849 France

Lour. Loureiro, João de 1710–1791 Portugal

Marsh. Marshall, Humphry 1722–1801 United States Student of American trees and shrubs

Maxim. Maximovich, Karl Johann 1829–1891 U.S.S.R.

Medik. Medikus, Friedrich Kasimir 1736–1808 Germany

Michx. Michaux, André 1746–1802 France Botanical collector and student of American trees

Michx.fil. Michaux, filius, François-André 1770–1855 France Son of André Michaux and author of works on American forest trees

Mill. Miller, Philip 1691–1771 England Horticulturist and gardener

Miq. Miquel, Friedrich Anton Wilhelm 1811–1871 Netherlands

Moench, Conrad 1744–1805 Germany

Muhl. Muhlenberg, Gotthilf Henry Ernest 1753–1815 United States Clergyman in Pennsylvania, educated in Germany

Munson, Thomas Volney 1843–1913 United States Nurseryman, horticulturist

Nakai, Takenoshin 1882–1952 Japan

Nees, Christian Gottfried Daniel von Esenbeck 1776–1858 Germany

Nutt. Nuttall, Thomas 1786–1859 United States Botanical collector and explorer, author of *Genera of North American Plants*

Oeder, Georg Christian von 1728–1791 Denmark

Pall. Pallas, Peter Simon 1741–1811 Germany Explorer in Russia and Siberia

Peck, Charles Horton 1833–1917 United States Student of *Rubus*

Pers. Persoon, Christian Hendrik 1761–1836 France Author of important works in mycology

Planch. Planchon, Jules Émile 1823–1888 France Professor at the University of Montpellier

Poir. Poiret, Jean Louis Marie 1755–1834 France Botanical collector in Africa, co-author (with Lamarck) of *Encyclopédie méthodique*

Polunin, Nicholas 1909– England Student of circumpolar flora, editor

Porter, Thomas Conrad 1822–1901 United States Author of *Flora of Pennsylvania* (1903)

Pringle, James S. 1937– Canada Taxonomist at the Royal Botanical Gardens, Hamilton

Pursh, Frederick Traugott 1774–1820 United States Collector and student of North American flora, author of *Flora Americae Septentrionalis*

R. & S. Roemer and Schultes
Roemer, Johann Jakob 1763–1819 Switzerland
Schultes, Josef August 1773–1831 Austria

Raf. Rafinesque-Schmaltz, Constantine Samuel 1783–1840 United States Eccentric naturalist, author of numerous publications on the flora of southern United States

Regel, Eduard August von 1815–1892 U.S.S.R.

Rehd. Rehder, Alfred 1863–1949 United States Authority on horticulture, especially trees and shrubs

Richards. Richardson, John 1787–1865 England Scottish naturalist with one of Sir John Franklin's expeditions in search of the Northwest Passage

Robins. Robinson, Benjamin Lincoln 1864–1935 United States Professor at Harvard, co-author (with Fernald) of the 1908 revision (7th edition) of Gray's *Manual of Botany*

Rouleau, Joseph Albert Ernest 1916– Canada Curator of Herbier Marie-Victorin, Montreal, and student of flora of eastern Canada, especially Newfoundland

Rowlee, Willard Winfield 1861–1923 United States

Rupr. Ruprecht, Franz Joseph 1814–1870 U.S.S.R. Curator of herbarium at St. Petersburg

Rydb. Rydberg, Per Axel 1860–1931 United States Author of *Flora of the Rocky Mountains and Adjacent Plains*

St.Hil. St. Hilaire, Auguste François César Prouvençal de 1779–1853 France Botanical collector and explorer in South America

Salisb. Salisbury, Richard Anthony 1761–1829 England

Sarg. Sargent, Charles Sprague 1841–1927 United States Student of American trees, especially *Crataegus*

Scheele, Georg Heinrich Adolf 1808–1864 Germany

Schneid. Schneider, Camillo Karl 1876–1951 Germany Dendrologist

Seem. Seemann, Berthold Carl 1825–1871 Germany Botanist on the voyage of H.M.S. *Herald*, 1845–1851

Sieb. Siebold, Philipp Franz von 1796–1866 Germany

S.J.Smith Smith, Stanley Jay 1915-1977 United States Curator at New York State Museum Herbarium

Sm. Smith, (Sir) James Edward 1759-1828 England Purchased the Linnaean Herbarium (1784); founder of The Linnaean Society

Small, John Kunkel 1869–1938 United States Curator at New York Botanical Garden Herbarium, author of *Flora of the Southeastern United States*

Spach, Édouard 1801–1879 France

Spreng. Sprengel, Curt Polycarp Joachim 1766–1833 Germany

Steud. Steudel, Ernst Gottlieb 1783–1856 Germany Physician and botanist

Suk. Sukatschev, Vladimir Nikolajevich 1880–1967 U.S.S.R.

T. & G. Torrey, John, (see Torr.) and Asa Gray (see Gray), co-authors of *A Flora of North America*

Tausch, Ignaz Friedrich 1793 - 1848 Bohemia

Thaxter, Roland 1858 - 1932 United States Pioneer mycologist

Thunb. Thunberg, Carl Pehr 1743 - 1828 Sweden Professor at the University of Uppsala

Torr. Torrey, John 1796–1873 United States Physician and botanist, co-author (with Asa Gray) of *A Flora of North America*

Trautv. Trautvetter, Ernst Rudolph von 1809–1899 U.S.S.R.

Trel. Trelease, William 1857–1945 United States

Trew, Christoph Jakob 1695–1769 Germany Physician and botanist

Turrill, William Bertram 1890–1961 England Student of the British flora

Vahl, Martin Hendriksen 1749–1804 Denmark

Vent. Ventenat, Étienne Pierre 1757–1808 France

Vill. Villars, Dominique 1745–1814 France Physician and Professor of Botany at the University of Strasbourg

Wahl. Wahlenberg, Göran 1780–1851 Sweden Professor at the University of Uppsala

Wall. Wallich, Nathaniel 1786–1854 Denmark Physician, Superintendent of Calcutta Botanic Garden, 1815–1846

Walt. Walter, Thomas 1740–1789 United States Student of flora of southeastern United States, author of *Flora Caroliniana*

Wang. Wangenheim, Friedrich Adam Julius von 1749–1800 Germany

Weigel, Christian Ehrenfried von 1748–1831 Germany

Wendl. Wendland, Heinrich Ludolph 1791–1869 Germany

Wieg. Wiegand, Karl McKay 1873–1942 United States Co-author (with A.J.Eames) of *Flora of the Cayuga Lake Basin, New York*

Willd. Willdenow, Karl Ludwig 1765–1812 Germany Professor of Botany at the Botanical Garden, Berlin; his herbarium contains type specimens of many North American plants

Wilson, Ernest Henry 1876–1930 United States Dendrologist, horticulturist, and plant explorer at the Arnold Arboretum of Harvard University; his expeditions to China resulted in many new and interesting plant introductions into American gardens

Wimmer, Christian Friedrich Heinrich 1803–1868 Germany

Wormsk. Wormskjold, Morten 1783–1845 Denmark Collected plants on Otto von Kotzebue's expedition to Alaska (1815–1818)

Young, Steven Burr 1938– United States Professor at Ohio State University, Columbus

Zab. Zabel, Hermann 1832–1912 Germany

Zucc. Zuccarini, Joseph Gerhard 1797–1848 Germany

GLOSSARY

achene a small dry indehiscent single-seeded fruit
acuminate tapering to a prolonged point
acute sharp-pointed but not long-tapering
adventitious arising from unusual positions, as buds from roots, roots from stems or leaves
aggregate collected together: as in the fruit of *Rubus*, with several to many drupelets on a common receptacle
alluvial (soils) deposited by flood along the banks of streams and rivers
alternate attached singly along the stem or axis at the nodes; not opposite nor whorled
ament a scaly spike bearing inconspicuous and usually unisexual flowers; catkin
angled with obvious longitudinal angles or ridges
anther the pollen-bearing portion of the stamen, usually sac-shaped; pollen-sac
anthesis the act of flowering or the time of expansion when the flower becomes fully functional
apetalous without petals
apex the tip or summit
apical occurring at the apex
apomict offspring resulting from apomixis
apomictic reproducing by apomixis
apomixis development of the ovule into a seed without fertilization; reproduction by non-sexual methods
appressed lying close to another part of the plant
arborescent treelike in size and habit
aril an appendage or an additional covering on some seeds, often pulpy or coloured or both
armed bearing spines or thorns
aromatic fragrant; with spicy aroma
ascending arising from the ground at an angle or curving upward
asexual non-sexual or purely vegetative
astringent causing a puckering effect on the mouth when eaten or tasted
attenuate gradually tapering to a long slender tip
awl-shaped narrowly sharp-pointed, tapering gradually to a stiff or slender point
awn a slender terminal bristle or appendage
awned provided with one or more awns
axil the angle formed by a leaf or a branch with a stem, or by a vein branching off a larger vein or off the midvein of a leaf

balsamic with characteristics of balsam, e.g., odour
beaked ending in a firm, rather prolonged tip
bearded bearing long stiff hairs
berry a fleshy fruit with a thin skin, the seeds embedded in the pulp
bilabiate two-lipped; as in the corolla of certain flowers divided into two lips, e.g., snapdragon
blade the expanded part of a leaf, a petal or a sepal
bloom a whitish or bluish-white fine powdery coating, as on some leaves and fruits
bract a reduced or modified leaf subtending a flower or associated with an inflorescence or flower stalk
bristle a stiff hair on the epidermis of a leaf or stem

caducous falling off very early
calcareous limy or containing lime
calciphile preferring soil with a high lime content
callus a hard prominence or protuberance
calyx sepals considered collectively; the outermost part of the floral envelope
campanulate bell-shaped
capitate headlike or knoblike
capsule a dry fruit produced from more than one carpel, releasing its seeds at maturity through slits, pores, or by splitting into valves
carpel a simple pistil or one component of a compound pistil
cartilaginous firm, tough, and flexible, like cartilage
catkin see ament
caudate with a tail or tail-like appendage
caudex the persistent more or less woody base of an otherwise annual stem
channelled conspicuously grooved longitudinally
ciliate fringed with hairs
clavate club-shaped; thicker at the distal end and tapering to the base
clawed with a claw; the long narrow base of some sepals and petals

coetaneous flowering at the same time as the leaves expand, as in some willows

colonial forming colonies of plants with underground connections

compound as in leaves, comprised of two or more leaflets

conic, conical cone-shaped

coniferous bearing cones

connate fused or united; connate-perfoliate leaves are united at the base to form a continuous tissue surrounding the stem

cordate heart-shaped

coriaceous leathery in texture

corolla petals considered collectively; the inner set(s) of floral leaves

corymb an inflorescence which is short and broad, more or less flat-topped, with the outer flowers opening first

corymbose in a corymb

crenate with rounded or blunt teeth

crenulate finely crenate

cultivar a variety or race that has originated and persisted under cultivation

cuneate wedge-shaped or narrowly triangular

cupule the cup (involucre) of an acorn or fruit of the oak

cuspidate tipped with a cusp or sharp firm point

cylindric, cylindrical shaped like a cylinder

cyme a broad, often flat-topped inflorescence with central flowers opening first

cymose cymelike

deciduous not persistent; falling at maturity or at the end of the growing season

decumbent lying on the ground; reclining but with the tips ascending

decurrent prolonged below the attachment and joined to the stem

decurved curved downward

decussate with each pair at right angles to the pair above and below, forming four rows in the longitudinal axis

dehiscence the act of opening at maturity, as in an anther releasing pollen or capsules releasing seeds

deltoid shaped like an equilateral triangle

dentate toothed, with sharp spreading teeth

denticulate with fine teeth

depressed pressed down

diabase a dark gray to black, fine-grained, basaltic igneous rock

dichasial indicating a cyme with two main branches

dichotomous forking into two branches of about equal size

digitate compound, with the parts spreading like the fingers of a hand

dioecious with staminate and pistillate flowers on different individuals

diploid having twice the basic chromosome number for the species

disc (floral) outgrowth from the receptacle around the base of the ovary

divaricate widely spreading from an axis, as branches from a trunk

divergent inclining away from each other

dolomite a limestone rich in magnesium

downy with a covering of soft hairs

drupe a fruit with a pulpy or fleshy layer surrounding the one to several seeds, each of which is enclosed in a stony layer

drupelet a small drupe, as one in a cluster forming the fruit of a raspberry

ellipsoid solid but with an elliptical outline

ellipsoidal with the shape of an ellipsoid

elliptic oval in outline and widest at or about the middle

emarginate with a shallow notch or cleft at the apex

entire without toothing or divisions; having an even margin

epicalyx an extra whorl of sepal-like appendages on the outside of the calyx

escape a plant, not part of the native flora, that has become established outside cultivation

evergreen holding the leaves over winter or longer

exfoliating peeling off in thin layers, as the bark on paper birch

exserted protruding beyond, as the stamens out of the corolla tube

falcate sickle-shaped

fascicle a cluster or bundle of similar parts, as needles of conifers

fastigiate parallel, clustered, and erect branches

fertile capable of producing fruit; of stamens, producing functional pollen

filament the part of the stamen which supports the anther, often threadlike

filiform threadlike

flexuous curved alternately in opposite directions

floccose coated with soft woolly hairs that rub off easily

floret a small flower, usually one of a cluster

floricane (in *Rubus*) the flowering cane or stem, the second year's growth of the usually non-flowering primocane

floriferous flower-bearing

foliaceous leaflike

follicle a dry dehiscent fruit developed from a single carpel and opening usually along one side
free not adhering to another organ
fruit a mature ovary or ovaries with or without associated parts of the flower or inflorescence

glabrate nearly glabrous
glabrescent becoming almost or quite smooth; becoming glabrous
glabrous smooth; lacking hairs
gland a secreting surface or structure, often a hair-shaped appendage with a swollen tip like a pinhead, usually producing nectar, oil, or waxy material
glandular with glands
glaucescence presence of a bloom
glaucescent more or less glaucous
glaucous covered with a whitish or bluish bloom that rubs off easily
globose spherical or nearly so
globular spherical
glutinous sticky
gynoecium the female part of the flower; the pistils, collectively

head a short dense spike
herb a non-woody plant that dies down to the ground at the end of the growing season
herbaceous with the characteristics of an herb
hermaphroditic having both stamens and pistils in the same flower
heterostylous having styles of different lengths
hirsute with rather rough or coarse hairs
hispid beset with stiff rough hairs or bristles
hoary covered with a close white or whitish pubescence
hybrid the progeny of a cross between two taxa, as in an inter-specific hybrid between two species
hypanthium a cup or ring around the ovary, usually formed from the union of calyx, corolla, and basal portion of stamens

impressed furrowed, as with veins below the level of the leaf surface
incised sharply and deeply cut, more or less irregularly
included not exserted, such as in stamens which do not project beyond the end of the corolla tube
indehiscent not splitting open
inferior relating to an organ borne below another one; inferior ovary has the ovary below the perianth
inflorescence a flower cluster
infrastipular borne below the stipules
internode the portion of an axis between two nodes
involucre an outer or accessory covering; a set of bracts surrounding a flower cluster or group of florets
irregular (flowers) with some parts different from others in the same series; not divisible into two equal halves

keeled ridged, like the bottom (keel) of a boat

lacerate with a jagged margin as if torn or slashed
laciniate cleft into narrow irregular segments
lanate clothed with woolly and interwoven hairs
lanceolate considerably longer than broad, tapering upwards from below the middle; shaped like a lance-head
lance-oblong intermediate in shape between lanceolate and oblong
lance-ovoid solid but with an outline between lanceolate and egg-shaped
lax loose or not closely spaced
lenticel a small corky spot, dot, or line on the bark of young branchlets
liana a woody vine
linear long and narrow, with parallel sides
lip one of the parts of an unequally divided calyx or corolla
littoral growing on, or pertaining to, shores
lobe a segment or division, often rounded, as of a leaf which is cleft or divided
locular pertaining to a locule or the number of chambers or cavities
locule a chamber or cavity within an anther, ovary, or fruit
lustrous glossy, shiny

marly abounding with marl, an earthy deposit of clay mixed with calcium carbonate
mealy comprised of fine flakes or granules
membranaceous thin, pliable, and more or less translucent
microspecies as used here, a "species" of apomictic origin; all progeny resembling the original mother plant
midrib the main vein of a leaf or leaflet running down the middle of the blade
midvein the central vein or midrib of a leaf or leaflike structure
monoecious with staminate and pistillate flowers on the same individual

monotypic with a single type, as in a genus with only one species
moraine an accumulation of earth, stones, and boulders carried and finally deposited by a glacier
mucro a short sharp point
mucronate abruptly tipped with a mucro
mycorrhiza a symbiotic association of living fungal organisms with the roots of certain plants
mycorrhizal pertaining to mycorrhiza

nectary a gland secreting nectar
nerved with veins or slender ribs
node the place along a stem at which a leaf or bud normally develops; leaves may arise singly (alternate), in pairs (opposite), or three or more (whorled) at the node
nutlet a small nut, like an achene with a thick wall

obcordate deeply lobed at the apex; the opposite of cordate
oblanceolate lance-shaped but broader upwards (away from the base)
oblique (leaves) slanting or unequal-sided at the base
oblong as of a leaf, two or three times longer than wide, with the sides nearly parallel
obovate inversely ovate, broadest beyond the middle
obtuse blunt
odd-pinnate pinnately compound with a terminal leaflet
opposite occurring in twos, one on each side of the axis at a node (paired)
orbicular spherical, circular in outline
oval broadly elliptical
ovary the basal enlarged part of a carpel or pistil containing the ovules
ovate egg-shaped, or like the longitudinal section of an egg, broadest below the middle
ovoid solid with an ovate outline
ovule the organ which contains the egg and which, after fertilization, becomes the seed

palmate having three or more lobes radiating from a common point, resembling a hand with the fingers spread
panicle a branched or compound inflorescence of the racemose type
paniculate resembling a panicle
papillose bearing minute roundish projections which produce a roughish surface
parti-coloured of variegated colour
pedicel the support of a single flower in a cluster
peduncle the primary flower stalk, supporting either a solitary flower or a cluster of flowers
pellucid clear, transparent
peltate shield-shaped, as a leaf with the blade attached to the petiole by the lower surface rather than by the margin of the blade
pendent handing down
perfect (flowers) with both stamens and pistils present
perfoliate (leaves) with the basal part of the leaf blade, or the united bases of opposite leaves, surrounding the stem
perianth the floral envelope, consisting of the calyx (sepals) and the corolla (petals)
pericarp the wall of a mature ovary; wall of the fruit
persistent remaining attached; not falling off at the end of the growing season
petal a member of the inner set(s) of floral leaves, just inside the calyx, either white or coloured
petiolate having a petiole
petiole the stalk of a leaf
petiolule the stalk of a leaflet
pilose covered or provided with soft hairs
pin (flowers) with relatively long style and short stamens (compare with thrum)
pinnate featherlike, as with the leaflets of a compound leaf on both sides of the rachis
pinnatifid cleft or divided in a pinnate fashion
pistil the central organ of a flower containing the ovules which become the seeds
pistillate unisexual flowers having pistils and lacking stamens
pit the central stone or bony-covered seed within a fleshy fruit such as a drupe
pith the spongy centre of a stem
plaited folded like a fan
pod a dehiscent dry fruit
pollen the dustlike grains produced in the anther
pollination the transfer of pollen from the stamen to the stigma (receptive part of the pistil)
polygamous bearing perfect and unisexual flowers on the same individual
polyploidy having three or more sets (genomes) of chromosomes in each cell
pome a fleshy fruit, like an apple
precocious developing early; expanding before the leaves, as in the catkins of some willows
prickle a small weak spinelike body arising from the epidermis of a stem or leaf
primocane (in *Rubus*) the stem of the first year, usually without flowers, from which the second year's flowering stems (floricanes) develop
procumbent prostrate, trailing, or lying flat on the ground

prostrate lying flat on the ground
protandrous the stamens shedding pollen before the stigmas are receptive
protogynous the stigma becoming receptive before the anthers of the same flower release their pollen
pruinose having a waxy powdery bloom
pseudogamous pollination necessary to initiate the development of apomictic seeds; very rarely fertilization occurs
pseudoterminal appearing as if at the end of a stem or branch but actually attached a short distance from the end
puberulent finely pubescent
pubescence a covering of hairs of any kind
pubescent clothed with hairs of any kind
pyramidal shaped like a pyramid
pyriform pear-shaped

raceme an elongate indeterminate inflorescence with each flower on a stalk along the stem
racemose racemelike or in racemes
rachis an axis bearing flowers or leaflets
receptacle the enlarged end of a stem (flower stalk) to which the flower parts are attached
recurved curved downward or backward
reflexed abruptly bent downward or backward
regular (flowers) with the parts so arranged as to be divisible into two equal halves by two or more planes
reniform kidney-shaped
resiniferous producing resin
resinous resinlike
reticulate netted: with a network of veins
retuse a shallow notch in a rounded apex, as in a leaf
revolute with the margin rolled towards the lower surface, as in a leaf
rhizome a horizontal underground stem that bears roots and leafy shoots
rhombic broadly diamond-shaped
rib a main vein of a leaf
rooting producing roots
rotate saucer-shaped
rotund rounded
rugose wrinkled

samara a winged indehiscent fruit; the "key" of maples
scabrous rough to the touch
scalloped with small or low semi-circular lobes
scale a thin, often membranaceous reduced leaf or bract; any small, thin, or flat structure
scape a leafless or bracted flowering stem arising from ground level
schizocarp a fruit which splits into one-seeded portions; mericarps
scurfy with scalelike particles
sedge a coarse grasslike plant often forming mats or clumps in damp soil
seed a mature ovule containing a minute embryonic plant with or without food tissue
self-incompatible the pollen of a flower not capable of fertilizing an egg in the pistil of the same flower
sepal one of the leaflike or bractlike parts of the outer floral envelope, sometimes petal-like
sericeous silky, with closely appressed soft straight hairs
serotinous late in the season, appearing after the leaves have expanded, as in the catkins of some willows
serrate with teeth pointing forward like those of a saw
serrulate with fine teeth or serrations
sessile not stalked; attached directly to the axis or receptacle
sheathing surrounding a stem or other organ with a more or less tubular structure (sheath)
shrub a woody plant with branches at or near the base and without a distinct single trunk
simple as of leaves, with a single blade; not compound
sinus a space between two lobes
spatulate rounded at the end and with a long narrowed base, like a spatula
sphagnum a moss typically found in wet acid areas where the accumulated remains become compacted with other plant debris to form peat
spike an elongate inflorescence with many sessile or subsessile flowers along the stem
spine a sharp woody outgrowth from the stem, usually a modified branch but sometimes a stipule or petiole
spinulose with diminutive spines
spur shoot a short compact stem or branch with little development of the internodes
stamen one of the pollen-bearing organs of the flower
staminate a unisexual flower with stamens but no carpels or pistils
staminode a sterile stamen
stellate star-shaped; with rays, like the points of a star
sterile infertile; unable to function in the fertilization or reproduction process
stigma the part of the pistil or style that receives the pollen
stigmatic relating to the stigma
stipe the stalk of a pistil or fruit
stipel the stipule of a leaflet
stipitate having a stipe

stipitate-glandular with stalked glands
stipule a leaflike or scaly appendage, one at each side at the base of the petiole of a leaf
stolon an elongate stem creeping over the surface of the ground
stoloniferous producing stolons
striate marked with longitudinal streaks or furrows (striae)
strobilus a cone; a stem with decussate or spirally arranged, usually overlapping scales that produce seeds from ovules borne on their surface
style the narrowed portion of the pistil between the ovary and the stigma
sub a Latin prefix meaning "nearly" or "more or less"
subtend to stand below or close to, as a bract below a flower, hence the flower is in the axil of the bract
subulate awl-shaped
sucker an adventitious shoot arising from a root
superior above; concerning the ovary, when attached above the perianth
surculose producing suckers
symbiosis the living together of dissimilar organisms with benefit to one or to both members of the association
symbiotic relating to symbiosis

talus an accumulation of rock debris at the base of a cliff, often in the form of a slope as a result of gravitational roll or slide
taxon any taxonomic unit or category
tendril a slender twining or clasping outgrowth from a stem or leaf
terete circular in transverse section
terminal attached to, or belonging at, the end of an axis or organ
ternate in threes
thorn a spine
thrum (flowers) with a short style and long stamens (compare with pin)
tomentose densely woolly or covered with a dense coating of matted wool-like hairs
tomentulose covered with minute tangled woolly hairs
tomentum a coating of tangled woolly hairs
trailing prostrate but not rooting
translucent transmitting light but obscuring the details of objects viewed through the structure or substance
tree a perennial woody plant with one main trunk, the branches forming a more or less distinct and elevated head or crown
trifoliate three-leaved
trifoliolate with three leaflets
truncate ending abruptly, as if cut off
tubercle minute tuber or swollen structure; any small excrescence
tuberculate furnished with tubercles
tubular tubelike
two-ranked arranged or held in two rows or series

umbel an inflorescence with the flower stalks arising from a common point and reaching about the same height; shaped like an inverted open umbrella, i.e., an umbrella blown inside out by the wind.
unarmed lacking thorns or spines
undulate wavy
unisexual a flower with one sex only, stamens or pistils but not both
urceolate urn-shaped and contracted at the mouth

venation the system or arrangement of veins
vestigial rudimentary or imperfectly developed; much reduced and not functional
villous covered with long straight silky hairs
vine a herbaceous or woody plant with trailing, running, or climbing stems

whorl a group of three or more similar organs radiating around a node
winged provided with a thin or membranaceous expansion
witches'-broom tufts of shoots or stems resulting from an infection by fungi or an infestation by insects
withe a tough flexible stem

INDEX TO BOTANICAL AND COMMON NAMES